GRE®

Graduate
Record
Examination

MATH WORKBOOK

Course Edition

Special thanks to the team that made this book possible:

Arthur Ahn, Matthew Belinkie, Shannon Berning, Lauren T. Bernstein, Kim Bowers, Gerard Cortinez, Elisa Davis, Lola Disparte, Boris Dvorkin, John Evans, Paula Fleming, Darcy Galane, Joanna Graham, Adam Grey, Allison Harm, Jack Hayes, Adam Hinz, Gar Hong, Sunny Hwang, Cinzia Iacono, Xandi Kagstrom, Avi Lidgi, Kate Lopaze, Keith Lubeley, TJ Mancini, Jennifer Moore, Jason Moss, Walt Niedner, Robert Reiss, Shmuel Ross, Derek Rusnak, Emily Sachar, Stephanie Schrauth, Sheryl Stebbins, Glen Stohr, Sascha Strelka, Gene Suhir, Martha Torres, Liza Weale, Lee A. Weiss, Emily West, and many others who have contributed materials and advice over the years.

Published by Kaplan Publishing, a division of Kaplan, Inc.
750 Third Avenue
New York, NY 10017

Printed in the United States of America

10 9 8 7 6

ISBN: 978-1-5062-0091-0

Important: Please Read

YOUR KAPLAN GRE COURSE BOOKS

1. ***GRE Math Workbook:*** This will be your book for in-class work. Please bring this book to your GRE Math class sessions (In Person and Live Online students) or use it when completing your Lessons on Demand (Self-Paced students). We recommend you do *not* begin working in this book before your class sessions—we will give you guidance for how best to use this material.
2. ***GRE Verbal Workbook:*** This will be your book for in-class work. Please bring this book to your GRE Verbal class sessions (In Person and Live Online students) or use it when completing your Lessons on Demand (Self-Paced students). We recommend you do *not* begin working in this book before your class sessions—we will give you guidance for how best to use this material.
3. ***GRE Premier:*** This book is a resource for homework and additional study. If you're eager to begin, start learning about the Kaplan Methods and Strategies and practicing GRE questions in this book.

YOUR KAPLAN GRE ONLINE RESOURCES

Go to **www.kaptest.com**, and log in to your Online Center using the username and password you received upon enrollment. Watch the Orientation Video, and explore the wealth of resources available to you as a Kaplan GRE student.

YOUR DIAGNOSTIC EXAM

Take your Diagnostic Exam as soon as possible, optimally before your first class session. You can find the Diagnostic in your Online Center. Taking the Diagnostic—a full-length practice exam—will help you establish a baseline score and familiarize yourself with the GRE. Review your Diagnostic Performance Summary, powered by Smart Reports™, Kaplan's adaptive learning technology, to identify your strengths and areas of opportunity.

Table of Contents

HOW TO USE THIS BOOK . . . xi

PART ONE: Getting Started

CHAPTER 1: Introduction to GRE Math . . . 3

Understanding the GRE Quantitative Reasoning Sections . . . 3

MST Mechanics . . . 3

Question Types . . . 5

Quantitative Comparison . . . 5

Problem Solving . . . 5

Data Interpretation . . . 6

PART TWO: Quantitative Reasoning

CHAPTER 2: Quantitative Comparison . . . 9

Quantitative Comparison . . . 9

Quantitative Comparison Questions . . . 10

Quantitative Comparison Practice Set . . . 12

Quantitative Comparison Practice Set Answer Key . . . 22

Quantitative Comparison Practice Set Answers and Explanations . . . 23

CHAPTER 3: Problem Solving . . . 31

Problem Solving . . . 31

Problem Solving Questions (Single Answer) . . . 31

Problem Solving (All-That-Apply) . . . 34

Problem Solving Questions (Numeric Entry) . . . 35

Problem Solving Practice Set . . . 38

Problem Solving Practice Set Answer Key . . . 47

Problem Solving Practice Set Answers and Explanations . . . 48

CHAPTER 4: Data Interpretation . . . 57

Data Interpretation . . . 57

Data Interpretation Questions (One Answer) . . . 58

Data Interpretation Questions (All-That-Apply) . . . 60

Data Interpretation Questions (Numeric Entry) . . . 62

Data Interpretation Practice Set . . . 65

Data Interpretation Practice Set Answer Key . . . 78

Data Interpretation Practice Set Answers and Explanations . . . 79

CHAPTER 5: Quantitative Reasoning Practice 87
Quantitative Reasoning Practice Set 1 88
Quantitative Reasoning Practice Set 1 Answer Key 95
Quantitative Reasoning Practice Set 1 Answers and Explanations 96
Quantitative Reasoning Practice Set 2 103
Quantitative Reasoning Practice Set 2 Answer Key 110
Quantitative Reasoning Practice Set 2 Answers and Explanations 111
Quantitative Reasoning Practice Set 3 116
Quantitative Reasoning Practice Set 3 Answer Key 123
Quantitative Reasoning Practice Set 3 Answers and Explanations 124
Quantitative Reasoning Practice Set 4 131
Quantitative Reasoning Practice Set 4 Answer Key 138
Quantitative Reasoning Practice Set 4 Answers and Explanations 139
Quantitative Reasoning Practice Set 5 146
Quantitative Reasoning Practice Set 5 Answer Key 153
Quantitative Reasoning Practice Set 5 Answers and Explanations 154
Quantitative Reasoning Practice Set 6 159
Quantitative Reasoning Practice Set 6 Answer Key 166
Quantitative Reasoning Practice Set 6 Answers and Explanations 167

PART THREE: Math Content Review

CHAPTER 6: Arithmetic 177
Real Numbers 177
Number Operations 178
Number Operations Exercises 186
Number Operations Answer Key 188
Number Properties 193
Number Properties Exercises 199
Number Properties Answer Key 201
Averages (Arithmetic Means) 205
Averages (Arithmetic Means) Exercises 207
Averages (Arithmetic Means) Answer Key 210
Ratios 217
Ratios Exercises 222
Ratios Answer Key 225
Percents 232
Percents Exercises 237
Percents Answer Key 240
Powers and Roots 243
Powers and Roots Exercises 246
Powers and Roots Answer Key 248

CHAPTER 7: Algebra 253
Understanding Algebra 253
Operations with Algebraic Expressions 254

Operations with Algebraic Expressions Exercises 258
Operations with Algebraic Expressions Answer Key 260
Rules of Exponents 264
Rules of Exponents Exercises 265
Rules of Exponents Answer Key 266
Solving Linear Equations 271
Solving Linear Equations Exercises 275
Solving Linear Equations Answer Key 276
Quadratic Equations. 283
Quadratic Equations Exercises 285
Quadratic Equations Answer Key. 286
Solving Inequalities 295
Solving Inequalities Exercises 296
Solving Inequalities Answer Key 297
Translation into Algebra. 301
Translation into Algebra Exercises 305
Translation into Algebra Answer Key. 308
Functions 317
Functions Exercises 319
Functions Answer Key 323
Coordinate Geometry 328
Coordinate Geometry Exercises 334
Coordinate Geometry Answer Key 339
Graphs of Functions 344
Graphs of Functions Exercises 347
Graphs of Functions Answer Key 353

CHAPTER 8: Geometry 363
Understanding Geometry 363
Lines and Angles. 364
Lines and Angles Exercises 369
Lines and Angles Answer Key. 373
Triangles and Pythagorean Theorem. 377
Triangles and Pythagorean Theorem Exercises 384
Triangles and Pythagorean Theorem Answer Key. 387
Polygons 392
Polygons Exercises 396
Polygons Answer Key 401
Circles. 407
Circles Exercises 410
Circles Answer Key. 414
Multiple Figures 420
Multiple Figures Exercises 423
Multiple Figures Answer Key 430
Three-Dimensional Figures (Uniform Solids) 435

Three-Dimensional Figures (Uniform Solids) Exercises 438
Three-Dimensional Figures (Uniform Solids) Answer Key 442

CHAPTER 9: Data Interpretation 449
Understanding Data Analysis 449
Counting Methods 449
Counting Methods Exercises 453
Counting Methods Answer Key 456
Descriptive Statistics 461
Descriptive Statistics Exercises 465
Descriptive Statistics Answer Key 469
Data Graphs and Tables 475
Data Graphs and Tables Exercises 480
Data Graphs and Table Answers Key 488
Probability 492
Probability Exercises 494
Probability Answer Key 497

PART FOUR: Advanced Math Practice

CHAPTER 10: High-Difficulty Question Sets 507
Introduction 507
Advanced Math Questions Practice Set 1 508
Advanced Math Questions Practice Set 1 Answer Key 517
Advanced Math Questions Practice Set 1 Answers and Explanations 518
Advanced Math Questions Practice Set 2 532
Advanced Math Questions Practice Set 2 Answer Key 542
Advanced Math Questions Practice Set 2 Answers and Explanations 543
Advanced Math Questions Practice Set 3 561
Advanced Math Questions Practice Set 3 Answer Key 569
Advanced Math Questions Practice Set 3 Answers and Explanations 570
Advanced Math Questions Practice Set 4 587
Advanced Math Questions Practice Set 4 Answer Key 596
Advanced Math Questions Practice Set 4 Answers and Explanations 597

PART FIVE: GRE Resources

APPENDIX: Math Reference 615

Getting the Most Out of Your Kaplan GRE Course

WELCOME TO YOUR KAPLAN GRE COURSE

Congratulations. By starting this Kaplan GRE course, you've just taken a big step toward your goal of a higher GRE score. At Kaplan, we've been working with students in your position for the past 75 years, so we know it can be daunting to contemplate mastering such a high-stakes test as the GRE. Don't worry; we'll help you build GRE expertise throughout this course. To help you get started, we'd like to share some advice for structuring your study and reaching your score goals.

Schedule Your Test Day

The GRE is required for admission to graduate programs in everything from anthropology to zoology. All these programs have different admissions deadlines. Check out the websites of the schools you're interested in, and make sure you take the GRE at least three weeks before the deadline to ensure that schools receive your score on time.

While you can take the GRE multiple times (although no more than once every 21 days), it's best to test once and to test confidently, so you want to give yourself enough time to get as much as possible out of your Kaplan course. We recommend that you take the GRE about two to four weeks after finishing your final class session (longer if you're enrolled in a course that meets more than once a week). Doing so gives you time to complete all of the required assignments, continue taking practice tests, and take full advantage of your Kaplan resources for personalized study. But don't wait so long that you lose momentum from the course. Three months of dedicated and consistent study works well for most test takers. As soon as you've given some thought to your goals and your available study time, schedule your GRE so that you have a clear target to work toward. Students who set their test date in advance tend to get the most out of the course.

Set Your Score Goals

As you begin your GRE preparation, spend some time setting your score goals. Base your goal score, above all, on the requirements of your target graduate programs. Some programs focus more on one section than another. Others require a minimum combined score before they'll even consider an applicant for admission or for scholarships. A good score is one that puts you in the 25th to 75th percentile of the students who attend your target program; many schools provide this data. An excellent score gives you a competitive advantage over other applicants to the programs you're considering. School websites or program-ranking sites can help you determine a solid target score. You can also call programs that interest you and speak with someone from the admissions department.

Once you've completed the Kaplan Diagnostic Test, you'll have a good sense of your starting point. Where you end up will depend less on where you start than on how actively you practice and how willing you are to try the new techniques you'll learn to improve your score.

Establish a Baseline Score

The first step of the process is to log in to your Online Center and complete the Diagnostic Test (MST 1) before your first GRE class session. Plan on taking about 3 hours and 45 minutes to complete the test. Sit down with your computer in a quiet place with a stable Internet connection where you can concentrate without interruption, and do the best you can.

Don't worry too much about your exact Diagnostic score. The Diagnostic is, as the name suggests, a diagnosis. It's *not* a prognosis for how much you can improve or how well you'll score on Test Day. The most important use of the Diagnostic is to focus your attention on the areas of the GRE that are your relative strengths and those in which you have opportunity for improvement. Knowing these areas will help you organize your preparation for the actual test. Analyze your Diagnostic results with the help of your Performance Summary (found in your Online Center).

Structure Your GRE Preparation

The best time to create a study plan is at the beginning of your GRE course. First, write into your calendar the dates of your class sessions. Then, schedule your practice tests, which we call MSTs. (MST stands for multi-stage test, the test format of the GRE; you'll learn how an MST works in class.) After you've attended two class sessions, begin taking and reviewing one Kaplan MST per week. Setting dates and times for your MSTs will better simulate Test Day and will help you fit all seven of them in. One week before Test Day, take your final MST (ideally *exactly* a week before: same time, same day) and spend the rest of the week reviewing the explanations and reinforcing your strengths.

There is homework to complete between class sessions, too, all of which you will find organized by session in your online center. Do your best to block out short, frequent periods of study time throughout the week. The amount of time you spend studying each week will depend on your schedule and your test date, but your course is structured with the expectation that you'll study outside of class for seven to twelve hours each week.

As you move through your GRE course, your study plan should be a living document that is molded each week based on your available study time and the results of your

most recent MST. Your strengths and weaknesses are likely to change over the course of your class sessions and out-of-class practice. Keep checking your Performance Summary and personalized recommendations to make sure you are addressing the areas that are most important to *your* score.

Raise Your Score

The big question everyone has when beginning a GRE course is how large a score increase to expect. It's difficult to make an exact score promise. We've seen consistently that the students who achieve the greatest success are the ones who participate actively in class, complete the required homework, and take practice tests at scheduled intervals. Our most successful students also make full use of the wide range of course materials, understand their areas of strength and opportunity, and check in with us with questions or to get advice.

Signing up for a GRE course is like joining a gym. Kaplan has great GRE gym equipment, and your teacher is your personal trainer. But you'll see your GRE test-taking skills get into shape to the extent that you put in the effort. Merely signing up for a gym membership doesn't guarantee results, but it *can* give you the tools and motivation you need to succeed. Your Kaplan GRE course works the same way. Here are a few additional tips for success:

Participate actively in your class sessions. Whether your course is In Person, Live Online, or Self-Paced, active learning will pay off as you apply what you learn in class to the homework assignments.

Know that you're going to make mistakes as you practice inside and outside of class. It's normal. Consider this: every wrong answer you choose, *and then learn from*, reduces the chance that you'll get a similar question wrong on the one and only day when wrong answers matter. So make mistakes willingly and even happily now, while they don't count. Just resolve to learn from every one of them.

Have patience with your progress. As you take more MSTs, it's common to make progress in fits and starts and even to experience some dips along the way. Don't be alarmed—score improvements hardly ever follow a perfectly smooth upward trajectory. But if you're studying according to our recommendations, you can trust that you're moving in the right direction to get the score you want on Test Day.

Stick with the new methods you'll be learning. You will be learning and assimilating many new techniques and strategies throughout your Kaplan course. The more consistently you practice these, the more they will become second nature, thereby eliminating uncertainty and saving you valuable time on Test Day. When you are starting out, however, new techniques often take longer or feel more cumbersome than the way you might otherwise have approached the question. The best analogy may be learning to type. When you start out, you may be typing effectively with two or three fingers, keeping a decent pace. But once you start learning touch typing, what happens? You slow down, and errors increase—until you've mastered this faster, more efficient way. Then, you breeze along more expertly than ever. As you become the GRE equivalent of a 100-words-per-minute typist, resist the temptation to dwell on short-term trends between one MST and the next; focus instead on practicing well. Aim for consistency and accuracy at first, then emphasize pacing as you get closer to Test Day and grow more confident with applying the methods.

Make Time to Review

You lead a busy life in addition to preparing for the GRE, and it can often feel difficult to fit in much study time outside of class. It may be tempting to push ahead and cover new material as quickly as possible, but we recommend strongly that you schedule ample time for review.

This may seem counterintuitive—after all, your time is limited, so why spend it going over stuff you've already done? But the brain rarely remembers *anything* it sees or does only once. When you build a connection in the brain and then don't follow up on it, that knowledge may still be "in there" somewhere, but it is not accessible the way you need it to be on Test Day. When you carefully review notes you've taken or problems you've solved (and the explanations for them, as well), the process of retrieving that information reopens and reinforces the connections you've built in your brain. This builds long-term retention and repeatable skill sets—exactly what you need to beat the GRE. Focus your review on the following areas:

Review the topics you've covered in class. Go session by session—read the notes you took, watch the Lesson on Demand for that session (in your Online Center), remind yourself of what you did to answer questions successfully, and work through the review assignments to reinforce those skills.

Review your MSTs and quizzes by working through the answers and explanations and noticing trends in your performance. Review the questions you answered incorrectly *and* those you got right. Rework the questions you got wrong so you can be sure to know exactly what to do when you see similar ones in the future. For the questions you got correct, did you answer correctly for the right reasons, or did you follow a lucky hunch? Does the explanation offer an alternative, more efficient path to the right answer that you can add to your arsenal? Or maybe you aced the problem and should take note of your process so that you can repeat it on similar problems!

Come to Us for Help

Take this advice to heart as you embark on your GRE preparation with Kaplan. Reach out to your teacher or to our team of GRE experts at **KaplanGREFeedback@kaplan.com** with any questions you have about your Kaplan GRE course. We are here to support your efforts every step of the way, and we back up our promises with the Higher Score Guarantee. For more details and eligibility requirements, please see **www.kaptest.com/hsg.**

Thanks for choosing Kaplan. We wish you the best of luck on your journey to graduate school success.

PART ONE

Getting Started

CHAPTER 1

Introduction to GRE Math

UNDERSTANDING THE GRE QUANTITATIVE REASONING SECTIONS

The Quantitative Reasoning sections of the GRE test your ability to reason quantitatively—to read a math problem, understand what it is asking, and solve it. The basic math disciplines covered on the GRE are arithmetic, algebra, geometry, and data interpretation. In this way, the GRE has topics similar to those covered on the SAT. There is no trigonometry or calculus on the GRE. The aim of the test is to provide an accurate indication of your ability to use your knowledge of the fundamental topics and apply reasoning skills to the various question types. The goal is to make the test a true indicator of your ability to apply given information, think logically, and draw conclusions. These are skills you will need at the graduate level of study.

The GRE contains two math sections, each having 20 questions. The allotted time for each section is 35 minutes. To perform well on the Quantitative Reasoning sections, you want to perform the fundamental math skills efficiently and apply your reasoning skills at the same time to answer as many questions correctly as possible. The various question types will be explained in detail in part 2 of this book. The foundations of arithmetic, algebra, geometry, and data interpretation will be reviewed in part 3 of this book. In every chapter, there will be plenty of opportunities to practice and check your answers.

MST MECHANICS

The GRE is a multi-stage test (MST). While working within a section of the test during the time allotted, you are allowed to skip questions and return to them as long as time remains for the section. The test is computer-based with built-in capabilities such as the "Mark" button to indicate a question you want to examine later (within the time allowed for that section) and the "Review" button to see your progress on the entire set of questions in a section, an optional time display, and an on-screen calculator. As you prepare for Test Day, always consider how you plan to manage your time for each section and how these computer capabilities can help you.

On the GRE, every question has the same value. So you do not want to waste time dwelling on one question while you may be able to answer several others correctly. This is one way the Mark and Review buttons can assist you in managing your time. You can turn the time display *on* or *off* as you wish. Some test takers will find it helpful; others will find it annoying.

The availability of an on-screen calculator is something you will want to consider as you practice for the test. Questions in the Quantitative Reasoning sections are not designed to be calculator-intensive. If you find yourself doing a long calculation to answer a question, you may be missing the reasoning aspect completely. That means most questions on the GRE are written to be answered without a calculator. You will want to reserve the calculator for an isolated calculation or a quick check of an answer, but not use it as your main resource for the questions. To help you plan for the best use of the on-screen calculator, an image of it is shown below. Notice that it is a simple, four-function calculator with a square root key and change of sign key.

As you review mathematical foundations and practice answering the various question types in this workbook, use your own calculator strategically. This will lead you to rely on your critical thinking skills.

QUESTION TYPES

A Quantitative Reasoning section with 20 questions will be composed of three question types:

- Quantitative Comparison
- Problem Solving
- Data Interpretation

Each question type will require you to draw upon your ability to combine your knowledge of mathematical concepts with your reasoning skills in a particular way. However, the question types are not distributed evenly within a section. The chart below shows how many questions you can expect of each type, as well as the average amount of time you should spend on each question type.

	Quantitative Comparison	**Problem Solving**	**Data Interpretation**
Number of Questions	Approx. 7–8	Approx. 9–10	Approx. 3
Time per Question	1.5 minutes	1.5–2 minutes	2 minutes

QUANTITATIVE COMPARISON

A Quantitative Comparison question will not ask you to identify a particular value as the answer, but it will ask you to make a comparison between two quantities and identify a relationship between them. The four answer choices for a Quantitative Comparison are always the same—same wording, same order. Because the answer choices are consistent, you will become familiar with them quickly and be able to concentrate on the comparison at hand.

The Kaplan Method for answering Quantitative Comparison questions and strategies to help solve them efficiently can be found in chapter 2.

PROBLEM SOLVING

A Problem Solving question may appear as a math problem, or it may appear as a word problem dealing with a real-world situation. These questions deal with percents, ratios, and other proportions; linear and quadratic equations; basic probability and statistics; and two- and three-dimensional geometry.

The Kaplan Method for answering Problem Solving questions and strategies to help solve them efficiently can be found in chapter 3.

DATA INTERPRETATION

Questions that require you to interpret data will be based on information located in graphs or tables. You will see a set of three or four questions based on each data presentation. Frequently, you will have to use information from more than one data source to answer a question. Data Interpretation questions need to be answered thoughtfully, only after you have taken time to analyze the contents of each graph or table of data.

The Kaplan Method for answering Data Interpretation questions and strategies to help solve them efficiently can be found in chapter 4.

PART TWO

Quantitative Reasoning

CHAPTER 2

Quantitative Comparison

QUANTITATIVE COMPARISON

A Quantitative Comparison question asks you to compare two mathematical expressions; one is Quantity A, the other Quantity B. Sometimes additional information is given centered above the quantities. The additional information applies to both quantities and is needed to make the comparison. The question asks you to compare the relative values of the quantities or to tell whether there is enough information to make a comparison. This type of question is generally about the *relative* values of the two quantities, so you won't need the on-screen calculator for the most part.

The four answer choices for Quantitative Comparison questions are always the same. Choices (A), (B), and (C) represent definite relationships between the quantities. Choice (D) represents a relationship that cannot be determined, based on the information given. Choice (D) means that more than one relationship is possible, depending on the numbers chosen for the variable(s) in the question. When there is at least one variable in a Quantitative Comparison problem, using the Kaplan Picking Numbers strategy is one way to answer the question quickly. If the numbers chosen demonstrate more than one possible relationship between the quantities, you know that the answer is choice (D).

The directions for Quantitative Comparison questions will look like this:

Directions: Select the correct answer.

QUANTITATIVE COMPARISON QUESTIONS

THE KAPLAN METHOD FOR QUANTITATIVE COMPARISON

STEP 1 Analyze the centered information and quantities.

STEP 2 Approach strategically.

$$h > 1$$

Quantity A	Quantity B
$\frac{60}{h}$	The number of minutes in h hours

Ⓐ Quantity A is greater.
Ⓑ Quantity B is greater.
Ⓒ The two quantities are equal.
Ⓓ The relationship cannot be determined from the information given.

APPLY THE KAPLAN METHOD FOR QUANTITATIVE COMPARISON

Now let's apply the Kaplan Method to a Quantitative Comparison question:

$$h > 1$$

Quantity A	Quantity B
$\frac{60}{h}$	The number of minutes in h hours

Ⓐ Quantity A is greater.
Ⓑ Quantity B is greater.
Ⓒ The two quantities are equal.
Ⓓ The relationship cannot be determined from the information given.

STEP 1

Analyze the centered information and quantities.

The centered information tells you that the variable h is greater than 1. In Quantity A, the variable appears in the denominator of a fraction. So, the variable h is a divisor in Quantity A. In Quantity B, you are told that the variable h represents a number of hours. To find the number of minutes in h hours, the variable will be used as a multiplier: the number of minutes in h hours $= 60h$.

STEP 2

Approach strategically.

The restriction that h is greater than 1 is critical to determining the relationship between the quantities. A whole number (60 in this case) divided by a number greater than 1 will always be less than the whole number (60) multiplied by the same number greater than 1. A quick check using the Picking Numbers strategy verifies the reasoning. Let $h = 2$: $\frac{60}{h} = \frac{60}{2} = 30$ and $60h = 60 \times 2 = 120$. Quantity B is greater and the answer is **(B)**.

QUANTITATIVE COMPARISON PRACTICE SET

Try the following Quantitative Comparison questions using the Kaplan Method.

Basic

1.

Quantity A	Quantity B
The number of edges on a cube	Twice the number of faces on a cube

(A) Quantity A is greater.
(B) Quantity B is greater.
(C) The two quantities are equal.
(D) The relationship cannot be determined from the information given.

2.

Quantity A	Quantity B
The length of the hypotenuse of a right triangle with legs of lengths 5 and 12	The length of a leg of a right triangle with a hypotenuse length of 17 and the other leg of length 8

(A) Quantity A is greater.
(B) Quantity B is greater.
(C) The two quantities are equal.
(D) The relationship cannot be determined from the information given.

3.

Quantity A	Quantity B
The number of degrees in the largest angle of a triangle inscribed in a circle, in which the diameter of the circle is one side of the triangle	The number of degrees in a right angle

(A) Quantity A is greater.
(B) Quantity B is greater.
(C) The two quantities are equal.
(D) The relationship cannot be determined from the information given.

4.
$$\frac{a}{b} = \frac{3}{7} = \frac{c}{d}$$

Quantity A	Quantity B
$a + d$	$b + c$

Ⓐ Quantity A is greater.
Ⓑ Quantity B is greater.
Ⓒ The two quantities are equal.
Ⓓ The relationship cannot be determined from the information given.

5.
$$12x - 46 = -18 + 5x$$

Quantity A	Quantity B
x	5

Ⓐ Quantity A is greater.
Ⓑ Quantity B is greater.
Ⓒ The two quantities are equal.
Ⓓ The relationship cannot be determined from the information given.

6. In a three-digit number n, the hundreds digit is 3 times the units digit.

Quantity A	Quantity B
The units digit of n	4

Ⓐ Quantity A is greater.
Ⓑ Quantity B is greater.
Ⓒ The two quantities are equal.
Ⓓ The relationship cannot be determined from the information given.

7.

Quantity A	Quantity B
The number of days in 17 weeks	The number of minutes in 2 hours

Ⓐ Quantity A is greater.
Ⓑ Quantity B is greater.
Ⓒ The two quantities are equal.
Ⓓ The relationship cannot be determined from the information given.

8. $$2x + y = z$$

Quantity A	Quantity B
The value of z when $x = -1$ and $y = 3$	The value of z when $x = -2$ and $y = 2$

Ⓐ Quantity A is greater.
Ⓑ Quantity B is greater.
Ⓒ The two quantities are equal.
Ⓓ The relationship cannot be determined from the information given.

9. $$5p + 6q = 74$$
$$q = 8$$

Quantity A	Quantity B
p	5

Ⓐ Quantity A is greater.
Ⓑ Quantity B is greater.
Ⓒ The two quantities are equal.
Ⓓ The relationship cannot be determined from the information given.

10. $$f(x) = 3x + 4$$

Quantity A	Quantity B
Slope of $f(x)$	y-intercept of $f(x)$

Ⓐ Quantity A is greater.
Ⓑ Quantity B is greater.
Ⓒ The two quantities are equal.
Ⓓ The relationship cannot be determined from the information given.

Intermediate

11. One marble is randomly selected from a bag that contains only 4 black marbles, 3 red marbles, 5 yellow marbles, and 4 green marbles.

Quantity A	Quantity B
The probability of selecting either a black marble or a red marble	The probability of selecting either a yellow marble or a green marble

Ⓐ Quantity A is greater.
Ⓑ Quantity B is greater.
Ⓒ The two quantities are equal.
Ⓓ The relationship cannot be determined from the information given.

12. $$f(x) = x^2 - 6x + 8$$

Quantity A	Quantity B
$f(-6)$	$f(6)$

Ⓐ Quantity A is greater.
Ⓑ Quantity B is greater.
Ⓒ The two quantities are equal.
Ⓓ The relationship cannot be determined from the information given.

13. A bag has 20 marbles that are either black or white.

Quantity A	Quantity B
The number of times one must randomly draw a marble from the bag, without replacing it, to ensure that at least 4 black marbles are selected	The number of times one must randomly draw a marble from the bag, without replacing it, to ensure that at least 4 white marbles are selected

Ⓐ Quantity A is greater.
Ⓑ Quantity B is greater.
Ⓒ The two quantities are equal.
Ⓓ The relationship cannot be determined from the information given.

14.

$$\frac{a}{b} = \frac{3}{4}$$

Quantity A	Quantity B
$\frac{2a - b}{b}$	$\frac{a}{a + b}$

Ⓐ Quantity A is greater.
Ⓑ Quantity B is greater.
Ⓒ The two quantities are equal.
Ⓓ The relationship cannot be determined from the information given.

15. A line is represented by the equation $4x + 3y = 12$.

Quantity A	Quantity B
The value of the x-intercept of the line	The value of the y-intercept of the line

Ⓐ Quantity A is greater.
Ⓑ Quantity B is greater.
Ⓒ The two quantities are equal.
Ⓓ The relationship cannot be determined from the information given.

16. The positive integer x is odd, and the positive integer y is even.

Quantity A	Quantity B
$(-1)^{2x + y}$	$\frac{1}{2}$

Ⓐ Quantity A is greater.
Ⓑ Quantity B is greater.
Ⓒ The two quantities are equal.
Ⓓ The relationship cannot be determined from the information given.

17.

Data set $M = \{60, 9, 10, 20, 12, 7, 10, 8\}$

Quantity A	Quantity B
The mode of data set M	The mean of data set M

Ⓐ Quantity A is greater.
Ⓑ Quantity B is greater.
Ⓒ The two quantities are equal.
Ⓓ The relationship cannot be determined from the information given.

18.

$$\sqrt{x^2 + 39} = 8$$

Quantity A	Quantity B
x	4

Ⓐ Quantity A is greater.
Ⓑ Quantity B is greater.
Ⓒ The two quantities are equal.
Ⓓ The relationship cannot be determined from the information given.

19.

Quantity A	Quantity B
The sum of the coordinates of a point in the fourth quadrant of an xy-coordinate plane	The product of the coordinates of a point in the first quadrant of an xy-coordinate plane

Ⓐ Quantity A is greater.
Ⓑ Quantity B is greater.
Ⓒ The two quantities are equal.
Ⓓ The relationship cannot be determined from the information given.

20.

$$4 < x < 10$$
$$3 < y < 5$$

Quantity A	Quantity B
$x + y$	14

Ⓐ Quantity A is greater.
Ⓑ Quantity B is greater.
Ⓒ The two quantities are equal.
Ⓓ The relationship cannot be determined from the information given.

Advanced

21. The length of a rectangular canvas is increased by x percent, and the width of the canvas is decreased by x percent.

Quantity A	Quantity B
The area of the new canvas if $x = 20$	The area of the new canvas if $x = 40$

Ⓐ Quantity A is greater.
Ⓑ Quantity B is greater.
Ⓒ The two quantities are equal.
Ⓓ The relationship cannot be determined from the information given.

22.

$$|4x + 24| = 96$$
$$|4x| = 120$$

Quantity A	Quantity B
x	18

Ⓐ Quantity A is greater.
Ⓑ Quantity B is greater.
Ⓒ The two quantities are equal.
Ⓓ The relationship cannot be determined from the information given.

23.

Quantity A	Quantity B
The number of miles driven at 50 miles per hour for 2 hours	The shortest distance between the starting point and the ending point of a trip if a vehicle is driven 60 miles north and 80 miles east

Ⓐ Quantity A is greater.
Ⓑ Quantity B is greater.
Ⓒ The two quantities are equal.
Ⓓ The relationship cannot be determined from the information given.

24.

Quantity A	Quantity B
The area of a triangle with sides 6, 8, and 10	The area of an equilateral triangle with side 8

Ⓐ Quantity A is greater.
Ⓑ Quantity B is greater.
Ⓒ The two quantities are equal.
Ⓓ The relationship cannot be determined from the information given.

25.

$x < 0 < y$

Quantity A	Quantity B
$-2(x + y)$	$-xy$

Ⓐ Quantity A is greater.
Ⓑ Quantity B is greater.
Ⓒ The two quantities are equal.
Ⓓ The relationship cannot be determined from the information given.

26.

$$f(x) = 2x$$
$$g(x) = \frac{1}{2}x$$

Quantity A	Quantity B
$f(g(3))$	$g(f(3))$

Ⓐ Quantity A is greater.
Ⓑ Quantity B is greater.
Ⓒ The two quantities are equal.
Ⓓ The relationship cannot be determined from the information given.

27.

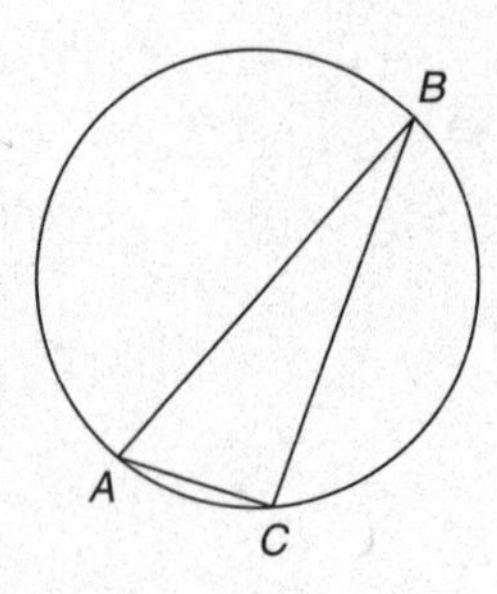

AB is a diameter of the circle.

$\angle ABC = 20°$

Quantity A	Quantity B
The measure of $\angle BAC$	70°

Ⓐ Quantity A is greater.
Ⓑ Quantity B is greater.
Ⓒ The two quantities are equal.
Ⓓ The relationship cannot be determined from the information given.

28. Five consecutive even integers have a sum of –20.

Quantity A	Quantity B
The greatest of the five even integers	0

Ⓐ Quantity A is greater.
Ⓑ Quantity B is greater.
Ⓒ The two quantities are equal.
Ⓓ The relationship cannot be determined from the information given.

29.

$$s = (t + r)^2$$
$$s = 4$$

Quantity A	Quantity B
$2 - r$	t

Ⓐ Quantity A is greater.
Ⓑ Quantity B is greater.
Ⓒ The two quantities are equal.
Ⓓ The relationship cannot be determined from the information given.

30. The numbers in a data set have a mean (arithmetic average) of 0.

Quantity A	Quantity B
Number of data elements below the mean	Number of data elements above the mean

Ⓐ Quantity A is greater.
Ⓑ Quantity B is greater.
Ⓒ The two quantities are equal.
Ⓓ The relationship cannot be determined from the information given.

QUANTITATIVE COMPARISON PRACTICE SET ANSWER KEY

1. C
2. B
3. C
4. D
5. B
6. B
7. B
8. A
9. A
10. B
11. B
12. A
13. D
14. A
15. B
16. A
17. B
18. D
19. D
20. D
21. A
22. B
23. C
24. B
25. D
26. C
27. C
28. C
29. D
30. D

QUANTITATIVE COMPARISON PRACTICE SET ANSWERS AND EXPLANATIONS

Basic

1. C

A question like this one shows the value of spending some time with common solids (cubes/rectangular solids—boxes; right cylinders—soup cans) before Test Day. In this way, you can familiarize yourself with the properties of these solids so you can visualize them when asked a question like this one.

A cube has 12 edges: 4 edges on top, 4 edges on the bottom, and 4 edges connecting the top to the bottom. A cube has 6 faces: top, bottom, back, front, right, and left: $2 \times 6 = 12$. The correct answer is **(C)**.

2. B

Quantity A is a 5:12:13 right triangle. The hypotenuse is 13. Quantity B is a right triangle with hypotenuse 17 and one of the legs 8. You may have this Pythagorean triplet memorized, too; it's an 8:15:17 right triangle. If not, the other leg can be found using the Pythagorean theorem. Let b be the length of the missing leg.

$8^2 + b^2 = 17^2$. Simplifying this equation: $64 + b^2 = 289$. Therefore $b^2 = 225$. Since the question asks for the length, we can disregard the negative value, so $b = 15$.

Quantity B is larger, so the correct answer is **(B)**.

3. C

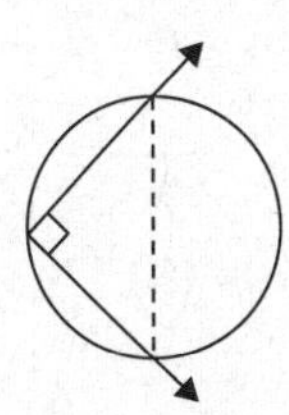

In the figure shown, the dashed line is the diameter of the circle.

When a triangle inscribed in a circle has the circle's diameter as one side, the angle on the circle's circumference intercepts a 180° arc. The measure of such an angle is always half the number of degrees of the intercepted arc, so the angle is a right angle: $180 \div 2 = 90$. Because a triangle has 180° total, the remaining angles have $180 - 90 = 90$ degrees between them; neither can be as large as the angle on the circumference, so it is the largest angle of the triangle. Quantity A equals 90°, which is the number of degrees in a right angle. The correct answer is **(C)**.

4. D

To simplify this Quantitative Comparison, use the strategy of Picking Numbers.

Let $a = 3$ and $b = 7$. Let $c = 3$ and $d = 7$.

$a + d = 3 + 7 = 10$
$b + c = 7 + 3 = 10$

In this case, the quantities are equal.

Repeat using a different set of numbers. Let $a = -3$ and $b = -7$. Let $c = 3$ and $d = 7$.

$a + d = -3 + 7 = 4$
$b + c = -7 + 3 = -4$

Now Quantity A is greater. When you pick a second set of numbers in a Quantitative Comparison, always consider the potential role of negatives and nonintegers. The correct answer is **(D)**.

5. B

Solve the equation for *x*. Subtract 5*x* from both sides:

$$\begin{aligned} 7x - 46 &= -18 \\ 7x &= 28 \\ x &= 4 \end{aligned}$$

Quantity A is 4, and Quantity B is 5. Therefore, the correct answer is **(B)**.

6. B

Start by trying to set the quantities equal. If the units digit of the number is 4, then the hundreds digit must be 3 times 4, based on the centered information in the problem. But that is impossible; the number 12 will not fit in the hundreds place. That means the units digit must be less than 4. Therefore, Quantity A is less than Quantity B. Choose **(B)** as the answer.

7. B

The number of days in 17 weeks is $17 \times 7 = 119$. The number of minutes in 2 hours is $2 \times 60 = 120$. The correct answer is **(B)**.

8. A

Before you take time to calculate, look closely at the equation in the centered information and the values given for *x* and *y* in the quantities. The value of *z* is twice *x* plus *y*. The value of *x* in Quantity A, –1, is greater than the value of *x* in Quantity B, –2. Therefore, 2*x* in Quantity A is greater than 2*x* in Quantity B. Likewise, the value of *y* in Quantity A, 3, is greater than the value of *y* in Quantity B, 2. Both values in Quantity A are greater than the corresponding values in Quantity B. The correct answer is **(A)**.

9. A

Use the two equations to determine the value of the variable *p*. Substitute 8 for *q* in the first equation. Then $5p + 6(8) = 74$, $5p = 26$, and $p = \frac{26}{5}$, or $5\frac{1}{5}$. Quantity A is $5\frac{1}{5}$, and Quantity B is 5. The correct answer is **(A)**.

10. B

The function $f(x) = 3x + 4$ is in slope-intercept form, where $y = mx + b$ and m is the slope and b is the y-intercept. $f(x)$ is another way of writing y.

The slope is 3 and the y-intercept is 4. The correct answer is **(B)**.

Intermediate

11. B

Use the probability formula to compare the quantities.

$$Probability = \frac{Number\ of\ desired\ outcomes}{Number\ of\ total\ possible\ outcomes}$$

The number of total possible outcomes is the same for both quantities. You only need to compare the numerators. The number of desired outcomes for Quantity A is $4 + 3 = 7$. The number of desired outcomes for Quantity B is $5 + 4 = 9$. The correct answer is **(B)**.

12. A

Look at the function $f(x) = x^2 - 6x + 8$. No matter what value is plugged in for x, you will add 8 to it; this is the same for both quantities. Also, when you evaluate $f(-6)$ and $f(6)$, $x^2 = 36$ in both cases. But $-6x = 36$ when $x = -6$, and $-6x = -36$ when $x = 6$. The correct answer is **(A)**.

13. D

You can't tell how many marbles you need to draw from the bag to have at least 4 black marbles or 4 white marbles, unless you know how many of each kind there are. The correct answer is **(D)**.

14. A

Try Picking Numbers here, making sure to pick numbers that conform to the centered information ($\frac{a}{b}$ must equal $\frac{3}{4}$). Also, make sure to try two sets of numbers.

Substitute $a = 3$ and $b = 4$ into both expressions:

$$\frac{2a - b}{b} = \frac{2(3) - 4}{4} = \frac{6 - 4}{4} = \frac{2}{4} = \frac{1}{2}$$

$$\frac{a}{a + b} = \frac{3}{3 + 4} = \frac{3}{7}$$

Now, let's try $a = -6$ and $b = -8$:

$$\frac{2a - b}{b} = \frac{2(-6) - (-8)}{-8} = \frac{-4}{-8} = \frac{1}{2}$$

$$\frac{a}{a + b} = \frac{-6}{-6 + -8} = \frac{-6}{-14} = \frac{3}{7}$$

Quantity A will always be larger than Quantity B, so the correct answer is **(A)**.

15. B

The x-intercept is where the line crosses the x-axis; at this point, the value of y is zero. To find the x-intercept, set $y = 0$ and solve for x.

$$\begin{aligned} 4x + 3y &= 12 \\ 4x + (3)(0) &= 12 \\ 4x + 0 &= 12 \\ 4x &= 12 \\ x &= 3 \end{aligned}$$

The y-intercept is where the line crosses the y-axis; at this point the value of x is zero. To find the y-intercept, set $x = 0$ and solve for y.

$$\begin{aligned} 4x + 3y &= 12 \\ (4)(0) + 3y &= 12 \\ 0 + 3y &= 12 \\ 3y &= 12 \\ y &= 4 \end{aligned}$$

The correct answer is **(B)**.

16. A

Since x is an integer, $2x$ must be even because 2 times any integer is even. Since $2x$ is even and y is even, $2x + y$ is even because an even plus an even is even. Then $(-1)^{2x+y} = 1$, because -1 is being raised to an even exponent. Quantity A is greater, so the correct answer is **(A)**.

17. B

Whenever a problem calls for the median or mode, put the numbers in order first: 7, 8, 9, 10, 10, 12, 20, 60. The mode—the number that occurs most often—is 10. Use reasoning, not calculating, to determine that the mean must be more than 10. Reason like this: if the mean were 10, then the sum of all eight numbers would be $8 \times 10 = 80$. However, you know the sum of the numbers is greater than 80 because the sum of just the last three numbers $(12 + 20 + 60)$ is greater than 80. So, the mean is greater than 10. The correct answer is **(B)**.

18. D

Eliminate the square root in the equation by squaring both sides; $x^2 + 39 = 64$. Then $x^2 = 25$. So $x = 5$ or $x = -5$. If $x = 5$, then Quantity A is greater. If $x = -5$, then Quantity B is greater. Because more than one relationship between the quantities is possible, the relationship between the quantities cannot be determined. The correct answer is **(D)**.

19. D

Use the strategy of Picking Numbers.

For Quantity A, pick $(10, -2)$; for Quantity B, pick $(1, 2)$. $10 + (-2) = 8$ and $1 \times 2 = 2$. In this case, Quantity A is greater.

Next, for Quantity A, pick (1, –2); for Quantity B, pick (1, 2). $1 + (-2) = -1$ and $1 \times 2 = 2$. In this case, Quantity B is greater.

The correct answer is **(D)**.

20. D

Since $4 < x$ and $3 < y$, the value of $x + y$ must be greater than 7. Since $x < 10$ and $y < 5$, the value of $x + y$ must be less than 15. So $7 < x + y < 15$. Quantity B is 14. Since $x + y$ can be any value between 7 and 15, it could be less than 14, or greater than 14. You can use Picking Numbers to see this. If $x = 9$ and $y = 4$, Quantity A is $9 + 4 = 13$ and Quantity B is greater. However, if $x = 9.5$ and $y = 4.5$, then Quantity A is $9.5 + 4.5 = 14$ and the quantities are the same. There are different possible relationships between Quantity A and Quantity B, so the correct answer is **(D)**.

Advanced

21. A

First, recall the formula for the area of a rectangle: $A = lw$. Then consider how to express the length after an increase and the width after a decrease. In Quantity A, the value of x is 20. Add 0.20 to l for the increase and subtract 0.20 from w for the decrease. That makes the new length $1.2l$ and the new width $0.8w$. So, if $x = 20$, $Area = (1.2l)(0.8w) = 1.20 \times 0.80(lw)$.

For $x = 40$ under Quantity B, add 0.40 to l and subtract 0.40 from w. That gives you a new length of $1.40l$ and a new width of $0.60w$. For Quantity B, $Area = (1.4l)(0.6w) = 1.40 \times 0.60(lw)$.

To compare the quantities in each column, ignore the piece in each expression that is the same (lw) and just compare the product 1.20×0.80 to the product 1.40×0.60.

Quantity A: $1.20 \times 0.80 = 0.96$

Quantity B: $1.40 \times 0.60 = 0.84$

Because $0.96 > 0.84$, the answer is **(A)**.

Picking Numbers is also a great strategy here. Try picking 10 for both l and w. Then Quantity A is $12 \times 8 = 96$, and Quantity B is $14 \times 6 = 84$. Then try picking 20 for l and 30 for w. Now, Quantity A is $24 \times 24 = 576$ and Quantity B is $28 \times 18 = 504$. Since Quantity A is still bigger than Quantity B, it's probably safe to assume that the relationship is consistent, and the answer is, again, **(A)**.

22. B

If $|4x + 24| = 96$, then $4x + 24$ equals either 96 or –96. If the expression equals 96, then $4x + 24 = 96$, $4x = 72$, and $x = 18$. If the expression equals –96, then $4x + 24 = -96$, $4x = -120$, and $x = -30$. So for the first equation, x is either 18 or –30.

For the second equation, if $|4x| = 120$, then $4x$ equals either 120 or –120. If the expression equals 120, then $4x = 120$, and $x = 30$. If the expression equals –120, then $4x = -120$, and $x = -30$.

Since the same variable x is used in both equations, it must have the same value. Therefore, $x = -30$. Quantity A is -30, and Quantity B is 18. The correct answer is **(B)**.

23. C

A vehicle driven 50 miles per hour for 2 hours travels $2 \times 50 = 100$ miles in that time.

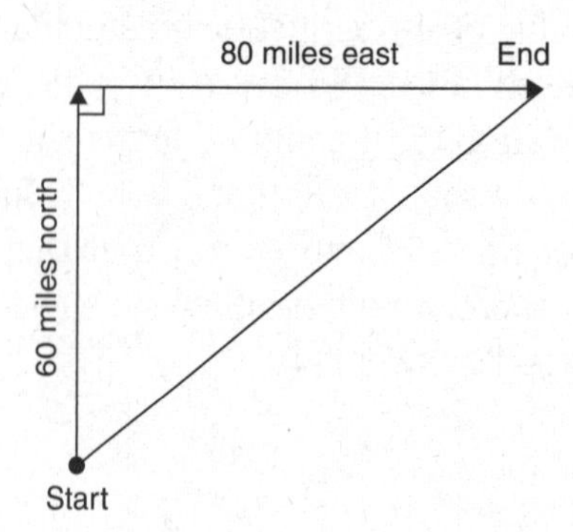

If a vehicle goes 60 miles north and 80 miles east, the distance from the starting point to the ending point is the hypotenuse of a right triangle with sides 60 and 80. The triangle formed is a 3:4:5 triangle because $60 = 3 \times 20$ and $80 = 4 \times 20$. The hypotenuse of the triangle is $5 \times 20 = 100$. Therefore, the quantities are equal. The correct answer is **(C)**.

24. B

A 6:8:10 triangle is a multiple of a 3:4:5 special right triangle. Because this is a right triangle with legs of 6 and 8, the base and height of the triangle are 6 and 8, and you can plug these into the area formula: $A = \frac{1}{2}(6)(8) = 24$. The area of the 6:8:10 right triangle is $A = \frac{1}{2}(6)(8) = 24$.

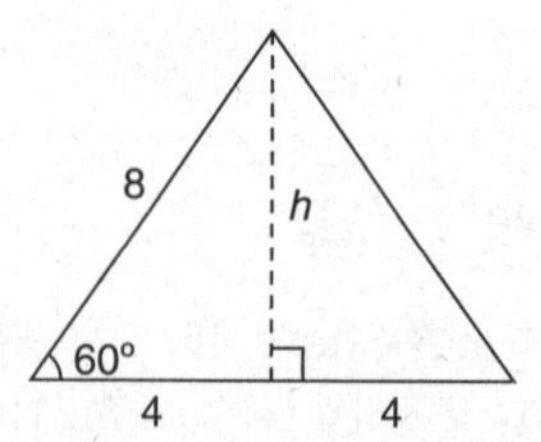

An equilateral triangle has three equal angles of 60 degrees each. Hence, when you drop the height as shown in the figure, you bisect the angle at the top, creating two 30°-60°-90° triangles. The ratios of a 30°-60°-90° triangle are $1{:}\sqrt{3}{:}2$, so the height, h, of the equilateral triangle is $4\sqrt{3}$. The area is $A = \frac{1}{2}(8)(4\sqrt{3}) = 16\sqrt{3}$. You can divide both quantities by 8 so that Quantity A is equal to 3 and Quantity B is equal to $2\sqrt{3}$. If you square both sides to make them look more alike, then Quantity A is 3^2, which equals 9. Quantity B is 4×3, which equals 12. Quantity B is larger, so the correct answer is **(B)**.

25. D

The centered information tells you that x is negative and y is positive. That means that Quantity B is always positive because it is the product of –1, a negative number, and a positive number. What about Quantity A? It can be positive or negative depending on the values chosen for x and y. Use the strategy of Picking Numbers to see whether it is possible to demonstrate that more than one relationship is possible, which is often the case when variables appear in both quantities.

Let $x = -1$ and $y = 2$. Then $-2(x + y) = -2(-1 + 2) = -2(1) = -2$ in Quantity A and $-xy = -1(-1)(2) = 2$ in Quantity B. In this case, Quantity B is greater.

Now, use reasoning before picking another set of numbers. What would it take for Quantity A to be greater? You need the sum of x and y to be a negative number with a large absolute value. That is, there must be a large negative difference between x and y.

Let $x = -51$ and $y = 1$. Then $-2(x + y) = -2(-51 + 1) = -2(-50) = 100$ in Quantity A and $-xy = -1(-51)(1) = 51$ in Quantity B. In this case, Quantity A is greater.

Two relationships have been demonstrated and the correct answer is **(D)**.

26. C

Quantity A: Take $g(x)$ first and then plug the result into $f(x)$.

$$g(x) = g(3) = \frac{1}{2}(3) = \left(\frac{3}{2}\right), \text{ and then } f(g(x)) = f\left(\frac{3}{2}\right) = 2\left(\frac{3}{2}\right) = 3.$$

Now, Quantity B: Take $f(x)$ first and then plug the result into $g(x)$.

$$f(x) = f(3) = 2(3) = 6, \text{ and then } g(f(x)) = g(6) = \frac{1}{2}(6) = 3.$$

The result is the same both times. The correct answer is **(C)**.

27. C

At first it may seem that there is not enough information to compare the measure of $\angle BAC$ to 70°. But the centered information tells you that AB is a diameter of the circle and provides a way to get started with the comparison.

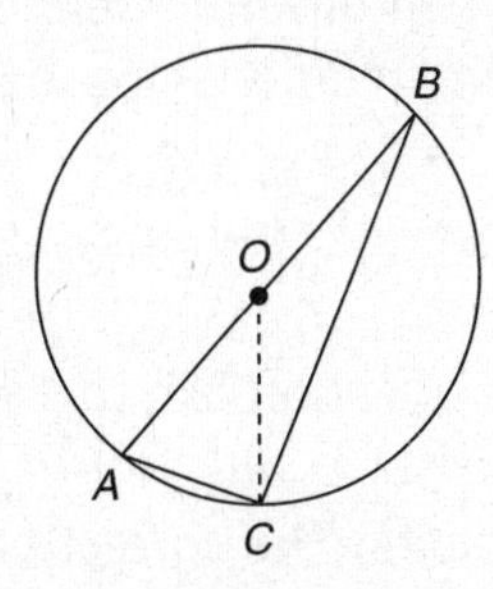

Because one side of the triangle is the diameter of the circle, this is a right triangle with $\angle BCA$ measuring 90 degrees. Since the centered information says that $\angle ABC = 20°$, $\angle BAC$ must equal 70 degrees. The two quantities are equal, so the answer is **(C)**.

28. C

Set x as the largest of the five consecutive even integers. Then, the second-largest even integer will be $x - 2$, the third-largest even integer will be $x - 4$, the fourth-largest will be $x - 6$, and the smallest of the five even integers will be $x - 8$. The sum of these five expressions is $(x) + (x - 2) + (x - 4) + (x - 6) + (x - 8)$, which equals $5x - 20$. Therefore $5x - 20 = -20$ and $x = 0$. This question may have been confusing at first because zero seems like a neutral number that is neither even nor odd, but zero is an even number because it is an integer multiple of 2 $(2 \times 0 = 0)$.

Alternatively, use the fact that, for a set of consecutive or evenly-spaced integers, the median is equal to the mean (average); and, as always, the average equals the sum divided by the number of terms. So median (the third term) = mean = $\frac{-20}{5} = -4$. To find the greatest term, just count up by 2s to the fifth term: $-4, -2, 0$.

The correct answer is **(C)**.

29. D

The question asks for a comparison of $2 - r$ and t. You know that the centered information in a GRE question is important, so begin by plugging the given value for s into the top equation.

If $s = 4$, then $4 = (t + r)^2$, and you now have a quadratic equation with a perfect square on both sides of the equal sign. A quadratic equation can have two solutions, one positive and one negative. So, $(t + r)$ may be $+2$ or -2. Write out the equations for each case:

If $t + r = 2$, then $t = 2 - r$

If $t + r = -2$, then $t = -2 - r$

There is not enough information to make a comparison, so the answer is **(D)**.

Another way to make short work of this one is to pick numbers that satisfy the centered information. If $t = 3$ and $r = -1$, for example, then the two quantities are equal: $2 - (-1) = 3$. If, on the other hand, $t = -1$ and $r = -1$, then Quantity A has the greater value: $2 - (-1) > -1$. When different sets of numbers satisfy the centered information but produce different relationships between the two quantities, the answer is always **(D)**.

30. D

Be careful when the GRE poses a question like this without telling you the elements in the set. For example, the data set could be $\{-2, 0, 2\}$ or $\{-4, 0, 2, 2\}$. It could also be a set of many more elements. The correct answer is **(D)**.

CHAPTER 3

Problem Solving

PROBLEM SOLVING

You will find about nine Problem Solving questions per Quantitative Reasoning section on the actual GRE. Problem Solving questions may be pure math or they may involve a real-world situation. The GRE tests algebraic, arithmetic, and geometric concepts. The questions will test your ability to reason mathematically using your knowledge of the various topics.

You will see three formats for answering Problem Solving questions on the GRE. You may be required to select one answer from the five choices given, select one or more choices from the choices given, or enter your answer in an on-screen box.

PROBLEM SOLVING QUESTIONS (SINGLE ANSWER)

THE KAPLAN METHOD FOR PROBLEM SOLVING

STEP 1 **Analyze the question.**

STEP 2 **Identify the task.**

STEP 3 **Approach strategically.**

STEP 4 **Confirm your answer.**

The directions for a Problem Solving question requiring a single answer will look like this:

> **Directions:** Select one answer choice.

A Problem Solving question requiring you to select a single answer will look like this:

> A retailer charges 25% more than his purchase price for any appliance he sells. When the retailer has a clearance sale, all appliances are marked 10% off. If the dealer sells a vacuum cleaner during a clearance sale, his profit (selling price minus purchase price) is what percent of his purchase price of the vacuum cleaner?
>
> (A) 10%
> (B) 12.5%
> (C) 15%
> (D) 17.5%
> (E) 20%

Apply the Kaplan Method for Problem Solving

Now let's apply the Kaplan Method to a Problem Solving question requiring a single answer.

A retailer charges 25% more than his purchase price for any appliance he sells. When the retailer has a clearance sale, all appliances are marked 10% off. If the retailer sells a vacuum cleaner during a clearance sale, his profit (selling price minus purchase price) is what percent of his purchase price of the vacuum cleaner?

(A) 10%
(B) 12.5%
(C) 15%
(D) 17.5%
(E) 20%

STEP 1

Analyze the question.

This is a real-world situation involving a retailer's percent markup to obtain a selling price and the retailer's percent markdown of items for a clearance sale. The question asks for the percent profit the retailer makes on an item he sells during a clearance sale. The answer choices are given as percents; however, no specific price is given for the item sold.

STEP 2

Identify the task.

The task is to calculate the percent profit, using the percent markup of 25% and the percent markdown of 10% for every item sold.

STEP 3

Approach strategically.

Although this problem could be solved using algebra, it may not be the fastest way for you to arrive at an answer. Instead, try the strategy of Picking Numbers for the initial purchase price of the vacuum cleaner and go from there. Suppose the retailer purchased the vacuum for $100.

A great number to pick for percent problems when no value is given is 100 because the calculations will be quite manageable.

First, determine the price assigned to the vacuum after a 25% markup:

$$125\% \text{ of } 100 = 1.25 \times 100 = 125$$

So the price assigned to the vacuum was $125, but it did not sell at that price. It was reduced 10%, so take 90% ($100 - 10 = 90$) of 125 to find the clearance sale price:

$$90\% \text{ of } 125 = 0.90 \times 125 = 112.50$$

When the vacuum sells at this price, the profit for the retailer is $\$112.50 - \$100 = \$12.50$.

Now you are ready to compare the profit of 12.50 to the original 100:

$$\frac{12.50}{100} = 0.125 = 12.5\%$$

The correct answer is **(B)**, 12.5%.

STEP 4

Confirm your answer.

There are two components to confirming your answer. First, does your answer make sense in the context of the question? In this case, a profit of 12.5% is less than the markup percent and more than the markdown percent, so it seems reasonable. Second, did you answer the question that was asked? In this question, you are asked *his profit is what percent of his purchase price?* Look back at the calculation to see that the correct comparison was made: dollar amount of profit compared to dollar amount of purchase price. Your answer is confirmed.

For more information on percents, see chapter 6.

PROBLEM SOLVING (ALL-THAT-APPLY)

The directions for a Problem Solving question requiring you to select one or more answers will look like this:

Directions: Select one or more answer choices.

A Problem Solving question requiring you to select one or more answers will look like this:

> The product of two integers is 14. Which of the following could be the average (arithmetic mean) of the two integers?
>
> Indicate <u>all</u> such averages.
>
> [A] –7.5
> [B] –6.5
> [C] –4.5
> [D] 4.5
> [E] 6.5

APPLY THE KAPLAN METHOD FOR PROBLEM SOLVING

Now let's apply the Kaplan Method to a Problem Solving question requiring you to select one or more answers.

The product of two integers is 14. Which of the following could be the average (arithmetic mean) of the two integers?

Indicate <u>all</u> such averages.

[A] –7.5
[B] –6.5
[C] –4.5
[D] 4.5
[E] 6.5

STEP 1

Analyze the question.

This is a pure math question involving integers. Whenever you see possible values described as integers, remember that the set of integers could include both positive and negative values and zero. In the question, you are given information about the product of two integers. Notice that the answer choices are not integers, but they do include both negative and positive values.

STEP 2

Identify the task.

The task is to select all the values that could be an average of the integers whose product is 14.

STEP 3

Approach strategically.

The best strategy here is to find all possible answers and then check which ones are given as answer choices. Start by writing the factors of positive 14: $1 \times 14 = 14$ and $2 \times 7 = 14$. Next consider the negative integers whose product is 14: $-1 \times (-14) = 14$ and $-2 \times (-7) = 14$.

Recall that the average of two numbers is the sum of the numbers divided by 2.

$$\frac{1+14}{2} = \frac{15}{2} = 7.5 \qquad \frac{-1+(-14)}{2} = \frac{-15}{2} = -7.5$$

$$\frac{2+7}{2} = \frac{9}{2} = 4.5 \qquad \frac{-2+(-7)}{2} = \frac{-9}{2} = -4.5$$

The correct answers are **(A)**, **(C)**, and **(D)**.

STEP 4

Confirm your answer.

For this question, check that you considered all the possible factors of 14. There were four possibilities because the only factors of 14 are 1 and 14, 2 and 7, –1 and –14, and –2 and –7. Therefore, it was reasonable to do all the calculations. The values chosen are the averages of the sets of factors of 14. The answer is confirmed.

For more information on averages, see chapter 6.

PROBLEM SOLVING QUESTIONS (NUMERIC ENTRY)

The directions for a Problem Solving question requiring you to make a Numeric Entry will look like this:

> **Directions:** Click in the box and type a number. Backspace to erase.

Enter your answer as an integer or decimal if there is one box or as a fraction if there are two boxes.

To enter an integer or decimal, type directly in the box or use the Transfer Display button on the calculator.

- Use backspace to erase.
- Use a hyphen to enter a negative sign; type a hyphen a second time to remove it. The digits will remain.

- Use a period for a decimal point.
- The Transfer Display button will enter your answer directly from the calculator.
- Equivalent forms of decimals are all correct. Example: $.14 = 0.140$.
- Enter the exact answer unless the question asks you to round your answer.

To enter a fraction, type the numerator and denominator in the appropriate boxes.

- Use a hyphen to enter a negative sign.
- The Transfer Display button does not work for fractions.
- Equivalent forms of fractions are all correct. Example: $\frac{25}{15} = \frac{5}{3}$.
- If numbers are large, reduce fractions to fit in boxes.

A Problem Solving question with Numeric Entry will look like this:

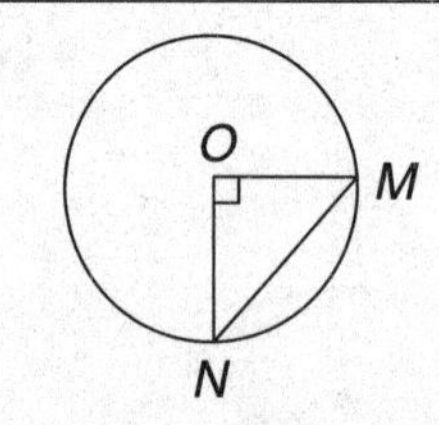

In the figure shown, the area of the circle whose center is *O* is 16π. What is the area of triangle *MNO*?

[ ] square units

Apply the Kaplan Method for Problem Solving

Now let's apply the Kaplan Method to a Problem Solving question involving Numeric Entry.

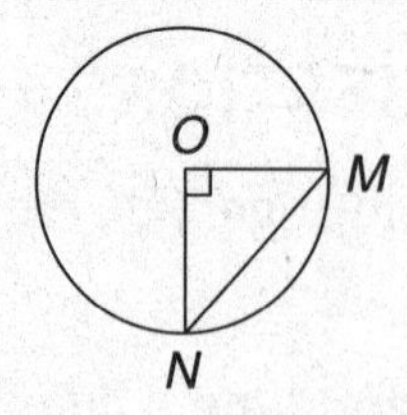

In the figure shown, the area of the circle whose center is *O* is 16π. What is the area of triangle *MNO*?

[ ] square units

STEP 1

Analyze the question.

You must use the figure shown and the information given to answer this geometry question. You have a circle with an inscribed triangle to deal with here. The center of the circle, *O*, is also a vertex of the right triangle. The question indicates that you will fill in a numerical answer for the square units.

STEP 2

Identify the task.

You must find the area of the triangle using the information given. That is, you will use the value given for the area of the circle to determine information about the dimensions of the triangle. Then you will be able to find the area of the triangle.

STEP 3

Approach strategically.

Start with what you know—the area of a circle is given by the equation $A = \pi r^2$. Solve for the length of the radius:

$$\begin{aligned} A &= \pi r^2 \\ 16\pi &= \pi r^2 \\ 16 &= r^2 \\ 4 &= r \end{aligned}$$

In the figure, the radius of the circle is also the base and height of the isosceles right triangle. The area of a triangle is given by the equation

$$A = \frac{1}{2}(\text{base})(\text{height}) = \frac{1}{2} \times 4 \times 4 = 8.$$

The area of the triangle is **8** square units.

STEP 4

Confirm your answer.

Do a quick check to confirm that your answer makes sense. If the area of the complete circle is 16π, then the area of one-fourth of the circle is 4π. Use 3 as an estimate for π and you have 4π is approximately $4 \times 3 = 12$. In the figure, the triangle takes up less than one-fourth of the circle, so the answer of **8** is reasonable.

For more information on circles and triangles, see chapter 8.

PROBLEM SOLVING PRACTICE SET

Try the following Problem Solving questions using the Kaplan Method.

Basic

1. What is the perimeter of a parallelogram with adjacent side lengths measuring $9a$ and $14b$?

   (A) $9a + 14b$
   (B) $18a + 28b$
   (C) $23(a + b)$
   (D) $36a$
   (E) $63ab$

2. What is the value of $-[(s + t)^0]$ if $s + t \neq 0$?

   (A) -1
   (B) 0
   (C) 1
   (D) $s + t$
   (E) $-(s + t)$

3. The data below show the monthly dollar amounts Marco spent on postage over the past 6 months. What is the average (arithmetic mean) of the data set?

   \$1.08, \$5.43, \$2.17, \$3.25, \$5.95, \$1.08

   \$ [ ]

4. What is the sixth term in a sequence in which the nth term is $n(n - 1)^2$?

   (A) 42
   (B) 150
   (C) 900
   (D) 1,225
   (E) 2,592

5. What is the value of $(2\sqrt{2})(\sqrt{6}) + 2\sqrt{3}$?

(A) 18
(B) $10\sqrt{3}$
(C) $6\sqrt{6}$
(D) $4\sqrt{2} + 2\sqrt{3}$
(E) $6\sqrt{3}$

6. If one of the angle measures of an equilateral triangle is given in degrees as $15n$, what is the value of n?

(A) 3
(B) 4
(C) 6
(D) 12
(E) 15

7. If $3x + y = -1$ and $y - 2x = 4$, what is the value of $x + 2y$?

(A) –5
(B) –3
(C) 0
(D) 3
(E) 5

8. The sum of the interior angles of a regular polygon is less than 540°. Which could be the polygon?

Indicate <u>all</u> such polygons.

[A] triangle
[B] quadrilateral
[C] pentagon
[D] hexagon

9. The ratio of $\frac{1}{2}$ to $\frac{3}{5}$ is the same as which of the following ratios?

(A) 1:5
(B) 3:10
(C) 2:3
(D) 5:6
(E) 3:2

10. What is the value of $(-8)^{-3}$?

(A) -512
(B) -24
(C) -2
(D) $-\frac{1}{2}$
(E) $-\frac{1}{512}$

Intermediate

11. A computer company's featured laptop cost \$800 last year. This year, the laptop sold for 15% less than it did last year. Next year, after updates are made to the model, there will be a 25% price increase over this year's price. What will be the price next year?

(A) \$810
(B) \$825
(C) \$840
(D) \$850
(E) \$880

12. If $x < -3$ and $x^2 + 5x + 12 = 8$, what is the value of $x + 2$?

(A) -4
(B) -2
(C) -1
(D) 0
(E) 3

13. What is the probability of rolling a number greater than 2 twice in a row on a fair six-sided die, with each of the numbers 1–6 on each side?

(A) $\frac{1}{4}$

(B) $\frac{5}{18}$

(C) $\frac{4}{9}$

(D) $\frac{5}{9}$

(E) $\frac{2}{3}$

14. If ♣$m = 3m$ and •$n = n + 7$, which of the following is the value of [(♣(•4) – 3(•2)]?

(A) –9
(B) 1
(C) 6
(D) 13
(E) 20

15. A flower shop sells flowers in a ratio of roses to carnations of 5:2. The ratio of carnations to tulips sold is 5:3. What is the ratio of roses to tulips?

(A) 2:3
(B) 3:2
(C) 5:3
(D) 25:6
(E) 25:9

16. A principal has four different trophies available for display in his two display cases. If only one trophy can fit in each display case, how many distinct ways are there to display two of the trophies in the cases at any given time?

(A) 16
(B) 12
(C) 9
(D) 8
(E) 6

17. Telephone company A charges $3.00 for the first minute of any long distance call and $0.50 for each additional minute. Telephone company B charges $2.00 for the first minute of any long distance call and $0.70 for each additional minute. If the cost of a call lasting x minutes, where x is a positive integer, is $15.00 more with telephone company B than with telephone company A, then what is the value of x?

Ⓐ 56
Ⓑ 60
Ⓒ 77
Ⓓ 80
Ⓔ 81

18. Find the measures, in degrees, of angles *GEF* and *DEG*.

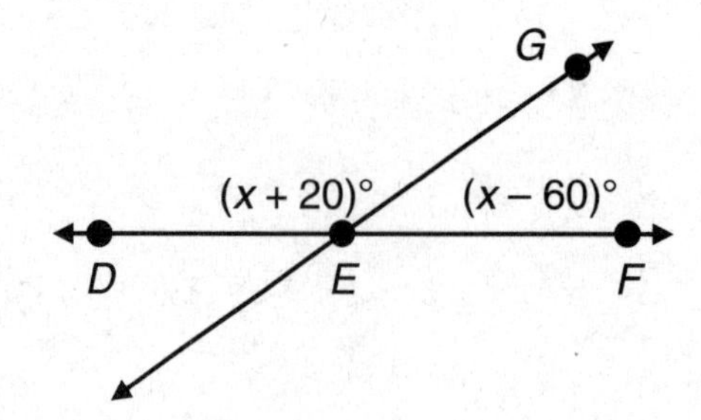

Ⓐ 30 and 150
Ⓑ 50 and 110
Ⓒ 50 and 130
Ⓓ 90 and 170
Ⓔ 110 and 130

19. The average (arithmetic mean) of a, b, and c is 70, and the average (arithmetic mean) of d and e is 120. What is the average (arithmetic mean) of a, b, c, d, and e?

Ⓐ 84
Ⓑ 90
Ⓒ 95
Ⓓ 96
Ⓔ 100

20. The length of one side of a triangle is 12. The length of another side is 18. Which of the following could be the perimeter of the triangle?

Indicate <u>all</u> such perimeters.

[A] 30
[B] 36
[C] 44
[D] 48
[E] 60

Advanced

21. The average of all the consecutive integers from a to b inclusive is 39. Which of the following could be a and b?

Indicate <u>all</u> such integers.

[A] 4 and 74
[B] 19 and 39
[C] 25 and 53
[D] 29 and 59
[E] 33 and 45

22. A square is inscribed inside a shaded circle, as shown. The circumference of the circle is $6\pi\sqrt{2}$. What is the area of the shaded region?

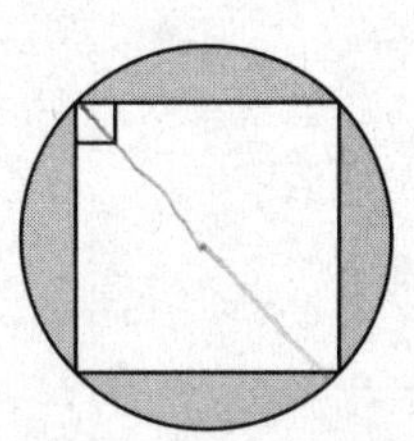

(A) $12\pi - 6\sqrt{2}$
(B) $12\pi - 18$
(C) $18\pi - 6$
(D) $18\pi - 36$
(E) $36\pi - 18$

23. Point A (4, 6) lies on a line with slope $-\frac{3}{4}$. Point B lies on the same line and is 5 units from Point A. Which of the following could be the coordinates of Point B?

    Indicate <u>all</u> such coordinates.

    [A] (−1, 1)
    [B] (−4, 12)
    [C] (8, 3)
    [D] (1, 10)
    [E] (0, 9)

24. The surface area of a cube with side length $(x + 4)$ is 294. What is the value of x?

    (A) −10
    (B) −1
    (C) 3
    (D) 7
    (E) 19

25. Pipe A can fill a tank in 3 hours. If pipe B can fill the same tank in 2 hours, how many minutes will it take both pipes to fill $\frac{2}{3}$ of the tank?

    (A) 30
    (B) 48
    (C) 54
    (D) 60
    (E) 72

26. If it takes Nathan 4 hours to unload a moving truck and it takes Iris 2 hours longer than Nathan to unload a moving truck, how long would it take the two of them, working together, to unload 2 moving trucks?

    (A) 5 hours
    (B) 4 hours, 48 minutes
    (C) 4 hours
    (D) 2 hours 24 minutes
    (E) 2 hours

27. The integer y is positive. If 6^y is a factor of $(2^{14})(3^{24})$, then what is the greatest possible value of y?

Ⓐ 7
Ⓑ 8
Ⓒ 14
Ⓓ 18
Ⓔ 36

28. Phillip has twice as many tropical fish as Jody. If Phillip gave Jody 10 of his tropical fish, he would have half as many as Jody. How many tropical fish do Phillip and Jody have together?

Ⓐ 10
Ⓑ 20
Ⓒ 30
Ⓓ 40
Ⓔ 60

29. When Dahlia's professor eliminated the lowest of her 4 quiz scores, her quiz average rose from 77 to 91. What was the score of the quiz that the professor eliminated?

Ⓐ 14
Ⓑ 21
Ⓒ 35
Ⓓ 42
Ⓔ 56

30. In company X, no employee is both a technician and an accountant. Also, in company X, $\frac{2}{5}$ of the employees are technicians, and $\frac{5}{16}$ of the remaining employees are accountants. What fraction of the total number of employees at company X are neither technicians nor accountants?

Ⓐ $\frac{23}{80}$

Ⓑ $\frac{33}{80}$

Ⓒ $\frac{3}{7}$

Ⓓ $\frac{47}{80}$

Ⓔ $\frac{57}{80}$

PROBLEM SOLVING PRACTICE SET ANSWER KEY

1. B
2. A
3. 3.16
4. B
5. E
6. B
7. D
8. A, B
9. D
10. E
11. D
12. B
13. C
14. C
15. D
16. B
17. E
18. C
19. B
20. C, D
21. A, C, E
22. D
23. C, E
24. C
25. B
26. B
27. C
28. C
29. C
30. B

PROBLEM SOLVING PRACTICE SET ANSWERS AND EXPLANATIONS

Basic

1. B

The perimeter of a parallelogram is the sum of the lengths of all four of its sides. Two side lengths are given, and a parallelogram has two pairs of equal opposite sides. So, the perimeter of the parallelogram is $2(9a) + 2(14b) = 18a + 28b$, and the answer is **(B)**.

2. A

For any non-zero base raised to the zero power, the value is equal to 1. (If the base and exponent are both zero, then the number is undefined.) Since the problem specifies that the base is not zero, it does not matter what *s* or *t* is. Their sum raised to the zero power will be 1. The expression asks for the negation of 1, which is –1, so the answer is **(A)**.

3. 3.16

Use the formula for average:

$$Average = \frac{Sum\ of\ terms}{Number\ of\ terms} = \frac{1.08 + 5.43 + 2.17 + 3.25 + 5.95 + 1.08}{6}$$

$$= \frac{18.96}{6} = 3.16$$

The answer is \$3.16.

4. B

Use the rule given, substituting in 6 for *n* to find the sixth term. Be sure to follow the order of operations (PEMDAS):

$$\begin{aligned} 6(6-1)^2 &= \\ 6(5)^2 &= \\ 6(25) &= 150 \end{aligned}$$

The correct choice is **(B)**.

5. E

Following PEMDAS, simplify the first part of the expression first. To multiply radical terms, deal with the inside and outside separately. Then, if the radical terms have the same number under the $\sqrt{\ }$, they can be added together.

$$\begin{aligned} \left(2\sqrt{2}\right)\left(\sqrt{6}\right) + 2\sqrt{3} &= \left(2\sqrt{2}\right)\left(\sqrt{2} \times \sqrt{3}\right) + 2\sqrt{3} \\ = (2)\left(\sqrt{2} \times \sqrt{2}\right)\left(\sqrt{3}\right) + 2\sqrt{3} &= 4\left(\sqrt{3}\right) + 2\sqrt{3} \\ &= 6\sqrt{3} \end{aligned}$$

The answer is **(E)**.

6. B

An equilateral triangle has three equal angles. Since the sum of the angles of a triangle always equals 180°, you can divide 180 by 3 to determine that the angles of an equilateral triangle are each 60°. So, $15n = 60$ and $n = 4$. The answer is **(B).**

7. D

The question asks for $x + 2y$, but not for x or y individually. In such situations, it is typically unnecessary to solve for the value of x and the value of y. Rearrange the equations so that like terms are on top of each other and then add the equations:

$$\begin{array}{r} 3x + y = -1 \\ +[-2x + y = 4] \\ \hline x + 2y = 3 \end{array}$$

Since $x + 2y = 3$, the answer is **(D).**

8. A, B

In a regular polygon with n sides, the formula for the sum of the interior angles is $(n - 2) \times 180$. This equation can be set up as an inequality to solve for n:

$(n - 2) \times 180 < 540°$

Dividing both sides by 180 gives us:

$n - 2 < 3$

$n < 5$

The polygons listed with fewer than 5 sides are the triangle and quadrilateral, so the answers are **(A)** and **(B).**

9. D

The ratio of two numbers a and b can be expressed as $\frac{a}{b}$. So the expression is the same as the value of $\frac{1}{2}$ divided by $\frac{3}{5}$. To divide by a fraction, you multiply by the reciprocal, $\frac{1}{2} \times \frac{5}{3} = \frac{5}{6}$. So the ratio of $\frac{1}{2}$ and $\frac{3}{5}$ is the same as $\frac{5}{6}$, which can be written as the ratio of 5:6. The answer is **(D).**

Alternatively, express both fractions in terms of a common denominator: $\frac{1}{2} = \frac{5}{10}$ and $\frac{3}{5} = \frac{6}{10}$. Now compare $\frac{5}{10}$ to $\frac{6}{10}$ and see that the ratio is 5:6. Again, the answer is **(D).**

10. E

A number raised to a negative exponent is equal to the reciprocal of that number raised to the positive exponent. So,

$$(-8)^{-3} = \frac{1}{(-8)^3} = \frac{1}{(-8)(-8)(-8)} = -\frac{1}{512}$$

The answer is **(E)**.

Intermediate

11. D

First, calculate the price of the laptop this year. Then, use that price to determine what the price will be next year. After a 15% decrease in price, the system would sell for 85% of \$800: $0.85 \times \$800 = \$680$. If there is a 25% increase next year, the system would sell for 125% of this year's price. That would be 125% of \$680: $1.25 \times \$680 = \$850$. The answer is **(D)**.

12. B

The equation is quadratic, so all terms should be moved to one side and set equal to zero to solve for x. The equation $x^2 + 5x + 12 = 8$ becomes $x^2 + 5x + 4 = 0$. This can be factored to $(x + 4)(x + 1) = 0$, so either $x + 4 = 0$ or $x + 1 = 0$, and solving for x gives $x = -1$ or $x = -4$. The problem states that $x < -3$, so the only valid solution is $x = -4$. Once the value of x is specified, you can substitute in the value for x to solve: $x + 2 = -4 + 2 = -2$. Always make sure to confirm your answer, because choice (A) is the trap that represents the value of x; but, remember, you're looking for the value of $x + 2$. The answer is **(B)**.

13. C

Use the probability formula:

$$\textit{Probability} = \frac{\textit{Number of desired outcomes}}{\textit{Number of total possible outcomes}}$$

There are 6 possible outcomes with each roll of a fair six-sided die, and 4 outcomes are greater than 2. The probability of rolling a number greater than 2 on the first roll is $\frac{4}{6} = \frac{2}{3}$.

The probability of rolling a number greater than 2 on the second roll is the same, $\frac{4}{6} = \frac{2}{3}$.

Multiply the two probabilities to find the chance of both events occurring: $\frac{2}{3} \times \frac{2}{3} = \frac{4}{9}$.

The correct choice is **(C)**.

14. C

The symbols represent operations on the numbers. So, to get the answer, apply the operation described in the problem statement each time the symbol appears. Apply PEMDAS and start with the interior parentheses. Since $\bullet n = n + 7$, $\bullet 4 = 4 + 7 = 11$ and $\bullet 2 = 2 + 7 = 9$. Now the expression is $[\clubsuit(11) - 3(9)] = (\clubsuit(11) - 27)$. Since $\clubsuit n = 3n$, $\clubsuit(11) = 3(11) = 33$. Finally, the expression is $33 - 27 = 6$. The answer is **(C)**.

15. D

This question is asking about a combined ratio. To compare the ratios, the number for carnations must be the same in each ratio. Since carnations are represented by 2 in one ratio and 5 in the other, convert the ratios so that both have carnations represented by 10. The number 10 is chosen because it is the least common multiple of 2 and 5. The ratio of roses to carnations is 5:2 = 25:10, and the ratio of carnations to tulips is 5:3 = 10:6. Now the value for carnations, 10, is the same in both ratios. The ratio of roses to tulips is 25:6, and the answer is **(D)**.

16. B

This is a permutations question. To understand why, first imagine that the two cases are in two different locations, say by the cafeteria and by the gymnasium. In this case, placing the hockey trophy in the cafeteria case and the debate team trophy in the gymnasium case is not the same as putting the debate trophy by the cafeteria and the hockey trophy by the gym. In other words, not only is the principal selecting two of four trophies, but he is also arranging them. Call the trophies A, B, C, and D. There are 3 ways for trophy A to be chosen first: AB, AC, AD. Likewise, there are 3 ways for Trophy B to be chosen first, 3 for trophy C, and 3 for trophy D. There are 12 total ways, so the answer is **(B)**.

You can also use the permutations formula to solve:

$$\frac{n!}{(n-k)!} = \frac{4!}{(4-2)!} = \frac{4 \times 3 \times 2 \times 1}{2 \times 1} = 12$$

17. E

The cost of a telephone call lasting x minutes with telephone company A is \$3.00 for the first minute and \$0.50 for each minute of the additional $x - 1$ minutes. So the cost of this telephone call with telephone company A is $3 + 0.5(x - 1)$ dollars. The cost of a telephone call lasting x minutes with telephone company B is \$2.00 for the first minute and \$0.70 for each minute of the additional $x - 1$ minutes. So the cost of this telephone call with telephone company B is $2 + 0.7(x - 1)$ dollars. Since the cost of a call lasting x minutes with telephone company B is \$15.00 more than the cost of a call lasting x minutes with telephone company A, we have this equation:

$$2 + 0.7(x - 1) = 3 + 0.5(x - 1) + 15$$

We can now solve this equation for x.

$$\begin{aligned} 2 + 0.7x - 0.7 &= 3 + 0.5x - 0.5 + 15 \\ 1.3 + 0.7x &= 17.5 + 0.5x \\ 0.2x &= 16.2 \end{aligned}$$

Therefore, $x = 81$. The answer is **(E)**.

18. C

From the general rules about relationships between angles formed by intersecting lines, you know that the sum of angles along a straight line is 180°. Angles *DEG* and *GEF* lie along a straight line. They are supplementary angles, so you can eliminate answer choices that do not sum to 180°; eliminate (B), (D), and (E). Now write an equation: $(x + 20) + (x - 60) = 180$. Simplify:

$$\begin{aligned} 2x - 40 &= 180 \\ 2x &= 220 \\ x &= 110 \end{aligned}$$

Since x is 110, the angle measures are $(x + 20) = (110 + 20) = 130$ and $(x - 60) = (110 - 60) = 50$, and the answer is **(C)**.

19. B

The average formula is $Average = \frac{Sum\ of\ terms}{Number\ of\ terms}$. Since the average of a, b, and c is 70, $\frac{a + b + c}{3} = 70$. Then $a + b + c = 3 \times 70$, or 210. Similarly, since the average of d and e is given as 120, $\frac{d + e}{2} = 120$. The sum of $d + e = 240$. Adding the corresponding sides of the equations $a + b + c = 210$ and $d + e = 240$, we have $(a + b + c) + (d + e) = 210 + 240$, and then $a + b + c + d + e = 450$. To find the average, divide 450 by 5. So the average of a, b, c, d, and e is 90. The correct answer is **(B)**.

20. C, D

The problem asks for the perimeter of the triangle. The triangle inequality theorem states that the length of the third side of a triangle must be between the positive difference and the sum of the other two sides. So, once you find the range of possible lengths for the third side, you can add the side lengths to find the range of possible perimeters.

The third side must be greater than the difference $18 - 12 = 6$ and less than the sum $12 + 18 = 30$. So, the perimeter must be greater than $12 + 18 + 6 = 36$ and less than $12 + 18 + 30 = 60$. Answer choices **(C)** and **(D)** are the only values that are greater than 36 and less than 60.

Advanced

21. A, C, E

The average of a group of consecutive integers is equal to the average of the smallest and largest integers, so any pair of numbers whose average is 39 could be a and b. Since the average of a and b is equal to their sum divided by 2, any pair of numbers whose sum is $2 \times 39 = 78$ is a valid answer choice. The answer choices with a pair of numbers that sum to 78 are **(A)**, **(C)**, and **(E)**.

22. D

The problem asks for the area of the shaded region, so you need to find the difference between the area of the circle and the area covered by the square.

For the area of the circle, first find the radius. The circumference of a circle is $2\pi r$. Since the circumference is given as $6\pi\sqrt{2}$, $2\pi r = 6\pi\sqrt{2}$, $2r = 6\sqrt{2}$, and $r = 3\sqrt{2}$. The area of the circle, then, is $\pi r^2 = \pi(3\sqrt{2})^2 = 18\pi$.

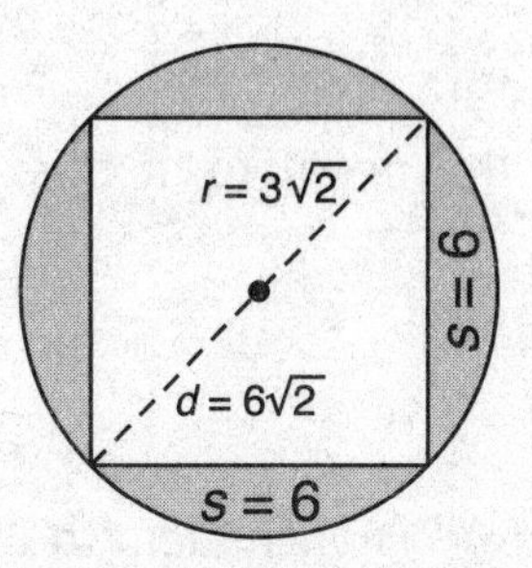

For the area of the square, find the side length. The square is inscribed in the circle, so the diagonal of the square is the diameter of the circle, or $6\sqrt{2}$. Since the diagonal of a square creates two 45°-45°-90° triangles, the sides of the square are the legs of a 45°-45°-90° triangle with a hypotenuse of $6\sqrt{2}$. Therefore, a side length is 6, and the area of the square is $6 \times 6 = 36$.

The difference between the area of the circle and the area of the square is $18\pi - 36$. The correct choice is **(D)**.

23. C, E

Using the definition of slope as $m = \frac{\text{rise}}{\text{run}} = -\frac{3}{4}$, you can plot point *A* and move vertically 3 and horizontally –4 to point $(4 - 4, 6 + 3) = (0, 9)$, which will also lie on the line with slope $-\frac{3}{4}$. This creates a 3:4:5 triangle, so the distance along the line from point *A* to the new point (0, 9) is 5 units, so (E) could be point *B*. You can also move vertically –3 and horizontally 4 to point $(4 + 4, 6 - 3) = (8, 3)$, which lies on the same line. Since the triangle formed is a 3:4:5 triangle again, this distance from point *A* to the new point (8, 3) is also 5. Choice (C) could also be point *B*. So, the answers are **(C)** and **(E)**.

24. C

First, consider that a cube has six faces and all edges of a cube are equal. So the total surface area is comprised of six squares. In this question, you're given the surface area as 294, so the area of one face is $\frac{294}{6} = 49$. The area of a square is equal to the square of its side, which is given as $x + 4$.

$(x + 4)^2 = 49$

$(x + 4)^2 = 7^2$

Remember, we can disregard the negative in this case, since distance won't be negative.

$$\begin{aligned} x + 4 &= 7 \\ x &= 3 \end{aligned}$$

Remember that you're solving for x, not the length of the side.

The answer is **(C)**.

25. B

Approach the problem strategically: find the number of hours it takes both pipes to fill the entire tank, multiply by $\frac{2}{3}$, and then convert hours to minutes.

To solve a combined work problem where the information is given in hours to complete the work, use the combined work formula. The time it takes pipe A and pipe B to fill the tank together is the product of their individual times to do the work divided by the sum of their individual times. In this case, that is $\left(\frac{2 \times 3}{2+3}\right) = \frac{6}{5}$. It would take the two pipes $\frac{6}{5}$ hours to fill the entire tank. It will take $\frac{2}{3} \times \frac{6}{5} = \frac{2}{{}_1\cancel{3}} \times \frac{\cancel{6}^2}{5} = \frac{4}{5}$ hour to fill $\frac{2}{3}$ of the tank. The final step is to convert $\frac{4}{5}$ hour to minutes. There are 60 minutes in one hour: $\frac{4}{5} \times 60 = 48$.

The answer is **(B)**.

26. B

According to the combined work formula, the amount of time it takes two people to do a single task together is the product of their individual times to do the task divided by the sum of those times. It takes Nathan 4 hours, so it takes Iris $4 + 2 = 6$ hours. Then:

$$T_{together} = \frac{T_a \times T_b}{T_a + T_b}$$

$$T_{together} = \frac{4 \times 6}{4 + 6}$$

$$T_{together} = \frac{24}{10}$$

$$T_{together} = \frac{12}{5}$$

So it would take $\frac{12}{5}$ hours, or 2 hours 24 minutes, to unload the truck together. However, the question asks how long would it take to unload *two* moving trucks, so the answer is 4 hours 48 minutes, which is **(B)**.

27. C

Let's rewrite 6^y by using the prime factorization of 6. The prime factorization of 6 is 2×3. We know by the law of exponents that $(ab)^n = a^n b^n$. So we have $6^y = (2 \times 3)^y = 2^y \times 3^y$.

Since we have 14 factors of two and 24 factors of three, there can be up to 14 factors of 6, since you're limited by the smallest number of possible factors. You cannot have 15 factors of 6, since you would need 15 factors of 2. Therefore, the greatest possible value of *y* is 14. The answer is **(C)**.

28. C

First, note that since Phillip has twice as many tropical fish as Jody, this can be written as the ratio $p:j = 2:1$, where *p* represents Phillip's original number of fish and *j* represents Jody's original number of fish. Then, because there are $2 + 1 = 3$ total parts in this ratio, the total number of fish that Phillip and Jody have must be a multiple of 3. Eliminate choices (A), (B), and (D). With just two possibilities to choose from, Backsolving would be a very efficient strategy here. However, this is also a great exercise in systems of linear equations, so here's the algebraic solution:

If Phillip has twice as many tropical fish as Jody, you can write $p = 2j$.

If Phillip gives Jody 10 fish, then he will have 10 fewer, or $p - 10$, and Jody will have 10 more, or $j + 10$. In this case Phillip would have half as many as Jody, so $p - 10 = \frac{1}{2}(j + 10)$.

Now you have two equations to describe the situation:

$$p = 2j$$
$$p - 10 = \frac{1}{2}(j + 10)$$

Solve this system of equations using substitution. Replace p in the second equation with $2j$ and solve:

$$p - 10 = \frac{1}{2}(j + 10)$$
$$2j - 10 = \frac{1}{2}(j + 10)$$
$$4j - 20 = j + 10 \qquad \text{Multiply both sides by 2.}$$
$$3j = 30 \qquad \text{Collect like terms.}$$
$$j = 10$$

Jody has 10 fish and Phillip has twice as many, 20. The total they have together is 30 fish. The answer is **(C)**.

29. C

The score that was deleted is equal to the difference between the original sum of the scores and the new sum of the scores. Since the average score is equal to the sum of the scores divided by the number of scores, use the average and number of scores given to find the sum before and after the lowest score was deleted. The original sum is equal to the original average score times the number of scores, or $77 \times 4 = 308$. The new sum is equal to the new average score times the number of scores, which is 3, since one score was eliminated. So, the new sum is $91 \times 3 = 273$. The score that was eliminated is $308 - 273 = 35$. The answer is **(C)**.

30. B

This is a good question for using the strategy of Picking Numbers. Start with picking a number for the total number of employees in the company. Since the denominators that appear in the fractions of the question stem contain the denominators 5 and 16, pick a total number of employees that works for both denominators. Since $5 \times 16 = 80$, start with 80. If $\frac{2}{5}$ of the 80 employees are technicians, then there are 32 technicians. This leaves $80 - 32 = 48$ employees who are not technicians. Then $\frac{5}{16}$ of these remaining 48 employees are accountants: $\frac{5}{16} \times 48 = 15$. Therefore, $48 - 15 = 33$ employees are neither technicians nor accountants. So $\frac{33}{80}$ of the employees are neither technicians nor accountants, and the answer is **(B)**.

CHAPTER 4

Data Interpretation

DATA INTERPRETATION

Data Interpretation questions are based on information presented to you in the form of graphs or tables of data. Typically, the data are presented in more than one graph or table, and you will be required to extract the data needed to answer the question from one or more of them. Questions range from those requiring simple arithmetic calculations to those that are more statistics-oriented. Typically, three or four questions will be associated with each data presentation.

You will see three formats for answering Data Interpretation questions on the GRE. You may be required to select one answer from the five choices given, select one or more choices from the choices given, or enter your answer in an on-screen box. These three formats for Data Interpretation questions are similar to those for Problem Solving questions.

THE KAPLAN METHOD FOR DATA INTERPRETATION

STEP 1 Analyze the tables and graphs.

STEP 2 Approach strategically.

DATA INTERPRETATION QUESTIONS (ONE ANSWER)

A Data Interpretation question requiring you to select a single answer will look like this:

Questions 1–3 are based on the following graphs.

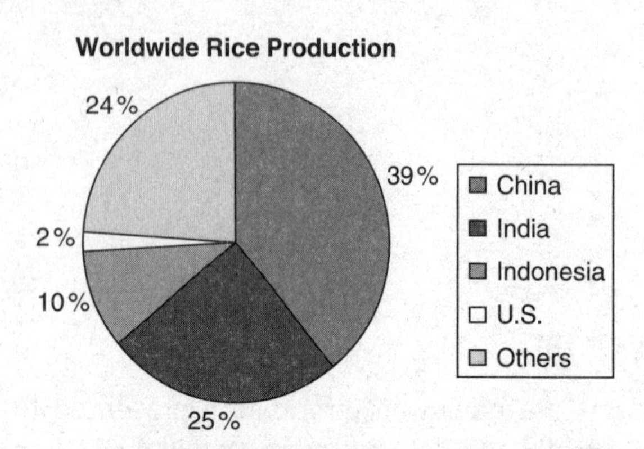

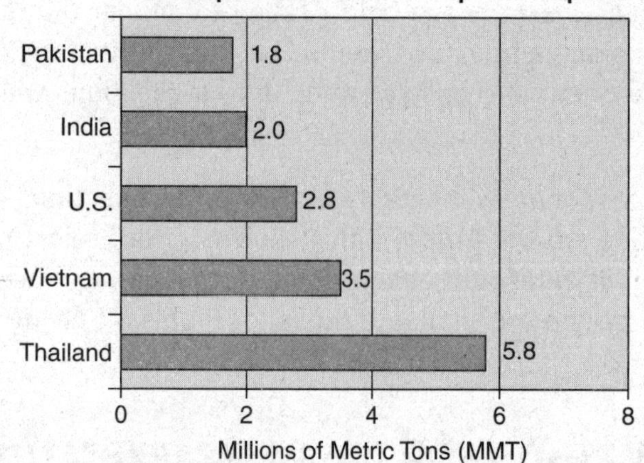

If the U.S. exports 50% of its crop one year, which is the best estimate, in MMT, of total worldwide production for that year?

(A) 56
(B) 112
(C) 200
(D) 280
(E) 400

APPLY THE KAPLAN METHOD FOR DATA INTERPRETATION

Now let's apply the Kaplan Method to a Data Interpretation question requiring a single answer.

If the U.S. exports 50% of its crop one year, which is the best estimate, in MMT, of total worldwide production for that year?

(A) 56
(B) 112
(C) 200
(D) 280
(E) 400

STEP 1

Analyze the tables and graphs.

The top graph is a pie chart showing a breakdown of worldwide rice production by country. Notice that the percents are given for each slice of the pie. The bottom bar graph shows data, given in MMT, concerning exporters of rice.

STEP 2

Approach strategically.

To approach a Data Interpretation question strategically, you must identify the information that is needed to answer the question and the calculation to be done. All other information presented in the data set is "put on hold" until that one question is answered. For this question, the amount 2.8 MMT is given for U.S. exports, and the question states that the number of MMT exported represents 50% of the U.S. production. You must use these pieces of information to estimate worldwide production. That requires going to the pie chart to see the percent assigned to the U.S. That is just 2% of worldwide production.

First, consider the amount exported, 2.8. If 2.8 represents 50% (half) of U.S. production, then the amount produced must be twice 2.8: $2 \times 2.8 = 5.6$.

Now use the percent formula: *Part* = *Percent* × *Whole*. You know the *Part*; it is 5.6. You also know the *Percent*; it is 2%. Set up the formula and solve for the *Whole*.

5.6 is 2% of what number?

$$5.6 = 2\left(\frac{1}{100}\right) \times \textit{Whole}$$

$$5.6 = \frac{2}{100} \times \textit{Whole}$$

$$5.6\left(\frac{100}{2}\right) = \left(\frac{2}{100}\right)\left(\frac{100}{2}\right)\textit{Whole}$$

$$5.6 \times 50 = \textit{Whole}$$

$$280 = \textit{Whole}$$

The solution to the equation indicates that 280 is the number of MMT produced worldwide. Choice **(D)** is correct.

For more information on percent equations, see chapter 6.

DATA INTERPRETATION QUESTIONS (ALL-THAT-APPLY)

The directions for a Data Interpretation question requiring you to select one or more answers will look like this:

> **Directions:** Select one or more answer choices.

A Data Interpretation question requiring you to select one or more answers will look like this:

Questions 1–3 are based on the following graphs.

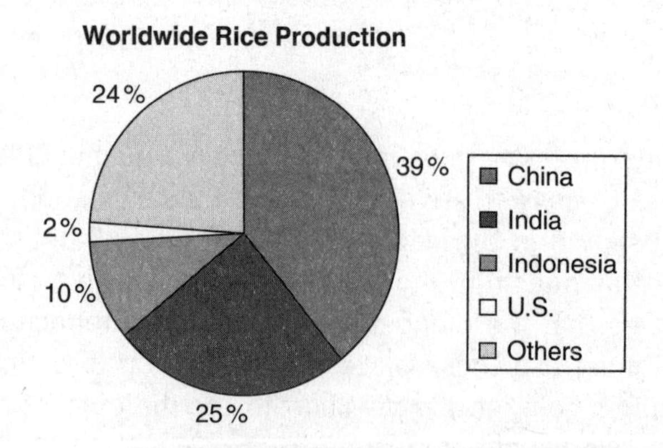

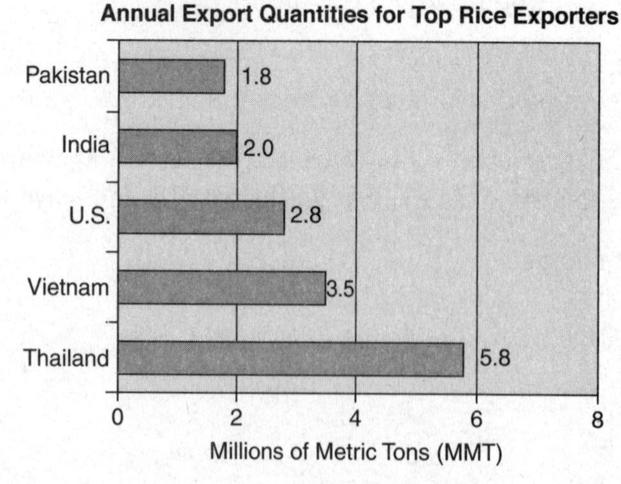

Which countries exported less than the average number of MMT for all exporting countries listed?

Indicate all such countries.

A Thailand
B Vietnam
C U.S.
D India
E Pakistan

Apply the Kaplan Method for Data Interpretation

Now let's apply the Kaplan Method to a Data Interpretation question requiring you to select one or more answers.

Which countries exported less than the average number of MMT for all exporting countries listed?

Indicate all such countries.

A Thailand
B Vietnam
C U.S.
D India
E Pakistan

STEP 1

Analyze the tables and graphs.

This question is based on the bottom graph dealing with exports.

STEP 2

Approach strategically.

To answer the question, the average number of MMT per country must be found. Read the export quantities for each country from the graph and use the equation for finding the average (arithmetic mean):

$$Average = \frac{Sum\,of\,terms}{Number\,of\,terms} = \frac{5.8 + 3.5 + 2.8 + 2 + 1.8}{5} = \frac{15.9}{5} \approx 3$$

There is no need to be concerned about decimal places; an estimate of 3 is sufficient to answer the question. The U.S., India, and Pakistan exported less than the average number of MMT. The correct choices are **(C)**, **(D)**, and **(E)**.

For more information on finding an average, see chapter 6.

DATA INTERPRETATION QUESTIONS (NUMERIC ENTRY)

The directions for a Data Interpretation question requiring you to make a Numeric Entry will look like this:

Directions: Click on the box and type a number. Backspace to erase.

Enter your answer as an integer or decimal if there is one box or as a fraction if there are two boxes.

To enter an integer or decimal, type directly in the box or use the Transfer Display button on the calculator.

- Use the backspace to erase.
- Use a hyphen to enter a negative sign; type a hyphen a second time to remove it. The digits will remain.
- Use a period for a decimal point.
- The Transfer Display button will enter your answer directly from the calculator.
- Equivalent forms of decimals are all correct. Example: $.14 = 0.140$.
- Enter the exact answer unless the question asks you to round your answer.

To enter a fraction, type the numerator and denominator in the appropriate boxes.

- Use a hyphen to enter a negative sign.
- The Transfer Display button does *not* work for fractions.
- Equivalent forms of fractions are all correct. Example: $\frac{25}{15} = \frac{5}{3}$.
- If numbers are large, reduce fractions to fit in boxes.

A Data Interpretation question with Numeric Entry will look like this:

Questions 1–3 are based on the following graphs.

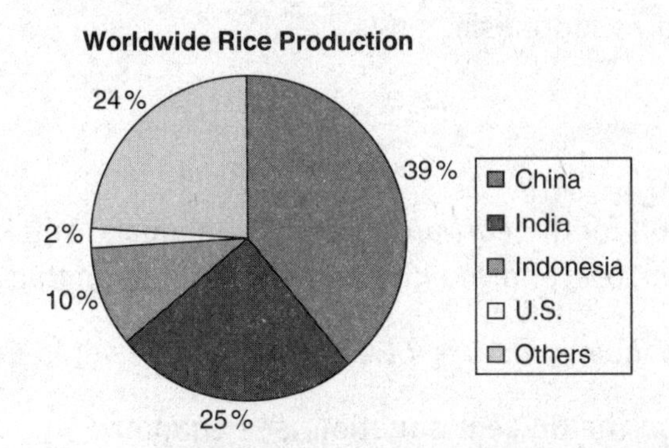

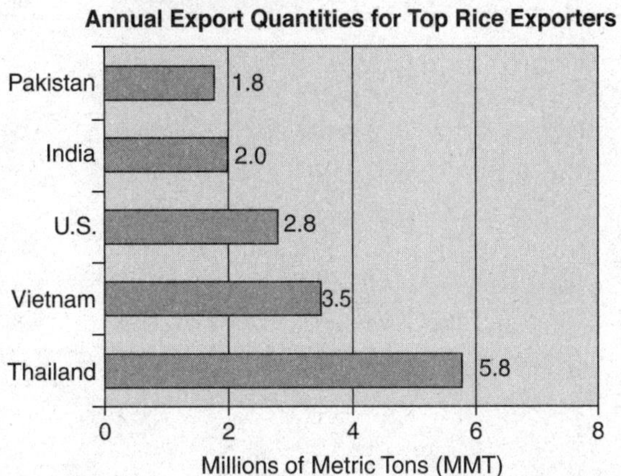

For a certain year, the worldwide production of rice was 600 MMT. Based on the data presented, what would Indonesia's production have been that year?

[ ] MMT

Apply the Kaplan Method for Data Interpretation

Now let's apply the Kaplan Method to a Data Interpretation question requiring you to make a Numeric Entry.

For a certain year, the worldwide production of rice was 600 MMT. Based on the data presented, what would Indonesia's production have been that year?

[ ] MMT

STEP 1

Analyze the tables and graphs.

The question requires you to use the top graph and find the percent of worldwide production contributed by Indonesia.

STEP 2

Approach strategically.

Solve a percent equation for the *Part* produced by Indonesia. You know from the pie chart that the percent is 10%, and you know from the question that the *Whole* is 600.

$Part = Percent \times Whole = 0.10 \times 600 = 60$. The answer is **60** MMT.

For more information on the percent equation, see chapter 6.

DATA INTERPRETATION PRACTICE SET

Try the following Data Interpretation questions using the Kaplan Method.

Basic

Questions 1–3 are based on the following graph.

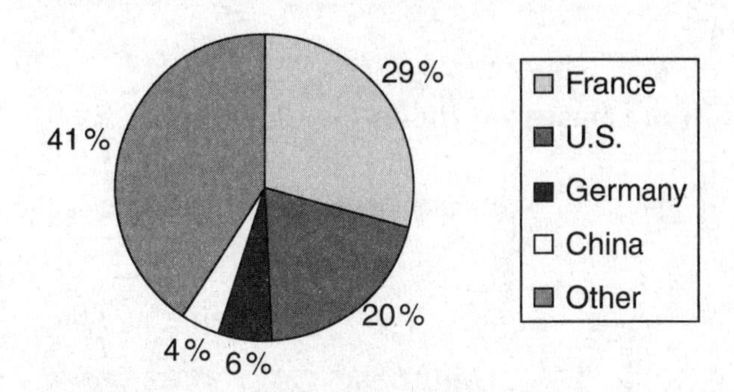

1. If Madagascar's exports totaled 1.3 billion dollars in 2009, approximately what was the value, in millions of dollars, of the country's exports to China?

   (A) 52
   (B) 78
   (C) 100
   (D) 325
   (E) 520

2. What is the approximate ratio of Madagascar's combined total exports to France, the United States, Germany, and China to Madagascar's exports to all other countries?

   (A) $\frac{2}{5}$
   (B) $\frac{1}{2}$
   (C) $\frac{2}{3}$
   (D) $\frac{3}{2}$
   (E) $\frac{2}{1}$

3. If Madagascar's exports to France increased to 33% of Madagascar's total exports in 2010, by approximately what percent did Madagascar's exports to France's increase from 2009–2010?

   Ⓐ 4%
   Ⓑ 14%
   Ⓒ 32%
   Ⓓ 88%
   Ⓔ 114%

Questions 4–6 are based on the following graph.

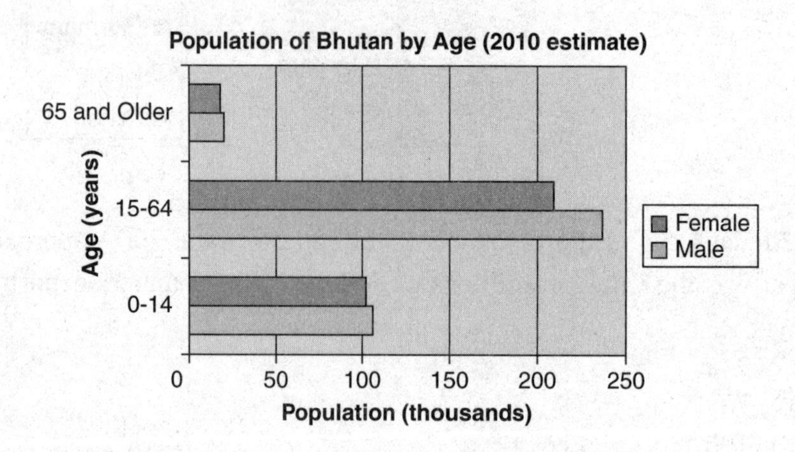

4. Approximately what percent of Bhutan's total population was between the ages of 0 and 14 years in 2010?

   Ⓐ 15%
   Ⓑ 25%
   Ⓒ 30%
   Ⓓ 40%
   Ⓔ 55%

5. What is the best estimate of the ratio of the female population age 65 and older to the female population age 0 to 14 years?

   Ⓐ 1:10
   Ⓑ 1:5
   Ⓒ 1:3
   Ⓓ 3:1
   Ⓔ 5:1

6. Which of the following ratios are greater than 1:2?

   Indicate all such ratios.

   A Males and females ages 0–14 to total population
   B Males ages 0–14 to males and females ages 0–14
   C Males ages 15–64 to total population
   D Males and females ages 0–14 to males and females ages 15–64
   E Males over 65 to males and females ages over 65

Questions 7–10 are based on the following graphs.

Distribution of Students at Shady Brook High School

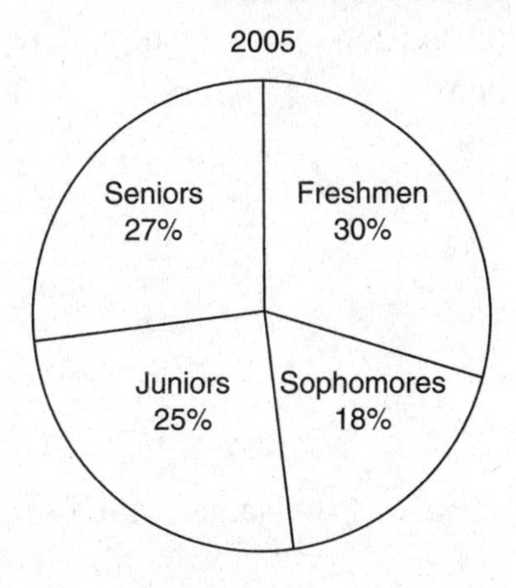

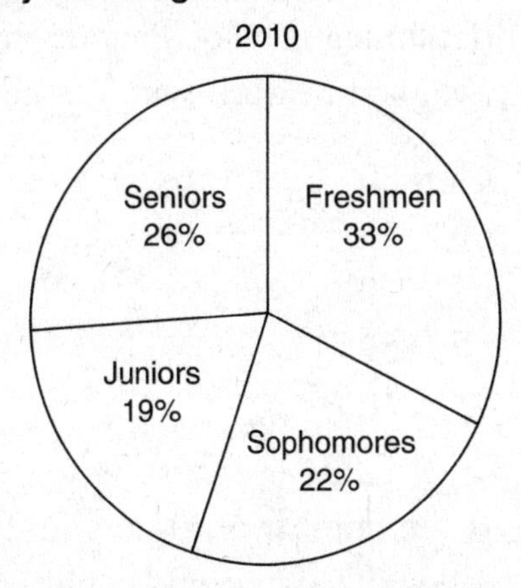

7. Suppose there were 1,100 students at Shady Brook High School in 2005 and 1,300 students in 2010. How many more freshmen attended the school in 2010 than in 2005?

   Ⓐ 429
   Ⓑ 200
   Ⓒ 99
   Ⓓ 33
   Ⓔ 3

8. Suppose 1,100 students attended Shady Brook High School in 2005. If two-thirds of the seniors were female, how many male seniors attended the school in 2005?

   (A) 99
   (B) 132
   (C) 216
   (D) 324
   (E) 400

9. Suppose 1,300 students attended Shady Brook High School in 2010. If the ratio of faculty assigned to work with underclassmen to underclassmen (freshmen and sophomores) was about 1:14, how many faculty were assigned to work with the underclassmen?

   (A) 30
   (B) 45
   (C) 50
   (D) 56
   (E) 100

10. If the number of juniors in 2005 and 2010 was the same, which of the following statements could be true?

    Indicate all such statements.

    [A] There were 988 students in 2005 and 1,300 students in 2010.
    [B] There were 952 students in 2005 and 1,300 students in 2010.
    [C] There were 987 students in 2005 and 750 students in 2010.
    [D] There were 1,158 students in 2005 and 875 students in 2010.
    [E] There were 1,064 students in 2005 and 1,400 students in 2010.

Intermediate

Questions 11–13 are based on the following graphs.

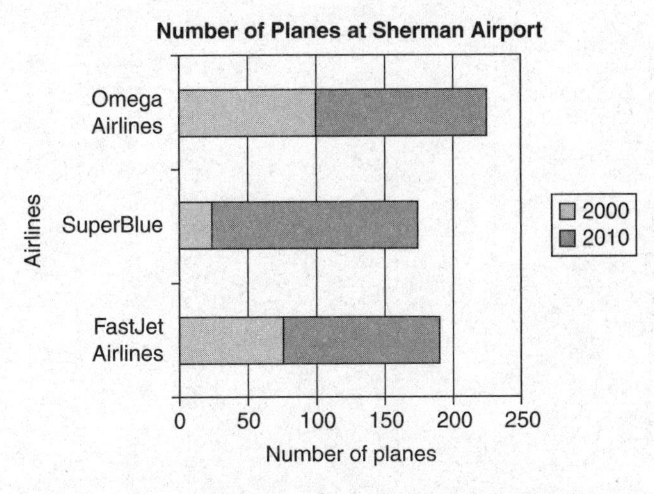

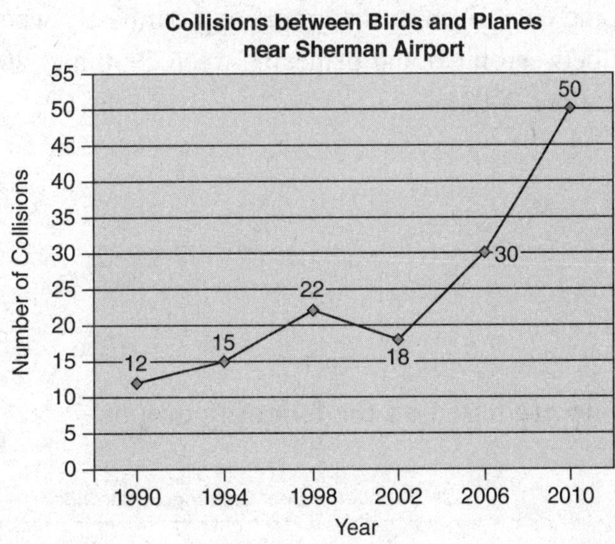

11. The number of collisions between birds and planes near Sherman Airport increased by approximately what percent between 1990 and 2010?

(A) 3%
(B) 37%
(C) 193%
(D) 245%
(E) 317%

12. What is the probability that a collision between a bird and a plane in 2010 involved a SuperBlue plane, assuming that collisions between birds and planes involved only Omega, SuperBlue, and FastJet airplanes?

Ⓐ $\frac{1}{15}$

Ⓑ $\frac{1}{3}$

Ⓒ $\frac{5}{13}$

Ⓓ $\frac{8}{11}$

Ⓔ $\frac{14}{15}$

13. Based on the data given in the graph, approximately what percentage of collisions between birds and planes between 1990 and 2010, inclusive, occurred before 2002?

Ⓐ 12%
Ⓑ 20%
Ⓒ 33%
Ⓓ 50%
Ⓔ 67%

Questions 14–16 are based on the following graphs.

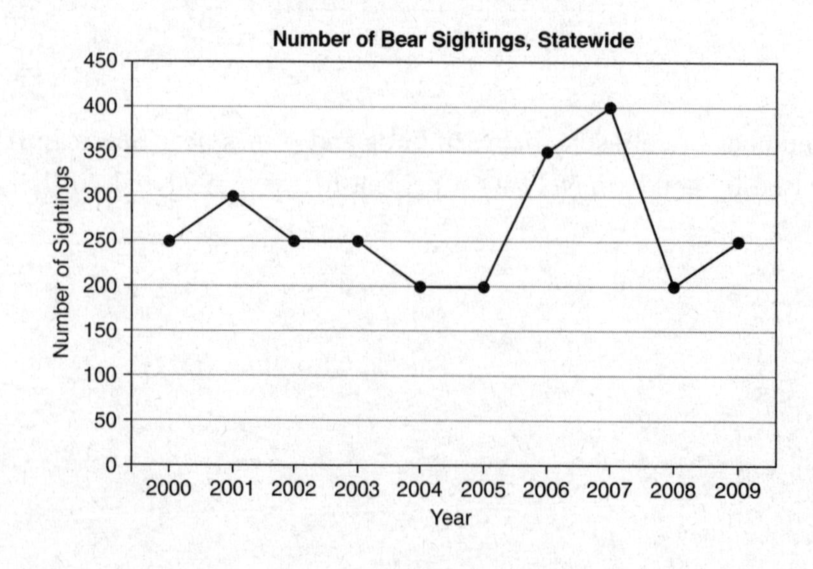

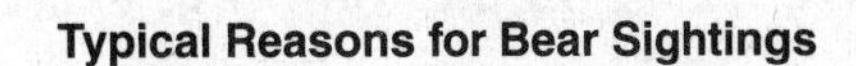

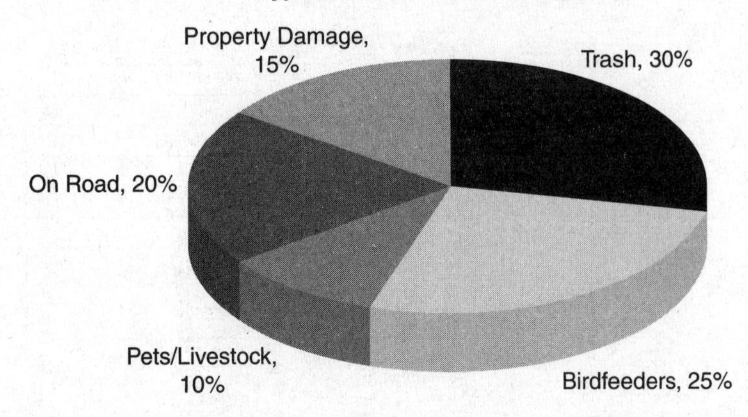

14. If Smithson County reported 20% of the bear sightings in the state in 2000, how many sightings were reported for that location?

Ⓐ 20
Ⓑ 25
Ⓒ 30
Ⓓ 40
Ⓔ 50

15. During the year that had the greatest increase in the number of bear sightings from the previous year, how many "on road" sightings were reported, assuming a typical distribution of bear-sighting types?

Ⓐ 20
Ⓑ 35
Ⓒ 40
Ⓓ 70
Ⓔ 80

16. According to the data given, in which year was there no change in the number of bear sightings from the previous year?

Select <u>all</u> that apply.

[A] 2002
[B] 2003
[C] 2004
[D] 2005
[E] 2006

Questions 17–20 are based on the following graphs.

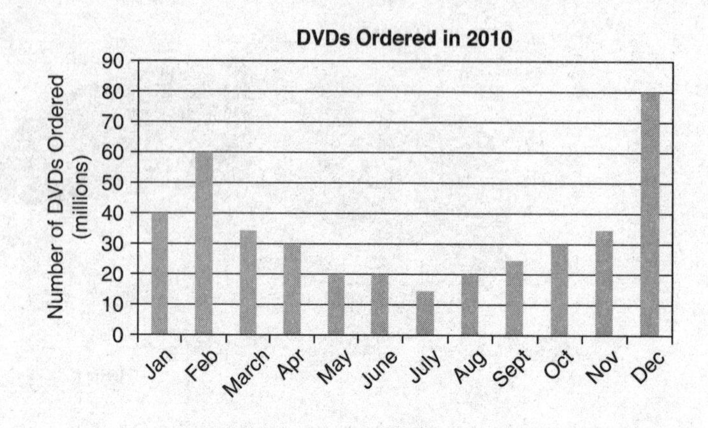

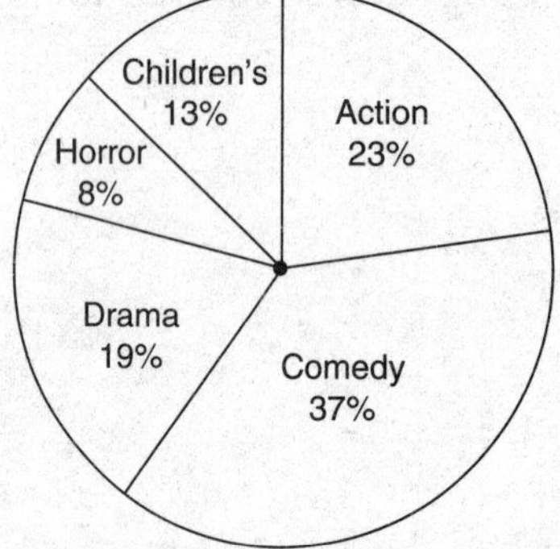

17. There was a $41\frac{2}{3}\%$ change in the number of DVDs ordered between which two consecutive months in 2010?

Ⓐ February to March
Ⓑ April to May
Ⓒ May to June
Ⓓ August to September
Ⓔ November to December

18. Which pairs of months in 2010 have a sales ratio of 2:3?

Select all such months.

[A] January:February
[B] April:May
[C] August:April
[D] October:November
[E] May:June

19. In December 2010, how many more comedy DVDs were ordered than drama DVDs?

(A) 4.2 million
(B) 6.3 million
(C) 8 million
(D) 14.4 million
(E) 18 million

20. Suppose 15% of the DVDs ordered in October 2010 were horror films. How many more horror films, in millions, were ordered in December than in October?

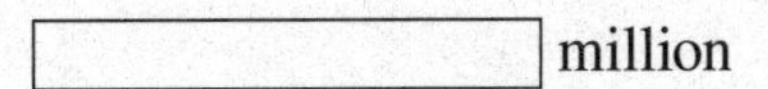 million

Advanced

Questions 21–24 are based on the following graphs.

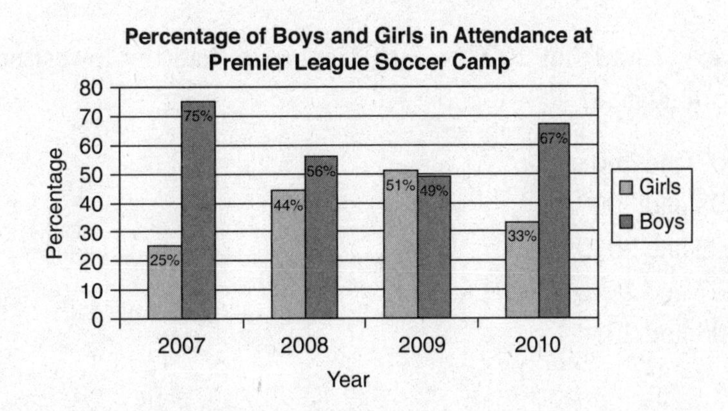

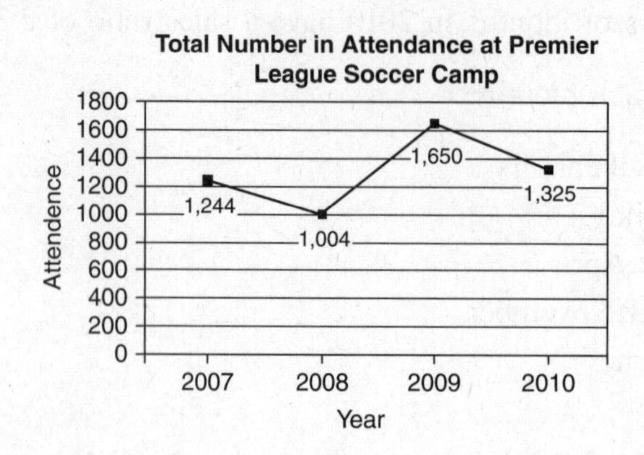

21. Approximately how many girls attended the 2010 Premier League Soccer Camp?

   Ⓐ 330
   Ⓑ 450
   Ⓒ 545
   Ⓓ 825
   Ⓔ 890

22. In which year(s) shown did approximately the same number of girls and boys attend the camp?

   Ⓐ 2007
   Ⓑ 2008
   Ⓒ 2009
   Ⓓ 2007 and 2008
   Ⓔ 2008 and 2009

23. Which two years from 2007 to 2010, inclusive, had the lowest numbers of boys in attendance?

   Ⓐ 2007 and 2008
   Ⓑ 2008 and 2009
   Ⓒ 2009 and 2010
   Ⓓ 2007 and 2010
   Ⓔ 2008 and 2010

24. Which is the best estimate of the total number of girls who attended in the two years that had the lowest total attendance?

Ⓐ 1,500
Ⓑ 1,275
Ⓒ 1,280
Ⓓ 750
Ⓔ 660

Questions 25–27 are based on the following graphs.

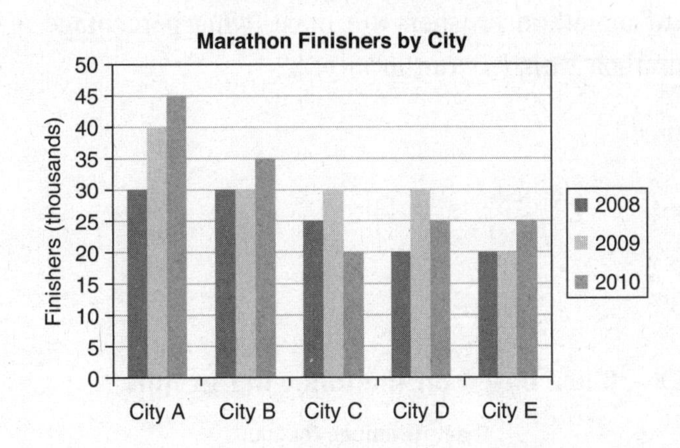

25. Which city showed the greatest percent increase in marathon finishers from 2009 to 2010?

Ⓐ City A
Ⓑ City B
Ⓒ City C
Ⓓ City D
Ⓔ City E

26. During the year in which there was no change in male marathon finishers from the previous year, what was the ratio of finishers in city A to finishers in city C?

  (A) 6:5
  (B) 4:3
  (C) 1:1
  (D) 8:5
  (E) 14:11

27. If 60% of marathon finishers are male, what percentage of the total 2009 U.S. marathon finishers ran in city A?

  (A) 5%
  (B) 10%
  (C) 25%
  (D) 30%
  (E) 40%

Questions 28–30 are based on the following graphs.

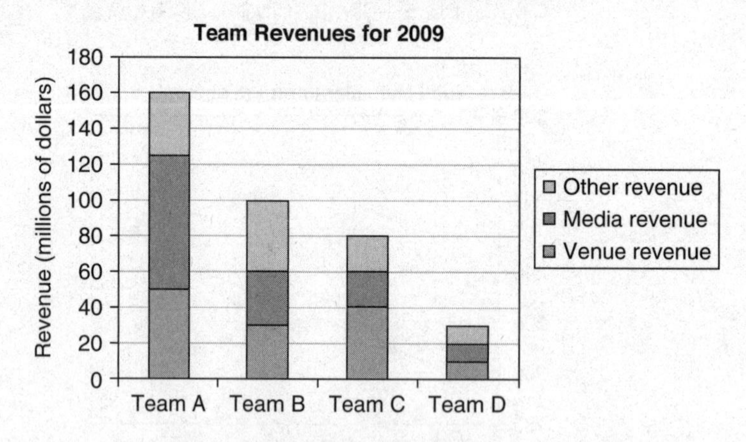

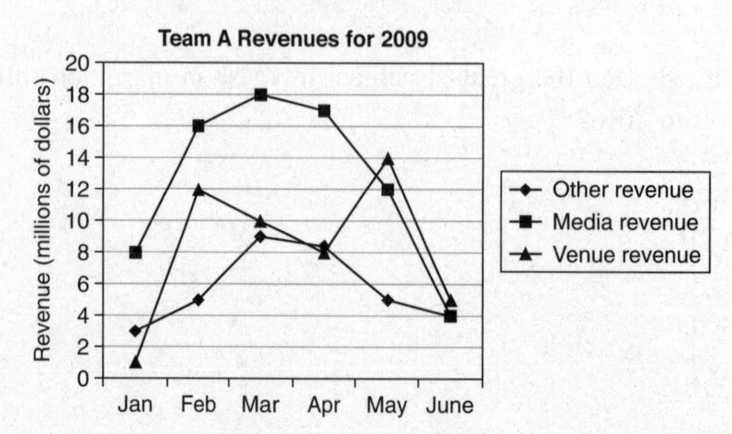

28. For the team that earned $20 million in media revenue in 2009, what percent of total revenue that year came from venue revenue?

[ ] percent

29. For the month in which team A showed the greatest media revenue, which statement(s) is (are) true?

Indicate all such statements.

- [A] Team A's venue was about the same as revenue from other sources.
- [B] Team A's revenue from media was more than $120 million.
- [C] Team A's media revenue for that month accounted for about one-fourth of the team's media revenue for the year.
- [D] In the same month, team A's revenue from other sources also showed its greatest amount.
- [E] Total revenues for team A for the month were less than $30 million.

30. What percent change did team A experience in venue revenue from May to June 2009?

- (A) Venue revenue decreased by 35.7%.
- (B) Venue revenue decreased by 64.3%.
- (C) Venue revenue increased by 66.6%.
- (D) Venue revenue increased by 80%.
- (E) Venue revenue stayed the same.

DATA INTERPRETATION PRACTICE SET ANSWER KEY

1. A
2. D
3. B
4. C
5. B
6. B, E
7. C
8. A
9. C
10. A, E
11. E
12. C
13. C
14. E
15. D
16. B, D
17. A
18. A, C
19. D
20. 1.9
21. B
22. C
23. B
24. D
25. E
26. A
27. B
28. 50
29. A, C, D
30. B

DATA INTERPRETATION PRACTICE SET ANSWERS AND EXPLANATIONS

Basic

1. A

Exports to China are 4% of all of Madagascar's exports. The dollar amount for Madagascar's exports is 1.3 billion dollars, but be aware that the question asks for millions of dollars. Multiply 1.3 billion by 1,000 to get the number of millions, or 1,300 million. Then, multiply 1,300 by 0.04 (4%) to get 52 million. The answer is choice **(A)**.

2. D

The percent shown for all other countries is 41%. So the percent for the remaining entries must be 100 − 41 = 59%. Write the ratio and then round:

$$\frac{59}{41} \approx \frac{60}{40} = \frac{3}{2}.$$

The correct answer is **(D)**.

3. B

The percent shown in the graph for France in 2009 is 29%. An increase to 33% would be an increase of 4 percentage points. Use the formula for percent increase:

$$\textit{Percent increase} = \frac{\textit{Amount of increase}}{\textit{Original whole}} \times 100\% = \frac{4}{29} \times 100\% \approx 14\%$$

The correct answer is **(B)**.

4. C

First, combine the approximate figures for males and females between the ages of 0 and 14 years. These are both just a little more than 100,000 for a total of about 210,000. Compare that figure to the total population. The total population is approximately 110,000 + 100,000 + 240,000 + 210,000 + 20,000 + 20,000 = 700,000. So, the percent of the population between 0 and 14 years (ignoring the thousands)

is about $\frac{210}{700} = \frac{\overset{3}{\cancel{21}}}{\cancel{70}_{10}} = \frac{3}{10} = 30\%$. The answer is **(C)**.

5. B

Compare the number of females in the oldest age group to the number of females in the youngest age group. Those numbers are approximately 20,000 and 100,000, respectively: 20:100 = 1:5. The answer is **(B)**.

6. B, E

(A) The population of males and females between 0 and 14 is about 210,000. That's less than half of the total population of about 700,000. Eliminate (A).

(B) For the age group 0–14, the length of the bar for the males is longer than the bar for females. Therefore, the males represent more than 50% of that group. Choice **(B)** is correct.

(C) There are approximately 240,000 males in the 15–64 age group. This is less than half of the total population of about 700,000. Eliminate (C).

(D) Do a quick estimate of the population numbers here: The total in the 0–14 age group is about 210,000. The total in the 15–64 age group is about 450,000. Eliminate (D) because twice 210,000 is less than 450,000.

(E) In the 65 and older age group, the bar for males is longer than the bar for females. Therefore, the males represent more than 50% of that group. Choice **(E)** is correct.

The correct choices are **(B)** and **(E)**.

7. C

Use the first graph to determine the number of freshmen in 2005 and use the second graph to determine the number of freshmen in 2010. Then subtract. The first graph shows that 30% of the students in 2005 were freshmen, so $0.30 \times 1{,}100 = 330$. The second graph shows that 33% of the students in 2010 were freshmen, so $0.33 \times 1{,}300 = 429$. Subtract: $429 - 330 = 99$. The answer is **(C)**.

8. A

Since the problem only involves the year 2005, only the first graph is needed. If two-thirds of the seniors were female, then the remaining one-third of the seniors were male. To find the number of male seniors, find one-third of the number of seniors. To find the number of seniors, find 27% of 1,100: $0.27 \times 1{,}100 = 297$. Then multiply by $\frac{1}{3}$: $297 \times \frac{1}{3} = 99$. The answer is **(A)**.

9. C

The percents given for the freshmen and sophomores in 2010 are 33% and 22% respectively. Use the sum of these percents ($33 + 22 = 55$) to find the number of underclassmen: $0.55 \times 1{,}300 = 715$. Let f represent the number of faculty in the proportion $\frac{1}{14} = \frac{f}{715}$. Cross multiply and solve for f:

$$175 = 14f$$
$$\frac{715}{14} = f \approx 51$$

The closest choice is **(C)**, 50.

10. A, E

In 2005, 25% of all the students were juniors. In 2010, 19% of all the students were juniors. For the number of juniors to be the same for both years, 25% of the 2005 student body must equal 19% of the 2010 student body, or 0.25 (number of students in 2005) = 0.19 (number of students in 2010). For this to be possible, the number of students in 2010 would have to be greater than the number of students in 2005. Eliminate any choices right away where this is not the case—eliminate choices (C) and (D).

Now check to see whether the numbers in choices (A), (B), and (E) make this a true statement. Answer choice **(A)** works: 0.25(988) = 247 and 0.19(1,300) = 247.

However, 0.25(952) = 238, and 0.19(1,300) = 247. Note that, since choice **(A)** works, choice (B) cannot possibly work, since (B) uses the same value for 2010, but a different value for 2005. Eliminate (B).

Try (E) next: 0.25(1,064) = 266 and 0.19(1,400) = 266. Choice **(E)** works.

The answers are **(A)** and **(E)** only.

Intermediate

11. E

The equation for percent increase is $\frac{\textit{Amount of increase}}{\textit{Original whole}} \times 100\%$. There were 50 collisions in 2010 and 12 collisions in 1990; there were 50 − 12 = 38 more collisions in 2010 than there were in 1990. To find the percent increase in collisions, divide 38 (the amount of increase) by 12 (the original whole) and multiply by 100. $\frac{38}{12} \times 100 \approx 317\%$. The answer is **(E)**.

12. C

From the data, in 2010 there were about 150 SuperBlue planes at the airport and a total of 390 planes at the airport. So, the probability that a collision involved a SuperBlue plane is $\frac{150}{390}$, which reduces to $\frac{5}{13}$. The answer is **(C)**.

13. C

In the years 1990, 1994, and 1998, there were about 12 + 15 + 22 = 49 collisions and a total of about 12 + 15 + 22 + 18 + 30 + 50 = 147 collisions during all the years shown on the graph. So, the percentage of collisions that occurred in the years 1990, 1994, and 1998 was $\frac{49}{147} \times 100 \approx 33\%$. The answer is **(C)**.

14. E

The question is asking about a percentage of the total bear sightings in 2000. The first graph shows total bear sightings, and in 2000, there were 250 sightings. Smithson County reported 20% of those, so 20% of 250 is 0.20 × 250 = 50. The answer is **(E)**.

15. D

The second graph shows the "on road" sightings as a percentage of the total sightings. So to find the number of "on road" sightings, you first need the total number of sightings. The year that had the greatest increase in the number of sightings from the previous year is the year with the steepest slope from the previous year on the first graph, which occurs in 2006. This graph shows 350 total sightings in 2006. The second graph shows that "on road" sightings account for 20% of the total: 20% of 350 = 0.20 × 350 = 70. The answer is **(D)**.

16. B, D

This question relates to the first graph. The two horizontal line segments represent no change in bear sightings. They occur from 2002 to 2003 and from 2004 to 2005. The question asks for the years that did not experience a change from the previous year, so choose the second year for each pair. The answers are 2003 and 2005, choices **(B)** and **(D)**.

17. A

This question involves a percent change, so the following formula can be used:

$$\frac{\textit{Amount of change}}{\textit{Original whole}} \times 100\% = \textit{percent change.}$$

Using the bar graph, look to see which answer choices seem unreasonable to avoid needless calculations.

(A) February to March showed a change from 60 to 35. That's a decrease of 25 from 60, which is a 41.67% or $41\frac{2}{3}\%$ change. This seems like the correct answer, but we can check the other answer choices to make sure.

(B) April to May was a change from 30 down to 20. That's a drop of 10 compared to 30. You know 10 divided by 30 is about 33%, so ignore this choice.

(C) May to June showed no real change, certainly not the percent you are looking for. Ignore this choice.

(D) August to September showed a change from 20 up to 25. When you compare 5 to 20, you get 25%. Not the percent you are looking for.

(E) The bar for December is more than twice as high as the bar for November. This increase is too high a percent.

The answer is **(A)**.

18. A, C

First, eliminate any ratios where the first month's sales are higher than or equal to the second month's sales, since you are looking for a ratio of 2:3. Then, for each remaining answer choice, write and simplify the ratio. If the answer choice simplifies to 2:3, then the answer is correct. The sales ratios are as follows:

(A) January:February → 40:60 = 2:3 → Correct.

(B) April:May → 30:20 → April larger than May. Eliminate.

(C) August:April → 20:30 = 2:3 → Correct.

(D) October:November → 30:35 = 6:7 → Incorrect.

(E) May:June → 20:20 → The two months are equal. Eliminate.

The answers are **(A)** and **(C)**.

19. D

To solve this problem, first use the bar graph to determine that 80 million DVDs were sold in December. Then, use the circle graph. To find the difference, subtract the percent of DVDs that were dramas from the percent that were comedies. Then, multiply that times the total number of DVDs to get the actual number of DVDs.

Percent comedies (37%) – *Percent dramas* (19%) = *Number of percentage points greater* (18%)

18% of 80 million = 14.4 million

The answer is **(D)**.

20. 1.9

To solve this problem, first use the bar graph to determine that there were 30 million DVDs sold in October and 80 million DVDs sold in December. From the question, you know that 15% of October's DVDs were horror films. From the circle graph, you know that 8% of December's DVDs were horror films. To find the difference, subtract the number of October horror DVDs from the number of December horror DVDs. In order to do this, you must use the percent formula to calculate each of these numbers.

Number of December horror films: 8% of 80 million = 0.08×80 million = 6.4 million

Number of October horror films: 15% of 30 million = 0.15×30 million = 4.5 million

Subtract: 6.4 million −4.5 million = 1.9 million.

The answer is 1.9 million. Since the numeric entry field is labeled with the word *million*, you only need to enter 1.9 for the answer.

Advanced

21. B

There were about 1,350 total children in attendance at the camp in 2010. Thirty-three percent of those attending that year were girls. Since 33% is close to $\frac{1}{3}$, divide 1,350 by 3, which equals 450. Therefore, the answer is **(B)**.

22. C

Examine the bar graph to find two bars that are closest to the same height. The bars are very nearly the same height in 2009. Therefore, in 2009, about the same number of boys and girls attended the camp. The answer is **(C)**.

23. B

Examine both graphs to see which years are the most likely years to have the least number of boys in attendance. If the bar that represents boys shows a high percentage and the corresponding year on the line graph shows a high total attendance, that year will have a high number of boys that attended. The opposite is also true:

if the bar that represents boys shows a low percentage and the corresponding year on the line graph shows a low total attendance, that year will have a low number of boys that attended. You can also find the number of boys that attended each year by multiplying the percentage of boys attending (taken from the bar chart) by the total number of children that attended (taken from the line graph). The calculations using rounded numbers would look like this:

Year 2007: $0.75 \times 1{,}250 \approx 938$

Year 2008: $0.55 \times 1{,}000 = 550$

Year 2009: $0.50 \times 1{,}650 = 825$

Year 2010: $0.65 \times 1{,}350 \approx 878$

The years 2008 and 2009 had the lowest numbers of boys attending. Therefore, the answer is **(B)**.

24. D

Estimate the total number of girls that attended in the two lowest years, 2007 and 2008, by multiplying the percentage of girls attending (taken from the bar chart) by the total number of children that attended (taken from the line graph). The answer choices are far enough apart that it is safe to round the numbers taken from the line graph.

$$(0.25 \times 1{,}200) + (0.45 \times 1{,}000) = 300 + 450 = 750$$

Therefore, the answer is **(D)**.

25. E

Only three cities showed an increase in the number of finishers between 2009 and 2010, so you only need to consider the data for cities A, B, and E. All three of these cities had the same increase in finishers, 5,000 more in 2010 than in 2009. You can avoid calculating the percent increase if you recall the percent increase formula:

$$\textit{Percent increase} = \frac{\textit{Amount of increase}}{\textit{Original whole}} \times 100\%$$

In the three options you are considering, the numerator is the same; it is 5,000. So, the city with the smallest number in the denominator will yield the greatest fraction. That is city E, making choice **(E)** the correct answer.

If you do compute the percent increase for each of the three cities, the results will look like this:

City A:

$$\textit{Percent increase} = \frac{\textit{Amount of increase}}{\textit{Original whole}} \times 100\% = \frac{5}{40} \times 100\% = \frac{1}{8} \times 100\% = 12.5\%$$

City B:

$$\text{Percent increase} = \frac{\text{Amount of increase}}{\text{Original whole}} \times 100\% = \frac{5}{30} \times 100\% = \frac{1}{6} \times 100\% = 16.7\%$$

City E:

$$\text{Percent increase} = \frac{\text{Amount of increase}}{\text{Original whole}} \times 100\% = \frac{5}{20} \times 100\% = \frac{1}{4} \times 100\% = 25\%$$

An important takeaway from this question is the property demonstrated about fractions; when the numerators are the same, the fraction with the smallest positive denominator has the greatest value.

26. A

To find the ratio of finishers in city A to finishers in city C, you need to know which year's data to use. The second graph shows the number of male marathon finishers. The year 2008 shows no increase or decrease in the number of male finishers from the previous year. Using the 2008 values, the first graph shows 30,000 city A marathon finishers and 25,000 city C marathon finishers. The ratio is 30:25, or 6:5, and the answer is **(A)**.

27. B

The second graph shows that in 2009 there were 240,000 male marathon finishers. The question states that this represents 60% of the total marathon finishers. Let T be total marathon finishers. Then $0.60 \times T = 240{,}000$. Solve for T by dividing both sides by 0.60 to give $T = 400{,}000$. The first graph shows that in 2009 there were 40,000 marathon finishers in city A. To find the percentage of the total that this represents, use the percent formula:

$$\begin{aligned}\text{Percent} &= \frac{\text{Part}}{\text{Whole}} \times 100\% \\ &= \frac{\text{2009 City A finishers}}{\text{2009 Total finishers}} \times 100\% \\ &= \frac{40{,}000}{400{,}000} \times 100\% \\ &= 10\%\end{aligned}$$

The answer is **(B)**.

28. 50

First, use the bar graph to determine that team C earned $20 million in media revenue in 2009. Then, find the requested percent by writing a fraction and renaming it as a percent. The venue revenue for team C is 40 million, and the total revenue for team C is 80 million. The fraction $\frac{40 \text{ million}}{80 \text{ million}} = \frac{40}{80} = \frac{1}{2}$, which equals 50%. Since the numeric entry field has the label *percent* after it, a percent sign is not needed.

The answer is **50**.

29. A, C, D

You really have no choice but to test each statement, but you can still approach the question strategically. First, in the bottom graph, locate the month described in the question: the month in which team A received the most revenue from media was March. In that month, the revenue figures (in millions of dollars) were media, 18; venue, 10; and other, 9. A tally of team A's monthly media revenues (in millions of dollars) shows that the total for January through June, inclusive, was $8 + 16 + 18 + 17 + 12 + 4 = 75$. Compare this total to team A's 2009 media revenues from the bar graph: $125 - 50 = 75$ million dollars. Therefore, all of team A's media revenues were generated in the January–June period. Now you are ready to evaluate each statement.

(A) Compare 10 for venue revenue to 9 for other revenue and conclude that these amounts are about the same. This is a true statement.

(B) Don't be fooled by looking at the top graph and reading the value at the top of the part for media. This is a false statement.

(C) Compare the amount given for media in March to the total for the year from media: $\frac{18}{75} = 0.24 = 24\%$. This is a true statement.

(D) The amount from other sources in March, 9, was the greatest amount shown for that category. This is a true statement.

(E) The total revenue (in millions of dollars) for March was $18 + 10 + 9$, which is greater than 30. This is a false statement.

The correct choices are **(A)**, **(C)**, and **(D)**.

30. B

According to the line graph, team A's venue revenue decreased from 14 million to 5 million from May to June, a change of 9 million. To find the percent decrease, you can use this formula:

$$\frac{\textit{Percent decrease}}{100} = \frac{\textit{Amount of decrease}}{\textit{Original whole}}$$

$$\frac{x}{100} = \frac{9}{14}$$

$$14x = 900$$

$$x = 64.3$$

The answer is **(B)**.

CHAPTER 5

Quantitative Reasoning Practice

In this section, you will work on practice sets composed of 20 questions each. This is the same number of questions you will see on each of the two sections of Quantitative Reasoning on the GRE. There is a diagnostic tool at the end of each practice set to help you learn from your mistakes and continue with more confidence as you prepare for the actual GRE.

On Test Day, you will have 35 minutes to complete 20 questions in a Quantitative Reasoning section. Keep the time allowed in mind as you work these practice sets.

QUANTITATIVE REASONING PRACTICE SET 1

Directions: Select the correct answer.

1.

Quantity A	Quantity B
40 percent of 0.75	$\frac{3}{5} \times \frac{3}{4}$

Ⓐ Quantity A is greater.
Ⓑ Quantity B is greater.
Ⓒ The two quantities are equal.
Ⓓ The relationship cannot be determined from the information given.

2. $m > 8$

Quantity A	Quantity B
$m + \sqrt{51}$	$\sqrt{64} + 7$

Ⓐ Quantity A is greater.
Ⓑ Quantity B is greater.
Ⓒ The two quantities are equal.
Ⓓ The relationship cannot be determined from the information given.

3. The side of equilateral triangle T is the same as the length of a side of square S.

Quantity A	Quantity B
The area of triangle T	The area of square S

Ⓐ Quantity A is greater.
Ⓑ Quantity B is greater.
Ⓒ The two quantities are equal.
Ⓓ The relationship cannot be determined from the information given.

4.

Quantity A	Quantity B
The number of distinct prime factors of 28	The number of distinct prime factors of 36

Ⓐ Quantity A is greater.
Ⓑ Quantity B is greater.
Ⓒ The two quantities are equal.
Ⓓ The relationship cannot be determined from the information given.

5.

$$4^5 = \frac{4^{20}}{4^x}$$

Quantity A	Quantity B
x	4

Ⓐ Quantity A is greater.
Ⓑ Quantity B is greater.
Ⓒ The two quantities are equal.
Ⓓ The relationship cannot be determined from the information given.

6.

$$g > 1$$

Quantity A	Quantity B
$\frac{g}{g+3} - 1$	$\frac{1}{g+3} - 1$

Ⓐ Quantity A is greater.
Ⓑ Quantity B is greater.
Ⓒ The two quantities are equal.
Ⓓ The relationship cannot be determined from the information given.

7. $♣j = j + 3$ and $◖j = \frac{j^2 + 2}{j}$

Quantity A	Quantity B
◖(♣2)	♣(◖2)

Ⓐ Quantity A is greater.
Ⓑ Quantity B is greater.
Ⓒ The two quantities are equal.
Ⓓ The relationship cannot be determined from the information given.

8. $a = bc \qquad c > 0$

Quantity A	Quantity B
a	b

Ⓐ Quantity A is greater.
Ⓑ Quantity B is greater.
Ⓒ The two quantities are equal.
Ⓓ The relationship cannot be determined from the information given.

9. The value of a certain stock rose by 40 percent from March to April, and then decreased by 30 percent from April to May. The stock's value in May was what percent of its value in March?

Ⓐ 90%
Ⓑ 98%
Ⓒ 110%
Ⓓ 130%
Ⓔ 142%

10. 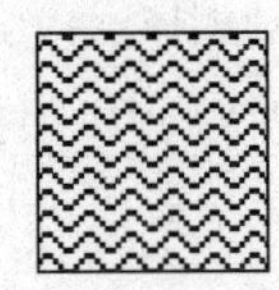

The figure shown represents a square garden. If each side is increased in length by 20 percent, by what percent is the area of the garden increased?

Ⓐ 44%
Ⓑ 50%
Ⓒ 125%
Ⓓ 144%
Ⓔ 150%

11. What is the value of a if $\frac{a+1}{a-3} - \frac{a+2}{a-4} = 0$?

Ⓐ –2
Ⓑ –1
Ⓒ 0
Ⓓ 1
Ⓔ 2

12. 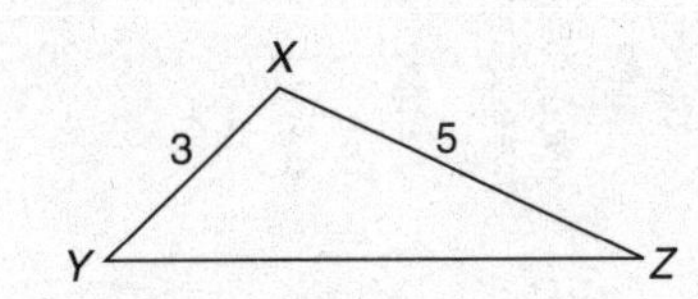

In the figure shown, which of the following could be the length of YZ?

Indicate all such lengths.

[A] 2
[B] 3
[C] 5
[D] 8
[E] 9

13. In a certain school, the ratio of boys to girls is 5:13. If there are 72 more girls than boys, how many boys are there?

Ⓐ 27
Ⓑ 36
Ⓒ 45
Ⓓ 72
Ⓔ 117

Questions 14–16 are based on the following graphs.

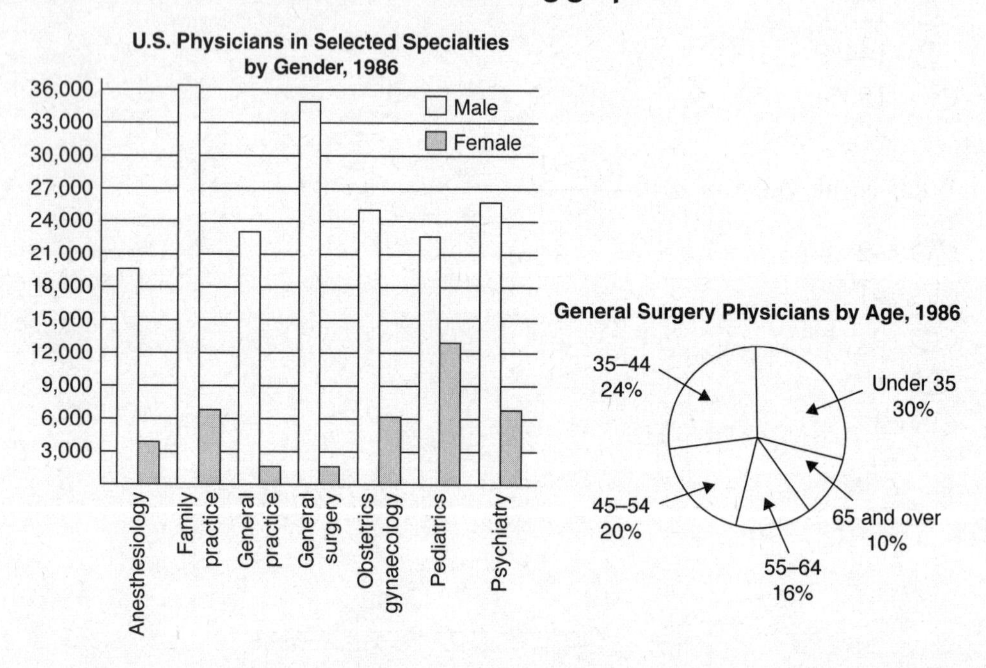

14. Approximately what percent of all general practice physicians in 1986 were female?

Ⓐ 8%
Ⓑ 23%
Ⓒ 75%
Ⓓ 82%
Ⓔ 90%

15. In 1986, approximately how many general surgery physicians were between the ages of 35 and 44, inclusive?

Ⓐ 5,440
Ⓑ 6,300
Ⓒ 7,350
Ⓓ 8,880
Ⓔ 10,200

16. If the number of female general surgery physicians in the under-35 category represented 8.5 percent of all the general surgery physicians, approximately how many male general surgery physicians were under 35 years of age?

Ⓐ 7,350
Ⓑ 7,960
Ⓒ 9,750
Ⓓ 10,260
Ⓔ 11,980

17.

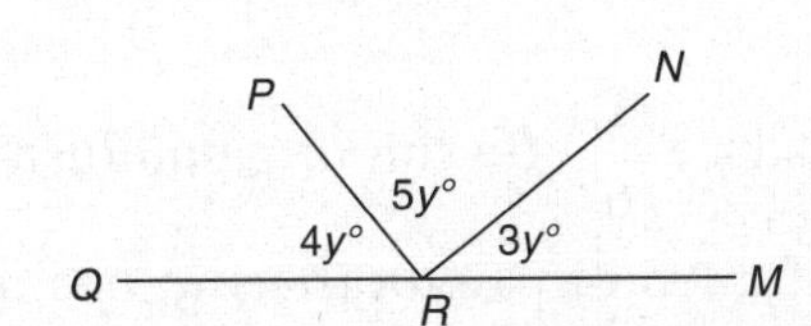

What is the degree measure of angle *PRM* shown?

[ ] degrees

18. Employee X is paid $19.50 an hour no matter how many hours he works per week. Employee Y is paid $18 an hour for the first 40 hours she works in a week and is paid 1.5 times the hourly rate for every additional hour she works. On a certain week, both employees worked the same number of hours and were paid the same amount. How many hours did each employee work that week?

Ⓐ 32
Ⓑ 36
Ⓒ 40
Ⓓ 42
Ⓔ 48

19.

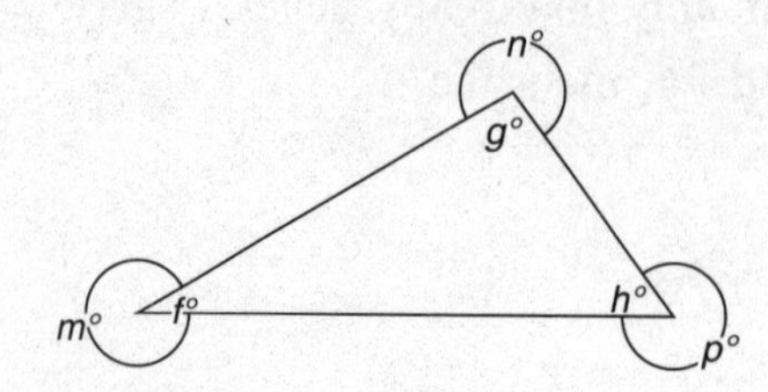

In the figure shown, what is $\frac{f + g + h}{m + n + p}$?

Ⓐ $\frac{1}{6}$

Ⓑ $\frac{1}{5}$

Ⓒ $\frac{1}{4}$

Ⓓ $\frac{1}{3}$

Ⓔ It cannot be determined from the information given.

20. Machine *A* can produce $\frac{1}{6}$ of a ton of paintbrushes in one hour. Machine *B* can produce $\frac{1}{14}$ of a ton of paintbrushes in one hour. Working together at their individual rates, how long would it take the two machines to produce 2 tons of paintbrushes?

Ⓐ 8 hours
Ⓑ 8 hours 24 minutes
Ⓒ 9 hours
Ⓓ 9 hours 46 minutes
Ⓔ 12 hours

QUANTITATIVE REASONING PRACTICE SET 1 ANSWER KEY

1. B
2. A
3. B
4. C
5. A
6. A
7. B
8. D
9. B
10. A
11. D
12. B, C
13. C
14. A
15. D
16. B
17. 120
18. E
19. B
20. B

QUANTITATIVE REASONING PRACTICE SET 1 ANSWERS AND EXPLANATIONS

1. B

Forty percent as a fraction is $\frac{4}{10}$. The decimal 0.75 as a fraction is $\frac{3}{4}$. We now have $\frac{4}{10} \times \frac{3}{4}$ for Quantity A. You know that $\frac{3}{5}$ is greater than $\frac{4}{10}$ because $\frac{3}{5} = \frac{6}{10}$. Therefore, $\frac{3}{5} \times \frac{3}{4}$ is greater than $\frac{4}{10} \times \frac{3}{4}$. Notice that you don't have to perform the multiplication because both sides are being multiplied by $\frac{3}{4}$. You only need to compare $\frac{3}{5}$ and $\frac{4}{10}$. In this case, Quantity B is greater than Quantity A, so the answer is **(B)**.

2. A

Compare each term in one quantity to see if it is larger than the corresponding term in the other quantity. Look at the first term in each quantity. Because *m* is larger than 8, you know that $m > \sqrt{64}$. Compare the next terms: $\sqrt{51} > 7$, since $\sqrt{49} = 7$. For every term of Quantity B, there is a larger term in Quantity A. Therefore, Quantity A is greater than Quantity B. The correct answer is **(A)**.

3. B

In geometry questions, if you are not given a diagram, then you should quickly sketch one on your scratch paper. If you draw a square where one side is also the side of an equilateral triangle, you should get something like this:

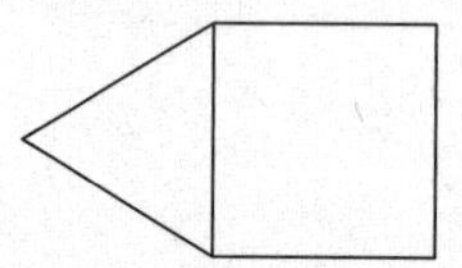

If you move the triangle over, it fits within the square.

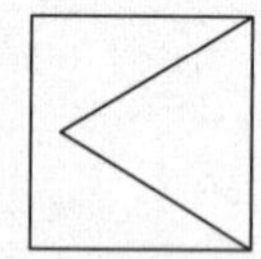

The square, or Quantity B, is larger. So the answer is **(B)**.

4. C

To find the number of distinct prime factors of a number, continue to break down factors of the number until you are left with only prime numbers. Remember that *distinct* simply means to count any repeated value only once. Therefore, 28 factors to 4×7, which further factors to $2 \times 2 \times 7$. The prime factors of 28 are 2, 2, and 7, and the distinct prime factors of 28 are 2 and 7. Likewise, 36 factors to 4×9, which further factors to $2 \times 2 \times 3 \times 3$. The prime factors of 36 are 2, 2, 3, and 3, and the distinct prime factors of 36 are 2 and 3. Each number has two distinct prime factors, so **(C)** is the correct choice.

5. A

When dividing values with the same base, you subtract the exponents. Look at the exponents in the given equation: $4^5 = \frac{4^{20}}{4^x}$. You know that $5 = 20 - x$. Some quick algebra shows you that $x = 15$. Quantity A is greater, so the correct choice is **(A)**.

6. A

Try simplifying each expression one step at a time. First, eliminate the 1 by adding 1 to both expressions:

Quantity A	Quantity B
$\frac{g}{g+3}$	$\frac{1}{g+3}$

Simplify by multiplying both expressions by $g + 3$. Remember, you can multiply both sides by the same positive value, and you know $(g + 3)$ is positive because you're given that $g > 1$.

Quantity A	Quantity B
$\frac{g}{g+3} \times (g+3) = g$	$\frac{1}{g+3} \times (g+3) = 1$

Now, compare g to 1. Given that $g > 1$, Quantity A is greater than Quantity B. The correct answer is **(A)**.

7. B

To compare these quantities, you will be applying the definitions of the symbols to the number 2. So everywhere you see j in the definitions, you will plug in the number 2. It also helps to say the definition of the symbol to yourself. For example, "club j" means to add 3 to j, and "semicircle j" means to square j and add 2 before dividing by j. As with any algebraic expression, work within the parentheses first.

For Quantity A, to calculate ◖ (♣2), you do (♣2) first: ♣2 $= 2 + 3 = 5$.

Then ◖5 $= \frac{5^2 + 2}{5} = \frac{25 + 2}{5} = \frac{27}{5} = 5\frac{2}{5}$.

For Quantity B, to calculate ♣ (◖ 2), you do (◖ 2) first: ◖ 2 $= \frac{2^2 + 2}{2} = \frac{4 + 2}{2} = \frac{6}{2} = 3$.

Then, ♣3 $= 3 + 3 = 6$. The answer is choice **(B)**.

8. D

You are given three variables, so it is an excellent time to use the Picking Numbers strategy. Remember when picking numbers, you must pick numbers that conform to the given information; that is, $a = bc$ and $c > 0$. Because c must be greater than zero, let's pick $c = 1$. If $c = 1$, any values you pick for a and b must be equal.

Because you picked numbers and found that the quantities can be equal, you know the correct answer is either (C) or (D).

Now try to pick values that give you a different relationship. If $c = 2$ and $b = 1$, a must equal 2. In this case, Quantity A is greater. The relationship changes, so **(D)** is correct. Remember, when picking numbers on Quantitative Comparison questions, once you find two different relationships, stop picking numbers and choose **(D)**.

9. B

Notice that even though the situation talks about the value of the stock, that value is never given. If you pick a number for the value of the stock, you can see much more easily what is going on.

Think about what would be a good number to pick for the value of the stock. The best number to pick when dealing with percents is almost always 100 because it is easy to find any percentage of 100.

Apply the percent changes to 100.

Forty percent of 100 is 40. Therefore, if the stock's value rises by 40 percent, its new value is 140.

Next, decrease this number by 30 percent.

Thirty percent of 140 is $0.30 \times 140 = 42$. So a decrease in price of 30 percent is $140 - 42 = 98$.

The original value was 100, and the new value is 98. So the stock is 98 percent of its original value. The correct choice is **(B)**.

10. A

Like almost all percent problems where no actual values are given, this is a great question to solve with the Picking Numbers strategy. Because you are dealing with a square garden, both sides are the same, and the area is side $\times$ side. Make the original dimensions 10 ft $\times$ 10 ft for an area of 100 sq ft. (Always start with 100 in percents questions because it is easy to find any percentage of 100.) Thus, if you increase each side length by 20 percent, the new dimensions would be 12 ft $\times$ 12 ft, for a new area of 144 sq ft. Consequently, the increase is 144 sq ft $-$ 100 sq ft $=$ 44 sq ft, which is 44 percent of 100 sq ft.

(A) is the correct choice.

11. D

Use the Backsolving strategy. Substitute the values in the answer choices into the equation. The one that gives you 0 must be correct.

(B): $\frac{a+1}{a-3} - \frac{a+2}{a-4} = \frac{-1+1}{-1-3} - \frac{-1+2}{-1-4} = \frac{0}{-4} - \frac{1}{-5} \neq 0$. Eliminate this answer choice.

Try another answer choice that is easy to work with.

(D): $\frac{a+1}{a-3} - \frac{a+2}{a-4} = \frac{1+1}{1-3} - \frac{1+2}{1-4} = \frac{2}{-2} - \frac{3}{-3} = -1 - (-1) = 0.$

Because the equation is true for **(D)**, it must be the answer.

12. B, C

In a triangle, each side must be shorter than the sum of the other two sides, and each side must be longer than the difference of the other two sides. Therefore, $(5 - 3) < YZ < (3 + 5)$, or $2 < YZ < 8$. The only possible values for YZ here that fall between 2 and 8 are 3 and 5.

So, **(B)** and **(C)** are the answers.

13. C

Use Backsolving on this one. The correct answer will yield a ratio of boys to girls of 5:13 when there are 72 more girls than boys. Start with (B).

(B): If there are 36 boys, there are $36 + 72 = 108$ girls, so the ratio of boys to girls is $36{:}108 = 1{:}3$. This is not the same as the ratio of 5:13 that you want. Try the next answer choice, (C).

(C): If there are 45 boys, there are $45 + 72 = 117$ girls, so the ratio of boys to girls is $45{:}117 = 5{:}13$, which is just what you want. The answer is **(C)**.

14. A

The bar graph doesn't give you the total number of general practice physicians, but you can find that number by adding the number of males to the number of females. To find the percent that are female, you take the number of females and put it over the total number. There are about 2,000 women and about 23,000 men, making the total about 25,000. If there are around 25,000 general practice physicians altogether and 2,000 of them are female, then 8 percent of the general practice physicians are female. The answer is **(A)**.

15. D

To refer to the ages of physicians, you need to find the slice of the pie that goes from 35 to 44. It's 24 percent, but 24 percent of what? You're not looking for a percent—you're looking for a number of doctors. For general surgery, the male bar goes up to about 35,000, and the female bar goes up to about 2,000—about 37,000 total. So 24 percent of 37,000 is the number of general surgery physicians between ages of 35 and 44, inclusive. That's 0.24 times 37,000, or 8,880. Choice **(D)** is the correct answer.

16. B

How many male general surgery physicians were under the age of 35? The pie chart breaks down general surgery by age, so you'll be working with it. And because you are looking for a number of general surgery physicians, you know that you are going to have to find the total number of general surgery physicians and then break it down according to the percentages on the pie chart.

You are given that the number of female general surgery physicians in the under-35 category represented 8.5 percent of all the general surgery physicians. What this does is break down the under-35 slice into two smaller slices, one for men under 35 and one for women under 35. You know that the whole under-35 slice is 30 percent of the total, and you are given that the number of females under 35 is 8.5 percent of the total. Thus, $30\% - 8.5\% = 21.5\%$, the amount you'll need to multiply by the total number of general surgery physicians.

For general surgery, the male bar goes up to about 35,000, and the female bar goes up to about 2,000—about 37,000 total. 21.5 percent of 37,000 is 7,955, which is closest to 7,960, making **(B)** the correct answer.

17. 120

Remember that a straight line has 180 degrees. That means that if you add the measures of angles *QRP*, *PRN*, and *NRM*, you'll get 180. Therefore, $4y + 5y + 3y = 180$, or $12y = 180$. Dividing both sides by 12 yields $y = 15$. Be careful, however; that's not your answer. You need to find the measure of angle *PRM*, which is the sum of the measures of angles *PRN* and *NRM*. That's $5(15) + 3(15) = 8(15) = 120$.

18. E

Use the Backsolving strategy. Note that in order for both employees to make the same amount in a given week, they must work more than 40 hours to allow Employee *Y*'s overtime rate to kick in. That means that choices (A), (B), and (C) are all too small. Start with (D). If (D) works, it's the correct answer; if not, the answer must be (E).

In 42 hours, employee *X* earns $42 \times \$19.50 = \$819$.

In 42 hours, employee *Y* earns $(40 \times \$18) + (2 \times \$27) = \$774$.

(D) is still too small, so **(E)** must be correct.

19. B

Remember that there are 360° around any point and that the interior angles of a triangle add up to 180°. From the first fact, you know that $f + m = g + n = h + p = 360$. Therefore, $m = 360 - f$, $n = 360 - g$, and $p = 360 - h$. From the second fact, you know that $f + g + h = 180$. Substitute these into the given rational expression:

$$\begin{aligned}\frac{f+g+h}{m+n+p} &= \frac{f+g+h}{360-f+360-g+360-h}\\ &= \frac{f+g+h}{360+360+360-f-g-h}\\ &= \frac{f+g+h}{360+360+360-(f+g+h)}\\ &= \frac{180}{360+360+360-180}\\ &= \frac{180}{900}\\ &= \frac{1}{5}\end{aligned}$$

The answer is **(B)**.

20. B

First you need to find out how long it will take each machine to produce 2 tons of paintbrushes. If Machine *A* can produce $\frac{1}{6}$ of a ton in one hour, it will take 6 hours for it to produce a full ton and twice that, or 12 hours, to produce 2 tons. If Machine *B* can produce $\frac{1}{14}$ of a ton in one hour, it will take 14 hours for it to produce a full ton and 28 hours to produce 2 tons. Now, use the combined work formula:

$$\frac{1}{r}+\frac{1}{s}=\frac{1}{t}$$

where r and s are the number of hours it takes Machines *A* and *B*, respectively, to produce 2 tons working by themselves, and t is the time it would take them to produce 2 tons working together.

$$\begin{aligned}\frac{1}{12}+\frac{1}{28} &= \frac{1}{t}\\ \frac{7}{84}+\frac{3}{84} &= \frac{1}{t}\\ \frac{10}{84} &= \frac{1}{t}\\ t &= \frac{84}{10}\end{aligned}$$

So it would take them $\frac{84}{10}$ or $8\frac{2}{5}$ hours to produce 2 tons of paintbrushes. The answer is **(B)**.

Diagnostic Tool

Tally up your score and write the results below.

Total

Total Correct: _______ out of 20

Percentage Correct: # you got right $\times$ 100 $\div$ 20: _______

By Section:

Quantitative Comparison _______ out of 8

Problem Solving _______ out of 9

Data Interpretation _______ out of 3

Diagnose Your Results

Look back at the questions you got wrong and think about your experience answering them. Were you stymied by a particular question type, or by a certain math topic? If the latter, studying the relevant math content review in Part 3 of this book should help.

QUANTITATIVE REASONING PRACTICE SET 2

Directions: Select the correct answer.

1. $a > 1$

Quantity A	Quantity B
$3 + (-9) - (-7)$	$4a$

(A) Quantity A is greater.
(B) Quantity B is greater.
(C) The two quantities are equal.
(D) The relationship cannot be determined from the information given.

2. 1 foot = 12 inches

Quantity A	Quantity B
The area of a square with a side length of 3 feet	400 square inches

(A) Quantity A is greater.
(B) Quantity B is greater.
(C) The two quantities are equal.
(D) The relationship cannot be determined from the information given.

3.

Quantity A	Quantity B
The number of square units in the area of a circle with a radius of 4	The number of units in the circumference of a circle with a radius of 8

(A) Quantity A is greater.
(B) Quantity B is greater.
(C) The two quantities are equal.
(D) The relationship cannot be determined from the information given.

4. $\square n = 5n$

Quantity A	Quantity B
$15 + \square 3$	$(\square 12) \div 2$

Ⓐ Quantity A is greater.
Ⓑ Quantity B is greater.
Ⓒ The two quantities are equal.
Ⓓ The relationship cannot be determined from the information given.

5. $ab \neq 0$

Quantity A	Quantity B
$(a^{-2}b^{-2})^{-1}$	$(a^{-1}b^{-1})^{-2}$

Ⓐ Quantity A is greater.
Ⓑ Quantity B is greater.
Ⓒ The two quantities are equal.
Ⓓ The relationship cannot be determined from the information given.

6. An integer x is selected at random from the set {17, 21, 23, 25, 27, 30, 33}.

Quantity A	Quantity B
The probability that the average (arithmetic mean) of 8, 16, and x is at least 17	$\frac{1}{2}$

Ⓐ Quantity A is greater.
Ⓑ Quantity B is greater.
Ⓒ The two quantities are equal.
Ⓓ The relationship cannot be determined from the information given.

7. (5, 2), (m, 4), (3, 5), (n, 7) are points on the same line.

Quantity A	Quantity B
m	n

Ⓐ Quantity A is greater.
Ⓑ Quantity B is greater.
Ⓒ The two quantities are equal.
Ⓓ The relationship cannot be determined from the information given.

8. $x \geq 0$

Quantity A	Quantity B
5^{5x}	5^{x+5}

Ⓐ Quantity A is greater.
Ⓑ Quantity B is greater.
Ⓒ The two quantities are equal.
Ⓓ The relationship cannot be determined from the information given.

9. The manager of a local grocery store earns an hourly wage of $21.00. The assistant manager earns 25% less than the manager. The stocker earns 60% less than the assistant manager. How much more does the manager make per hour than the stocker?

Ⓐ $6.30
Ⓑ $8.40
Ⓒ $9.45
Ⓓ $12.60
Ⓔ $14.70

10. Joe has a collection of 280 sports cards. If 30% of them are baseball cards, 25% of them are football cards, and the rest are basketball cards, how many basketball cards does Joe have?

Ⓐ 70
Ⓑ 84
Ⓒ 126
Ⓓ 154
Ⓔ 196

11. If a fair six-sided die with faces numbered 1 through 6 is tossed 3 times, what is the probability of getting a 1 or a 2 on all three tosses?

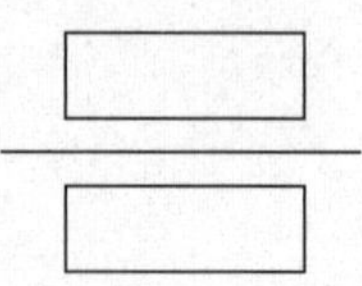

12. The perimeter of a square is 48 inches. The length, in inches, of its diagonal is

Ⓐ $6\sqrt{2}$
Ⓑ $8\sqrt{2}$
Ⓒ $12\sqrt{2}$
Ⓓ $24\sqrt{2}$
Ⓔ $48\sqrt{2}$

13. During a semester of her U.S. history class, Sophia received quiz scores of 85, 76, 98, 76, 100, and 75. How much greater is Sophia's average (arithmetic mean) quiz score than the mode of the set of scores?

Ⓐ 1
Ⓑ 5
Ⓒ 9
Ⓓ 10
Ⓔ 14

14. The ratio of girls to boys in a class is 6:7. If there are 18 girls, how many total students are in the class?

Ⓐ 18
Ⓑ 21
Ⓒ 27
Ⓓ 28
Ⓔ 39

15. If 12 is x percent of 60, what is 30 percent of x?

    (A) 6
    (B) 15
    (C) 18
    (D) 24
    (E) 66

16. 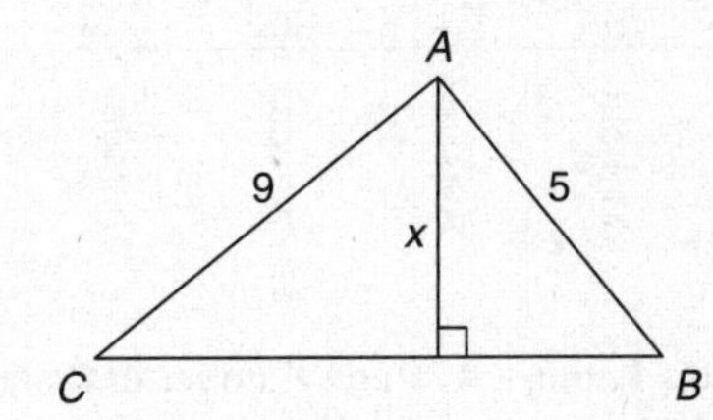

    The area of triangle ABC is $6x$. What is the perimeter of triangle ABC?

    (A) 12
    (B) 17
    (C) 20
    (D) 25
    (E) 26

17. Circle A has an area of 9π. Circle B has an area of 49π. If the circles intersect at exactly one point, which of the following could be the distance from the center of circle A to the center of circle B?

    (A) 6
    (B) 10
    (C) 21
    (D) 29
    (E) 58

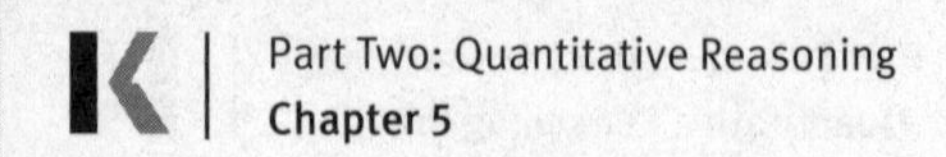

Questions 18–20 are based on the following graphs.

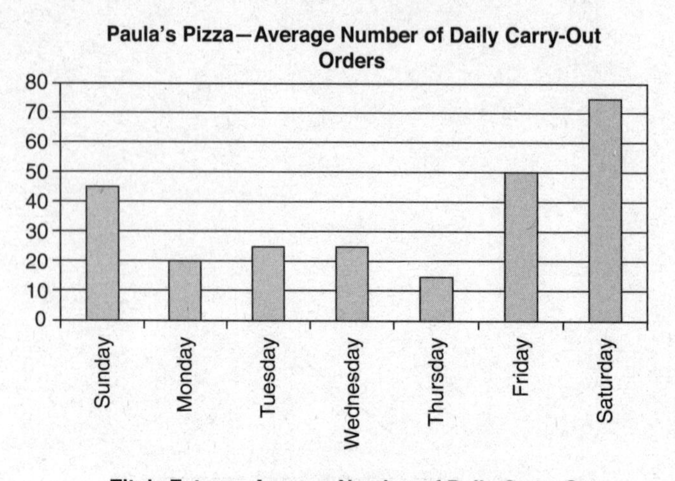

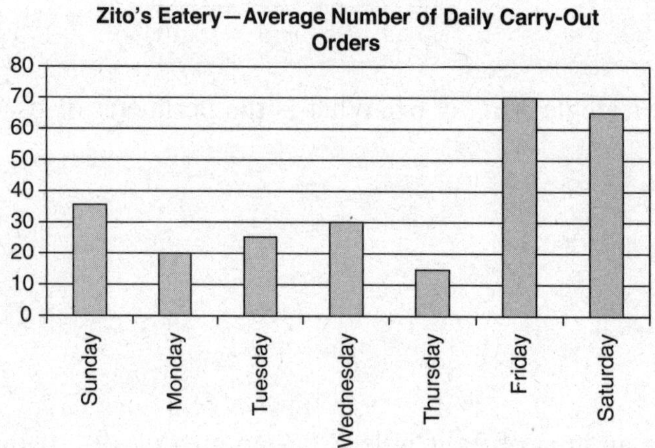

18. On which day of the week is the average number of carry-out orders the same for Paula's Pizza and Zito's Eatery?

    Indicate <u>all</u> such days.

    [A] Sunday
    [B] Monday
    [C] Tuesday
    [D] Wednesday
    [E] Thursday
    [F] Friday
    [G] Saturday

19. By what percent are Zito's Eatery's total average weekday (Monday through Friday) carry-out orders greater than its total average weekend (Saturday and Sunday) carry-out orders?

    (A) 12%
    (B) 18%
    (C) 50%
    (D) 60%
    (E) 63%

20. Between which two days does Paula's Pizza see the biggest drop in average carry-out orders?

    (A) Sunday to Monday
    (B) Wednesday to Thursday
    (C) Thursday to Friday
    (D) Friday to Saturday
    (E) Saturday to Sunday

QUANTITATIVE REASONING PRACTICE SET 2 ANSWER KEY

1. B
2. A
3. C
4. C
5. C
6. B
7. A
8. D
9. E
10. C
11. $\frac{1}{27}$
12. C
13. C
14. E
15. A
16. E
17. B
18. B, C, E
19. D
20. E

QUANTITATIVE REASONING PRACTICE SET 2 ANSWERS AND EXPLANATIONS

1. B

Quantity A contains all numbers, so it can be evaluated. Adding a negative number is the same as subtracting the positive of that number. So $3 + (-9)$ is the same as $3 - 9$, which is -6. Subtracting a negative number is the same as adding the positive of that number. So $-6 - (-7)$ is the same as $-6 + 7$, which is 1. Quantity B contains a variable, and its value is greater than 1. That means $4a$ is greater than 4 and Quantity B will always be greater than Quantity A. The correct answer is **(B).**

2. A

There are 12 inches in 1 foot. Therefore, a square that measures 3 feet by 3 feet is the same as a 36-inch by 36-inch square. So Quantity A equals $36 \times 36 = 1{,}296$ square inches, which is greater than 400 square inches. The correct answer is **(A).**

3. C

The area of a circle can be calculated by the formula $A = \pi r^2$. Quantity A is the area of a circle with a radius of 4, so its area is 16π. The circumference of a circle can be calculated by the formula $C = 2\pi r$. Quantity B is the circumference of a circle with a radius of 8, so its circumference is $2 \times \pi \times 8 = 16\pi$. The quantities are equal and, therefore, choice **(C)** is correct.

4. C

To apply the symbol in the centered information, multiply the number that follows it by 5. You must apply the symbol in both Quantity A and Quantity B. Look at Quantity A first:

$$15 + \square 3 = 15 + (5 \times 3) = 15 + 15 = 30$$

Now evaluate Quantity B:

$$(\square 12) \div 2 = 5 \times 12 \div 2 = 60 \div 2 = 30$$

Both values are the same, and choice **(C)** is correct.

5. C

When one exponent is raised to another, multiply the exponents. Simplifying Quantity A gives:

$$\left(a^{-2}b^{-2}\right)^{-1} = a^2b^2$$

Simplifying Quantity B gives:

$$\left(a^{-1}b^{-1}\right)^{-2} = a^2b^2$$

So the Quantities are equal, and the correct answer is **(C).**

6. B

The average of 8, 16, and x is $\frac{8 + 16 + x}{3}$. Since the average is to be at least 17, we can say that $\frac{8 + 16 + x}{3} \geq 17$. Solving the inequality for x by first multiplying both sides by 3, we have $24 + x \geq 51$. Subtracting 24 from both sides, we have $x \geq 27$. The set contains seven numbers, three of which are greater than or equal to 27. So the probability of choosing a number greater than or equal to 27 is $\frac{3}{7}$. Quantity A is $\frac{3}{7}$, and Quantity B is $\frac{1}{2}$. Since $\frac{3}{7}$ is a little less than $\frac{1}{2}$, Quantity B is larger. The correct answer is **(B)**.

7. A

Notice in the given points (5, 2) and (3, 5) that as the x-value decreases, the y-value increases. This indicates a line that has a negative slope. In the point (m, 4) the y-value, 4, is between the other given y-values, 2 and 5, so therefore the x-value, m, must be between the other given x-values, 5 and 3. In the other point (n, 7), the y-value, 7, is greater than the given y-value of 5, so therefore the x-value, n, must be less than the other given x-value, 3. So, $5 > m > 3$, and $n < 3$. The correct answer is **(A)**.

8. D

One way of solving this problem is to use the strategy of Picking Numbers for the variable, x. Substitute values in for x to determine if Quantity A or B is greater. Be sure that in doing so, enough values of x are evaluated to present a wide range of possibilities. For example:

If $x = 0$:

Quantity A: $5^{5x} = 5^0 = 1$

Quantity B: $5^{x+5} = 5^5$, which is greater than 1.

So when $x = 0$, Quantity B would be greater.

If $x = 1$:

Quantity A: $5^{5x} = 5^5$.

Quantity B: $5^{x+5} = 5^6$.

So when $x = 1$, Quantity B would still be greater.

If $x = 2$:

Quantity A: $5^{5x} = 5^{10}$.

Quantity B: $5^{x+5} = 5^7$.

So when $x = 2$, Quantity A would be greater.

Since the relationship varies depending on the value of x, the correct answer is **(D)**.

9. E

The manager of a grocery store makes $21.00 an hour. If the assistant manager makes 25% less than the manager, the assistant manager makes $21.00 \times 0.25 = 5.25$ less, or $15.75. The stocker makes 60% less than the assistant manager, so he makes $15.75 \times 0.60 = 9.45$ less, or $6.30. The question asks how much more the manager makes per hour than the stocker. Subtract the stocker's hourly wage from the manager's hourly wage: $21.00 − $6.30 = $14.70. The correct answer is **(E)**.

10. C

Joe has 280 sports cards in his collection. Of these, 30% are baseball cards and 25% are football cards. Therefore, 55% of the collection is baseball and football cards, and just a little less than half of the collection—45%—is basketball cards. Calculate 45% of the total: $280 \times 0.45 = 126$. The answer is **(C)**, 126 basketball cards.

11. $\mathbf{\frac{1}{27}}$

Since a die has 6 sides, the probability of getting a 1 or a 2 on the first toss is 2 out of 6, or $\frac{1}{3}$. This would be the same on the next two tosses. Because the probability of getting a 1 or a 2 is $\frac{1}{3}$ on each toss, the probability of getting a 1 or a 2 on all three tosses would be $\frac{1}{3} \times \frac{1}{3} \times \frac{1}{3} = \frac{1}{27}$.

12. C

If the perimeter of the square is 48 inches, each of the 4 sides would be 12 inches in length. The diagonal of the square would form a right triangle, with each of the sides being 12 inches long and the diagonal acting as the hypotenuse. The diagonal would split the right angle in half, forming a 45°-45°-90° triangle. In a 45°-45°-90° triangle, the hypotenuse is $\sqrt{2}$ times the side length. Therefore the answer is **(C)**, $12\sqrt{2}$.

13. C

The average quiz score is calculated by adding the quiz scores and dividing by the number of scores: $(85 + 76 + 98 + 76 + 100 + 75) \div 6 = 510 \div 6 = 85$. The mode of a data set is the value that is listed most often. In this list of U.S. history quiz scores, the score that is listed most often is 76. The difference between the average quiz score and the mode of the data set is $85 - 76 = 9$. The answer is **(C)**.

14. E

The ratio of girls to boys in a class is 6:7. There are 18 girls in the class ($6 \times 3 = 18$), so the number of boys can be determined by $7 \times 3 = 21$. Since there are 18 girls and 21 boys, the total number of students is $18 + 21 = 39$. The correct answer is **(E)**.

15. A

To calculate what percent 12 is of 60, divide: $12 \div 60 = 0.20$, which is 20%. That means $x = 20$. To calculate 30 percent of this value, multiply: $20 \times 0.30 = 6$. Therefore, the answer is **(A)**.

16. E

The area of a triangle can be calculated using the formula $A = \frac{1}{2}(base)(height)$. Given that the area of the triangle is $6x$, and the height of the triangle is x, the base must be 12, since $\frac{1}{2}(x)(12) = 6x$. To find the perimeter of the triangle, simply add the lengths of the three sides: $12 + 9 + 5 = 26$. The answer is **(E)**.

17. B

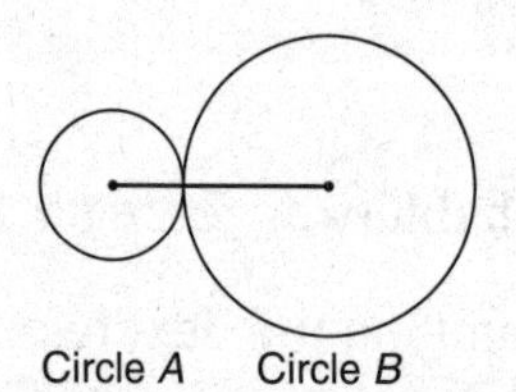

Draw a figure similar to the one shown to help visualize the information given. Recall that the area of a circle is πr^2. The area of circle A is 9π. So, $9 = r^2$, and $r = 3$. So the radius of circle A is 3. The area of circle B is 49π. So, $49 = r^2$, and $r = 7$. So the radius of circle B is 7. If the circles intersect at exactly one point, this means that the two circles are next to each other just touching each other (i.e., they are tangent to each other). To find the distance from the center of circle A to the center of circle B, simply add their radii: $3 + 7 = 10$. The correct answer is **(B)**.

18. B, C, E

Read each graph carefully to choose the days whose bars have the same height for each restaurant. The correct choices are Monday, Tuesday, and Thursday, or **(B)**, **(C)**, and **(E)**.

19. D

The total average number of daily carry-out orders at Zito's Eatery on the weekends is $65 + 35 = 100$. The total average number of daily carry-out orders at Zito's Eatery on the weekdays is $20 + 25 + 30 + 15 + 70 = 160$. So there are 60 more carry-out orders during the week than there are on the weekends. Since the weekend total is 100, it is easy to see that the weekday total is a 60% increase. Therefore, the answer is **(D)**.

20. E

Looking at the graph for Paula's Pizza, the biggest drop in average carry-out orders from one day to the next would either be between Saturday and Sunday or between Sunday and Monday. (The greatest difference is from Thursday to Friday, but that's an increase, and the question asks for a decrease.) From Saturday to Sunday, the average number of carry-out orders fell from 75 to 45, a drop of 30 orders. From Sunday to Monday, the average number of carry-out orders fell from 45 to 20, a drop of 25 orders. So the biggest drop occurs between Saturday and Sunday. Therefore, the answer is **(E)**.

Diagnostic Tool

Tally up your score and write your results below.

Total

Total Correct: _______ out of 20

Percentage Correct: # you got right $\times$ 100 $\div$ 20: _______

By Section:

Quantitative Comparison _______ out of 8

Problem Solving _______ out of 9

Data Interpretation _______ out of 3

DIAGNOSE YOUR RESULTS

Look back at the questions you got wrong and think about your experience answering them. Were you stymied by a particular question type, or by a certain math topic? If the latter, studying the relevant math content review in Part 3 of this book should help.

QUANTITATIVE REASONING PRACTICE SET 3

Directions: Select the correct answer.

1. $x < 0 < y$

Quantity A	Quantity B
x^2y^2	$(xy)^3$

Ⓐ Quantity A is greater.
Ⓑ Quantity B is greater.
Ⓒ The two quantities are equal.
Ⓓ The relationship cannot be determined from the information given.

2.

Quantity A	Quantity B
The hypotenuse of a right triangle with leg lengths 7 and 24	25

Ⓐ Quantity A is greater.
Ⓑ Quantity B is greater.
Ⓒ The two quantities are equal.
Ⓓ The relationship cannot be determined from the information given.

3.

Quantity A	Quantity B
$10^{-4} \times 10^3$	10^{-12}

Ⓐ Quantity A is greater.
Ⓑ Quantity B is greater.
Ⓒ The two quantities are equal.
Ⓓ The relationship cannot be determined from the information given.

4.

Quantity A	Quantity B
The area of a square with perimeter 36	3^4

Ⓐ Quantity A is greater.
Ⓑ Quantity B is greater.
Ⓒ The two quantities are equal.
Ⓓ The relationship cannot be determined from the information given.

5. a is a positive integer.

Quantity A	Quantity B
$-3a + 15$	$-3(a + 5)$

Ⓐ Quantity A is greater.
Ⓑ Quantity B is greater.
Ⓒ The two quantities are equal.
Ⓓ The relationship cannot be determined from the information given.

6.

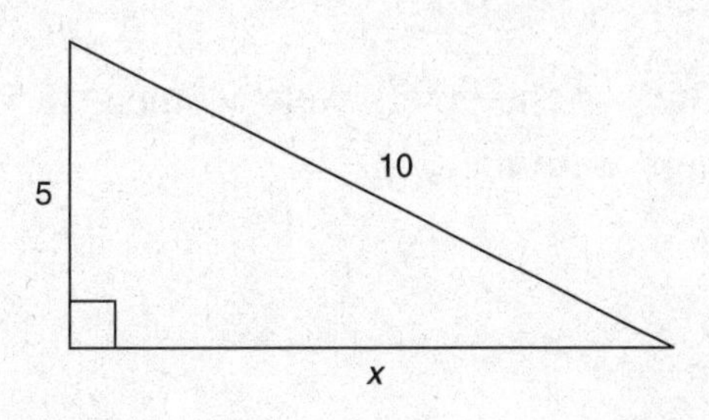

Quantity A	Quantity B
x	$5\sqrt{2}$

Ⓐ Quantity A is greater.
Ⓑ Quantity B is greater.
Ⓒ The two quantities are equal.
Ⓓ The relationship cannot be determined from the information given.

7. $\{-4, -3, 0, 2, 4, 6, 7, 9\}$

Quantity A	Quantity B
The average (arithmetic mean) of the set of numbers	The median of the set of numbers

Ⓐ Quantity A is greater.
Ⓑ Quantity B is greater.
Ⓒ The two quantities are equal.
Ⓓ The relationship cannot be determined from the information given.

8. $xy < 0$

Quantity A	Quantity B
$x - y$	$\frac{x}{y}$

(A) Quantity A is greater.
(B) Quantity B is greater.
(C) The two quantities are equal.
(D) The relationship cannot be determined from the information given.

9. In an election, 36% of the voters were women. If 48,000 people voted, how many of the voters were men?

(A) 12,000
(B) 17,280
(C) 28,520
(D) 30,720
(E) 35,520

10. If $a = 2b + \frac{1}{2}$ and $4b = 3$, what is the value of a?

(A) 1
(B) $1\frac{1}{2}$
(C) 2
(D) $2\frac{1}{2}$
(E) 3

11. If Sierra's scores on her first three tests were 90, 93, and 98, what must she score on the fourth test to have 95 as her test average?

Ⓐ 95
Ⓑ 96
Ⓒ 97
Ⓓ 98
Ⓔ 99

12. If a is an even integer and b is an odd integer, which of the following must be odd?

Indicate all such expressions.

[A] $a - b$
[B] $a + 2b$
[C] $3a + b$
[D] $2a - b$
[E] $a + b$

13. If Eugene can complete a project in 4 hours and Steve can complete the same project in 6 hours, how many hours will it take Eugene and Steve to complete the project if they work together?

Ⓐ 2
Ⓑ $2\frac{1}{4}$
Ⓒ $2\frac{2}{5}$
Ⓓ $2\frac{3}{4}$
Ⓔ 3

14. The product of two consecutive positive integers is 156. What is the larger of the two integers?

Ⓐ 11
Ⓑ 12
Ⓒ 13
Ⓓ 14
Ⓔ 15

15. How many milliliters of acid are there in 350 milliliters of a 4% acid solution?

Ⓐ 3.5
Ⓑ 10
Ⓒ 14
Ⓓ 35
Ⓔ 87.5

16. Maia packed three skirts, two hats, and four blouses for a trip. How many different outfits consisting of one skirt, one hat, and one blouse can she make with the clothes she has packed?

[ ] outfits

Questions 17–20 are based on the following graphs.

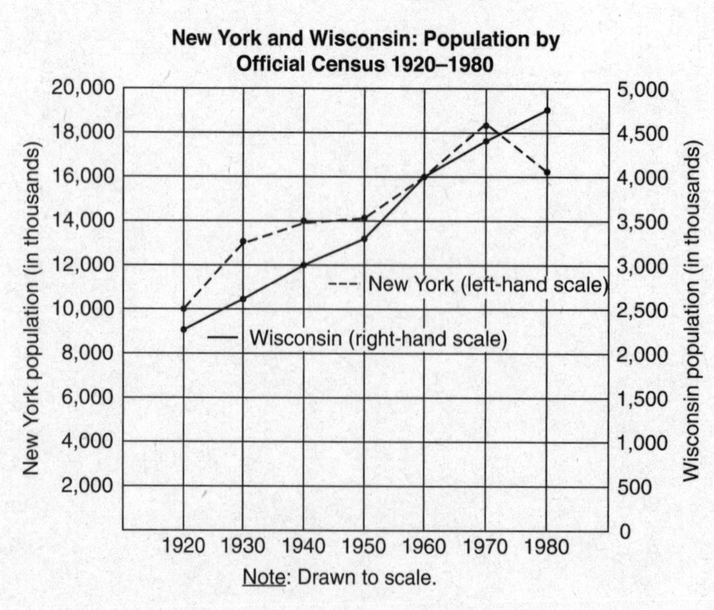

Note: Drawn to scale.

Density of Population by State
(per square mile, land area only)

State	1920	1960	1970	1980
Arkansas	33.4	34.2	37.0	43.9
Illinois	115.7	180.4	199.4	205.3
New York	217.9	350.6	381.3	370.6
Texas	17.8	36.4	42.7	54.3
Wisconsin	47.6	72.6	81.1	86.5

17. What was the least densely populated of the listed states in 1960?

Ⓐ Arkansas
Ⓑ Illinois
Ⓒ New York
Ⓓ Texas
Ⓔ Wisconsin

18. In 1930, what was the approximate ratio of the number of people living in Wisconsin to the number of people living in New York?

Ⓐ 1:6
Ⓑ 1:5
Ⓒ 5:6
Ⓓ 6:5
Ⓔ 4:1

19. If, in 1830, the population density of New York was 44 people per square mile, and the area of New York has stayed relatively constant since then, approximately what was the population of New York in 1830, in millions?

Ⓐ 1.2
Ⓑ 2.0
Ⓒ 3.5
Ⓓ 4.0
Ⓔ 4.8

20. Which of the following statements can be inferred from the information in the graph and table?

Indicate <u>all</u> such statements.

- [A] From 1920 through 1980, the population of New York has always been more than twice the population of Wisconsin.
- [B] Between 1960 and 1970, the percent increase in Wisconsin's population was greater than its percent increase in population between 1920 and 1930.
- [C] Of the five states listed, New York had the greatest increase in population between 1920 and 1960.
- [D] For the years 1920, 1960, 1970, and 1980, New York had a greater population density than each of the other states listed.
- [E] The population density of Texas decreased between 1960 and 1970.

QUANTITATIVE REASONING PRACTICE SET 3 ANSWER KEY

1. A
2. C
3. A
4. C
5. A
6. A
7. B
8. D
9. D
10. C
11. E
12. A, C, D, E
13. C
14. C
15. C
16. 24
17. A
18. B
19. B
20. A, D

QUANTITATIVE REASONING PRACTICE SET 3 ANSWERS AND EXPLANATIONS

1. A

Even though x is a negative number and y is a positive number, both x^2 and y^2 are positive. Since the product of two positive numbers is positive, x^2y^2 in Quantity A is positive. The product of a negative number and a positive number is negative, so xy is negative. It follows that $(xy)^3$ in Quantity B is negative because a negative number raised to an odd exponent is negative. A positive number is always greater than a negative number, so Quantity A is larger. The answer is **(A)**.

2. C

You can use the Pythagorean theorem, $a^2 + b^2 = c^2$, to find the hypotenuse of the right triangle. The letters a and b represent the legs, so $7^2 + 24^2 = c^2$.

$$\begin{aligned} 49 + 576 &= c^2 \\ 625 &= c^2 \\ 25 &= c \end{aligned}$$

Solving the equation for c, you get $c = 25$. So, Quantity A and Quantity B are equal. The answer is **(C)**.

3. A

Recall the rules for working with exponents to make this comparison. To multiply numbers with the same base, add the exponents:

$10^{-4} \times 10^3 = 10^{-4+3} = 10^{-1}$

That makes Quantity A equal to 10^{-1}. Now compare Quantity A to Quantity B:

$10^{-1} = \frac{1}{10}$ and $10^{-12} = \frac{1}{10^{12}} = \frac{1}{1{,}000{,}000{,}000{,}000}$. So, Quantity A is larger, much larger. The answer is **(A)**.

4. C

If the perimeter of a square is 36, then each side length is 9 because all four sides of a square are equal, and $36 \div 4 = 9$. The area of a square is given by the formula $A = s^2$, making the area of the square $9^2 = 81$. Likewise, $3^4 = 3 \times 3 \times 3 \times 3 = 81$. So, Quantity A and Quantity B are equal. The answer is **(C)**.

5. A

First, make the quantities look the same. Leave Quantity A alone for now and go to the expression in Quantity B. Distribute the –3, making the expression:

$$-3(a - 5) = -3a - 15$$

Now, compare the expressions in both sides of the Quantitative Comparison. Each has $-3a$, so you can eliminate those from each quantity. This leaves you comparing the value 15 under Quantity A with the value -15 in Quantity B. Quantity A is larger, so the answer is **(A)**.

6. A

A right triangle whose shorter leg is 5 and whose hypotenuse is 10 is a 30°-60°-90° right triangle. In a 30°-60°-90° triangle, the sides of the triangle are in the ratio of $x : x\sqrt{3} : 2x$. So, the other side of the triangle must be $5\sqrt{3}$. Quantity B is $5\sqrt{2}$. You do not need to calculate the individual square roots. The square root of 3 is greater than the square root of 2, so Quantity A is larger. The answer is **(A)**.

7. B

The mean of the set is $\frac{-4 + (-3) + 0 + 2 + 4 + 6 + 7 + 9}{8} = \frac{21}{8} = 2\frac{5}{8}$. The median is the middle value in a set of numbers arranged in order—when there is an odd number of terms in the set. This set has an even number of entries, so the median is the average of the two middle numbers, 2 and 4. That makes the median 3 since $(2 + 4) \div 2 = 3$. Quantity B is larger. The answer is **(B)**.

8. D

Because $xy < 0$, either x is positive and y is negative, or x is negative and y is positive. Use the Kaplan strategy of Picking Numbers for x and y to test the values of $x - y$ and $\frac{x}{y}$. If x is positive and y is negative, you can choose $x = 3$ and $y = -1$. Then, $x - y = 3 - (-1) = 3 + 1 = 4$ and $\frac{x}{y} = \frac{3}{-1} = -3$. So, Quantity A is larger. But if x is negative and y is positive, you can choose $x = -3$ and $y = 1$. Then, $x - y = -3 - 1 = -4$ and $\frac{x}{y} = \frac{-3}{1} = -3$. In this case, Quantity B is larger. So, the relationship cannot be determined from the given information. The answer is **(D)**.

9. D

If 36% of the voters were women, then 64% of the voters were men. To find 64% of the total number of voters, change 64% to its decimal form and multiply by 48,000: $0.64 \times 48{,}000 = 30{,}720$. So, 30,720 of the voters were men. This is choice **(D)**.

Alternatively, you can use a strategy to get the answer while lessening your chances of making a computation error—and without having to take the time to pull up the online calculator. If the percent of voters who are women is 36%, that means that a little more than one-third of the voters are women, leaving a little less than two-thirds of the voters to be men. Divide 48,000 by 3 to get 16,000, then double it to get two-thirds (32,000). Looking at the answer choices, only one answer, **(D)** 30,720, is a little less than 32,000, making it the only answer that can be correct. The answer is **(D)**.

10. C

First, solve the equation $4b = 3$ to find b. To solve for b, divide both sides of the equation by 4. So $b = \frac{3}{4}$. Then, substitute $\frac{3}{4}$ for b in the equation $a = 2b + \frac{1}{2}$. Simplify to get the value of a: $a = 2\left(\frac{3}{4}\right) + \frac{1}{2}$; $a = \frac{3}{2} + \frac{1}{2}$. Therefore, a is equal to 2. The answer is **(C)**.

11. E

To figure out what Sierra needs to score on her fourth test, use the variable x to represent the fourth test score. Add up all four test values, including x, and divide by 4. This equals the test average. Set this expression equal to 95 : $\frac{90 + 93 + 98 + x}{4} = 95$. Then, solve the equation for x.

$$\begin{aligned} 90 + 93 + 98 + x &= 4(95) \\ 281 + x &= 380 \\ x &= 380 - 281 \\ x &= 99 \end{aligned}$$

Sierra needs to score 99 on the fourth test to have a test average of 95, which is answer choice **(E)**.

An alternative approach, involving much less computation, is to use the Balancing method. An average is "balanced" in that the surpluses (amounts above the average) are equal to the deficits (amounts below the average). In Sierra's case, she has a 90 that's 5 points below her desired average of 95, a 93 that's 2 points below, and a 98 that's 3 points above. Overall, that's $(-5) + (-2) + (+3) = -4$. To balance that deficit, she needs a surplus of $+4$ on her last test; that's $95 + 4$, or 99: again, answer choice **(E)**.

12. A, C, D, E

Use the Picking Numbers strategy for a and b to test each answer choice. You can choose $a = 4$ and $b = 3$.

Then test each answer choice.

(A) $a - b = 4 - 3 = 1$, which is odd.

(B) $a + 2b = 4 + 2(3) = 4 + 6 = 10$, which is even.

(C) $3a + b = 3(4) + 3 = 12 + 3 = 15$, which is odd.

(D) $2a - b = 2(4) - 3 = 8 - 3 = 5$, which is odd.

(E) $a + b = 4 + 3 = 7$, which is odd.

So, $a - b$, $3a + b$, $2a - b$, and $a + b$ are odd. The answers are **(A)**, **(C)** , **(D)**, and **(E)**.

13. C

This is a combined work problem asking how long it will take two separate people (or machines, or whatever) to complete one job together, if we know how long it will take each person to do the same job separately. If you had no idea how to do this problem, (A) and (E) could still be eliminated straight off, leaving a 33% chance of guessing correctly, because (A) requires Steve to work at Eugene's rate and (E) requires Eugene to work at Steve's rate.

The generic equation for solving such a problem is $\frac{1}{a} + \frac{1}{b} = \frac{1}{t}$, where a is the time that it takes one person to complete the job alone, b is the the time that it takes the other person to complete the job alone, and t is the time the job will take when the two people work on it together.

$$\frac{1}{4} + \frac{1}{6} = \frac{1}{t}$$

$$\frac{3}{12} + \frac{2}{12} = \frac{1}{t}$$

$$\frac{5}{12} = \frac{1}{t}$$

To solve for t, find the reciprocal of the fraction on the left: $t = \frac{12}{5}$ or $2\frac{2}{5}$ hours.

Note that this problem is also set up in such a way that you can take advantage of the simpler "time per task" formula: the time it takes two people to complete a task together is equal to the product of their individual times divided by the sum of their individual times.

$$T_{together} = \frac{T_a \times T_b}{T_a + T_b}$$

$$T_{together} = \frac{4 \times 6}{4 + 6}$$

$$T_{together} = \frac{24}{10}$$

$$T_{together} = \frac{12}{5}$$

No matter how you approach it, the answer is **(C)**.

14. C

Kaplan's Backsolving strategy is a great approach to solving this problem. You are told that two consecutive integers, multiplied together, will produce 156, and you're asked to identify the larger of the two integers. In order to multiply to 156, the units digits of the integers will have to multiply to 6 (or some other number that has a units digit of 6).

Let's test the answer choices, which each represent the larger of the two integers, to see which one(s) multiply to have a units digit of 6:

(A) 11: 0×1 No

(B) 12: 1×2 No

(C) 13: 2×3 Yes!

(D) 14: 3×4 No

(E) 15: 4×5 No

Therefore, the two consecutive numbers end in a 2 and a 3, respectively. The question asks for the larger number, so the larger integer is 13. The answer is **(C)**.

15. C

The volume of acid in 350 milliliters of a 4% acid solution is 4% of 350 milliliters, or $0.04 \times 350 = 14$ milliliters. The answer is **(C)**.

16. 24

Since Maia has 3 skirts, 2 pairs of shoes, and 4 blouses, she can create $3 \times 2 \times 4 = 24$ different outfits. The answer is **24**.

17. A

You only need to use the table; that gives you information on population density. Look down the column for 1960 and find the state with the smallest density. It's Arkansas, with 34.2 people per square mile. The correct answer is **(A)**.

18. B

You need the ratio of the number in Wisconsin in 1930 to the number in New York in 1930. Because of the different scales, you can't estimate the ratios by comparing the respective heights of the two graphs—you need to find the actual figures. Because the populations for each state are shown in the thousands, you can compare the thousands in Wisconsin to the thousands in New York. The ratio of Wisconsin's population in 1930 to that of New York's in the same year is approximately 2,600:13,000, which reduces to a ratio of 1:5.

The answer is **(B)**.

19. B

You have to reason with proportions here while also using the given table. The question gives the population density of New York for 1830. The population density of the state is the ratio of the population to the amount of land; if the amount of land stays constant, then the density will increase at the same rate as the population. Now you could work with any of the four density figures given in the chart, but it's probably easiest to work with 1920's figure: New York's population in 1920 was approximately 10 million people, a nice round number to work with. In addition, the 1830 density was 44 people per square mile—very close to $\frac{1}{5}$ of the 1920 density, 217.9 people per square mile. That means the 1830 population must have been about one-fifth of the 1920 population. The 1920 population was about 10 million, so the 1830 population must have been approximately $\frac{1}{5}$ of 10 million, or 2 million people. The correct choice is **(B)**.

20. A, D

For a question of this type, you must test each statement.

(A) Rather than calculate the ratio for every year, look at the smallest population for New York, 10 million, and the largest population for Wisconsin, about 4.75 million. Therefore, this statement must be true. Choose **(A)**.

(B) There are two things to consider in order to evaluate this statement about the population changes in Wisconsin for the time periods given. Look at the solid line on the graph for the two periods and notice that the amount of change is the same. Recall the formula for percent change:

$$\textit{Percent change} = \frac{\textit{Amount of change}}{\textit{Original whole}}.$$

Now use logic rather than your calculator. The amount of change for both time periods is the same—that is the numerator of each fraction in the formula. However, the denominators are different. The population in 1920 was less than the population in 1960, so the original amount is smaller in that fraction, making the percent increase greater from 1920 to 1930 than it was from 1960 to 1970, not lesser.

(C) The table gives population density (number of people per square mile) for five states, but it does not give the number of square miles in the state. It is not possible to infer anything about the increase in population numbers of the five states from the table.

(D) This is a true statement and can be verified by reading down each column of the table of population densities. In each of the years shown, New York's population density is the greatest value. Choose **(D)**.

(E) In 1960, Texas's population density was 36.4. In 1970, it was 42.7. That's an increase, not a decrease, so (E) is not a true statement.

The correct answers are **(A)** and **(D)**.

Diagnostic Tool

Tally up your score and write the results below.

Total

Total Correct: _______ out of 20

Percentage Correct: # you got right × 100 ÷ 20: _______

By Section:

Quantitative Comparison _______ out of 8

Problem Solving _______ out of 8

Data Interpretation _______ out of 4

Diagnose Your Results

Look back at the questions you got wrong and think about your experience answering them. Were you stymied by a particular question type, or by a certain math topic? If the latter, studying the relevant math content review in Part 3 of this book should help.

QUANTITATIVE REASONING PRACTICE SET 4

Directions: Select the correct answer.

1. $m \neq 1$

Quantity A	Quantity B
m	m^3

Ⓐ Quantity A is greater.
Ⓑ Quantity B is greater.
Ⓒ The two quantities are equal.
Ⓓ The relationship cannot be determined from the information given.

2.

Quantity A	Quantity B
$(3^{-1} + 3^{-2})^{-1}$	1

Ⓐ Quantity A is greater.
Ⓑ Quantity B is greater.
Ⓒ The two quantities are equal.
Ⓓ The relationship cannot be determined from the information given.

3. $x^2 + 5x + 6 = 0$

Quantity A	Quantity B
The square of the sum of the roots of the equation	25

Ⓐ Quantity A is greater.
Ⓑ Quantity B is greater.
Ⓒ The two quantities are equal.
Ⓓ The relationship cannot be determined from the information given.

4.

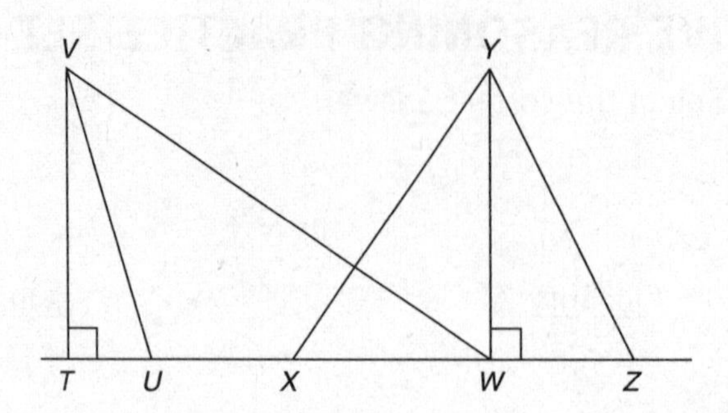

The area of triangle UVW is greater than the area of triangle XYZ.

$UX = WZ$

Quantity A	Quantity B
TV	YW

(A) Quantity A is greater.
(B) Quantity B is greater.
(C) The two quantities are equal.
(D) The relationship cannot be determined from the information given.

5.

Three circles have radii of x, y, and $x + y$ units, respectively.

Quantity A	Quantity B
The sum of the areas of the two smaller circles	The area of the largest circle

(A) Quantity A is greater.
(B) Quantity B is greater.
(C) The two quantities are equal.
(D) The relationship cannot be determined from the information given.

6.

A child ate 1 less than 25% of the 28 apples his father purchased.

Quantity A	Quantity B
The number of apples the child ate	8

(A) Quantity A is greater.
(B) Quantity B is greater.
(C) The two quantities are equal.
(D) The relationship cannot be determined from the information given.

7. Line m on a coordinate plane can be defined by the equation $-2x + 3y = 6$.

Quantity A	Quantity B
The slope of a line parallel to line m	The slope of a line perpendicular to line m

Ⓐ Quantity A is greater.
Ⓑ Quantity B is greater.
Ⓒ The two quantities are equal.
Ⓓ The relationship cannot be determined from the information given.

8. The price of a $240 coat was discounted by 25%. Then the coat was discounted by an additional 30% of the discounted price.

Quantity A	Quantity B
The final price of the coat	$120

Ⓐ Quantity A is greater.
Ⓑ Quantity B is greater.
Ⓒ The two quantities are equal.
Ⓓ The relationship cannot be determined from the information given.

9. Maria bought a new cell phone, cell phone case, and wall charger. The cell phone cost $149.99, the case cost $24.99, and the wall charger cost $29.99. If tax on each of these items was 9.5%, which of the following is closest to the amount Maria spent?

Ⓐ $175
Ⓑ $200
Ⓒ $210
Ⓓ $224
Ⓔ $250

10. Three consecutive even integers have a sum of 102. If x represents the least number in the set, what equation can be used to determine the value of the sum of the integers?

Ⓐ $x + (x + 1) + (x + 2) = 102$
Ⓑ $x + (x + 1) + (x + 3) = 102$
Ⓒ $x + (x + 2) + (x + 4) = 102$
Ⓓ $x + 2x + 4x = 102$
Ⓔ $x + x + x = 102$

11. Kourtland and Caleb share an apartment. If each month Caleb pays c dollars and Kourtland pays k dollars, what percent of the total cost does Kourtland pay?

Ⓐ $\frac{k}{c}\%$
Ⓑ $\frac{c}{k}\%$
Ⓒ $\frac{k}{c + k}\%$
Ⓓ $\frac{100k}{c}\%$
Ⓔ $\frac{100k}{c + k}\%$

12. Which of the following is a possible value of k for which $-\frac{24}{\sqrt{k}}$ is an integer?

Indicate all possible choices.

[A] 9
[B] 12
[C] 16
[D] 25
[E] 64

13. If $M = \frac{at^2}{h}$ and $M \neq 0$, what is the effect on M of doubling t, tripling a, and quadrupling h?

Ⓐ M is multiplied by 1.5.
Ⓑ M is multiplied by 3.
Ⓒ M is multiplied by 4.
Ⓓ M is multiplied by 4.5.
Ⓔ M is multiplied by 79.

14. If Caroline drove 211 miles between 9:30 a.m. and 12:45 p.m. of the same day, what was her approximate average speed in miles per hour?

Ⓐ 50
Ⓑ 55
Ⓒ 60
Ⓓ 65
Ⓔ 70

15. Each person in a room is a junior or a senior. The number of juniors in the room is 7 times the number of seniors. Which of the following could be the number of people in the room?

Indicate all such numbers.

[A] 14
[B] 26
[C] 35
[D] 48
[E] 49
[F] 63

16. Astrid wrote down all the different three-digit numbers that can be written using each of the numerals 1, 2, and 3 exactly once. What is the median of the numbers Astrid wrote down?

Ⓐ 213
Ⓑ 222
Ⓒ 223
Ⓓ 231
Ⓔ 233

17.

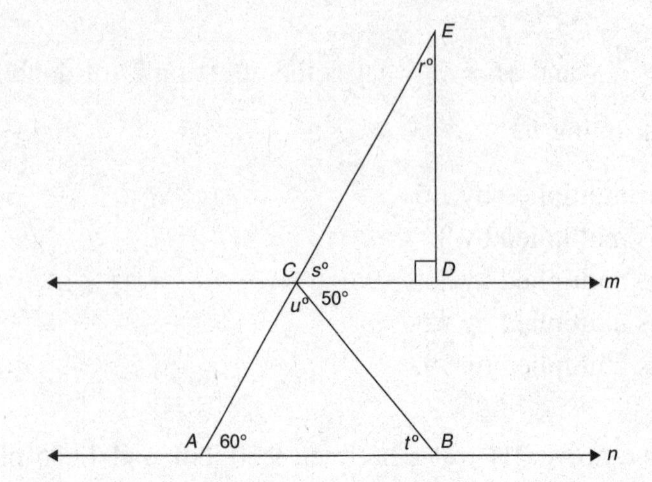

In the figure shown, line m is parallel to line n. What is the value of $s - 2r$?

Questions 18–20 refer to the following graphs.

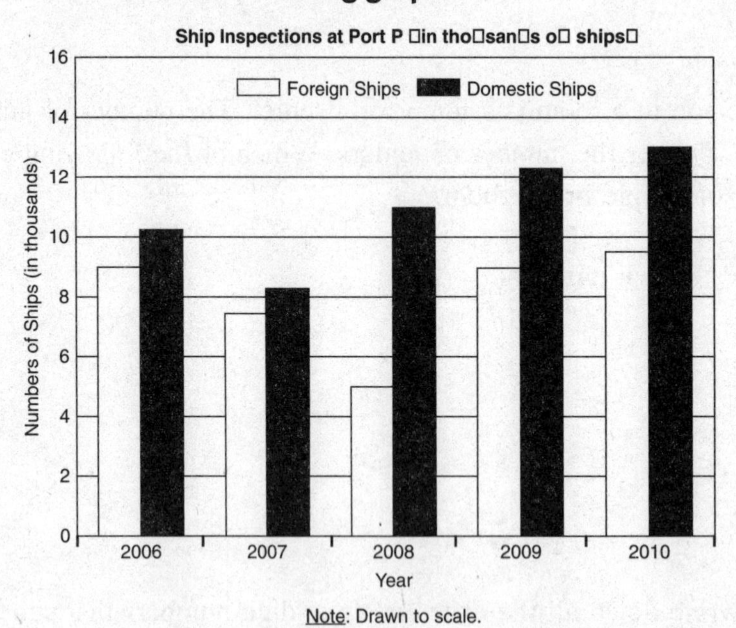

Note: Drawn to scale.

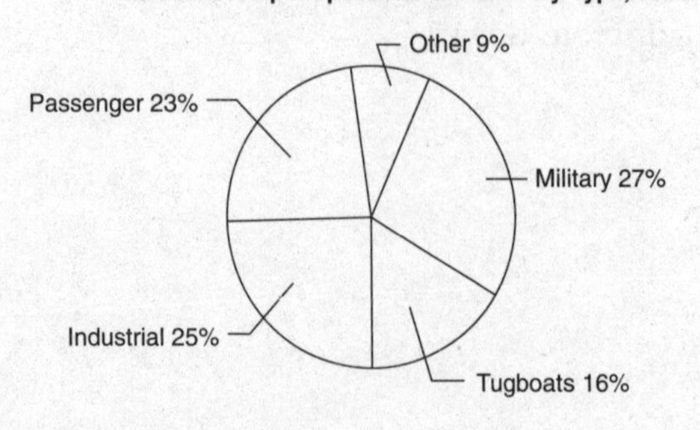

Note: Drawn to scale.

18. In 2009, the ratio of domestic ships inspected to foreign ships inspected was approximately

(A) $\frac{9}{4}$
(B) $\frac{2}{1}$
(C) $\frac{5}{3}$
(D) $\frac{8}{5}$
(E) $\frac{4}{3}$

19. If the average ship carries 500 tons of cargo, which of the following is closest to the number of tons of cargo inspected at Port P in 2008?

(A) 1.0 million
(B) 2.0 million
(C) 4.0 million
(D) 8.0 million
(E) 16.0 million

20. In 2010, approximately how many more domestic passenger ships were inspected than domestic tugboats at Port P?

(A) 820
(B) 855
(C) 890
(D) 910
(E) 955

QUANTITATIVE REASONING PRACTICE SET 4 ANSWER KEY

1. D
2. A
3. C
4. A
5. B
6. B
7. A
8. A
9. D
10. C
11. E
12. A, C, E
13. B
14. D
15. D
16. B
17. 0
18. E
19. D
20. D

QUANTITATIVE REASONING PRACTICE SET 4 ANSWERS AND EXPLANATIONS

1. D

For a question like this, with the variable in both quantities, it is best to use the strategy of Picking Numbers and try several values for the variable. If $m = 0$, both quantities equal 0. If $m = 2$, Quantity B is greater. Just testing these two options is enough to conclude that **(D)** is the answer.

2. A

This question requires that you apply the law of exponents dealing with negative exponents. Recall that $n^{-1} = \frac{1}{n}$ and $n^{-2} = \frac{1}{n^2}$ and so on, provided n does not equal 0. To simplify the expression in Quantity A, work inside the parentheses first.

$$\begin{aligned}\left(3^{-1} + 3^{-2}\right)^{-1} &= \left(\frac{1}{3} + \frac{1}{3^2}\right)^{-1} \\ &= \left(\frac{3}{9} + \frac{1}{9}\right)^{-1} \\ &= \left(\frac{4}{9}\right)^{-1} \\ &= \frac{9}{4}\end{aligned}$$

Notice that in the last step of work shown, the fraction is inverted. The result is a number greater than 1, and the answer is **(A)**.

3. C

Begin by finding the roots of the quadratic equation, using the reverse of the FOIL process. Find two numbers whose product is positive 6 and whose sum is positive 5. The numbers are 2 and 3; use these numbers in the factors and solve the equation.

$$\begin{aligned}x^2 + 5x + 6 &= 0 \\ (x + 2)(x + 3) &= 0 \\ x = -2 \text{ or } x &= -3\end{aligned}$$

To find the value of Quantity A, add the roots and then square the result:

$-2 + -3 = -5$; $(-5)^2 = 25$. Therefore, the quantities are equal. Choose **(C)**.

4. A

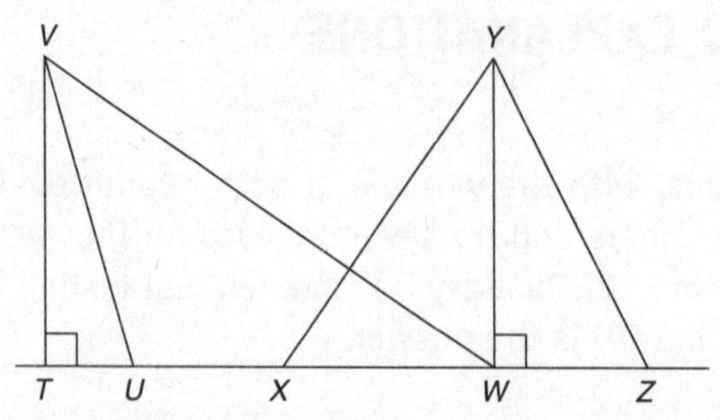

The area of triangle *UVW* is greater than the area of triangle *XYZ*.

$UX = WZ$

You are given that the area of triangle *UVW* is greater than the area of triangle *XYZ*. Make note of the other information conveyed by the diagram. The height of each triangle is indicated by the right angle box in each triangle, but no other information is given about the heights. Notice that the bases of the triangles are overlapping in the diagram; they have segment *XW* in common. Segment *UW* is the base of triangle *UVW*, and segment *XZ* is the base of triangle *XYZ*.

It may be helpful to draw the triangles separately to keep the information given about triangle *UVW* and triangle *XYZ* organized as you apply the area formula to each one.

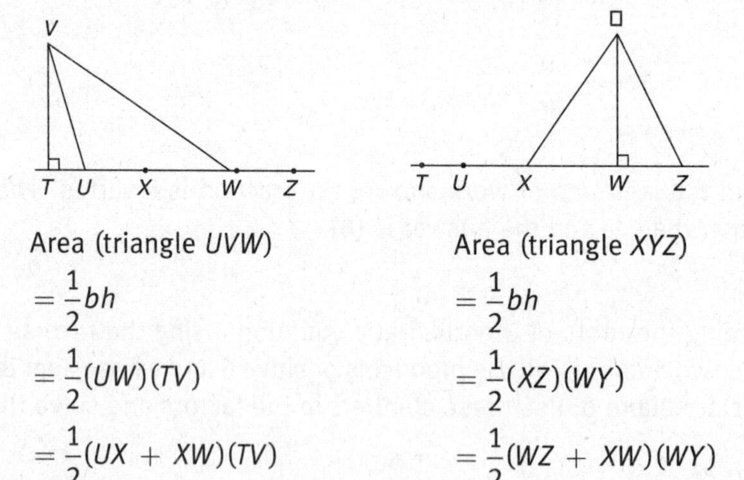

Area (triangle *UVW*)	Area (triangle *XYZ*)
$= \frac{1}{2}bh$	$= \frac{1}{2}bh$
$= \frac{1}{2}(UW)(TV)$	$= \frac{1}{2}(XZ)(WY)$
$= \frac{1}{2}(UX + XW)(TV)$	$= \frac{1}{2}(WZ + XW)(WY)$

The key to making the correct comparison is recognizing that the bases of the triangles are the same length. You are given that $UX = WZ$ and certainly $XW = XW$. If the area of triangle *UVW* is greater, as the question states, it is because its height is greater. Therefore, $TV > YW$ and the answer is **(A)**.

5. B

Use the Picking Numbers strategy to make this question more concrete. Let $x = 2$ and $y = 3$. The two smaller circles will have radii of 2 and 3, and their areas will be 4π and 9π, respectively. Quantity A would be $4\pi + 9\pi = 13\pi$ in this case. Using the same values for *x* and *y*, the radius of the largest circle would be $2 + 3 = 5$, and its area would be 25π. Therefore, Quantity B is greater. The correct choice is **(B)**.

6. B

Find 25% of 28 first and then subtract 1: $0.25 \times 28 = 7$ and $7 - 1 = 6$. The child ate 6 apples. Quantity B is greater. The correct choice is **(B)**.

7. A

To compare these quantities, you need to determine the slope of a parallel and perpendicular line to the linear function $-2x + 3y = 6$. First, solve the equation for y in order to put the equation in the form of $y = mx + b$, where m represents the slope of the line and b represents the y-intercept:

$$\begin{aligned} -2x + 3y &= 6 \\ 3y &= 2x + 6 \\ y &= \frac{2}{3}x + 2 \end{aligned}$$

This is the slope-intercept form of the equation of the line. Parallel lines have the same slope, so the value of Quantity A is $\frac{2}{3}$. The slopes of two perpendicular lines are negative reciprocals of each other, so the value of Quantity B is $-\frac{3}{2}$. Therefore, you can conclude that Quantity A is greater than Quantity B. The correct choice is **(A)**.

8. A

The original price of the coat was \$240. After the first discount of 25%, the coat cost $\$240 - (\$240 \times 0.25) = \$240 - \$60 = \$180$. After the second discount of 30%, the coat cost $\$180 - (\$180 \times 0.30) = \$180 - \$54 = \$126$. Therefore, Quantity A, the final cost of the coat, is \$126. This is larger than Quantity B, \$120. The correct choice is **(A)**.

9. D

First, estimate the total cost: $\$150 + \$25 + \$30 = \$205$. To find the estimated tax, multiply the estimated total by 10%, or 0.1. Ten percent of \$205 is \$20.50. Adding the two amounts together will give the approximate amount she spent: \$225.50, which is closest to \$224. The correct choice is **(D)**.

10. C

The question states that the least number in the set is x. Therefore, the next even number following x will be $x + 2$, and the next even number will be $x + 4$. Since the sum is equal to 102, you set $x + (x + 2) + (x + 4) = 102$. The correct choice is **(C)**.

11. E

The question asks you to find the percent that a part (Kourtland's share) is of a whole (Kourtland and Caleb's total cost). The formula for finding the answer is $\frac{part}{whole} \times 100\%$. In this case, Kourtland's part is k, and the whole is the sum of Kourtland and Caleb's cost, which is $c + k$. $\frac{Kourtland's\ share}{total\ amount} = \frac{k}{c+k} \times 100\% = \frac{100k}{c+k}\%$. The correct choice is **(E)**.

12. A, C, E

The keyword in the question is *integer.* The set of *integers* is composed of all whole numbers: positive, negative, and zero. Therefore, you are looking for $-\frac{24}{\sqrt{k}}$ to equal a whole number. Begin by Backsolving and substituting each answer choice into the expression.

(A) $-\frac{24}{\sqrt{9}} = -\frac{24}{3} = -8$. Since −8 is an integer, choice **(A)** works.

(B) 12 is not a perfect square. Only a perfect square will work here; choice (B) can be eliminated. For the record:

$$-\frac{24}{\sqrt{12}} = -\frac{24}{\sqrt{3 \times 4}} = -\frac{24}{2\sqrt{3}} = -\frac{12}{\sqrt{3}}.$$

(C) $-\frac{24}{\sqrt{16}} = -\frac{24}{4} = -6$. Since −6 is an integer, choice **(C)** works.

(D) $-\frac{24}{\sqrt{25}} = -\frac{24}{5} = -4.8$. Since −4.8 is not an integer, choice (D) can be eliminated.

(E) $-\frac{24}{\sqrt{64}} = -\frac{24}{8} = -3$. Since −3 is an integer, choice **(E)** works.

The correct choices are **(A)**, **(C)**, and **(E)**.

13. B

Doubling t is equal to $2t$, tripling a is equal to $3a$, and quadrupling h is equal to $4h$. Substitute these new terms into the expression and simplify:

$$M = \frac{(3a)(2t)^2}{4h}$$

$$M = \frac{(3a)(4t^2)}{4h}$$

$$M = \frac{3at^2}{h}$$

This expression is 3 times the original expression of $M = \frac{at^2}{h}$. The correct choice is **(B)**.

14. D

The key here is *miles per hour.* You need to determine the time, in hours, between 9:30 a.m. and 12:45 p.m. It is 3 hours and 15 minutes or 3.25 hours. To find miles per hour, divide the miles by the amount of time: $\frac{211 \text{ miles}}{3.25 \text{ hours}} \approx 64.9 \approx 65$ mph.

The correct choice is **(D)**.

15. D

If the number of seniors in the room is x, then the number of juniors is $7x$. The total number of people in the room is $7x + x = 8x$. So the number of people in the room, which is $8x$, must be a multiple of 8. Let's look at the answer choices to see which ones are multiples of 8.

Choice **(D)**, 48, is the only answer choice that is a multiple of 8. The correct answer is **(D)**.

16. B

Start by determining the different three-digit numbers that can be written using 1, 2, and 3. They are 123, 132, 213, 231, 312, and 321.

The *median* of a set is the middle value when the set is listed from least to greatest.

There are two middle values in this set: 213 and 231. To find the median, add them and divide by 2. The correct choice is **(B)**.

17. 0

You are given that lines m and n are parallel. When a transversal cuts across two parallel lines, the corresponding angles formed are congruent. Line AE is a transversal. Therefore, angle ECD is congruent to angle CAB, so $s = 60°$.

You know that angle CDE is 90° and the sum of angles in a triangle is 180°.

$$\begin{aligned} s + r + 90 &= 180 \\ 60 + r + 90 &= 180 \\ r + 150 &= 180 \\ r &= 30 \end{aligned}$$

Since you know the values of s and r, substitute them into the expression:

$$s - 2r = 60 - 2(30) = 60 - 60 = 0$$

The correct answer is 0.

18. E

The bars start at zero on the graph, so it is possible to do some estimation to narrow down the choices. In 2009, more domestic ships were inspected than foreign ships, so the ratio is greater than 1. However, the ratio is not greater than 2 based on the heights of the bars. Choices (A) and (B) can be eliminated. Reading across from the vertical scale to the height of the bar for domestic inspections, you see that a little more than 12,000 domestic ships were inspected and about 9,000 foreign ships were inspected. So the answer is a little more than the ratio of 12 to 9. Since $\frac{12}{9}$ can be simplified to $\frac{4}{3}$, the answer is **(E)**.

19. D

Add the number of domestic ships inspected in 2008 to the number of foreign ships, then multiply the total number of ships by 500 tons to get the total tonnage. There were approximately 11,000 domestic ships inspected and 5,000 foreign ships inspected for a total of 16,000 ships. Multiply that by 500 tons per ship to get an estimate of 8 million tons. The correct answer is **(D)**.

20. D

You need both the graph and the pie chart to answer the question. The simplest approach is this: You are taking the percents of the same whole; therefore, the difference in the two amounts is the same as the difference in the percents, multiplied by the whole. The whole here is the number of 2010 domestic ship inspections, or 13,000 ships. The percents are 23% and 16%, for a difference of 7%. Take 7% of 13,000:

$$0.07 \times 13{,}000 = 910$$

The answer is **(D)**.

Diagnostic Tool

Tally up your score and write the results below.

Total

Total Correct: _______ out of 20

Percentage Correct: # you got right × 100 ÷ 20: _______

By Section:

Quantitative Comparison _______ out of 8

Problem Solving _______ out of 9

Data Interpretation _______ out of 3

DIAGNOSE YOUR RESULTS

Look back at the questions you got wrong and think about your experience answering them. Were you stymied by a particular question type, or by a certain math topic? If the latter, studying the relevant math content review in Part 3 of this book should help.

QUANTITATIVE REASONING PRACTICE SET 5

Directions: Select the correct answer.

1. $$y = 7x - 14$$
$$x > 4$$

Quantity A	Quantity B
y	14

Ⓐ Quantity A is greater.
Ⓑ Quantity B is greater.
Ⓒ The two quantities are equal.
Ⓓ The relationship cannot be determined from the information given.

2. Lines m and n are parallel lines cut by a transversal, l.

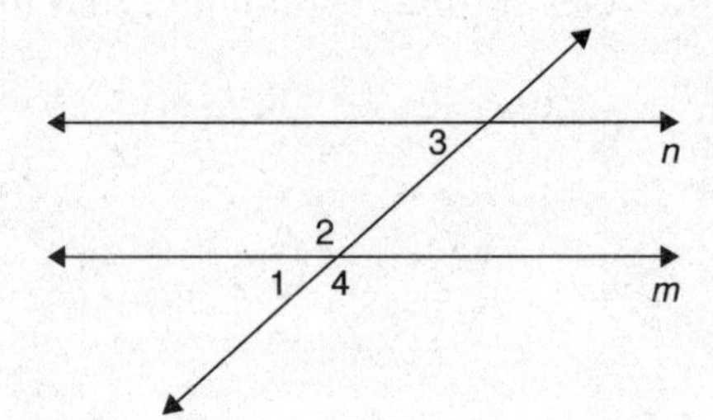

Quantity A	Quantity B
$\angle 1 + \angle 2$	$\angle 3 + \angle 4$

Ⓐ Quantity A is greater.
Ⓑ Quantity B is greater.
Ⓒ The two quantities are equal.
Ⓓ The relationship cannot be determined from the information given.

3.

Quantity A	Quantity B
0.10% of 0.15	1.5×10^{-5}

Ⓐ Quantity A is greater.
Ⓑ Quantity B is greater.
Ⓒ The two quantities are equal.
Ⓓ The relationship cannot be determined from the information given.

4. Use the following data set to answer the question.

3 3 4 4 4 5

Quantity A	Quantity B
The mode of the data set	The average (arithmetic) mean of the data set

(A) Quantity A is greater.
(B) Quantity B is greater.
(C) The two quantities are equal.
(D) The relationship cannot be determined from the information given.

5. $2 \leq x \leq 10$

$y > 3$

Quantity A	Quantity B
$x - y$	xy

(A) Quantity A is greater.
(B) Quantity B is greater.
(C) The two quantities are equal.
(D) The relationship cannot be determined from the information given.

6.

Quantity A	Quantity B
$\frac{2\sqrt{3}}{6}$	$\frac{1}{\sqrt{3}}$

(A) Quantity A is greater.
(B) Quantity B is greater.
(C) The two quantities are equal.
(D) The relationship cannot be determined from the information given.

7. The ratio of y to c is equal to the ratio of x to b.

x, y, b, and c are positive integers.

$0 < y < 4$ and $x > 4$.

Quantity A	Quantity B
$3c$	$2b$

(A) Quantity A is greater.
(B) Quantity B is greater.
(C) The two quantities are equal.
(D) The relationship cannot be determined from the information given.

8. There are 60% fewer parking spots than cars.
There is one car for every 3 people.

Quantity A	Quantity B
the number of parking spots	the number of people

(A) Quantity A is greater.
(B) Quantity B is greater.
(C) The two quantities are equal.
(D) The relationship cannot be determined from the information given.

9. Which of the following has a value that is greater than 25?

Indicate all such values.

[A] $\sqrt{525}$

[B] $10\sqrt{6}$

[C] $\sqrt{625}$

[D] $2\sqrt{169}$

[E] $10\sqrt{7}$

10. At a sandwich shop, there are 2 breads, 3 kinds of meat, 4 types of cheeses, and 6 different sauces to choose from. Customers can choose only one of each. Assuming a customer has already chosen one kind of bread and one kind of meat for a sandwich, how many different sandwiches can be made from the other ingredients?

(A) 36
(B) 24
(C) 18
(D) 10
(E) 2

11. Which of the following is less than the sum of all the prime factors of 330?

Indicate <u>all</u> such values.

- [A] 15
- [B] 17
- [C] 19
- [D] 21

12. There are at least 150 jars of jelly in a grocery store. The ratio of jars of jelly to jars of peanut butter is 5:6. Which of the following could be the number of jars of peanut butter in the store?

Indicate <u>all</u> such numbers.

- [A] 115
- [B] 125
- [C] 180
- [D] 192
- [E] 225

13.

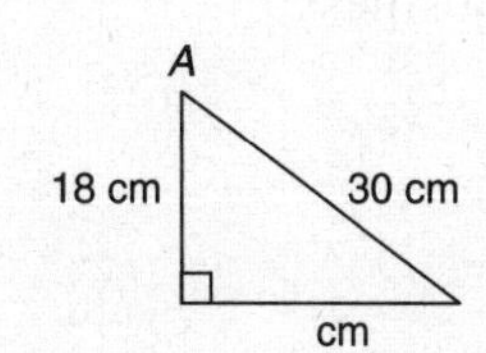

What is the area of triangle ABC in square centimeters?

[ ] square centimeters

Questions 14–16 are based on the following data.

Sales Convention Facts	
Total businesses	226
Business Types:	
Sole Proprietorships (single owner)	134
Male owners	61
Female owners	73
Joint Ventures (multiple owners)	92
Two Owners	33
Three Owners	22
Four or more Owners	37
Business Fields:	
Automotive Sales Business	72
Agriculture Sales Business	18
Electronics Sales Business	66
Household Sales Business	39
Other Businesses	31
Revenues:	
\$25,000 – \$49,999	87
\$50,000 – \$99,999	94
\$100,000 – \$249,999	39
\$250,000 or higher	6

14. All electronics sales businesses are conducted as sole proprietorships. If $\frac{1}{3}$ of electronic sales businesses have female owners, how many electronic sales businesses have male owners?

Ⓐ 73
Ⓑ 66
Ⓒ 61
Ⓓ 44
Ⓔ 22

15. Approximately what percentage of the total number of businesses at the sales convention generate less than $100,000 in annual sales?

   Ⓐ 80%
   Ⓑ 52%
   Ⓒ 42%
   Ⓓ 20%
   Ⓔ 18%

16. Of the household sales businesses present at the convention, there are 7 female-owned sole proprietorships and 4 male-owned sole proprietorships. Approximately what percentage of the household sales businesses are joint ventures?

   Ⓐ 12%
   Ⓑ 28%
   Ⓒ 31%
   Ⓓ 42%
   Ⓔ 72%

17.

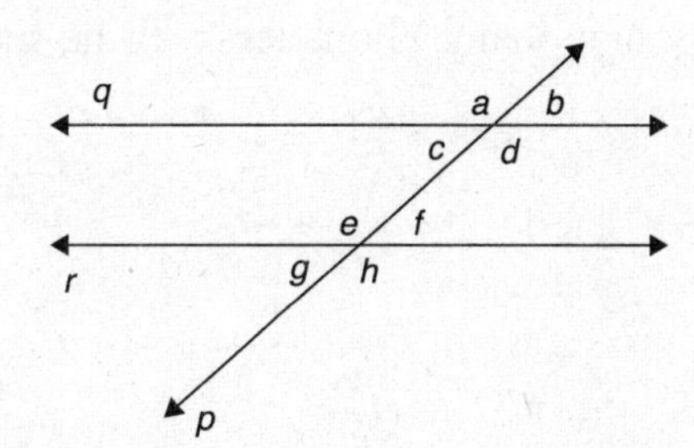

Lines q and r in the figure shown are parallel lines cut by transversal p. What is the sum of the measure of angles a, d, f, and g in degrees?

[ ] degrees

18. The average of John's test scores in his economics class is 80%. Only one test remains, and it is weighted at 25% of the final grade. If his final grade is based entirely on his test scores, what is the lowest percent John can receive on the final to earn an 85% in the class?

(A) 96%
(B) 97%
(C) 98%
(D) 99%
(E) 100%

19. What is the total number of unique prime factors of 450?

(A) 2
(B) 3
(C) 4
(D) 5
(E) 6

20. Pedro traveled 6,252 kilometers last year. If he continues to travel at the same rate, approximately how many kilometers will he travel in the next 3 months?

(A) 500
(B) 1,500
(C) 2,000
(D) 2,500
(E) 3,000

QUANTITATIVE REASONING PRACTICE SET 5 ANSWER KEY

1. A
2. C
3. A
4. A
5. B
6. C
7. B
8. B
9. D, E
10. B
11. A, B, C
12. C, D
13. 216
14. D
15. A
16. E
17. 360
18. E
19. B
20. B

QUANTITATIVE REASONING PRACTICE SET 5 ANSWERS AND EXPLANATIONS

1. A

For y to be less than or equal to 14, x has to be less than or equal to 4. Since x is greater than 4, it follows that y is greater than 14. This relationship can also be determined by using the Picking Numbers strategy. Since x is greater than 4, a substitution of any value greater than 4 for x will result in a value of y that is greater than 14. For example, let $x = 5$. That gives you $y = 7x - 14 = 7(5) - 14 = 35 - 14 = 21$. So, Quantity A is greater; the answer is **(A)**.

2. C

$m\angle 1 + m\angle 2 = 180°$ and $m\angle 1 + m\angle 4 = 180°$ because they are supplementary angles. Lines m and n are parallel lines cut by a transversal, l. They create corresponding angles that are congruent, so $m\angle 1 = m\angle 3$. Substituting $m\angle 3$ for its equivalent angle, $m\angle 1$, results in the equation $m\angle 3 + m\angle 4 = 180°$. Therefore, Quantity A is equal to Quantity B; the answer is **(C)**.

3. A

For Quantity A, write the percent in decimal form and then multiply: $0.10\% = 0.0010$; $0.001 \times 0.15 = 0.00015$. Quantity B $= 1.5 \times 10^{-5} = 1.5 \times 0.00001 = 0.000015$. So, Quantity A is greater than Quantity B; the answer is **(A)**.

4. A

The mode, or data item that occurs most frequently, of 3, 3, 4, 4, 4, 5 is 4. The mean, or average of the data, is $\frac{3 + 3 + 4 + 4 + 4 + 5}{6} = 3.83$. So, Quantity A is greater than Quantity B; the answer is **(A)**.

5. B

In this problem, x can be any number from 2 to 10 inclusive, and y can be any number greater than 3. Substituting possible values for x and y, xy is always greater than $x - y$. For example, let $x = 3$ and $y = 4$. Then $x - y = 3 - 4 = -1$; $xy = 3 \times 4 = 12$. Try a different set of numbers to be certain: let $x = 9$ and $y = 9$. Then $x - y = 9 - 9 = 0$; $xy = 9 \times 9 = 81$. You can conclude that Quantity B is greater than Quantity A; the correct answer is **(B)**.

Note that you could also approach this problem strategically by thinking like this: Because x and y are both positive, $x - y$ will always be less than x. Because x and y are both greater than 1, xy will always be greater than x.

6. C

Quantity A can be simplified as follows: $\frac{2\sqrt{3}}{6} = \frac{\sqrt{3}}{3}$.

Quantity B can be rationalized as follows: $\frac{1}{\sqrt{3}} = \frac{1}{\sqrt{3}} \times \frac{\sqrt{3}}{\sqrt{3}} = \frac{\sqrt{3}}{3}$.

So, Quantity A is equal to Quantity B; the correct answer is **(C)**.

7. B

This problem can be solved by Picking Numbers. You are given that y is less than 4 and greater than 0; x is greater than 4. Since x and y are whole numbers, y cannot be greater than 3. Representative values can be assigned to each variable. For example, $y = 3$ and $x = 5$. The variables c and b are directly related to each other. For example, if c were 7 times the amount of y, then b would be 7 times the amount of x. Using these values, $y = 3$, $x = 5$, $c = 21$ ($y \times 7$), and $b = 35$ ($x \times 7$). Substituting these values into the expressions for Quantity A and Quantity B, $3c = 63$ and $2b = 70$. $2b$ is greater than $3c$. So, Quantity B is greater than Quantity A; the correct answer is **(B)**.

8. B

At a glance, you can see that there are fewer parking spots than cars and that there are fewer cars than people. Thus Quantity B must be larger than Quantity A.

The easiest way to see this is simply to Pick Numbers. If there are 30 people, there are ten cars (one car for every three people). That would mean that there are four parking spots (60% of 10 is 6 and $10 - 6 = 4$).

To see the algebra, let the number of parking spots be represented by s, the number of cars be represented by c, and the number of people be represented by p. s equals c minus 60% of c and c equals $\frac{p}{3}$. Substitute $\frac{p}{3}$ for c in the first equation, $s = c - 0.6c$, resulting in $s = \frac{p}{3} - 0.6\left(\frac{p}{3}\right)$.

Next, simplifying the equation results in $3s = 0.4p$. Dividing each side by 0.4 yields $7.5s = p$. Plug in the numbers from the earlier Picking Numbers approach (four parking spots for every 30 people) and you'll see that this equation is correct. So, the number of people (Quantity B) is greater than the number of parking spots (Quantity A); the answer is **(B)**.

9. D, E

Notice that all the answers involve a square root in one way or another. So, first consider the given number, 25, and its square: $25 \times 25 = 625$. A number greater than 25 must be greater than $\sqrt{625}$. Choices (A) and (C) have values that are less than or equal to 25. In choice (B), $10\sqrt{6}$ is equal to $\sqrt{600}$, which is less than 25.

Choice **(D)** is correct because $\sqrt{169}$ is 13 and $13 \times 2 = 26$. Choice **(E)** is also correct because $10\sqrt{7}$ is equal to $\sqrt{700}$. So, the correct choices are **(D)** and **(E)**.

10. B

Since the bread and meat have been decided, the remaining options consist of 4 cheeses and 6 sauces. Remember, each can only be used once. Using the multiplication principle, the 4 cheeses and the 6 sauces allow for 24 different sandwiches ($6 \times 4 = 24$). So, the answer is **(B)**.

11. A, B, C

The prime factorization of 330 is $2 \times 3 \times 5 \times 11$. The sum of the prime factors is $2 + 3 + 5 + 11 = 21$. So, the correct choices are **(A)**, **(B)**, and **(C)**.

12. C, D

Using the ratio of 5:6, having at least 150 jars of jelly indicates that there are at least 180 jars of peanut butter. Since the number of jars of jelly can be 150 or greater, the number of jars of peanut butter can be 180 or greater. Adhering to the ratio of 5:6, the number of jars of peanut butter must be a multiple of 6 greater than or equal to 180. Note that 225 is greater than 180 but it's not divisible by 6; therefore, it cannot form a ratio of 5:6 with the number of jars of jelly. So, 180 and 192 are the only choices that meet those criteria; the correct answers are **(C)** and **(D)**.

13. 216

Triangle *ABC* is a right triangle. Its side lengths consist of three positive integers a, b, and c such that $a^2 + b^2 = c^2$. 18 and 30 are multiples of 6 and the representatives of the 3 and 5 in the 3:4:5 ratio. So, $x = 6 \times 4 = 24$, or 24 cm.

Apply the formula for area: $\frac{1}{2} \times \text{base} \times \text{height} = \frac{1}{2} \times 24 \times 18 = 216$, or 216 cm^2.

This can also be solved using the Pythagorean theorem, $a^2 + b^2 = c^2$. Substitute the values into the equation and simplify.

$$\begin{aligned}(18)^2 + (x)^2 &= (30)^2 \\ 324 + x^2 &= 900 \\ x^2 &= 576 \\ x &= 24\end{aligned}$$

Use $x = 24$ to find the area of the triangle as shown. The correct answer is **216**.

14. D

One-third of the 66 electronics sales businesses have female owners: $\frac{1}{3} \times 66 = 22$.

The female-owned electronics sales businesses, 22, subtracted from the total electronics sales businesses, 66, leaves 44 male-owned electronics sales businesses. So, the answer is **(D)**.

15. A

There are 87 businesses at the convention that generate less than $50,000 annually and 94 businesses at the convention that generate between $50,000 and $99,999 annually. So, the total number of businesses that generate less than $100,000 is 181.

These 181 businesses account for approximately 80% of the 226 businesses at the convention: $\frac{181}{226} = 80.09\%$. So, the answer is **(A)**.

16. E

There are 39 household sales businesses. Subtracting the 7 female-owned sole proprietorships and 4 male-owned sole proprietorships, the remaining 28 household sales businesses are joint ventures. About 72% of household sales businesses are joint ventures: $\frac{28}{39} = 71.79\%$. So, the answer is **(E)**.

17. 360

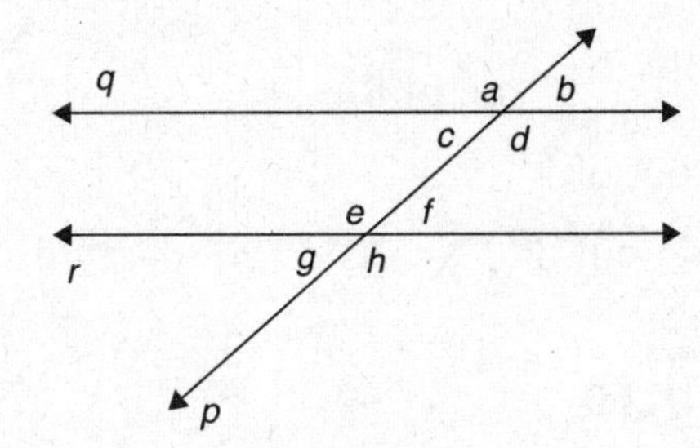

In this problem, every obtuse angle is equal to every other obtuse angle, and every acute angle is equal to every other acute angle. Also, any obtuse angle plus any acute angle equals 180 degrees; a and d are obtuse angles, and f and g are acute angles.

So, the correct answer is **360**.

18. E

Use the Picking Numbers strategy here, and start with 100 for the total number of possible points to score in the class. The tests taken so far weigh 75%, or 75 points out of 100 points. John has achieved a score of 80% so far. This means he has earned $0.75 \times 80 = 60$, or 60 points out of 100 points in the class.

To earn 85%, or 85 points, in the class, John needs $85 - 60 = 25$, or 25 points more. John would have to get a score x such that $0.25x = 25$, or $x = 100$, to earn an 85% in the class. Thus, John has to get 100% on his last test to pull his average up to 85% for the class. The correct answer is **(E)**.

19. B

The prime factorization of 450 is $2 \times 3 \times 3 \times 5 \times 5$. This factorization consists of 3 unique prime factors: 2, 3, and 5. So, there are 3 unique prime factors; the answer is **(B)**.

20. B

Three months is one fourth of a year. Thus, Pedro will travel $6{,}252 \div 4 = 1{,}563$ miles in 3 months if he continues traveling at the same rate. The closest answer is 1,500, choice **(B)**.

Diagnostic Tool

Tally up your score and write the results below.

Total

Total Correct: ______ out of 20

Percentage Correct: # you got right × 100 ÷ 20: ______

By Section:

Quantitative Comparison ______ out of 8

Problem Solving ______ out of 9

Data Interpretation ______ out of 3

DIAGNOSE YOUR RESULTS

Look back at the questions you got wrong and think about your experience answering them. Were you stymied by a particular question type, or by a certain math topic? If the latter, studying the relevant math content review in Part 3 of this book should help.

QUANTITATIVE REASONING PRACTICE SET 6

Directions: Select the correct answer.

1.

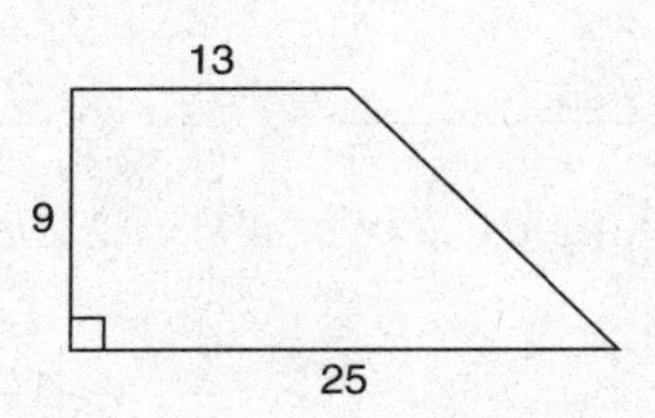

Quantity A	Quantity B
Three times the number of units in the perimeter of the figure	The number of square units in the area of the figure

Ⓐ Quantity A is greater.
Ⓑ Quantity B is greater.
Ⓒ The two quantities are equal.
Ⓓ The relationship cannot be determined from the information given.

2.

$$5 < c < 8$$
$$7 < d < 10$$

Quantity A	Quantity B
$c + d$	16

Ⓐ Quantity A is greater.
Ⓑ Quantity B is greater.
Ⓒ The two quantities are equal.
Ⓓ The relationship cannot be determined from the information given.

3.

The surface area of a cube is 96 square units.

Quantity A	Quantity B
The volume of the cube	256

Ⓐ Quantity A is greater.
Ⓑ Quantity B is greater.
Ⓒ The two quantities are equal.
Ⓓ The relationship cannot be determined from the information given.

4.

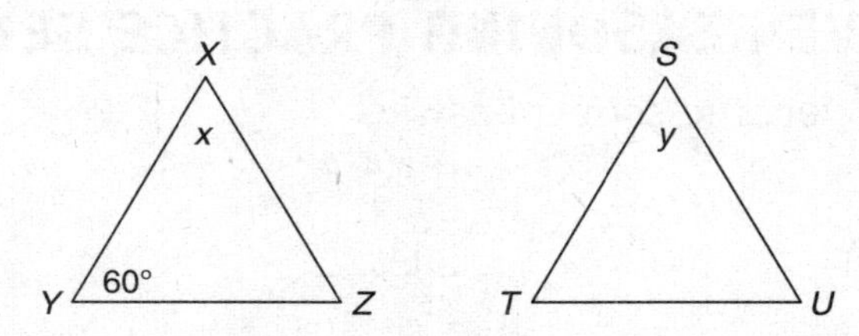

In the triangles above, $XY = XZ$ and $ST = TU = SU$.

Quantity A	Quantity B
x	y

(A) Quantity A is greater.
(B) Quantity B is greater.
(C) The two quantities are equal.
(D) The relationship cannot be determined from the information given.

5. The price of an item was increased by 20 percent. The new price of the item was decreased by x percent, resulting in a final price that is 28 percent less than the original price.

Quantity A	Quantity B
x	43

(A) Quantity A is greater.
(B) Quantity B is greater.
(C) The two quantities are equal.
(D) The relationship cannot be determined from the information given.

6. $$2x^2 + 4x - 30 = 0$$

Quantity A	Quantity B
The product of the roots of the equation	The sum of the roots of the equation

(A) Quantity A is greater.
(B) Quantity B is greater.
(C) The two quantities are equal.
(D) The relationship cannot be determined from the information given.

7.

$$⌂c = -c + \frac{1}{c}$$

Quantity A	Quantity B
The value of $⌂c$ if $c = 4$	The value of $⌂c$ if $c = 3$

Ⓐ Quantity A is greater.
Ⓑ Quantity B is greater.
Ⓒ The two quantities are equal.
Ⓓ The relationship cannot be determined from the information given.

8.

Quantity A	Quantity B
$9^2 + 8^2 + 7^2$	$49 + 63 + 80$

Ⓐ Quantity A is greater.
Ⓑ Quantity B is greater.
Ⓒ The two quantities are equal.
Ⓓ The relationship cannot be determined from the information given.

9. Which of the following is equivalent to the expression $\frac{(ab)^3c^0}{a^3b^4}$, where $abc \neq 0$?

Ⓐ c
Ⓑ ab
Ⓒ abc
Ⓓ $\frac{1}{b}$
Ⓔ $\frac{c}{ab}$

10. Jayson has 172 ounces of sports drink in a cooler. If he pours an equal amount into 8 bottles with 4 ounces left over, approximately what percent of the total does he pour into each bottle?

Ⓐ 4%
Ⓑ 8%
Ⓒ 12%
Ⓓ 20%
Ⓔ 32%

11. Given that $f(x) = (x - 4)^2$ and $g(x) = x^2 - 5$, what is the value of $f(2) - g(2)$?

   (A) −1
   (B) 1
   (C) 3
   (D) 5
   (E) 9

12. Out of every 500 picture frames shipped, there are always exactly 20 damaged frames. Which of the following could be the number of damaged and undamaged frames, respectively, within a given shipment?

   Indicate all such numbers.

   [A] 5 and 100
   [B] 5 and 120
   [C] 10 and 250
   [D] 20 and 480
   [E] 40 and 800

13. Keep Cool Air-Conditioning charges $60 for the first 30 minutes of every house call and $15 for each additional quarter hour. In addition, the customer has to pay for any required parts needed for repairs. If Mrs. Lewis's bill was $225, which included $90 for parts, how long did the air-conditioning repair technician work?

   (A) 1 hour and 15 minutes
   (B) 1 hour and 45 minutes
   (C) 2 hours
   (D) 2 hours and 15 minutes
   (E) 3 hours

14. Darion has at least one quarter, one dime, one nickel, and one penny in his pocket. He has three times as many pennies as nickels, three times as many nickels as dimes, and three times as many dimes as quarters. If he has 120 coins, how much money does he have?

Ⓐ $1.20
Ⓑ $3.81
Ⓒ $4.06
Ⓓ $4.07
Ⓔ $7.75

15.

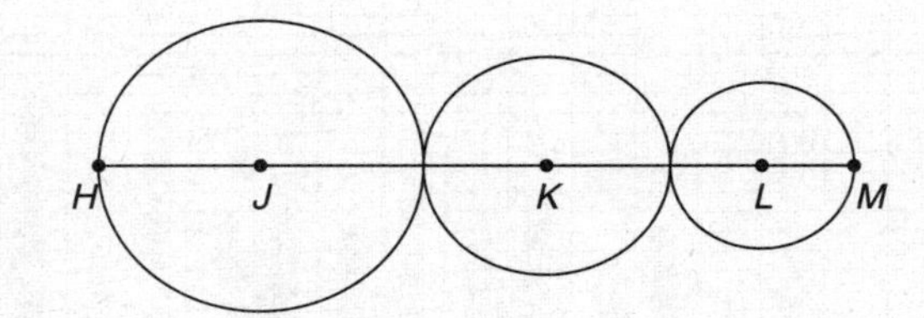

In the figure shown, J, K, and L are the centers of the three circles. The radius of the circle with center J is four times the radius of the circle with center L, and the radius of the circle with center J is two times the radius of the circle with center K. If the sum of the areas of the three circles is 525π square units, what is the measure, in units, of JL?

Ⓐ 35
Ⓑ 45
Ⓒ 50
Ⓓ 65
Ⓔ 70

16. A line whose slope is $-\frac{1}{4}$ passes through the points (4, 3) and (x, 1). What is the value of x?

Ⓐ −12
Ⓑ −4
Ⓒ 0
Ⓓ 8
Ⓔ 12

17. Full Force's annual dance recital was attended by 205 people, each of whom purchased a ticket. Children and youth tickets cost $3.50 and adult tickets cost $5.00. If Full Force collected $828.50 in ticket sales, how many adults attended?

[ ] adults

Questions 18–20 are based on the following graph:

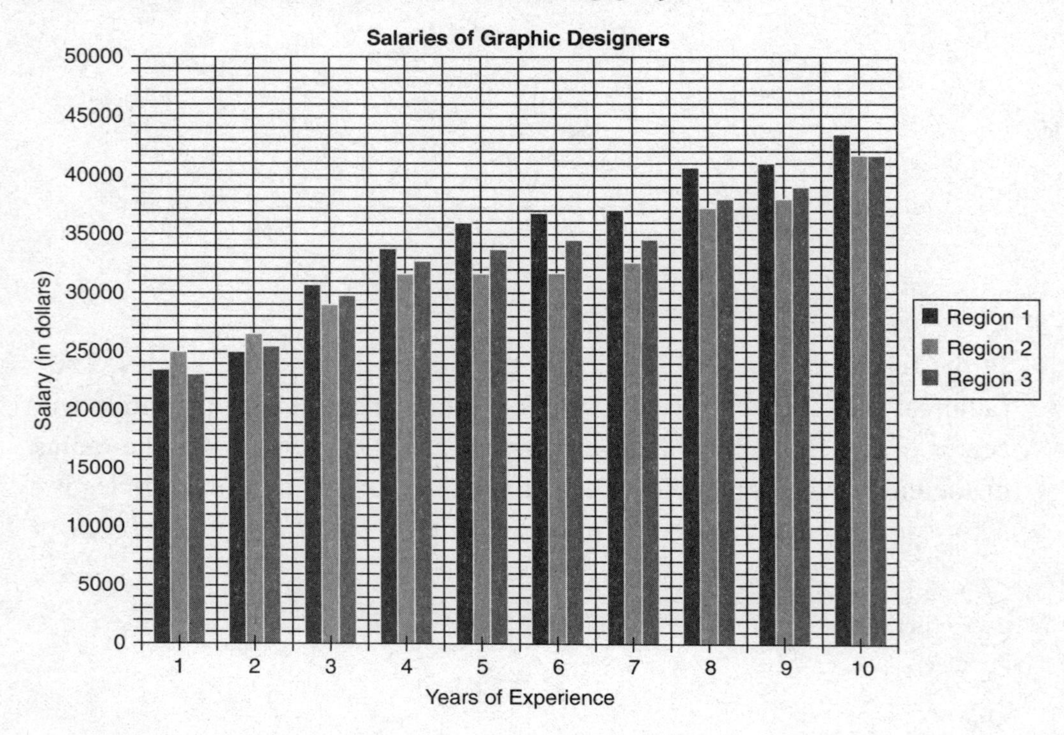

18. How many years of experience does a graphic designer in Region 2 need in order to earn the same salary as a graphic designer in Region 1 with 7 years of experience?

(A) 5
(B) 6
(C) 7
(D) 8
(E) 9

19. Richard, who lives in Region 2, just graduated from college and has zero years of experience. His starting salary is $23,900 per year. Suppose the correlation between experience and pay shown in the graph remains the same over the next 10 years. What is the approximate percent increase in annual pay over his starting salary that he can expect to receive if he works 10 years?

Ⓐ 42%
Ⓑ 45%
Ⓒ 55%
Ⓓ 58%
Ⓔ 74%

20. What is the approximate difference, in dollars, between the median of the region with the highest median salary for the ten different possible years shown and the median of the region with the lowest median salary for the ten different possible years shown?

Ⓐ 1,000
Ⓑ 2,000
Ⓒ 2,500
Ⓓ 3,000
Ⓔ 5,000

QUANTITATIVE REASONING PRACTICE SET 6 ANSWER KEY

1. A
2. D
3. B
4. C
5. B
6. B
7. B
8. A
9. D
10. C
11. D
12. B, D
13. B
14. B
15. B
16. E
17. 74
18. D
19. E
20. E

QUANTITATIVE REASONING PRACTICE SET 6 ANSWERS AND EXPLANATIONS

1. A

To compare the quantities, you must first determine the perimeter and the area of the figure. Break up the complex figure into two smaller figures:

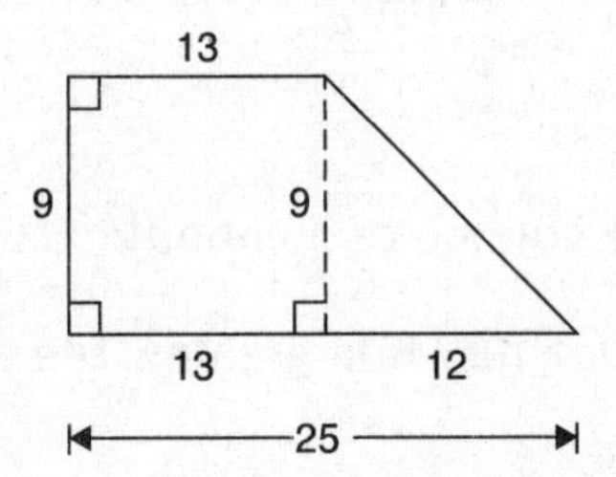

Look at the triangle created by drawing the dashed segment. The dashed segment is both a leg of a right triangle and a side of a 9×13 rectangle. To calculate the perimeter of the entire figure, you have to find the hypotenuse of the triangle created. The legs of the triangle are 9 and 12, so the triangle is a 3:4:5 triangle—$3 \times 3 = 9$, $3 \times 4 = 12$—whose hypotenuse is 15, because $3 \times 5 = 15$.

The perimeter of the figure can be found now: $9 + 13 + 15 + 25 = 62$. That means Quantity A is $3 \times 62 = 186$.

The area of the entire figure is the sum of the area of the rectangle and the area of the triangle:

Rectangle: $A = l \times w = 9 \times 13 = 117$

Triangle: $A = \frac{1}{2}bh = \frac{1}{2}(12 \times 9) = 54$

Complete figure: $117 + 54 = 171$

Compare 186 for Quantity A to 171 for Quantity B. The answer is **(A)**.

2. D

Consider Picking Numbers to test the relationship between $c + d$ and 16. It is not necessary that you choose whole numbers for c and d, but it is easier that way. Suppose $c = 6$ and $d = 8$. Then Quantity A is less: $6 + 8 = 14$; $14 < 16$.

But don't jump to conclusions yet; try another set of values. Suppose $c = 7$ and $d = 9$. Then Quantity A equals Quantity B: $7 + 9 = 16$. So, more than one relationship is possible, and the correct choice is **(D)**.

3. B

The key to solving this problem is knowing the formula for the surface area of a cube, knowing the formula for the volume of a cube, and knowing that all the sides of a cube are equal. The formula for the surface area of a cube is $6s^2$, where s is the side length. To find the side length, solve the equation for s:

$$\begin{aligned} 6s^2 &= 96 \\ s^2 &= 16 \\ s &= 4 \end{aligned}$$

Now, use the formula for the volume of a cube: $V = s^3$, where s is the side length.

$V = s^3 = 4 \times 4 \times 4 = 64$. Quantity B is greater. The answer is **(B)**.

4. C

The sum of the interior angles in a triangle is 180°. The given information indicates that the triangle on the left is isosceles. Since its two top sides are equal, the two base angles are also equal. If the two base angles are equal, then the value of x is also 60°. The given information indicates that the triangle on the right is equilateral. Since all three sides are equal, all three interior angles are also equal, and the value of y is 60°. Therefore, the two quantities are equal, and the correct choice is **(C)**.

5. B

If the original price of the item was 100 dollars and the price was increased by 20%, the new price, in dollars, was $100 + (0.20 \times 100) = 120$. The final price was 28% less than the original price, so the final price, in dollars, was $100 - (0.28 \times 100) = 100 - 28 = 72$.

Use the percent decrease formula to find what x, the percent decrease, is from 120 to 72.

$$\textit{Percent decrease} = \frac{\textit{Amount of decrease}}{\textit{Original whole}} \times 100\%$$

$\frac{120 - 72}{120} \times 100\% = \frac{48}{120} \times 100\%$. $\frac{48}{120}$ simplifies to $\frac{2}{5}$. So $\frac{2}{5} \times 100\% = 40\%$.

Thus, $x = 40$. Quantity A has a value of 40, and Quantity B has a value of 43. The correct answer is **(B)**.

6. B

To compare the quantities, you have to solve the equation first. Start by factoring out the greatest common factor, 2.

$$\begin{aligned} 2x^2 + 4x - 30 &= 0 \\ 2(x^2 + 2x - 15) &= 0 \end{aligned}$$

Divide both sides by 2 and factor the expression inside the parentheses into two binomials:

$$(x + 5)(x - 3) = 0$$

To solve the equation, set each binomial equal to 0 and solve for x:

$$x + 5 = 0 \qquad x - 3 = 0$$

So $x = -5$ or $x = 3$. The product of the two possible values of x is –15, and the sum of the two possible values of x is –2. The correct choice is **(B)** because $-2 > -15$.

7. B

The operation indicated by the symbol says to take negative c and then add the reciprocal of c. For $c = 4$, you have

$$-4 + \frac{1}{4} = \frac{-16 + 1}{4} = -\frac{15}{4} = -3\frac{3}{4}.$$

For $c = 3$, you have

$$-3 + \frac{1}{3} = \frac{-9 + 1}{3} = -\frac{8}{3} = -2\frac{2}{3}.$$

On a number line, Quantity B would be to the right of Quantity A. Quantity B is greater, so the answer is **(B)**.

8. A

There is no need to do a lot of calculating for this question—use reasoning instead. Compare the pieces in the quantities. On the left there is 7^2, and on the right there is 49. Cross these off because they are the same. Now compare the pieces that are left. Because 9^2 is greater than 80 and 8^2 is greater than 63, Quantity A is greater.

9. D

To simplify the expression, you will use the rules of exponents. $(ab)^3$ can be rewritten as a^3b^3 by applying the power rule, and c^0 should be rewritten as 1 because any nonzero number to the zero power equals 1. That gives you $\frac{a^3b^3}{a^3b^4}$. Although you are not finished simplifying, you can eliminate choices (A), (C), and (E) because they contain c. Next, simplify the expression to $\frac{1}{b}$.

The correct choice is **(D)**.

10. C

If there were 4 ounces left over, then $172 - 4 = 168$ ounces were poured. If 8 bottles were filled, then each bottle got $168 \div 8 = 21$ ounces. Now calculate the percentage that 21 is of the total 172: $\frac{21}{172} \approx 0.12$, or 12%. The correct answer is **(C)**.

11. D

Start by evaluating each function for $x = 2$.

$$f(x) = (x - 4)^2 = (2 - 4)^2 = (-2)^2 = 4$$
$$g(x) = x^2 - 5 = (2)^2 - 5 = 4 - 5 = -1$$

The question asks for the difference: $4 - (-1) = 5$. The correct answer is **(D)**.

12. B, D

The key to determining the ratio is to calculate the number of picture frames that are undamaged, which is $500 - 20 = 480$ for every 500 frames. The ratio of damaged frames to undamaged frames is $\frac{20}{480} = \frac{10}{240} = \frac{5}{120} = \frac{1}{24}$. Choices **(B)** and **(D)** are multiples of this ratio: $\frac{5}{5} \times \frac{1}{24} = \frac{5}{120}$ and $\frac{20}{20} \times \frac{1}{24} = \frac{20}{480}$.

The correct choices are **(B)** and **(D)**.

13. B

One way to solve this problem is to set up an equation, where x represents the number of quarter hours the repair technician worked:

$$\begin{aligned} \$60 + \$15x + \$90 &= \$225 \\ \$150 + \$15x &= \$225 \\ \$15x &= \$75 \\ x &= 5 \end{aligned}$$

The repair technician worked for 5 quarter hours (in addition to the first 30 minutes). A quarter of an hour is 15 minutes, so 5 quarter hours is equal to 1 hour and 15 minutes. Add in the first 30 minutes, and the answer is **(B)**, 1 hour and 45 minutes.

14. B

This problem can be solved with algebra, but it can also be solved by Picking Numbers. The total number of coins is 120, so pick a number for the number of quarters and go from there. The number of coins are all based on the number of quarters as shown. Suppose Darion has 2 quarters:

pennies	$3 \times 18 = 54$
nickels	$3 \times 6 = 18$
dimes	$3 \times 2 = 6$
quarters	2
Total coins	80 (too few)

That is not enough coins, so try 3 quarters:

pennies	$3 \times 27 = 81$
nickels	$3 \times 9 = 27$
dimes	$3 \times 3 = 9$
quarters	3
Total coins	120

Now that you have the correct number of coins, their value may be found:

$0.81 + (27 \times 0.05) + (9 \times 0.10) + (3 \times 0.25) = 3.81$. The correct answer is **(B)**.

15. B

To solve this problem, use the formula for the area of a circle, $A = \pi r^2$, where r is the radius. Before you find the length of JL, you have to find the length of the radius of each circle. Since the radii of circles J and K are four and two times the radius of circle L, respectively, you can set up the equation, where r represents the radius of circle L:

$$\begin{aligned} \pi(4r)^2 + \pi(2r)^2 + \pi(r)^2 &= 525\pi \\ 16r^2\pi + 4r^2\pi + r^2\pi &= 525\pi \\ 21r^2\pi &= 525\pi \\ r^2 &= 25 \\ r &= 5 \end{aligned}$$

If the radius of circle L is 5 units, the radius of circle K is 10 units, and the radius of circle J is 20 units. So the length of line segment JL is composed of the radius of circle L, the diameter of circle K, and the radius of circle J. This equals $20 + 10 + 10 + 5$ or 45 units. The correct choice is **(B)**.

16. E

For this question, use the formula for the slope of a line: $Slope = \dfrac{y_2 - y_1}{x_2 - x_1}$.

$$\begin{aligned} -\frac{1}{4} &= \frac{3-1}{4-x} \\ -\frac{1}{4}(4-x) &= \frac{3-1}{4-x}(4-x) \\ -1+\frac{x}{4} &= 2 \\ \frac{x}{4} &= 3 \\ x &= 12 \end{aligned}$$

The answer is **(E)**.

17. 74

To solve this problem, set up a system of equations where a represents the number of adults and c represents the number of children and youths:

$$a + c = 205$$
$$3.50c + 5a = \$828.50$$

Solve for a in the first equation and then substitute it into the second equation: $a = 205 - c$.

$$\begin{aligned} 3.50c + 5(205 - c) &= 828.50 \\ 3.50c + 1{,}025 - 5c &= 828.50 \\ -1.50c &= -196.50 \\ c &= 131 \end{aligned}$$

Since there were 131 children and youths, there were $205 - 131 = 74$ adults. The correct answer is **74**.

18. D

First, locate the bar of a graphic designer in Region 1 with 7 years of experience. Since Region 1 remains highest compared to all other bars from 1 to 7 years, you can eliminate any number of years that are 7 or less in the answer choices. Then, locate a bar that has the same height as the bar you found from 7 years and beyond. A graphic designer in Region 2 with 8 years of experience has the same salary. The correct choice is **(D)**.

19. E

The key to finding the percent increase is to compare the increase to the original amount. If Richard works 10 years, he can expect to earn \$41,500 per year at the end of that time. The amount of increase is $\$41{,}500 - \$23{,}900 = \$17{,}600$.

$$\frac{17{,}600}{23{,}900} \times 100\% = 73.6\% \approx 74\%$$

The percent increase is about 74%. The correct choice is **(E)**.

20. E

Start by determining the medians for the regions you'll want to compare. Since there are ten data values in each set, the median for each region will be the average of the fifth and sixth data values. It's clear from the graph that Region 1 will have the highest median value and Region 2 will have the lowest median value, so those are the ones to calculate:

Region 1: $(\$36{,}000 + \$37{,}000) \div 2 = \$36{,}500$
Region 2: $(\$31{,}500 + \$31{,}500) \div 2 = \$31{,}500$

The approximate difference is \$5,000. The correct choice is **(E)**.

Diagnostic Tool

Tally up your score and write the results below.

Total

Total Correct: ______ out of 20

Percentage Correct: # you got right $\times$ 100 $\div$ 20: ______

By Section:

Quantitative Comparison ______ out of 8

Problem Solving ______ out of 9

Data Interpretation ______ out of 3

Diagnose Your Results

Look back at the questions you got wrong and think about your experience answering them. Were you stymied by a particular question type, or by a certain math topic? If the latter, studying the relevant math content review in Part 3 of this book should help.

PART THREE

Math Content Review

CHAPTER 6

Arithmetic

Most of the problems on the GRE involve arithmetic to some extent. Among the most important topics are number properties, ratios, and percents. You should know most of the definitions of different types of numbers such as what an integer is, what even numbers are, etc.

Not only do arithmetic topics covered in the unit themselves appear on the GRE, but they are also essential for understanding some of the more advanced topics that will be covered later. For instance, many of the rules covering arithmetic operations, such as the commutative law, will be important when we discuss variables and algebraic expressions. In addition, the concepts we cover here will be needed for solving problems.

REAL NUMBERS

Number Types

The set of real numbers consists of all rational and irrational numbers, including integers, fractions, and decimals. This number tree is a visual representation of the types of numbers and their relationships:

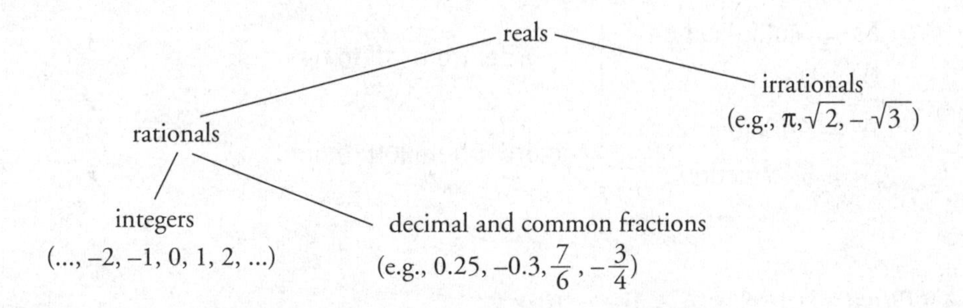

Real Numbers: All numbers on the number line; all the numbers on the GRE are real

Rational Numbers: All numbers that can be expressed as the ratio of two integers (all integers and fractions)

Irrational Numbers: All real numbers that are not rational, both positive and negative (e.g., π, $-\sqrt{3}$)

Integers: All numbers, including zero, with no fractional or decimal parts

A number line can be used to represent the set of real numbers. Every real number corresponds to a point on the number line, and every point on the number line corresponds to a real number. Numbers to the left of 0 on the number line are negative and numbers to the right of 0 are positive. Zero is neither negative nor positive.

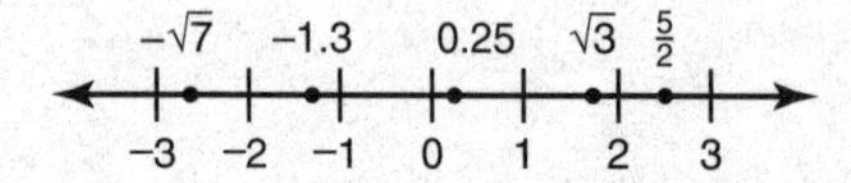

The number line can be used to compare real numbers. For real numbers a and b, if a is to the left of b, then a is less than b, $a < b$. If a is to the right of b, then a is greater than b, $a > b$. So, $-\sqrt{7} < \sqrt{3}$, $\sqrt{3} > -1.3$, and $0.25 < \frac{5}{2}$.

NUMBER OPERATIONS

Working efficiently and confidently with numeric expressions can save you time on the GRE. The following operations may be just one important step in solving a problem.

Order of Operations

PEMDAS = **P**lease **E**xcuse **M**y **D**ear **A**unt **S**ally—This mnemonic will help you remember the order of operations.

P = Parentheses
E = Exponents
M = Multiplication } in order from left to right
D = Division
A = Addition } in order from left to right
S = Subtraction

Example: $30 - 5 \times 4 + (7 - 3)^2 \div 8$

First perform any operations within **Parentheses**. (If the expression has parentheses within parentheses, work from the innermost out.) $\quad 30 - 5 \times 4 + 4^2 \div 8$

Next, raise to any powers indicated by **Exponents**.	$30 - 5 \times 4 + 16 \div 8$
Then do all **Multiplication** and **Division** in order from left to right.	$30 - 20 + 2$
Last, do all **Addition** and **Subtraction** in order from left to right.	$10 + 2$ 12

LAWS OF OPERATIONS

Commutative law: Addition and multiplication are both **commutative**; it doesn't matter **in what order** the operation is performed.

Example: $5 + 8 = 8 + 5$; $2 \times 6 = 6 \times 2$

Division and subtraction are **not** commutative.

Example: $3 - 2 \neq 2 - 3$; $6 \div 2 \neq 2 \div 6$

Associative law: Addition and multiplication are also **associative**; the terms can be **regrouped** without changing the result.

Example:

$$\begin{aligned} (a + b) + c &= a + (b + c) \\ (3 + 5) + 8 &= 3 + (5 + 8) \\ 8 + 8 &= 3 + 13 \\ 16 &= 16 \end{aligned} \qquad \begin{aligned} (a \times b) \times c &= a \times (b \times c) \\ (4 \times 5) \times 6 &= 4 \times (5 \times 6) \\ 20 \times 6 &= 4 \times 30 \\ 120 &= 120 \end{aligned}$$

Distributive law: The **distributive law** of multiplication allows us to "distribute" a factor among the terms being added or subtracted. In general, $a(b + c) = ab + ac$.

Example:

$$\begin{aligned} 4(3 + 7) &= 4 \times 3 + 4 \times 7 \\ 4 \times 10 &= 12 + 28 \\ 40 &= 40 \end{aligned}$$

Division can be distributed in a similar way.

Example:

$$\begin{aligned} \frac{3 + 5}{2} &= \frac{3}{2} + \frac{5}{2} \\ \frac{8}{2} &= 1\frac{1}{2} + 2\frac{1}{2} \\ 4 &= 4 \end{aligned}$$

Don't get carried away, though. When the sum or difference is in the **denominator**, no distribution is possible.

Example: $\frac{9}{4+5}$ is NOT equal to $\frac{9}{4} + \frac{9}{5}$.

FRACTIONS

$4 \leftarrow$ numerator

$- \leftarrow$ fraction bar (means "divided by")

$5 \leftarrow$ denominator

Equivalent fractions: The value of a number is unchanged if you multiply the number by 1. In a fraction, multiplying the numerator and denominator by the same nonzero number is the same as multiplying the fraction by 1; the fraction is unchanged. Similarly, dividing the top and bottom by the same nonzero number leaves the fraction unchanged.

Example: $\frac{1}{2} = \frac{1 \times 2}{2 \times 2} = \frac{2}{4}$

$\frac{5}{10} = \frac{5 \div 5}{10 \div 5} = \frac{1}{2}$

Canceling and reducing: Generally speaking, when you work with fractions on the GRE, you'll need to put them in **lowest terms**. That means that the numerator and the denominator are not divisible by any common integer greater than 1. For example, the fraction $\frac{1}{2}$ is in lowest terms, but the fraction $\frac{3}{6}$ is not, since 3 and 6 are both divisible by 3.

The method we use to take such a fraction and put it in lowest terms is called **reducing**. That simply means to divide out any common multiples from both the numerator and denominator. This process is also commonly called **canceling**.

Example: Reduce $\frac{15}{35}$ to lowest terms.

First, determine the largest common factor of the numerator and denominator. Then, divide the top and bottom by that number to reduce.

$\frac{15}{35} = \frac{3 \times 5}{7 \times 5} = \frac{3 \times 5 \div 5}{7 \times 5 \div 5} = \frac{3}{7}$

Addition and subtraction: We can't add or subtract two fractions directly unless they have the same denominator. Therefore, before adding or subtracting, we must find a common denominator. A common denominator is just a **common multiple** of the denominators of the fractions. The **least common denominator** is the **least common multiple** (the smallest positive number that is a multiple of all the terms).

Example: $\frac{3}{5}+\frac{2}{3}-\frac{1}{2}$

Denominators are 5, 3, 2. $\quad$ $LCM = 5 \times 3 \times 2 = 30 = LCD$

Multiply numerator and denominator of each fraction by the value that raises each denominator to the LCD.

$$\left(\frac{3}{5}\times\frac{6}{6}\right)+\left(\frac{2}{3}\times\frac{10}{10}\right)-\left(\frac{1}{2}\times\frac{15}{15}\right)$$

$$=\frac{18}{30}+\frac{20}{30}-\frac{15}{30}$$

Combine the numerators by adding or subtracting and keep the LCD as the denominator.

$$=\frac{18+20-15}{30}=\frac{23}{30}$$

Multiplication:

Example: $\frac{10}{9}\times\frac{3}{4}\times\frac{8}{15}$

First, reduce (cancel) diagonally and vertically.

$$\frac{{}^{2}\cancel{10}}{{}_{3}\cancel{9}}\times\frac{{}^{1}\cancel{3}}{{}_{1}\cancel{4}}\times\frac{\cancel{8}^{2}}{\cancel{15}_{3}}$$

Then multiply numerators together and denominators together.

$$\frac{2\times1\times2}{3\times1\times3}=\frac{4}{9}$$

Division: Dividing is the same as multiplying by the **reciprocal** of the divisor. To get the reciprocal of a fraction, just invert it by interchanging the numerator and the denominator. For example, the reciprocal of the fraction $\frac{3}{7}$ is $\frac{7}{3}$.

Example: $\frac{4}{3}\div\frac{4}{9}$

To divide, invert the second term (the divisor), and then multiply as above.

$$\frac{4}{3}\div\frac{4}{9}=\frac{4}{3}\times\frac{9}{4}=\frac{{}^{1}\cancel{4}}{{}_{1}\cancel{3}}\times\frac{\cancel{9}^{3}}{\cancel{4}_{1}}=\frac{1\times3}{1\times1}=3$$

Complex fractions: A complex fraction is a fraction that contains one or more fractions in its numerator or denominator. There are two ways to simplify complex fractions.

Method I: Use the distributive law. Find the least common multiple of all the denominators, and multiply all the terms in the top and bottom of the complex fraction by the LCM. This will eliminate all the denominators, greatly simplifying the calculation.

Example: $$\frac{\frac{7}{9}-\frac{1}{6}}{\frac{1}{3}+\frac{1}{2}} = \frac{18 \times \left(\frac{7}{9}-\frac{1}{6}\right)}{18 \times \left(\frac{1}{3}+\frac{1}{2}\right)}$$ LCM of all the denominators is 18.

$$= \frac{\frac{^{2}\cancel{18}}{1} \times \frac{7}{\cancel{9}_{1}} - \frac{^{3}\cancel{18}}{1} \times \frac{1}{\cancel{6}_{1}}}{\frac{^{6}\cancel{18}}{1} \times \frac{1}{\cancel{3}_{1}} + \frac{^{9}\cancel{18}}{1} \times \frac{1}{\cancel{2}_{1}}}$$

$$= \frac{2 \times 7 - 3 \times 1}{6 \times 1 + 9 \times 1}$$

$$= \frac{14 - 3}{6 + 9} = \frac{11}{15}$$

Method II: Treat the numerator and denominator separately. Combine the terms in each to get a single fraction on top and a single fraction on bottom. We are left with the division of two fractions, which we perform by multiplying the top fraction by the reciprocal of the bottom one. This method is preferable when it is difficult to get an LCM for all the denominators.

Example: $$\frac{\frac{7}{9}-\frac{1}{6}}{\frac{1}{3}+\frac{1}{2}} = \frac{\frac{14}{18}-\frac{3}{18}}{\frac{2}{6}+\frac{3}{6}} = \frac{\frac{11}{18}}{\frac{5}{6}} = \frac{11}{18} \div \frac{5}{6} = \frac{11}{\cancel{18}_{3}} \times \frac{\cancel{6}^{1}}{5} = \frac{11}{15}$$

Example: $$\frac{\frac{5}{11}-\frac{5}{22}}{\frac{7}{16}+\frac{3}{8}} = \frac{\frac{10}{22}-\frac{5}{22}}{\frac{7}{16}+\frac{6}{16}} = \frac{\frac{5}{22}}{\frac{13}{16}} = \frac{5}{\cancel{22}_{11}} \times \frac{\cancel{16}^{8}}{13} = \frac{40}{143}$$

Comparing positive fractions: If the numerators are the same, the fraction with the smaller denominator will have the larger value, since the numerator is divided into a smaller number of parts.

Example: $\frac{4}{5} > \frac{4}{7}$ i.e.: >

If the denominators are the same, the fraction with the larger numerator will have the larger value.

Example: $\frac{5}{8} > \frac{3}{8}$ i.e.: >

If neither the numerators nor the denominators are the same, express all of the fractions in terms of some common denominator. The fraction with the largest numerator will be the largest.

Example: Compare $\frac{11}{15}$ and $\frac{13}{20}$.

$$\frac{11}{15} = \frac{11 \times 20}{15 \times 20} \qquad \frac{13}{20} = \frac{13 \times 15}{20 \times 15}$$

$$= \frac{220}{15 \times 20} \qquad = \frac{195}{20 \times 15}$$

Since $220 > 195$, $\frac{11}{15} > \frac{13}{20}$.

Notice that it is not necessary to calculate the denominators. A shorter version of this method is to multiply the numerator of the left fraction by the denominator of the right fraction and vice versa (cross multiply). Then compare the products obtained this way. If the left product is greater, then the left fraction was greater to start with.

Example: Compare $\frac{5}{7}$ and $\frac{9}{11}$.

$5 \times 11 \ ? \ 9 \times 7$

$55 < 63$ so $\frac{5}{7} < \frac{9}{11}$

Sometimes it is easier to find a common **numerator**. In this case, the fraction with the **smaller** denominator will be the **larger** fraction.

Example: Compare $\frac{22}{19}$ and $\frac{11}{9}$.

Multiply $\frac{11}{9} \times \frac{2}{2}$ to obtain a common numerator of 22.

$$\frac{11}{9} = \frac{11 \times 22}{9 \times 2} = \frac{22}{18}$$

Since $\frac{22}{19} < \frac{22}{18}$, $\frac{22}{19} < \frac{11}{9}$.

As before, the comparison can also be made by cross multiplying.

$22 \times 9 < 11 \times 19$, so $\frac{22}{19} < \frac{11}{9}$

Mixed Numbers: Mixed numbers are numbers consisting of an integer and a fraction. For example, $3\frac{1}{4}$, $12\frac{2}{5}$, and $5\frac{7}{8}$ are all mixed numbers. Fractions whose numerators are greater than their denominators may be converted into mixed numbers, and vice versa.

Example: Convert $\frac{23}{4}$ to a mixed number.

$$\frac{23}{4} = \frac{20}{4} + \frac{3}{4} = 5\frac{3}{4}$$

Example: Convert $2\frac{3}{7}$ to a fraction.

$$2\frac{3}{7} = 2 + \frac{3}{7} = \frac{14}{7} + \frac{3}{7} = \frac{17}{7}$$

Decimal Fractions

Decimal fractions are just another way of expressing common fractions; they can be converted to common fractions with a power of ten in the denominator.

Example: $0.053 = \frac{53}{10^3}$ or $\frac{53}{1{,}000}$

Each position, or **digit**, in the decimal has a name associated with it. The GRE occasionally contains questions on digits, so you should be familiar with this naming convention:

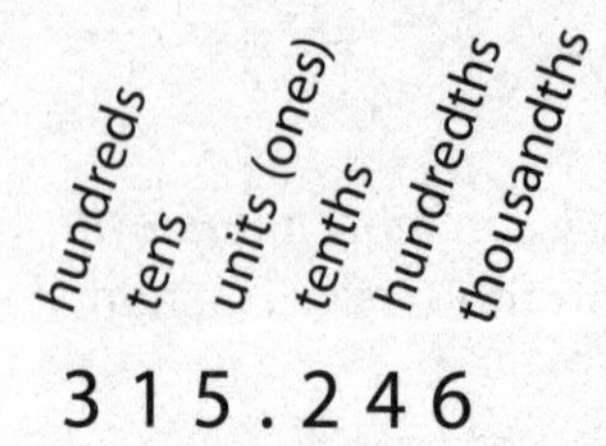

Comparing decimal fractions: To compare decimals, add zeros to the decimals (after the last digit to the right of the decimal point) until all the decimals have the same number of digits. Since the denominators of all the fractions are the same, the numerators determine the order of values.

Example: Arrange in order from smallest to largest: 0.7, 0.77, 0.07, 0.707 and 0.077.

$$0.7 = 0.700 = \frac{700}{1{,}000}$$

$$0.77 = 0.770 = \frac{770}{1{,}000}$$

$$0.07 = 0.070 = \frac{70}{1{,}000}$$

$$0.707 = 0.707 = \frac{707}{1{,}000}$$

$$0.077 = 0.077 = \frac{77}{1{,}000}$$

$70 < 77 < 700 < 707 < 770$; therefore, $0.07 < 0.077 < 0.7 < 0.707 < 0.77$

Addition and subtraction: When adding or subtracting one decimal to or from another, make sure that the decimal points are lined up, one under the other. This will ensure that tenths are added to tenths, hundredths to hundredths, etc.

Example: $0.6 + 0.06 + 0.006 =$

$$\begin{array}{r} 0.6 \\ 0.06 \\ +0.006 \\ \hline 0.666 \end{array}$$

Answer: 0.666

Example: $0.72 - 0.072 =$

$$\begin{array}{r} 0.72 \\ -0.072 \\ \hline \end{array} = \begin{array}{r} 0.720 \\ -0.072 \\ \hline 0.648 \end{array}$$

Answer: 0.648

Multiplication and division: To multiply two decimals, multiply them as you would integers. The number of decimal places in the product will be the total number of decimal places in the factors that are multiplied together.

Example: $0.675 \times 0.42 =$

$$\begin{array}{rl} 0.675 & \text{(3 decimal places)} \\ \times 0.42 & +\text{(2 decimal places)} \\ \hline 1350 & \\ 2700 & \\ \hline 0.28350 & \text{(5 decimal places)} \end{array}$$

Answer: 0.2835

When dividing a decimal by another decimal, multiply each by a power of 10 such that the divisor becomes an integer. (This doesn't change the value of the quotient.) Then carry out the division as you would with integers, placing the decimal point in the quotient directly above the decimal point in the dividend.

Example: $0.675 \div 0.25$

Multiply each decimal by 100 by moving the decimal point two places to the right (since there are two zeros in 100).

$67.5 \div 25$

$$\begin{array}{r} 2.7 \\ 25\overline{)67.5} \\ 50 \\ \hline 175 \\ 175 \\ \hline 0 \end{array}$$

Answer: 2.7

NUMBER OPERATIONS EXERCISES

BASIC

1. Is $6.205 > 6.250$?

2. According to the order of operations, what will be evaluated first?
$3^2 - (6 + 2)$

3. Does $(6 \times 12) \times 8$ equal $6 + (12 \times 8)$?

4. Evaluate and write as a mixed number: $\frac{2}{7} - \frac{3}{21} + 2\frac{4}{14}$.

5. Convert $\frac{17}{4}$ to a mixed number.

6. Reduce $\frac{6}{42}$ to simplest terms.

7. Simplify $\frac{3}{5} \times \frac{3}{8} \times \frac{2}{3}$.

Evaluate Exercises 8-10.

8. $\frac{1}{3} + \frac{1}{5}$

9. $8.84 \div 5.2$

10. 2.67×4.08

INTERMEDIATE

11. Evaluate $\frac{4}{11} + \frac{11}{12}$.

12. Which operation will be evaluated third? $6 \times \left(4.02 - 7^2\right) + 5.15$

13. True or false? $\frac{3}{7 + 10} = \frac{3}{7} + \frac{3}{10}$

14. Write in simplest form: $\frac{3}{8} \times \frac{7}{3} \times 2\frac{1}{4}$.

15. Evaluate $3\frac{2}{7} \div \frac{1}{3}$.

16. Simplify the complex fraction: $\dfrac{\frac{7}{12} - \frac{1}{4}}{\frac{4}{9} + \frac{1}{3}}$.

17. Fill in the box with <, >, or = : $\frac{3}{13} \square \frac{7}{28}$.

Evaluate Exercises 18–20.

18. $4.875 \div 6.5$

19. $\frac{3}{2} - \left(\frac{1}{6}\right)^2$

20. $5\left(\frac{9}{8} - \frac{2}{3}\right)$

ADVANCED

21. Evaluate $7.02^2 + \left(3 - 4.3^2\right) \times \frac{3}{4}$.

22. True or false: $\frac{\frac{1}{5} - \frac{3}{8}}{3} = \frac{3}{5} - \frac{9}{8}$?

23. Evaluate and write in simplest terms: $\frac{2}{5} \times \frac{7}{10} + \left(\frac{11}{5}\right)^2$.

24. Evaluate the expression: $\frac{\frac{3}{8} + \frac{5}{6} - 3}{\frac{9}{2} - \frac{3}{8} + 6}$.

25. Fill in the box with <, >, or =: $\frac{\frac{2}{5}}{\frac{2}{3}} \square \frac{5}{3}$

26. Write 3.067 as a mixed number fraction.

27. Order from least to greatest: 1.085, $1\frac{1}{7}$, 1.850, $\frac{10,805}{10,000}$.

Evaluate Exercises 28–30.

28. $3\frac{2}{7} + 1\frac{1}{5} - \frac{4}{3}$

29. $\frac{(2.2 - 0.3)^2}{0.95} - (1.1 - 7.08)$

30. $(14.08 - 5.78) \times 3.28 \div 1.2$

NUMBER OPERATIONS ANSWER KEY

Basic

1. No

Since both numbers start with 6, you need only compare the decimals. Since 250 is greater than 205, 6.250 is greater than 6.205.

2. (6 + 2)

Actions inside parentheses must be completed first according to the order of operations.

3. No

According to the rules of multiplication, $(a \times b) \times c = a \times (b \times c)$, but plus signs and multiplication signs are not interchangeable. So, $(6 \times 12) \times 8$ is not equal to $6 + (12 \times 8)$.

4. $2\frac{3}{7}$

First, convert to an improper fraction and reduce: $\frac{2}{7} - \frac{1}{7} + \frac{16}{7}$. Then, add and subtract numerators to get $\frac{17}{7}$, and convert to a mixed number, $2\frac{3}{7}$.

5. $4\frac{1}{4}$

Divide 17 by 4 to find the whole number, 4. There is 1 left over. Write the remainder as a fraction, $\frac{1}{4}$.

6. $\frac{1}{7}$

To reduce, find the greatest common factor of 6 and 42, which is 6. Divide both the numerator and the denominator by that common factor. This gives us $\frac{1}{7}$.

7. $\frac{3}{20}$

Multiply across the numerators and the denominators, $\frac{18}{120}$. Then, reduce the result by dividing both by the greatest common factor of 6. However, a much simpler way is to cancel out the common factors on the top and bottom before multiplying: $\frac{3}{5} \times \frac{{}^{1}\not{3}}{{}_{4}\not{8}} \times \frac{\not{2}^{1}}{\not{3}_{1}} = \frac{3}{20}$.

8. $\frac{8}{15}$

Since 3 and 5 have no factors in common, multiply them together to find the Least Common Multiple. Then, change each fraction to have a denominator with this common multiple: $\frac{5}{15} + \frac{3}{15}$. Then, add. Since 8 and 15 have no common factors, the result cannot be reduced.

9. 1.7

When dividing by a number with a decimal place, first move the decimal point over by the same number of places in the dividend and divisor. So you are dividing 88.4 by 52:

$$5.2\overline{)8.84} = 52\overline{)88.4}$$

```
     1.7
52)88.4
   52
   364
   364
     0
```

10. 10.8936

First, multiply the two numbers as though they have no decimals. Then, count the number of decimal places in the two numbers that are multiplied; in this case, there are 4. Count 4 places from the rightmost digit in the product and place the decimal there. In this case, that means putting the decimal between the 0 and the 8, so the answer is 10.8936.

```
    2.67
  × 4.08
   21 36
 10 68
 10.8936
```

Intermediate

11. $1\frac{37}{132}$

First, find a common denominator. Since 11 and 12 have no common factors, multiply them together to get 132, which is the Least Common Multiple: $\frac{48}{132}+\frac{121}{132}$. Then, add the numerators, $\frac{169}{132}$, and convert to a mixed number, $1\frac{37}{132}$.

12. Multiplication

According to the order of operations, the parentheses must be cleared first, and this includes evaluating the exponent and performing subtraction. Next, this result is multiplied by 6.

13. False

Be careful! If you have addition or subtraction in the numerator, you can split the fraction into two fractions with the same denominator. The same does not hold true if you have addition or subtraction in the denominator. In that situation, you first have to complete the operation(s) in the denominator.

14. $\mathbf{1\frac{31}{32}}$

First, convert to an improper fraction: $\frac{3}{8} \times \frac{7}{3} \times \frac{9}{4}$. Next, cancel common factors in the numerator and denominator wherever possible. In this case, the 3s can be cancelled, giving you a product of $\frac{63}{32}$, which can then be converted to a mixed number, $1\frac{31}{32}$.

15. $\mathbf{9\frac{6}{7}}$

To evaluate, first convert the mixed number to an improper fraction, $\frac{23}{7}$. Next, convert the division to multiplication by multiplying the first fraction by the reciprocal of the second: $\frac{23}{7} \times \frac{3}{1} = \frac{69}{7}$. Convert back to a mixed number by dividing by 7: $9\frac{6}{7}$.

16. $\mathbf{\frac{3}{7}}$

First, evaluate the expression in the numerator, $\frac{7}{12} - \frac{1}{4}$ or $\frac{1}{3}$. Then, evaluate the expression in the denominator: $\frac{4}{9} + \frac{1}{3}$, or $\frac{7}{9}$. Finally, perform the division: $\frac{1}{3} \div \frac{7}{9} = \frac{1}{3} \times \frac{9}{7} = \frac{9}{21} = \frac{3}{7}$.

17. $<$

Express both fractions with common denominators to determine which fraction is greater. The least common denominator of $\frac{3}{13}$ and $\frac{7}{28} = \frac{1}{4}$ is 52 (13×4).

Thus $\frac{3}{13} = \frac{3 \times 4}{13 \times 4} = \frac{12}{52}$ and $\frac{1}{4} = \frac{1 \times 13}{4 \times 13} = \frac{13}{52}$. Then $\frac{12}{52} < \frac{13}{52}$, so $\frac{3}{13}$ is less than $\frac{7}{28}$.

18. 0.75

Dividing 4.875 by 6.5 gives:

$$\begin{array}{r} 0.75 \\ 65\overline{)48.75} \\ \underline{455} \\ 325 \\ \underline{325} \\ 0 \end{array}$$

19. $\mathbf{1\frac{17}{36}}$

First, evaluate $\left(\frac{1}{6}\right)^2$, or $\frac{1}{36}$. Then, convert the fractions using a common denominator of 36 and subtract: $\frac{54}{36} - \frac{1}{36} = \frac{53}{36}$. Finally, convert to a mixed number, $1\frac{17}{36}$.

20. $2\frac{7}{24}$

Begin by converting the fractions in parentheses to a common denominator of 24: $5\left(\frac{27}{24} - \frac{16}{24}\right)$. Evaluate the expression inside the parentheses: $5\left(\frac{11}{24}\right)$. Next, multiply: $\frac{5}{1} \times \frac{11}{24} = \frac{55}{24}$. Finally, convert to a mixed number, $2\frac{7}{24}$.

Advanced

21. 37.6629

First, evaluate the expression inside the parentheses: $7.02^2 + (-15.49) \times \frac{3}{4}$. Next, calculate $7.02^2 = 49.2804$. Multiplying the second term by $\frac{3}{4}$ then gives $49.2804 + (-11.6175)$. Performing the addition gives 37.6629.

22. False

Dividing by 3 is the same as multiplying by $\frac{1}{3}$: $\left(\frac{1}{5} - \frac{3}{8}\right) \times \frac{1}{3}$. Distributing $\frac{1}{3}$ across the parentheses gives $\frac{1}{15} - \frac{3}{24}$ or $\frac{1}{15} - \frac{1}{8}$. Also, $\frac{3}{5} - \frac{9}{8}$ can be restated as $\frac{9}{15} - \frac{9}{8}$. Comparing piece by piece, it can be seen without further calculation that these expressions are not equal.

23. $5\frac{3}{25}$

To evaluate, first clear the exponent: $\left(\frac{11}{5}\right)^2 = \frac{121}{25}$. Then, multiply: $\frac{2}{5} \times \frac{7}{10} = \frac{\cancel{2}^{1}}{5} \times \frac{7}{\cancel{10}_{5}} = \frac{7}{25}$. Since the two terms have the same denominator, you can simply add them together: $\frac{7}{25} + \frac{121}{25} = \frac{128}{25} = 5\frac{3}{25}$.

24. $-\frac{43}{243}$

First, evaluate the expression in the numerator:

$$\frac{3}{8} + \frac{5}{6} - 3 = \frac{9}{24} + \frac{20}{24} - \frac{72}{24} = -\frac{43}{24}.$$

Then, evaluate the expression in the denominator:

$\frac{9}{2} - \frac{3}{8} + 6 = \frac{36}{8} - \frac{3}{8} + \frac{48}{8} = \frac{81}{8}$. Finally, perform the division:

$$-\frac{43}{24} \div \frac{81}{8} = -\frac{43}{24} \times \frac{8}{81} = -\frac{43}{\cancel{24}_{3}} \times \frac{\cancel{8}^{1}}{81} = -\frac{43}{243}.$$

25. $<$

Simplify $\frac{\frac{2}{5}}{\frac{2}{3}}$. Dividing by $\frac{2}{3}$ is the same as multiplying by $\frac{3}{2}$: $\frac{\not{2}^{1}}{5} \times \frac{3}{\not{2}_{1}} = \frac{3}{5}$.

Finally, compare: $\frac{3}{5} < \frac{5}{3}$.

26. $\mathbf{3\frac{67}{1,000}}$

The decimal part is 67 thousandths. Write as a mixed number. $3 + \frac{67}{1,000} = 3\frac{67}{1,000}$.

27. $\mathbf{\frac{10,805}{10,000}}$**, 1.085,** $\mathbf{1\frac{1}{7}}$**, 1.850**

Convert the fractions to decimals to compare: $\frac{10,805}{10,000} = 1.0805$ and $1\frac{1}{7} = 1.14286$.

Compare the ones place, then the tenths, then the hundredths, and then the thousandths.

28. $\mathbf{3\frac{16}{105}}$

First, convert mixed numbers to improper fractions and find a common denominator: $\frac{345}{105} + \frac{126}{105} - \frac{140}{105}$. Then, add and subtract numerators, $\frac{331}{105}$, and convert to a mixed number, $3\frac{16}{105}$.

29. 9.78

Evaluate the expressions inside the parentheses: $\frac{(1.9)^2}{0.95} - (-5.98)$. Calculate the exponent: $\frac{3.61}{0.95} - (-5.98)$. Divide: 3.8 – (–5.98). Finally, subtract: 9.78.

30. 22.6867

Evaluate the expression inside the parentheses: $(8.3) \times 3.28 \div 1.2$. Then, multiply and divide from left to right: $(8.3) \times 3.28 \div 1.2 = 27.224 \div 1.2 = 22.6867$.

NUMBER PROPERTIES

NUMBER LINE AND ABSOLUTE VALUE

A **number line** is a straight line that extends infinitely in either direction, on which real numbers are represented as points.

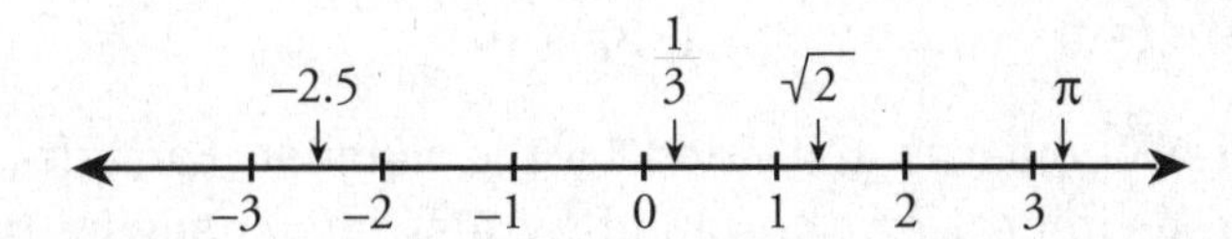

As you move to the right on a number line, the values increase.

Conversely, as you move to the left, the values decrease.

Zero separates the positive numbers (to the right of zero) and the negative numbers (to the left of zero) along the number line. Zero is neither positive nor negative.

The **absolute value** of a number is just the number without its sign. It is written as two vertical lines.

Example: $|-3| = |+3| = 3$

The absolute value can be thought of as the number's distance from zero on the number line; for instance, both +3 and –3 are 3 units from zero, so their absolute values are both 3.

PROPERTIES OF –1, 0, 1, AND NUMBERS IN BETWEEN

Properties of zero: Adding or subtracting zero from a number does not change the number.

Example: $0 + x = x$; $2 + 0 = 2$; $4 - 0 = 4$

Any number multiplied by zero equals zero.

Example: $z \times 0 = 0$; $12 \times 0 = 0$

Division by zero is **undefined**. When given an algebraic expression, be sure that the denominator is not zero. $\frac{0}{0}$ is also undefined.

Properties of 1 and –1: Multiplying or dividing a number by 1 does not change the number.

Example: $x \div 1 = x$; $4 \times 1 = 4$; $-3 \times 1 = -3$

Multiplying or dividing a number by –1 changes the sign.

Example: $y \times (-1) = -y$; $6 \times (-1) = -6$; $-2 \div (-1) = -(-2) = 2$; $(x - y) \times (-1) = -x + y$

Note: The sum of a number and –1 times that number is equal to zero.

Example: $a + (-a) = 0$; $8 + (-8) = 0$;

The **reciprocal** of a number is 1 divided by the number. For a fraction, as we've already seen, the reciprocal can be found by just interchanging the denominator and the numerator. The product of a number and its reciprocal is 1. Zero has no reciprocal, since $\frac{1}{0}$ is undefined.

Properties of numbers between –1 and 1: The reciprocal of a number between 0 and 1 is greater than the number.

Example: The reciprocal of $\frac{2}{3} = \frac{1}{\frac{2}{3}} = \frac{3}{2} = 1\frac{1}{2}$, which is greater than $\frac{2}{3}$.

The reciprocal of a number between –1 and 0 is less than the number.

Example: The reciprocal of $-\frac{2}{3} = \frac{1}{\left(-\frac{2}{3}\right)} = -\frac{3}{2} = -1\frac{1}{2}$, which is less than $-\frac{2}{3}$.

The square of a number between 0 and 1 is less than the number.

Example: $\left(\frac{1}{2}\right)^2 = \frac{1}{2} \times \frac{1}{2} = \frac{1}{4}$, which is less than $\frac{1}{2}$.

Multiplying any positive number by a fraction between 0 and 1 gives a product smaller than the original number.

Example: $6 \times \frac{1}{4} = 1\frac{1}{2}$, which is less than 6.

Multiplying any negative number by a fraction between 0 and 1 gives a product greater than the original number.

Example: $-3 \times \frac{1}{6} = -\frac{1}{2}$, which is greater than –3.

All these properties can best be seen by observation rather than by memorization.

Operations with Signed Numbers

The ability to add and subtract signed numbers is best learned by practice and common sense.

Addition: Like signs: Add the absolute values and keep the same sign.

Example: $(-6) + (-3) = -9$

Unlike signs: Take the difference of the absolute values and keep the sign of the number with the larger absolute value.

Example: $(-7) + (+3) = -4$

Subtraction: Subtraction is the inverse operation of addition; subtracting a number is the same as adding its inverse. Subtraction is often easier if you change to addition by changing the sign of the number being subtracted. Then use the rules for addition of signed numbers.

Example: $(-5) - (-10) = (-5) + (+10) = +5$

Multiplication and division: The product or the quotient of two numbers with the same sign is positive.

Example: $(-2) \times (-5) = +10; \frac{-50}{-5} = +10$

The product or the quotient of two numbers with opposite signs is negative.

Example: $(-2)(+3) = -6; \frac{-6}{2} = -3$

Odd and Even

Odd and even apply only to integers. There are no odd or even noninteger numbers. Put simply, even numbers are integers that are divisible by 2, and odd numbers are integers that are not divisible by 2. Another easy way to classify an integer as even or odd is to check its last digit. If an integer's last digit is either 0, 2, 4, 6, or 8, the integer is even; if its last digit is 1, 3, 5, 7, or 9, it is odd. Odd and even numbers may be negative.

A number needs just a single factor of 2 to be even, so the product of an even number and **any** integer will always be even.

Rules for Odds and Evens:

Odd $\pm$ Odd $=$ Even	Odd $\times$ Odd $=$ Odd
Even $\pm$ Even $=$ Even	Even $\times$ Even $=$ Even
Odd $\pm$ Even $=$ Odd	Odd $\times$ Even $=$ Even

You can easily establish these rules when you need them by picking sample numbers.

Example: $3 + 5 = 8$, so the sum of any two odd numbers is even.

Example: $\frac{4}{2} = 2$, but $\frac{6}{2} = 3$, so the quotient of two even numbers could be odd or even (or a fraction!)

Factors, Primes, and Divisibility

Multiples: An integer that is divisible by another integer is a **multiple** of that integer.

Example: 12 is a multiple of 3, since 12 is divisible by 3; $3 \times 4 = 12$.

Remainders: The remainder is what is left over in a division problem. A remainder is always smaller than the number we are dividing by.

Example: 17 divided by 3 is 5, with a remainder of 2.

Factors: The **factors**, or **divisors**, of a number are the positive integers that evenly divide into that number.

Example: 36 has nine factors: 1, 2, 3, 4, 6, 9, 12, 18, and 36.

We can group these factors in pairs:

$$1 \times 36 = 2 \times 18 = 3 \times 12 = 4 \times 9 = 6 \times 6$$

The **greatest common factor**, or **greatest common divisor**, of a pair of numbers is the largest factor shared by the two numbers.

Divisibility tests: There are several tests to determine whether a number is divisible by 2, 3, 4, 5, 6, and 9.

A number is divisible by 2 if its last digit is divisible by 2.

Example: 138 is divisible by 2 because 8 is divisible by 2.

A number is divisible by 3 if the **sum** of its digits is divisible by 3.

Example: 4,317 is divisible by 3 because $4 + 3 + 1 + 7 = 15$, and 15 is divisible by 3.

239 is **not** divisible by 3 because $2 + 3 + 9 = 14$, and 14 is not divisible by 3.

A number is divisible by 4 if its last two digits are divisible by 4.

Example: 1,748 is divisible by 4 because 48 is divisible by 4.

A number is divisible by 5 if its last digit is 0 or 5.

Example: 2,635 is divisible by 5. 5,052 is **not** divisible by 5.

A number is divisible by 6 if it is divisible by both 2 and 3.

Example: 4,326 is divisible by 6 because it is divisible by 2 (last digit is 6) and by 3 ($4 + 3 + 2 + 6 = 15$).

A number is divisible by 9 if the sum of its digits is divisible by 9.

Example: 22,428 is divisible by 9 because $2 + 2 + 4 + 2 + 8 = 18$, and 18 is divisible by 9.

A number is divisible by 10 if its last digit is 0.

Example: 790 is divisible by 10. 8,431 is not.

A number is divisible by 12 if it is divisible by both 3 and 4.

Example: 21,528 is divisible by 12 because it is divisible by 3 ($2 + 1 + 5 + 2 + 8 = 18$) and by 4 (last two digits are 28).

Prime number: A **prime** number is an integer greater than 1 that has no factors other than 1 and itself. The number 1 is not considered prime. The number 2 is the first prime number and the only even prime. (Do you see why? Any other even number has 2 as a factor, and therefore is not prime.) The first ten prime numbers are 2, 3, 5, 7, 11, 13, 17, 19, 23, 29.

Prime factorization: The **prime factorization** of a number is the expression of the number as the product of its prime factors. No matter how you factor a number, its prime factors will always be the same.

Example: $36 = 6 \times 6 = 2 \times 3 \times 2 \times 3$ or $2 \times 2 \times 3 \times 3$ or $2^2 \times 3^2$

Example:
$$\begin{aligned} 480 &= 48 \times 10 = 8 \times 6 \times 2 \times 5 \\ &= 2 \times 4 \times 2 \times 3 \times 2 \times 5 \\ &= 2 \times 2 \times 2 \times 2 \times 3 \times 2 \times 5 \\ &= 2^5 \times 3 \times 5 \end{aligned}$$

The easiest way to determine a number's prime factorization is to figure out a pair of factors of the number, and then determine their factors, continuing the process until you're left with only prime numbers. Those primes will be the prime factorization.

Example: Find the prime factorization of 1,050.

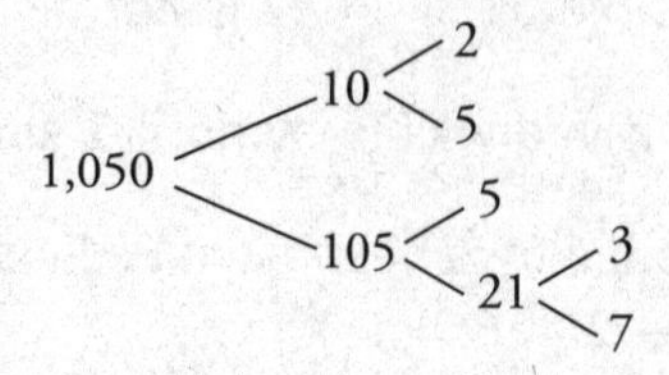

So the prime factorization of 1,050 is $2 \times 3 \times 5^2 \times 7$.

CONSECUTIVE NUMBERS

A list of numbers is **consecutive** if the numbers either occur at a fixed interval, or exhibit a fixed pattern. All the consecutive numbers you will encounter on the GRE are integers. Consecutive numbers could be in ascending **or** descending order.

Example: 1, 2, 3, 4, 5, 6 . . . is a series of consecutive positive integers.

Example: –6, –4, –2, 0, 2, 4 . . . is a series of consecutive even numbers.

Example: 5, 7, 11, 13, 17, 19 . . . is a series of consecutive prime numbers.

NUMBER PROPERTIES EXERCISES

Basic

1. Evaluate (–7) + (–4 + 3).

2. Which of the following numbers is divisible by 7: 43, 51, 56, 58?

3. Which is greater: |–6| or –6?

For Exercises 4 and 5, use your knowledge of number properties to determine if the values will be odd or even.

4. The product of an even number and an odd number.

5. The sum of two odd numbers.

6. What is the greatest integer that will divide evenly into both 48 and 60?

7. The following list contains all the factors of what number: 1, 2, 3, 5, 6, 10, 15, 30?

8. The prime factorization of a number contains exactly five 2s and one 3 and no other numbers. What is the number?

9. What is the smallest positive integer divisible by both 18 and 7?

10. What are the next 3 consecutive integers in this arithmetic sequence: –5, 5, 15, 25?

Intermediate

11. Which is greater: |6| or |–7|?

12. Find all numbers between 70 and 90 that are evenly divisible by 8.

For Exercises 13 and 14, use your knowledge of number properties to determine if the values will be odd or even.

13. The sum of four odd numbers.

14. The product of two even numbers multiplied by one odd number.

15. What is the greatest number that is less than 153 and is also divisible by 3 and 7?

16. The sum of a set of 3 consecutive even integers is 102. What is the sum of the 3 consecutive odd integers that precede the lowest number in that set?

17. True or false? If n is an integer, $2n + 2$ will always be even.

18. Which of the following numbers are divisible by 6: 168, 273, 348, 434?

For Exercises 19 and 20, find the number's prime factorization.

19. 208

20. 52^2

ADVANCED

21. Which expression is greater: $|(-6) + (-3)|$ or $(-3)(-3)(-1)$?

22. $200 \leq x \leq 300$. How many values of x are divisible by both 5 and 8?

For Exercises 23 and 24, use your knowledge of number properties to determine if the values will be odd or even.

23. The difference between an even number and an odd number, squared.

24. The sum of two odd numbers plus the product of an even number times an odd number.

25. What number between 70 and 75, inclusive, has the greatest number of factors?

26. What are the smallest three prime numbers greater than 65?

27. What is the smallest number that can be written as the product of three distinct prime factors?

28. True or false? The product of three consecutive integers is always even.

29. How many positive integers between 21 and 59 are equal to the product of a multiple of 4 and an odd number?

30. What is the prime factorization of $18^4 \times 6$?

NUMBER PROPERTIES EXERCISES ANSWER KEY

BASIC

1. –8

First, evaluate the expressions inside parentheses from left to right: $(-7) + (-1)$. Then, add.

2. 56

The product of 7 and 8 is 56, so 56 is divisible by 7.

3. |–6|

The absolute value of -6 is 6. Since 6 is greater than -6, $|-6| > -6$.

4. Even

The product of an even number and an odd number is an even number.

5. Even

The sum of two odd numbers is an even number.

6. 12

Use prime factorization to determine the greatest common factor of two numbers: $48 = 2 \times 2 \times 2 \times 2 \times 3$ and $60 = 2 \times 2 \times 3 \times 5$. Their common factors are 2, 2, and 3, so the greatest common factor is $2 \times 2 \times 3 = 12$.

7. 30

Every number is a factor of itself, and by definition a number itself is its greatest factor. Since this list contains every factor of the number, the largest number in the list must be the right answer to the question. The largest number in this list is 30.

8. 96

The prime factorization is the expression of a number as the product of its prime factors. A prime factorization that contains five 2s and one 3 is the expression $2 \times 2 \times 2 \times 2 \times 2 \times 3$, which equals 96.

9. 126

Find the least common multiple, or LCM, of the numbers. In this case the LCM is the product of the two factors, because 7 is a prime number and it isn't a factor of 18.

10. 35, 45, 55

This sequence is created by adding 10 to the previous number. Following this pattern, the next 3 integers will be 35, 45, and 55.

INTERMEDIATE

11. |−7|

The absolute value of 6 is 6, and the absolute value of −7 is 7. Since 7 is greater than 6, $|6| < |-7|$.

12. 72, 80, 88

To be divisible by 8, the quotient must divide evenly. Start at 70 and increase by 2 (as any number evenly divisible by an even number must be even), and determine if the number is divisible by 8. Since 72 is divisible by 8, start here. Then, the next numbers will each be 8 greater than the last: $72 + 8 = 80$, and $80 + 8 = 88$.

13. Even

Two odd numbers added together produce an even number. Twice that amount of odd numbers (or four odd numbers) will also produce an even number. In fact, any even number of odd numbers, added together, will yield an even number.

14. Even

The product of two even numbers is an even number. Then the product of an odd number and an even number will be an even number.

15. 147

A number that is divisible by both 3 and 7 is divisible by $3 \times 7 = 21$. Find the largest number less than 153 that is divisible by 21. It is 147.

16. 87

If numbers are evenly spaced and added together, the average is the median of those numbers. 102 divided by 3 is 34, so the three numbers in the original list are $32 + 34 + 36 = 102$. The three consecutive odd integers that come before 32 are 27, 29 and 31, which add up to 87.

Alternatively, use algebra to find the least of the even integers. If n is the least of the even integers, then the next two even integers are $n + 2$ and $n + 4$. $n + n + 2 + n + 4 = 102$. Combine like terms and solve for n: $3n + 6 = 102$, and $n = 32$.

17. True

The product of an even number and any other number is also even. Therefore, the product of $2n$ will always be even. Then, the sum of two even numbers is even, so adding 2 to that even number will result in an even number.

18. 168, 348

Use the divisibility rules to find the numbers that are divisible by the prime factors of 6: 2 and 3. Of the numbers in the list, 168 and 348 are even numbers, so they are divisible by 2, and the sum of their digits produces a number that is divisible by 3.

19. $2 \times 2 \times 2 \times 2 \times 13$

The prime factorization is the expression of a number as the product of its prime factors.

20. $2 \times 2 \times 13 \times 2 \times 2 \times 13$

First, find the prime factorization of the base number, 52: $52 = 2 \times 2 \times 13$. The question asks for the prime factorization of 52^2, so square the prime factorization (repeat each number in it) to get the final answer.

Advanced

21. $|(-6) + (-3)|$

Evaluate each expression according to the rules of operations with signed numbers. The expression $|(-6) + (-3)| = |-9| = 9$, and the expression $(-3)(-3)(-1) = (9)(-1) = -9$. And 9 is greater than −9, so $|(-6) + (-3)| > (-3)(-3)(-1)$.

22. 3

Any number divisible by 8 and 5 is divisible by 8×5, since 5 is prime and not a factor of 8. Also, $8 \times 5 = 40$ is a factor of 200, the smallest number in *x*'s range. Add 40 to 200 to find the next number in the series, 240, then 40 again for 280. The next number that is a multiple of 40 after 280 is larger than 300, the top of the range. This means 200, 240, and 280 are divisible by 8 and 5.

23. Odd

The difference is an odd number because an even minus an odd is odd, and then an odd number raised to any power remains odd, so the square of an odd number is odd.

24. Even

The sum of two odd numbers is an even number, and the product of an even number and an odd number is an even number. Finally, the sum of two even numbers is an even number.

25. 72

There are 12 factors of 72: 1, 2, 3, 4, 6, 8, 9, 12, 18, 24, 36, and 72. Each of the other numbers in the list has fewer factors than does 72.

70: 1, 2, 5, 7, 10, 14, 35, 70

71: 1, 71

73: 1, 73

74: 1, 2, 37, 74

75: 1, 3, 5, 15, 25, 75

26. 67, 71, 73

A prime number is one that is divisible only by 1 and itself. The smallest three numbers greater than 65 that are prime are 67, 71, and 73.

27. 30

Consider the smallest prime numbers. The number 2 is the smallest (and the only even) prime number. So the smallest prime factors are 2, 3, and 5: $2 \times 3 \times 5 = 30$.

28. True

Three consecutive numbers must include at least one even number. If an even number is multiplied by any other integer, the product must be even. You can also Pick Numbers to see that this will always work.

29. 9

The first few multiples of 4 are 4, 8, 12, 16, and 20. The multiples of 4 that are between 21 and 59 are 24, 28, 32, 36, 40, 44, 48, 52, and 56. These can each be written as the product of a multiple of 4 and an odd number as follows:

$24 = 8 \times 3$

$28 = 4 \times 7$

$32 = 32 \times 1$

$36 = 12 \times 3 = 4 \times 9$

$40 = 8 \times 5$

$44 = 4 \times 11$

$48 = 16 \times 3$

$52 = 4 \times 13$

$56 = 8 \times 7$

30. $2 \times 2 \times 2 \times 2 \times 2 \times 3 \times 3 \times 3 \times 3 \times 3 \times 3 \times 3 \times 3 \times 3$, $2^5 \times 3^9$

The prime factorization of 18 is $2 \times 3 \times 3$. This shows up 4 times in the prime factorization of $18^4 \times 6$ because 18 is raised to the 4th power. Then, multiply this by the prime factorization of 6, 3×2, resulting in $(2 \times 3 \times 3) \times (2 \times 3 \times 3) \times (2 \times 3 \times 3) \times (2 \times 3 \times 3) \times (2 \times 3)$.

AVERAGES (ARITHMETIC MEANS)

The average (arithmetic mean) of a group of numbers is defined as the sum of the values divided by the number of values.

$$\text{Average value} = \frac{\text{Sum of values}}{\text{Number of values}}$$

Example: Henry buys three items costing \$2.00, \$0.75, and \$0.25. What is the average price?

$$\text{Average price} = \frac{\text{Sum of prices}}{\text{Number of prices}} = \frac{\text{Total price}}{\text{Total items}} = \frac{\$2.00 + \$0.75 + \$0.25}{3}$$

$$= \frac{\$3.00}{3} = \$1.00$$

On the GRE you might see a reference to the median. If a group of numbers is arranged in numerical order, the median is the middle value. For instance, the median of the numbers 4, 5, 100, 1, and 6 is 5. The median can be quite different from the average. For instance, in the example shown, the average was \$1.00, while the median is simply the middle of the three prices given, or \$0.75.

If we know the average of a group of numbers, and the number of numbers in the group, we can find the **sum** of the numbers. It's as if all the numbers in the group have the average value.

$$\text{Sum of values} = \text{Average value} \times \text{Number of values}$$

Example: The average daily temperature for the first week in January was 31 degrees. If the average temperature for the first six days was 30 degrees, what was the temperature on the seventh day?

The sum for all 7 days $= 31 \times 7 = 217$ degrees.

The sum of the first six days $= 30 \times 6 = 180$ degrees.

The temperature on the seventh day $= 217 - 180 = 37$ degrees.

For evenly spaced numbers, the average is the middle value. The average of consecutive integers 6, 7, and 8 is 7. The average of 5, 10, 15, and 20 is $12\frac{1}{2}$ (midway between the middle values 10 and 15).

It might be useful to try and think of the average as the "balanced" value. That is, all the numbers below the average are less than the average by an amount that will "balance out" the amount that the numbers above the average are greater than the average. For example, the average of 3, 5, and 10 is 6. 3 is 3 less than 6 and 5 is 1 less than 6. This, in total, is 4, which is the same as the amount that 10 is greater than 6.

Example: The average of 3, 4, 5, and x is 5. What is the value of x?

Think of each value in terms of its position relative to the average, 5.

3 is 2 less than the average.

4 is 1 less than the average.

5 is at the average.

So these 3 terms together are $1 + 2 + 0$, or 3, less than the average. Therefore, x must be 3 **more** than the average, to restore the balance at 5. So x is $3 + 5$ or 8.

AVERAGE RATE (AVERAGE *A* PER *B*)

$$\textbf{Average } A \textbf{ per } B = \frac{\textbf{Total } A}{\textbf{Total } B}$$

Example: John travels 30 miles in 2 hours and then 60 miles in 3 hours. What is his average speed in miles per hour?

$$\begin{aligned}\text{Average miles per hour} &= \frac{\text{Total miles}}{\text{Total hours}} \\ &= \frac{(30 + 60)\text{ miles}}{(2 + 3)\text{ hours}} \\ &= \frac{90\text{ miles}}{5\text{ hours}} \\ &= 18\text{ miles/hour}\end{aligned}$$

AVERAGES (ARITHMETIC MEANS) EXERCISES

BASIC

1. What is the average of $\frac{1}{3}$ and $\frac{1}{2}$?

2. What is the average of 5, 10, 50, and 100?

3. What is the value of x if the average of 3, 7, 8, and x is 6?

4. What is the value of x if the average of –6, 4, 10, and x is 2?

5. If the average of 7, 5, 14, 9, and x is 9, what is the value of x?

6. What is the average of $\frac{1}{4}, \frac{1}{5}, \frac{1}{20}$, and $\frac{1}{2}$?

7. The average of six numbers is 8. If 2 is subtracted from three of the numbers, what is the new average?

8. Deonte travels 72 miles in 3 hours and then 84 miles in 3 hours. What is his average speed for the whole trip in miles per hour?

9. What is the average of $\frac{4}{3}, \frac{2}{3}$, and 3?

10. What is the average of –12, 4, 15, and –25?

INTERMEDIATE

11. What is the sum of the five consecutive even numbers whose average (arithmetic mean) is 20?

12. What is the average of y, $y + 2$, $y + 4$, and $y + 6$ in terms of y?

13. If the average of 32, 5, 11, and n is 21, then what is the sum of $32 + 5 + 11 + n$?

14. What is the sum of the five consecutive even numbers whose average is 12?

15. Five bakeries sell an average of 300 muffins per bakery per day. If two of the bakeries stop making muffins, but the total number of muffins sold stays the same, what is the average number of muffins sold per bakery among the remaining bakeries?

16. What is the sum of the five consecutive odd numbers whose average is 75?

17. The average of r and s is 40. If $t = 10$, what is the average of r, s, and t?

18. If the average of 5 consecutive integers is 14, what is the sum of the least and greatest of the 5 integers?

19. If the average of 7 consecutive even integers is 44, what is the sum of the integers?

20. On three consecutive passes, a football team gains 5 yards, loses 33 yards, and gains 25 yards. How many yards must the team gain on the next pass to have an average gain of 3 yards?

ADVANCED

21. If 20 students in one class had an average grade of 94% and 18 students from another class had an average grade of 92%, what is the average grade for all 38 students across both classes? Round to the nearest hundredth.

22. Ashley's scores on five math tests are 84, 100, 88, 95, and 92. What does she need to score on the next test for her final average on all the tests to be at least 92?

23. Rob has received scores of 96, 89, and 85 on 3 quizzes. If the exam is weighted twice as heavily as each of the three quizzes, what is the lowest score Rob can get on the exam to have a final average of at least 90?

24. If the average of 5 consecutive integers is 15, what is the sum of the least and greatest of the 5 integers?

25. The Ryder family traveled at an average rate of 60 miles per hour (mph), and it took 10 hours to complete their trip. If they traveled for 4 of their 10 hours at a constant speed of 70 miles per hour, at what average speed did they travel for the remaining 6 hours to obtain the 60 miles per hour average for the entire trip?

26. If the average of 10 consecutive odd integers is 224, what is the least of these integers?

27. The average of five numbers is 30. After one of the numbers is removed, the average arithmetic mean of the remaining numbers is 32. What number was removed?

28. The table shows the closing price of a stock during one week. If the average (arithmetic mean) closing price for the five days was $32.88, what was the closing price on Friday?

Day	Price
Monday	$32.59
Tuesday	$33.62
Wednesday	$30.78
Thursday	$35.23
Friday	?

29. Two airplanes leave the same airport at the same time, one traveling west and the other east. Their average speeds differ by 10 miles per hour. After 1.5 hours, they are 520 miles apart. What is the approximate average speed of each plane over the 1.5 hours? Round to the nearest tenth.

30. If the average (arithmetic mean) of $a + b$ is n, what is the average of a, b, and c in terms of n and c?

AVERAGES (ARITHMETIC MEANS) EXERCISES ANSWER KEY

BASIC

1. $\frac{5}{12}$

Find the sum of the values: $\frac{1}{2} + \frac{1}{3} = \frac{3}{6} + \frac{2}{6} = \frac{5}{6}$. Then divide by the number of values: $\frac{5}{6} \div 2 = \frac{5}{6} \times \frac{1}{2} = \frac{5}{12}$.

2. 41.25 or $41\frac{1}{4}$

Find the sum of the values: $5 + 10 + 50 + 100 = 165$. Then divide by the number of values: $165 \div 4 = 41.25$ or $41\frac{1}{4}$.

3. 6

Think of each value in terms of its position relative to the average, 6:

3 is 3 less than the average.
7 is 1 more than the average.
8 is 2 more than the average.

So these 3 terms together are $-3 + 1 + 2$, or 0 more than the average. Therefore x must equal the average. So x is 6.

Alternatively, if the average of the four numbers is 6, then their sum must be $4 \times 6 = 24$. Then $24 - (3 + 7 + 8) = 6$. The unknown value, x, equals 6.

4. 0

Think of each value in terms of its position relative to the average, 2:

-6 is 8 less than the average.
4 is 2 more than the average.
10 is 8 more than the average.

So, these 3 terms together are $-8 + 2 + 8$, or 2 more than the average. Therefore, x must be 2 less than the average to restore the balance at 2. So x is $2 - 2$, or 0.

Alternatively, if the average of the four numbers is 2, then their sum must be $4 \times 2 = 8$. Then $8 - (-6 + 4 + 10) = 0$. The unknown value, x, equals 0.

5. 10

Think of each value in terms of its position relative to the average, 9:

7 is 2 less than the average.
5 is 4 less than the average.

14 is 5 more than the average.

9 is 0 more than the average.

So, these 4 terms together are $-2 + -4 + 5 + 0$, or 1 less than the average. Therefore, x must be 1 more than the average to restore the balance at 9. So x is $1 + 9$ or 10.

Alternatively, if the average of the five numbers is 9, then their sum must be $5 \times 9 = 45$. Then $45 - (7 + 5 + 14 + 9) = 10$. The unknown value, x, equals 10.

6. $\frac{1}{4}$

Find the sum of the values: $\frac{1}{4} + \frac{1}{5} + \frac{1}{20} + \frac{1}{2} = \frac{5}{20} + \frac{4}{20} + \frac{1}{20} + \frac{10}{20} = \frac{20}{20} = 1$.

Then divide by the number of values: $1 \div 4 = \frac{1}{4}$.

7. 7

The average of six numbers is 8. While we don't know the individual values, we can multiply $6 \times 8 = 48$ to get the original total of the numbers. If 2 is subtracted from each of three numbers, $2 \times 3 = 6$ was subtracted from the original total, leaving $48 - 6 = 42$. The new average is $42 \div 6 = 7$.

8. 26

Use the equation total miles $\div$ total hours $=$ average miles per hour. Total miles $= 72 + 84 = 156$ miles and total hours $= 3 + 3 = 6$ hours. Divide: 156 miles $\div$ 6 hours $=$ 26 miles per hour. Deonte's average speed is 26 miles per hour.

9. $1\frac{2}{3}$

Find the sum of the values: $\frac{4}{3} + \frac{2}{3} + 3 = \frac{4}{3} + \frac{2}{3} + \frac{9}{3} = \frac{15}{3} = 5$. Then divide 5 by the number of values: $5 \div 3 = \frac{5}{3} = 1\frac{2}{3}$.

10. -4.5 **or** $-4\frac{1}{2}$

Find the sum of the values: $-12 + 4 + 15 + -25 = -18$. Then divide by the number of values: $-18 \div 4 = -4.5$.

Intermediate

11. 100

If the average is 20 and there are 5 terms, the sum of the terms is $20 \times 5 = 100$. There is no need to determine all the individual values in the set.

12. $y + 3$

Find the sum of the values: $y + y + 2 + y + 4 + y + 6 = 4y + 12$. Then divide by the number of values: $\frac{4y + 12}{4} = y + 3$.

13. 84

If you know the average of a set of terms and you know how many terms there are, you can multiply the two together to get the total of all the terms: $21 \times 4 = 84$. In this case, the sum of the values is 84.

14. 60

If the average is 12 and there are 5 terms, the sum of the terms is $12 \times 5 = 60$. There is no need to determine all the individual values in the set.

15. 500

Originally, five bakeries sold an average of 300 muffins each day, which equals $5 \times 300 = 1{,}500$ total muffins per day.

If two bakeries drop out but the total number of muffins stays the same, the 1,500 muffins will be made by only $5 - 2 = 3$ bakeries. Divide the 1,500 by 3 to find the average for the 3 bakeries: $1{,}500 \div 3 = 500$.

So the average number of muffins sold per bakery among the remaining bakeries is 500.

16. 375

If the average is 75 and there are 5 terms, the sum of the terms is $75 \times 5 = 375$. There is no need to determine all the individual values in the set.

17. 30

If you know the average of a set of terms and you know how many terms there are, you can multiply the average and the number of terms together to get the total. In this case, the average is 40 and the number of terms is 2, so the total for $r + s$ is $40 \times 2 = 80$. If $t = 10$, then $r + s + t = 80 + 10 = 90$. Therefore, the average of r, s, and t is $\frac{r + s + t}{3} = \frac{90}{3} = 30$.

18. 28

The arithmetic mean of 5 evenly spaced integers is the middle number in the set. The sum of the least and the greatest integers is twice the average, or in this case, $2 \times 14 = 28$. There is no need to know the specific numbers in the set.

19. 308

If the average is 44 and there are 7 terms, the sum of the terms is $44 \times 7 = 308$. There is no need to determine all the individual values in the set.

20. 15

To have an average of 3 yards over 4 plays, the total yards gained must equal 12 yards ($3 \times 4 = 12$). If $y =$ the number of yards gained on the last play, $5 - 33 + 25 + y = 12$. Solving for y, $-3 + y = 12$, and $y = 15$. So the team must gain 15 yards on the next pass to have an average gain of 3 yards.

Alternatively, you can use the balancing method:

To get to the average:

5 yards must lose 2

–33 must gain 36

25 must lose 22

Add $(-2 + 36 - 22)$ to the average of 3 to find that the next pass must gain 15 yards to get an average of 3.

Advanced

21. 93.05%

If 20 students averaged 94, then the total number of points they scored is $20 \times 94 = 1{,}880$. If 18 students averaged 92, then the total number of points they scored is $18 \times 92 = 1{,}656$. So, the total number of points scored for the group is $1{,}880 + 1{,}656 = 3{,}536$. Divide the total number of points by the total number of students ($20 + 18 = 38$): $\frac{3{,}536}{38} \approx 93.05\%$. The average, rounded to the nearest hundredth, is 93.05%.

22. 93

The easiest way to solve a problem like this one is to use the balancing method. Determine how many points each of the actual scores must gain or lose to get to the average, then add that to the average to get your solution. In this case:

84 must gain 8

100 must lose 8

88 must gain 4

95 must lose 3

92 doesn't have to gain or lose

Add $(8 - 8 + 4 - 3)$ to 92, for a required score of 93 on the next test to have at least a 92 average.

Alternatively, call the missing score x and solve algebraically using the fact that the sum of the tests divided by the number of tests is the average.

$$\frac{84 + 100 + 88 + 95 + 92 + x}{6} \geq 92$$
$$\frac{459 + x}{6} \geq 92$$
$$(6)\left(\frac{459 + x}{6}\right) \geq 6(92)$$
$$459 + x \geq 552$$
$$x \geq 552 - 459$$
$$x \geq 93$$

Ashley needs to score at least a 93 on the next test to have at least a 92 average overall.

23. 90

The sum of the tests divided by the number of tests is the average. Because the exam counts twice as much as the quizzes, it is included two times.

The easiest way to solve a problem like this one is to use the balancing method. Determine how many points each of the actual scores must gain or lose to get to the average, then add that to the average to get your solution. In this case:

96 must lose 6

89 must gain 1

85 must gain 5

Since $-6 + 1 + 5 = 0$, Rob already has an average of 90, so he can get a score as low as 90 on his final exam to maintain at least that average.

If you prefer to "do the math":

$$\frac{96 + 89 + 85 + x + x}{5} = 90$$
$$\frac{270 + 2x}{5} = 90$$
$$(5)\left(\frac{270 + 2x}{5}\right) = 5(90)$$
$$270 + 2x = 450$$
$$2x = 450 - 270$$
$$2x = 180$$
$$x = 90$$

Rob needs to score at least 90 on the exam.

24. 30

The sum of two numbers at an equal distance from the middle number will be twice their average. In this case, the average is 15, so the sum of the smallest and largest numbers is $2 \times 15 = 30$. There is no need to know the specific numbers in the set.

25. $53\frac{1}{3}$ mph

Use the distance formula: distance = rate × time. Find how many miles the Ryder family traveled: total distance = 60 mph × 10 hours = 600 miles. Next find how many of the 600 miles were traveled at 70 mph: d = 70 mph × 4 hours = 280 miles. Subtract to find the remaining number of miles traveled at an unknown speed: 600 – 280 = 320. Now, use the rate formula to determine the miles per hour, $\text{rate} = \frac{\text{distance}}{\text{time}}$. In this case, $\text{mph} = \frac{320 \text{ miles}}{6 \text{ hours}} = 53\frac{1}{3}$ mph.

The Ryder family traveled at an average of $53\frac{1}{3}$ miles per hour for the remaining 6 hours of the trip.

26. 215

For a set of numbers that are evenly spaced, the average is the same as the middle (median) of the numbers. If 224 is the average of an even number of evenly spaced odd numbers, half the terms are less than the average. To find the smallest number in the 10-number set, count down 5 odd numbers below 224: 223, 221, 219, 217, 215. The least of these integers is 215.

27. 22

While you don't know the individual values, you can multiply the number of values times the average, or 5 × 30 = 150, to get the original total of the numbers. If one of the numbers is removed and the new average is 32, then the new total for the four remaining values is 4 × 32 = 128. Subtract the sum of four numbers from the sum of five numbers to find the value of the number that was removed: 150 – 128 = 22.

28. $32.18

Even with daunting numbers like these, the easiest way to solve a problem like this one is to use the balancing method. Determine how much money each of the prices must gain or lose to get to the average, then add that to the average to get your solution. In this case:

$32.59 must gain $0.29

$33.62 must lose $0.74

$30.78 must gain $2.10

$35.23 must lose $2.35

This means the final closing price must be (0.29 – 0.74 + 2.10 – 2.35) plus the average of $32.88, or $32.18.

To use algebra to solve the problem, let x = Friday's price. The sum of the values ÷ number of days = average price.

$$\frac{32.59 + 33.62 + 30.78 + 35.23 + x}{5} = 32.88$$

$$\frac{132.22 + x}{5} = 32.88$$

$$132.22 + x = 164.4$$

$$x = 32.18$$

The closing price on Friday was \$32.18.

29. 168.3 mph, 178.3 mph

To solve this problem, let x = speed of one plane and $x + 10$ = speed of other plane. To travel the distance the two planes are apart from each other, add the distance traveled by the slow one (1.5 hours times its rate, x) to the distance traveled by the faster plane (1.5 hours times its rate, $x + 10$).

$$1.5x + 1.5(x + 10) = 520$$

$$1.5x + 1.5x + 15 = 520$$

$$3x = 520 - 15$$

$$3x = 505$$

$$x = 168.3 \text{ mph}$$

$$x + 10 = 168.3 + 10 = 178.3 \text{ mph}$$

30. $\mathbf{\frac{2n + c}{3}}$

Find the average of a and b, and simplify to find the value of $a + b$:

$$\frac{a + b}{2} = n$$

$$a + b = 2n$$

Find the average of a, b, and c by substituting the value of $a + b$: $\frac{a + b + c}{3} = \frac{2n + c}{3}$.

RATIOS

A ratio is a comparison of two quantities by division.

Ratios may be written with a fraction bar $\left(\frac{x}{y}\right)$, a colon (x:y), or English terms (ratio of x to y). We recommend the first way, since ratios can be treated as fractions for the purposes of computation.

Ratios can (and in most cases, should) be reduced to lowest terms just as fractions are reduced.

Example: Joe is 16 years old and Mary is 12.

The ratio of Joe's age to Mary's age is $\frac{16}{12}$. (Read "16 to 12.")

$\frac{16}{12} = \frac{4}{3}$ or 4:3

In a ratio of two numbers, the numerator is often associated with the word *of*; the denominator with the word *to*.

The ratio **of** 3 **to** 4 is $\frac{\text{of } 3}{\text{to } 4} = \frac{3}{4}$.

$$\textbf{Ratio} = \frac{\textbf{of...}}{\textbf{to...}}$$

Example: In a box of doughnuts, 12 are sugar and 18 are chocolate. What is the ratio of sugar doughnuts to chocolate doughnuts?

$$\text{Ratio} = \frac{\text{of sugar}}{\text{to chocolate}} = \frac{12}{18} = \frac{2}{3}$$

We frequently deal with ratios by working with a **proportion**. A proportion is simply an equation in which two ratios are set equal to one another.

Ratios typically deal with "parts" and "wholes." The whole is the entire set; for instance, all the workers in a factory. The part is a certain section of the whole; for instance, the female workers in the factory.

The ratio of a part to a whole is usually called a fraction. "What fraction of the workers are female?" means the same thing as "What is the ratio of the number of female workers to the total number of workers?"

A fraction can represent the ratio of a part to a whole:

$\frac{\text{Part}}{\text{Whole}}$ or Part : Whole.

Example: There are 15 men and 20 women in a class. What fraction of the students are female?

$$\text{Fraction} = \frac{\text{Part}}{\text{Whole}}$$
$$= \frac{\text{Number of female students}}{\text{Total number of students}}$$
$$= \frac{20}{15 + 20}$$
$$= \frac{\cancel{20}^{4}}{\cancel{35}_{7}}$$
$$= \frac{4}{7}$$

This means that $\frac{4}{7}$ of the students are female, 4 out of every 7 students are female, or the ratio of female students to total students is 4:7.

Part:Part Ratios and Part:Whole Ratios

A ratio can compare either a part to another part or a part to a whole. One type of ratio can readily be converted to the other **if** all the parts together equal the whole and there is no overlap among the parts (that is, if the whole is equal to the sum of its parts).

Example: The ratio of domestic sales to foreign sales of a certain product is 3:5. What fraction of the total sales are domestic sales? (Note: This is the same as asking for the ratio of the amount of domestic sales to the amount of total sales.)

In this case, the whole (total sales) is equal to the sum of the parts (domestic and foreign sales). We can convert from a **part:part** ratio to a **part:whole** ratio.

Of every 8 sales of the product, 3 are domestic and 5 are foreign. The ratio of domestic sales to total sales is $\frac{3}{8}$ or 3:8.

Example: The ratio of domestic to foreign sales of a certain product is 3:5. What is the ratio of domestic sales to European sales?

Here we cannot convert from a **part:whole** ratio (domestic sales:total sales) to a **part:part** ratio (domestic sales:European sales) because we don't know if there are any other sales besides domestic and European sales. The question doesn't say that the product is sold only domestically and in Europe, so we cannot assume there are no African, Australian, Asian, etc., sales, and so the ratio asked for here cannot be determined.

Ratios with more than two terms: Ratios involving more than two terms are governed by the same principles. These ratios contain more relationships, so they convey more information than two-term ratios. Ratios involving more than two terms are usually ratios of various parts, and it is usually the case that the sum of these parts does equal the whole, which makes it possible to find **part:whole** ratios as well.

Example: Given that the ratio of men to women to children in a room is 4:3:2, what other ratios can be determined?

Quite a few. The **whole** here is the number of people in the room, and since every person is either a man, a woman, or a child, we can determine **part:whole** ratios for each of these parts. Of every nine ($4 + 3 + 2$) people in the room, 4 are men, 3 are women, and 2 are children. This gives us three **part:whole** ratios:

$$\text{Ratio of men:total people} = 4:9 \text{ or } \frac{4}{9}$$

$$\text{Ratio of women:total people} = 3:9 = 1:3 \text{ or } \frac{1}{3}$$

$$\text{Ratio of children:total people} = 2:9 \text{ or } \frac{2}{9}$$

In addition, from any ratio of more than two terms, we can determine various two-term ratios among the parts.

$$\text{Ratio of women:men} = 3:4$$
$$\text{Ratio of men:children} = 4:2 = 2:1$$

And finally, if we were asked to establish a relationship between the number of adults in the room and the number of children, we would find that this would be possible as well. For every 2 children there are 4 men and 3 women, which is $4 + 3$ or 7 adults. So:

$$\text{Ratio of children:adults} = 2:7 \text{ or}$$
$$\text{Ratio of adults:children} = 7:2$$

Naturally, a test question will require you to determine only one or at most two of these ratios, but knowing how much information is contained in a given ratio will help you to determine quickly which questions are solvable and which, if any, are not.

Ratio versus Actual Number

Ratios are always reduced to simplest form. If a team's ratio of wins to losses is 5:3, this does not necessarily mean that the team has won 5 games and lost 3. For instance, if a team has won 30 games and lost 18, the ratio is still 5:3. Unless we know the actual **number** of games played (or the actual number won or lost), we don't know the actual values of the parts in the ratio.

Example: In a classroom of 30 students, the ratio of the boys in the class to students in the class is 2:5. How many are boys?

We are given a part to whole ratio (boys:students). This ratio is a fraction. Multiplying this fraction by the actual whole gives the value of the corresponding part. There are 30 students; $\frac{2}{5}$ of them are boys, so the number of boys must be $\frac{2}{5} \times 30$.

$$\frac{2 \text{ boys}}{{}_1\cancel{5} \text{ }\cancel{\text{students}}} \times \cancel{30}^{6} \text{ }\cancel{\text{students}} = 2 \times 6 = 12 \text{ boys}$$

Picking Numbers

Ratio problems that do not contain any actual values, just ratios, are ideal for solving by Picking Numbers. Just make sure that the numbers you pick are divisible by both the numerator and denominator of the ratio.

Example: A building has $\frac{2}{5}$ of its floors below ground. What is the ratio of the number of floors above ground to the number of floors below ground?

(A) 5:2
(B) 3:2
(C) 4:3
(D) 3:5
(E) 2:5

Pick a value for the total number of floors, one that is divisible by both the numerator and denominator of $\frac{2}{5}$. Let's say 10. Then, since $\frac{2}{5}$ of the floors are below ground, $\frac{2}{5} \times 10$, or 4 floors, are below ground. This leaves 6 floors above ground.

Therefore, the ratio of the number of floors above ground to the number of floors below ground is 6:4, or 3:2, choice **(B)**.

We'll see more on ratios and how we can pick numbers to simplify things in the Problem Solving chapter.

RATES

A rate is a ratio that relates two different kinds of quantities. Speed, the ratio of distance traveled to time elapsed, is an example of a rate.

When we talk about rates, we usually use the word *per*, as in "miles per hour," "cost per item," etc. Since *per* means "for one" or "for each," we express the rates as ratios reduced to a denominator of 1.

Example: John travels 50 miles in two hours. His average rate is $\frac{\text{50 miles}}{\text{2 hours}}$ or 25 miles per hour.

Note: We frequently speak in terms of "average rate," since it may be improbable (as in the case of speed) that the rate has been constant over the period in question. See the Averages section for more details.

RATIOS EXERCISES

Basic

For exercises 1–5, reduce each ratio to simplest form.

1. 16:172

2. 15.5:3.1

3. $\frac{6}{8}:\frac{1}{4}$

4. 4.8:0.8:1.6

5. 27:3

6. If the ratio of the number of men to the number of women on a committee of 25 members is 2:3, how many members of the committee are women?

7. If the ratio of boys to girls in a class is 4:3 and there are 24 boys, how many girls are in the class?

8. Nicholas has 9 goldfish and 6 guppies in his fish tank. What is the ratio of goldfish to guppies in his fish tank?

9. After spending $\frac{1}{4}$ of her paycheck, Rachelle has $150 left. How much was Rachelle's paycheck?

10. In a local recreation center, votes were cast for a new yoga, spin, or Pilates exercise class in the ratio of 3:4:2. If there were 360 votes total, how many votes did yoga receive?

Intermediate

11. A punch recipe calls for 3 pints of ginger ale for every 2 quarts of juice. How many pints of ginger ale will be needed to mix with 10 quarts of juice?

12. The ratio of the ages of Anna and Emma is 3:5 and the ratio of the ages of Emma and Nicholas is 3:5. What is the ratio of Anna's age to Nicholas's age?

13. Selena collects football, baseball, and basketball cards. If the ratio of football to baseball cards is 4:2 and the ratio of baseball to basketball cards is 4:1, what is the ratio of football to basketball cards?

14. A certain juice is in the ratio of 2 parts concentrate to 1 part water. If there are currently 9 gallons of juice, how much water must be added to make the juice in the ratio of 1 part concentrate to 1 part water?

15. If the ratio of $3x$ to $7y$ is 3:4, what is the ratio of x to y?

16. Lin finishes the first half of an exam in one-third the time it takes her to finish the second half. If the whole exam takes her 60 minutes, how many minutes does she spend on the first half of the exam?

17. Bella's grade in a course is determined by 5 quizzes and 1 exam. If the exam counts twice as much as each of the quizzes, what fraction of the final grade is determined by the exam?

18. If red, blue, and yellow gravels are to be mixed in the ratio 4:5:2 respectively, and 12 pounds of red gravel are available, how many pounds of the colored gravel mixture can be made? Assume there is enough blue and yellow gravel available to use all the red gravel.

19. At a football game with 5,400 fans, two-thirds of the fans are rooting for the home team. If 800 of the home-team fans are students, what fraction of the home-team fans are not students?

20. In a certain year, California produced $\frac{7}{10}$ and South Carolina produced $\frac{1}{10}$ of all fresh peach crops in the United States. If all the other states combined produced 242 million pounds that year, how many million pounds did South Carolina produce?

ADVANCED

21. If the ratio of $2a - b$ to $a + b$ is 2:5, what is the ratio of a to b?

22. A moisturizer has 4 parts aloe for every 1 part glycerin. If 65 ounces of this moisturizer are made, how many ounces of glycerin are required?

23. A sportswear store ordered an equal number of blue and white jerseys. The jersey company delivered 22 extra blue jerseys, making the ratio of blue jerseys to white jerseys 7:5. How many jerseys of each color did the store originally order?

24. In a pet store, the ratio of the number of puppies to kittens is 4:7. When 7 more puppies are received, the ratio of the number of puppies to the number of kittens changes to 5:7. How many puppies does the pet store now have?

25. A soccer field has a total area of 7,700 square yards. If 1,100 pounds of grass seed are spread evenly across the entire field, approximately how many pounds of grass seed, to the nearest whole number, are spread over an area of the field totaling 2,500 square yards?

26. If $\frac{1}{2}$ of the number of white roses in a garden is $\frac{1}{8}$ of the total number of roses, and $\frac{1}{3}$ of the number of red roses is $\frac{1}{9}$ of the total number of roses, then what is the ratio of white roses to red roses?

27. The ratio of $5x$ to $2(x + y)$ is 2:3. What is the ratio of x to y?

28. Liquids A and B are in the ratio 2:1 in the first container, and 1:2 in the second container. In what ratio should the contents of the two containers be mixed to obtain a mixture of A and B in the ratio 1:1?

29. A full soap dispenser contains 400 milliliters of soap. After the first 100 milliliters of soap are used, the missing quantity is replaced with pure water, which is added to the dispenser. What is the ratio of soap to water in the dispenser after the water is added?

30. The ratio of the age of a woman to that of her husband was 5:7 when they were first married. After 20 years of marriage, this ratio will be 5:6. What were the ages of the husband and the wife at the time they were first married?

RATIOS EXERCISES ANSWER KEY

BASIC

1. 4:43

Divide both numbers by their greatest common factor, 4. 16:172 = 4:43

2. 5:1

First, multiply both numbers by ten, to compare whole numbers. Then, divide by their greatest common factor, 31. 155:31 = 5:1

3. 3:1

First, multiply both sides by the lowest common denominator, 8, to convert the fractions to whole numbers. Then, divide both sides by their greatest common factor.

$$8\left(\frac{6}{8}\right):8\left(\frac{1}{4}\right) = 6:2 = 3:1$$

4. 6:1:2

First, multiply all three numbers by 10, to compare whole numbers. Then, divide by their greatest common factor, 8:

$$4.8:0.8:1.6 = 48:8:16 = 6:1:2$$

5. 9:1

Divide both numbers by their greatest common factor, 3. Thus, 27:3 = 9:1.

6. 15

The ratio of men to women is 2:3. Now, convert to a part:whole ratio to find the actual number of women. The proportion of women to the whole is $\frac{3}{(2+3)} = \frac{3}{5}$.

Find $\frac{3}{5}$ of the total to calculate the number of women: $\frac{3}{5}(25) = 15$.

There are 15 women on the committee.

7. 18

The ratio of boys to girls is 4:3. Let x = number of girls. Set up a proportion and solve.

$$\begin{aligned}\frac{4}{3} &= \frac{24}{x}\\ 4x &= 72\\ x &= 18\end{aligned}$$

There are 18 girls.

8. 3:2

The ratio of goldfish to guppies is 9:6. Divide both numbers by their greatest common factor, 3, to reduce to simplest form, 3:2.

9. $200

Rachelle spent $\frac{1}{4}$ of her paycheck, so $150 is $1 - \frac{1}{4} = \frac{3}{4}$ of her paycheck. Let x equal her total paycheck.

$$\begin{aligned} \text{fraction} &= \frac{\text{part}}{\text{whole}} \\ \frac{3}{4} &= \frac{150}{x} \\ 3x &= 600 \\ x &= 200 \end{aligned}$$

Her paycheck is $200.

10. 120

Yoga accounts for 3 out of every 9 votes ($3 + 4 + 2 = 9$). Let x equal the number of votes for yoga. Set up a proportion and solve for x.

$$\begin{aligned} \frac{3}{9} &= \frac{x}{360} \\ 9x &= 1{,}080 \\ x &= 120 \end{aligned}$$

Yoga received 120 votes.

INTERMEDIATE

11. 15

Let x be the number of pints of ginger ale. Set up a proportion and solve for x.

$$\begin{aligned} \frac{3 \text{ pints ginger ale}}{2 \text{ quarts juice}} &= \frac{x \text{ pints ginger ale}}{10 \text{ quarts juice}} \\ \frac{3}{2} &= \frac{x}{10} \\ 2x &= 30 \\ x &= 15 \end{aligned}$$

There will be 15 pints of ginger ale needed for 10 quarts of juice.

12. 9:25

The ratio of the ages of Anna and Emma is 3:5. The ratio of the ages of Emma and Nicholas is 3:5. Emma's age is common to both ratios, so replace Emma's age with the least common multiple of 3 and 5, or 15. Anna's age to Emma's age is 9:15 (multiply both values by 3). Emma's age to Nicholas's age is 15:25 (multiply both values by 5 to get Emma's age to 15, as in the earlier ratio). Now that Emma's age is the same in each ratio, you can combine the two ratios. Anna:Emma:Nicholas = 9:15:25.

The ratio of Anna's age to Nicholas's age is 9:25.

13. 8:1

Baseball cards are common to both ratios. Multiply both ratios by appropriate factors so that the baseball cards are represented by the same number in each ratio. Football:baseball = 4:2 and baseball:basketball = 4:1. Multiply the first ratio by 2. Now, football:baseball = 8:4. So, football:baseball:basketball is 8:4:1; football:basketball is 8:1.

14. 3

The juice is 2 parts of concentrate and 1 part of water. Since there are 9 gallons of juice, there must be 6 gallons of concentrate and 3 gallons of water. In order to have the concentrate and water be the same, you must add 3 gallons of water.

15. 7:4

Set up a proportion using the given ratios and solve.

$$\frac{3x}{7y} = \frac{3}{4}$$

$$\frac{x}{y} = \frac{3}{4} \times \frac{7}{3}$$

$$\frac{x}{y} = \frac{21}{12} = \frac{7}{4}$$

The ratio of x to y is 7:4.

16. 15 minutes

Let x equal the number of minutes it took to complete the first half of the exam. Set up a proportion and solve.

$$\frac{1}{3} = \frac{x}{60 - x}$$

$$3x = 60 - x$$

$$4x = 60$$

$$x = 15$$

Lin finishes the first half of the exam in 15 minutes.

17. $\frac{2}{7}$

Each of the 5 quizzes counts once and the exam counts twice, which means there are $5 + 2 = 7$ total parts. The exam is 2 parts, so it counts as $\frac{2}{7}$ of the final grade.

18. 33 pounds

There are 4 parts of red. $4 \times 3 = 12$, so multiply each element in the ratio 4:5:2 by 3 to determine how many pounds of each color should be used: $(4 \times 3):(5 \times 3):(2 \times 3) = 12:15:6$. Add the amounts together to determine how many pounds of the mixture will be made: $12 + 15 + 6 = 33$.

19. $\frac{7}{9}$

First, find the number of home team fans: $\frac{2}{3} \times 5{,}400 = 3{,}600$ home team fans. Then, subtract the number of those fans who are students to find the number of home-team fans who are not students: 3,600 – 800 students = 2,800. Create a ratio of part to whole, and simplify:

$$\frac{2{,}800 \text{ home-team fans who are not students}}{3{,}600 \text{ home-team fans}} = \frac{7}{9}$$

Thus, $\frac{7}{9}$ of the home-team fans are not students.

20. 121

Together, California and South Carolina produced $\frac{7}{10} + \frac{1}{10}$ or $\frac{8}{10}$ of all the fresh peaches in the country. All the other states combined produced $1 - \frac{8}{10}$ or $\frac{2}{10}$ of all the fresh peaches. So, the ratio of California to South Carolina to other states is 7:1:2. The other states produced 242 million pounds. Let t = total million pounds of peaches. First, solve for t:

$$\begin{aligned} \frac{2}{10}t &= 242 \\ t &= 1{,}210 \end{aligned}$$

South Carolina produced $\frac{1}{10}$ of the total, or $\frac{1}{10}(1{,}210) = 121$.

Therefore, South Carolina produced 121 million pounds of peaches.

Advanced

21. 7:8

Set up a proportion using the given ratios and solve:

$$\begin{aligned} \frac{2a-b}{a+b} &= \frac{2}{5} \\ 5(2a-b) &= 2(a+b) \\ 10a-5b &= 2a+2b \\ 10a-2a &= 2b+5b \\ 8a &= 7b \\ \frac{a}{b} &= \frac{7}{8} \end{aligned}$$

The ratio of a to b is 7:8.

22. 13

4 parts aloe + 1 part glycerin = 5 total parts

$$\begin{aligned} \frac{\text{1 part glycerin}}{\text{5 parts}} &= \frac{x\text{ ounces glycerin}}{\text{65 ounces total}} \\ 5x &= 65 \\ x &= 13 \end{aligned}$$

13 ounces of glycerin are required.

23. 55

Let the initial number of each color of blue and white jerseys be x. The jersey company delivered 22 extra blue jerseys, so total blue jerseys $= x + 22$.

$$\begin{aligned} \frac{7}{5} &= \frac{x+22}{x} \\ 5x+110 &= 7x \\ 110 &= 2x \\ 55 &= x \end{aligned}$$

The store originally ordered 55 jerseys of each color.

24. 35

Let x be the initial scale of puppies to kittens: $\frac{4x}{7x}$. Add 7 more puppies: $\frac{4x+7}{7x}$.

The new ratio is 5:7. Use the new ratios to create a proportion and solve for x.

$$\begin{aligned} \frac{4x+7}{7x} &= \frac{5}{7} \\ 28x+49 &= 35x \\ 49 &= 7x \\ 7 &= x \end{aligned}$$

The original number of puppies is $4x$, so the current number of puppies is $4x + 7 = 4(7) + 7 = 35$. So, there are 35 puppies.

25. Approximately 357 pounds

Use the given information to set up a proportion and solve for the unknown.

$$\begin{aligned}\frac{7 \text{ square yards}}{1 \text{ pounds}} &= \frac{2{,}500 \text{ square yards}}{x \text{ pounds}}\\ 2{,}500 &= 7x\\ 357.14 &\approx 357 = x\end{aligned}$$

26. 3:4

Ratio of white roses to total roses: $\frac{1}{\frac{1}{2}}:\frac{1}{\frac{1}{8}} = 1:4 = \frac{1}{4}:1$. So white roses are $\frac{1}{4}$ of total roses.

Ratio of red roses to total roses: $\frac{1}{\frac{1}{3}}:\frac{1}{\frac{1}{9}} = 3:9 = 1:3 = \frac{1}{3}:1$. So red roses are $\frac{1}{3}$ of total roses.

Ratio of white roses to red roses: $\frac{1}{4}:\frac{1}{3} = 12 \times \frac{1}{4}:12 \times \frac{1}{3} = 3:4$.

27. 4:11

Set up a proportion using the given ratios and solve:

$$\begin{aligned}\frac{5x}{2(x+y)} &= \frac{2}{3}\\ 15x &= 4(x+y)\\ 15x &= 4x + 4y\\ 11x &= 4y\\ \frac{x}{y} &= \frac{4}{11}\end{aligned}$$

The ratio of x to y is 4:11.

28. 1:1

In the first container, the ratio of A to total liquid is $\frac{2}{2+1} = \frac{2}{3}$, and the ratio of B to total liquid is $\frac{1}{1+2} = \frac{1}{3}$. In the second container, the ratio of A to total liquid is $\frac{1}{1+2} = \frac{1}{3}$, and the ratio of B to total liquid is $\frac{2}{2+1} = \frac{2}{3}$. So the ratio of A to total liquid in both containers is 3:6 or 1:2, and the ratio of B to total liquid in both containers is 3:6 or 1:2. If equal parts from each container are mixed, the ratio of A to B will be 1:1.

29. 3:1

The amount of soap left is 300 milliliters. The amount of water is 100 milliliters. So the ratio of soap to water is 300:100, which reduces to 3:1.

30. The wife was 20, and the husband was 28.

Let the woman's age at marriage = $5x$ and the man's age at marriage = $7x$. Set up a proportion of their ages in 20 years and solve:

$$\begin{aligned} \frac{5x + 20}{7x + 20} &= \frac{5}{6} \\ 6(5x + 20) &= 5(7x + 20) \\ 30x + 120 &= 35x + 100 \\ 20 &= 5x \\ 4 &= x \end{aligned}$$

Woman's age at marriage = 5(4) = 20.

Man's age at marriage = 7(4) = 28.

PERCENTS

Percents are one of the most commonly used math relationships. Percents are also a popular topic on the GRE. *Percent* is just another word for *hundredth*. Therefore, 19% (19 percent) means

19 hundredths

or $\frac{19}{100}$

or 0.19

or 19 out of every 100 things

or 19 parts out of a whole of 100 parts.

They're all just different names for the same thing.

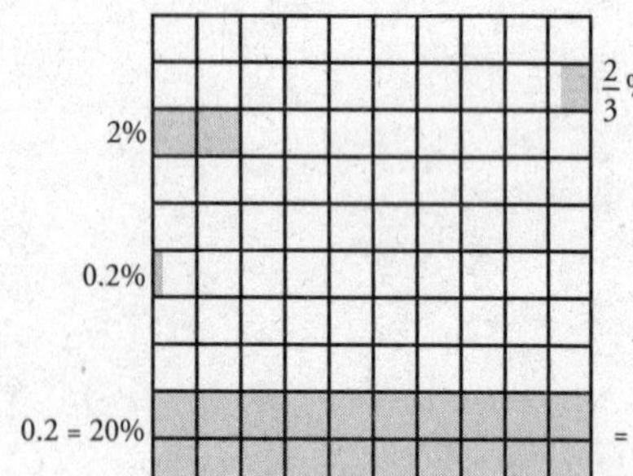

Each box at the left represents 1%. 100 boxes = (100)(1%) = 100% = 1 whole. Note that we have, in increasing order, 0.2%, $\frac{2}{3}$%, 2%, and 20%.

MAKING AND DROPPING PERCENTS

To make a percent, multiply by 100%. Since 100% means 100 hundredths or 1, multiplying by 100% will not change the value.

Example: $0.17 = 0.17 \times 100\% = 17.0\%$ *or* 17%

Example: $\frac{1}{4} = \frac{1}{4} \times 100\% = 0.25 \times 100\% = 25\%$

To drop a percent, divide by 100%. Once again, dividing by 100% will not change the value.

Example: $32\% = \frac{32\%}{100\%} = \frac{32}{100} = \frac{8}{25}$

Example: $\frac{1}{2}\% = \frac{\frac{1}{2}\%}{100\%} = \frac{1}{200}$

To change a percent to a decimal, just drop the percent and move the decimal point two places to the left. (This is the same as dividing by 100%.)

Example: $0.8\% = 0.00.8 = 0.008$

Example: $2\frac{1}{4}\% = 2.25\% = 0.02.25 = 0.0225$

Common Percent and Fractional Equivalents

$\frac{1}{20} = 5\%$	$\frac{1}{10} = 10\%$	$\frac{1}{8} = 12\frac{1}{2}\%$	$\frac{1}{6} = 16\frac{2}{3}\%$
$\frac{1}{5} = 20\%$	$\frac{1}{4} = 25\%$	$\frac{1}{3} = 33\frac{1}{3}\%$	$\frac{1}{2} = 50\%$

$10\% = \frac{1}{10}$	$12\frac{1}{2}\% = \frac{1}{8}$	$16\frac{2}{3}\% = \frac{1}{6}$
$20\% = \frac{2}{10} = \frac{1}{5}$	$25\% = \frac{2}{8} = \frac{1}{4}$	
$30\% = \frac{3}{10}$	$37\frac{1}{2}\% = \frac{3}{8}$	$33\frac{1}{3}\% = \frac{2}{6} = \frac{1}{3}$
$40\% = \frac{4}{10} = \frac{2}{5}$		
$50\% = \frac{5}{10} = \frac{1}{2}$	$50\% = \frac{4}{8} = \frac{2}{4} = \frac{1}{2}$	$50\% = \frac{3}{6} = \frac{1}{2}$
$60\% = \frac{6}{10} = \frac{3}{5}$	$62\frac{1}{2}\% = \frac{5}{8}$	$66\frac{2}{3}\% = \frac{4}{6} = \frac{2}{3}$
$70\% = \frac{7}{10}$	$75\% = \frac{6}{8} = \frac{3}{4}$	$83\frac{1}{3}\% = \frac{5}{6}$
$80\% = \frac{8}{10} = \frac{4}{5}$	$87\frac{1}{2}\% = \frac{7}{8}$	
$90\% = \frac{9}{10}$		
$100\% = \frac{10}{10} = 1$		

Being familiar with these equivalents can save you a lot of time on Test Day.

Percent Problems

Most percent problems can be solved by plugging into one formula:

Percent × Whole = Part

This formula has three variables: percent, whole, and part. In percent problems, generally, the **whole** will be associated with the word *of*; the **part** will be associated with the word *is*. The percent can be represented as the ratio of the part to the whole, or the *is* to the *of*.

Percent problems will usually give you two of the variables and ask for the third. See the examples of the three types of problems. On the GRE, it is usually easiest to change the percent to a common fraction and work it out from there.

Example: What is 25% of 36?

Here we are given the percent and the whole. To find the part, change the percent to a fraction, then multiply. Use the formula above.

$$\text{Percent} \times \text{Whole} = \text{Part}$$

Since $25\% = \frac{1}{4}$, we are really asking what one-fourth of 36 is.

$$\frac{1}{4} \times 36 = 9$$

Example: 13 is $33\frac{1}{3}\%$ of what number?

Here we are given the percent and the part and asked for the whole.

If Percent × Whole = Part, then

$$\text{Whole} = \frac{\text{Part}}{\text{Percent}}. \text{ Recall that } 33\frac{1}{3}\% = \frac{1}{3}.$$

$$= \frac{13}{\frac{1}{3}}$$

$$= 13 \times \frac{3}{1} = 39$$

We can avoid all this algebra. All we are asked is "13 is one-third of what number?" And 13 is one-third of 3 × 13 or 39.

Example: 18 is what percent of 3?

Here we are given the whole (3) and the part (18) and asked for the percent. If Percent × Whole = Part, then

$$\text{Percent} = \frac{\text{Part}}{\text{Whole}}$$

Since the part and the whole are both integers and we're looking for a percent, we're going to have to make our result into a percent by multiplying it by 100%.

$$\text{Percent} = \frac{18}{3}(100\%) = 6(100\%) = 600\%$$

Note here that we can find the percent as the "is" part divided by the "of" part:

What percent is 18 of 3?

$$\text{Percent} = \frac{\text{is}}{\text{of}} = \frac{18}{3} = 6 = 600\%$$

Alternative method: The base 3 represents 100%. Since 18 is 6 times as large, the percent equals $6 \times 100\% = 600\%$.

Percent increase and decrease:

$$\textbf{Percent increase} = \frac{\textbf{Amount of increase}}{\textbf{Original whole}} \times (\textbf{100\%})$$

$$\textbf{Percent decrease} = \frac{\textbf{Amount of decrease}}{\textbf{Original whole}} \times (\textbf{100\%})$$

$$\textbf{New whole} = \textbf{Original whole} \pm \textbf{Amount of change}$$

When dealing with percent increase and percent decrease, always be careful to put the amount of increase or decrease over the original whole, not the new whole.

Example: If a $120 dress is increased in price by 25%, what is the new selling price?

Our original whole here is $120, and the percent increase is 25%. Change 25% to a fraction, $\frac{1}{4}$, and use the formula.

$$\begin{aligned}\text{Amount of increase} &= \text{Percent increase} \times \text{Original whole} \\ &= 25\% \times \$120 \\ &= \frac{1}{4} \times \$120 \\ &= \$30\end{aligned}$$

To find the **new whole** (the new selling price):

New whole = Original whole + Amount of increase

New whole = $120 + $30 = **$150**

Combining percents: On some problems, you'll need to find more than one percent, or a percent of a percent. Be careful. You can't just add percents, unless you're taking the percents of the same whole. Let's look at an example.

Example: The price of an antique is reduced by 20 percent and then this price is reduced by 10 percent. If the antique originally cost \$200, what is its final price?

First, we know that the price is reduced by 20%. That's the same thing as saying that the price becomes 100% – 20%, or 80% of what it originally was. 80% of \$200 is equal to $\frac{8}{10} \times \$200$, or \$160. Then, *this* price is reduced by 10%. $10\% \times \$160 = \$16$, so the final price of the antique is $\$160 - \$16 = \$144$.

A common error in this kind of problem is to assume that the final price is simply a 30% reduction of the original price. That would mean that the final price is 70% of the original, or $70\% \times \$200 = \$140$. But, as we've just seen, this is *not* correct. Adding or subtracting percents directly only works if those percents are being taken of the same whole. In this example, since we took 20% of the original price, and then 10% of that reduced price, we can't just add the percents together.

For this type of question, the GRE will always provide the answer choice of simply adding or subtracting the percents, so beware of this trap. More practice with percent problems can be found in the Problem Solving chapter.

PERCENTS EXERCISES

BASIC

For Exercises 1–2, convert to a percent:

1. 0.087

2. $\frac{3}{20}$

For Exercises 3–4, convert to a fraction:

3. 65%

4. 232%

5. What is 28% of 70?

6. What is 125% of 48?

7. What percent of 40 is 22?

8. Twenty percent of 60 is 50% of what number?

9. Fifteen percent of 40% of 50 is what number?

10. What is 2.5% of $\frac{1}{4}$?

INTERMEDIATE

For Exercises 11–12, convert to a percent:

11. 1.675

12. 0.101

For Exercises 13–14, convert to a fraction:

13. $\frac{1}{8}\%$

14. 0.7%

15. What is 2.5% of 15?

16. What is $\frac{3}{25}$% of 100?

17. Of 25 students in a class, 15 have completed their tests. What percent of students in the class have not completed their tests?

18. Seventy percent of 125 is 80% of what number?

19. 60% of 180% of 40 is what number?

20. The price of a car accessory that originally costs $80 is discounted by 25%. What is the discounted price of the car accessory?

ADVANCED

For Exercises 21–22, convert to a percent:

21. 0.0003

22. 5.035

For Exercises 23–24, convert to a fraction:

23. 6.75%

24. $2\frac{3}{5}$%

25. What is 0.05% of 5,000?

26. What is $150\frac{2}{5}$% of 1,000?

27. A scientist is studying the population change in the number of foxes for a certain area. She observes a 25% increase in the population of foxes for a certain area. If the new population is 45 foxes, what was the previous population?

28. Jeff sold $650,000 in software in the fourth quarter. If this was a 40% increase over his third quarter sales, what were his third quarter sales? Round to the nearest dollar.

29. A hostess at an art gallery makes $100 for each exhibit that she works. She also receives $2\frac{1}{2}\%$ of the art sales. If she earned $900 for a single exhibit, how much were the art sales?

30. Gasoline at a certain station has increased from $2.98 to $3.07. If it then decreases by half the percent of the percent increase, what is the new price? Round to the nearest tenth of a percent at each step, then round the final price to the nearest cent.

PERCENTS EXERCISES ANSWER KEY

Basic

1. 8.7%

To convert to a percent, multiply by 100%. Move the decimal point two places to the right to reach 8.7.

2. 15%

To convert to a percent, multiply by 100%: $\frac{3}{20} \times 100\% = \frac{300}{20}\% = 15\%$.

3. $\frac{13}{20}$

To convert to a fraction, divide by 100%: $\frac{65\%}{100\%} = \frac{65}{100} = \frac{13}{20}$.

4. $2\frac{8}{25}$

To convert to a fraction, divide by 100%: $\frac{232\%}{100\%} = \frac{232}{100} = 2\frac{32}{100} = 2\frac{8}{25}$.

5. 19.6

Use the formula percent $\times$ whole $=$ part: $28\% \times 70 = 0.28 \times 70 = 19.6$.

6. 60

Use the formula percent $\times$ whole $=$ part: $125\% \times 48 = 1.25 \times 48 = 60$.

7. 55%

Use the formula percent $\times$ whole $=$ part: percent $\times 40 = 22$. Divide both sides by 40: percent $= 22 \div 40 = 0.55$. Multiply by 100 and add the % sign.

8. 24

Use the formula percent $\times$ whole $=$ part for both steps. First, find 20% of 60: $0.20 \times 60 = 12$. Then, determine what number 12 is 50% of: $50\% \times x = 12$.

$$0.50 \times x = 12$$

$$x = 12 \div 0.50 = 24$$

9. 3

Use the formula percent $\times$ whole $=$ part for both steps. First, find 40% of 50: $0.40 \times 50 = 20$. Then, find 15% of 20: $0.15 \times 20 = 3$.

10. 0.00625

Use the formula percent $\times$ whole $=$ part:

$$2.5\% \times \frac{1}{4} = 0.025 \times 0.25 = 0.00625.$$

Intermediate

11. 167.5%

To convert to a percent, multiply by 100%. Move the decimal point two places to the right to reach 167.5%.

12. 10.1%

To convert to a percent, multiply by 100%. Move the decimal point two places to the right to reach 10.1%.

13. $\mathbf{\frac{1}{800}}$

To convert to a fraction, divide by 100%: $\frac{\frac{1}{8}\%}{100\%} = \frac{1}{8} \times \frac{1}{100} = \frac{1}{800}$.

14. $\mathbf{\frac{7}{1000}}$

To convert to a fraction, divide by 100%: $\frac{0.7\%}{100\%} = \frac{0.7}{100} = \frac{7}{1{,}000}$.

15. 0.375

Use the formula percent $\times$ whole $=$ part: $2.5\% \times 15 = 0.025 \times 15 = 0.375$.

16. 0.12

Use the formula percent $\times$ whole $=$ part:

$$\frac{3}{25}\% \times 100 = 0.12\% \times 100 = 0.0012 \times 100 = 0.12.$$

17. 40%

First, find the number of students who have not completed the test: $25 - 15 = 10$. Then, divide that number by the total: $10 \div 25 = 0.4$ or 40%.

18. 109.375

Use the formula percent $\times$ whole $=$ part for both steps. First, find 70% of 125: $0.70 \times 125 = 87.5$. Then, determine what number 87.5 is 80% of: $80\% \times x = 87.5$

$$x = 87.5 \div 0.80 = 109.375$$

19. 43.2

Use the formula percent $\times$ whole $=$ part for both steps. First, find 180% of 40: $1.80 \times 40 = 72$. Then, find 60% of 72: $0.60 \times 72 = 43.2$.

20. $60

First, find the amount of the discount, or 25% of 80: $0.25 \times 80 = 20$. Then subtract the discount amount from the original price: $80 – $20 = $60.

ADVANCED

21. 0.03%

To convert to a percent, multiply by 100%. Move the decimal point two places to the right to reach 0.03%.

22. 503.5%

To convert to a percent, multiply by 100%. Move the decimal point two places to the right to reach 503.5%.

23. $\mathbf{\frac{27}{400}}$

To convert to a fraction, divide by 100%: $\frac{6.75\%}{100\%} = \frac{6.75}{100} = \frac{675}{10{,}000} = \frac{27}{400}$.

24. $\mathbf{\frac{13}{500}}$

To convert to a fraction, divide by 100%: $\frac{2\frac{3}{5}\%}{100\%} = \frac{13}{5} \times \frac{1}{100} = \frac{13}{500}$.

25. 2.5

Use the formula percent × whole = part: $0.05\% \times 5{,}000 = 0.0005 \times 5{,}000 = 2.5$.

26. 1,504

Use the formula percent × whole = part. Convert $150\frac{2}{5}\%$ to a decimal, 150.4%. Then, multiply. $150.4\% \times 1{,}000 = 1.504 \times 1{,}000 = 1{,}504$.

27. 36

Find the number that sums to 45 when 25% of itself is added to it. Use the formula $x + 0.25x = 45$ and solve for x. $1.25x = 45$, so $x = 36$.

28. $464,286

Find the number that sums to $650,000 when 40% of itself is added to it. Use the formula $x + 0.40x = \$650{,}000$ and solve for x. $1.40x = \$650{,}000$. Divide both sides by 1.40, so $x \approx \$464{,}286$.

29. $32,000

Set up a formula that models the situation, with x being the number of art sales. Convert $2\frac{1}{2}\%$ to a decimal, 0.025. $\$100 + 0.025x = \$900$. Solve for x. $0.025x = \$800$. Divide both sides by 0.025, so $x = \$32{,}000$.

30. $3.02

Set up a formula that models the situation and solve for x, where x = original increase in the price of gasoline. Thus, $\$2.98 + \$2.98x = \$3.07$; $\$3.07 - \$2.98 = \$0.09 = \$2.98x$. Then, divide by $2.98 to get an increase of approximately 0.03 or 3%. If the decrease is half the increase, then the decrease is 1.5%. Find 1.5% of $3.07: $0.015 \times \$3.07 \approx \$0.05$. Subtract: $\$3.07 - \$0.05 = \$3.02$.

POWERS AND ROOTS

RULES OF OPERATION WITH POWERS

In the term $3x^2$, 3 is the **coefficient**, x is the **base**, and 2 is the **exponent**. The exponent refers to the number of times the base is multiplied by itself, or how many times the base is a factor. For instance, in 4^3, there are 3 factors of 4 : $4^3 = 4 \times 4 \times 4 = 64$.

A number multiplied by itself twice is called the **square** of that number (e.g., x^2 is x squared).

A number multiplied by itself three times is called the **cube** of that number (e.g., 4^3 is 4 cubed).

To multiply two terms with the same base, keep the base and add the exponents.

Example: $$2^2 \times 2^3 = (2 \times 2)(2 \times 2 \times 2) = (2 \times 2 \times 2 \times 2 \times 2) = 2^5$$ or $$2^2 \times 2^3 = 2^{2+3} = 2^5$$

Example: $$x^4 \times x^7 = x^{4+7} = x^{11}$$

To divide two terms with the same base, keep the base and subtract the exponent of the denominator from the exponent of the numerator.

Example: $$4^4 \div 4^2 = \frac{4 \times 4 \times 4 \times 4}{4 \times 4} = \frac{4 \times 4}{1} = 4^2$$ or $$4^4 \div 4^2 = 4^{4-2} = 4^2$$

To raise a power to another power, multiply the exponents.

Example: $$\left(3^2\right)^4 = (3 \times 3)^4 = (3 \times 3)(3 \times 3)(3 \times 3)(3 \times 3) = 3^8$$ or $$(3^2)^4 = 3^{2 \times 4} = 3^8$$

Any nonzero number raised to the zero power is equal to 1. $a^0 = 1$ if $a \neq 0$, but 0^0 is undefined.

A negative exponent indicates a reciprocal. To arrive at an equivalent expression, take the reciprocal of the base and change the sign of the exponent.

$$a^{-n} = \frac{1}{a^n} \text{ or } \left(\frac{1}{a}\right)^n$$

Example: $2^{-3} = \left(\frac{1}{2}\right)^3 = \frac{1}{2^3} = \frac{1}{8}$

A fractional exponent indicates a **root**.

$(a)^{\frac{1}{n}} = \sqrt[n]{a}$ (read "the nth root of a." If no "n" is present, the radical sign means a square root.)

Example: $8^{\frac{1}{3}} = \sqrt[3]{8} = 2$

On the GRE you will probably only see the square root. The square root of a non-negative number x is equal to the number which, when multiplied by itself, gives you x. Every positive number has two square roots, one positive and one negative. The positive square root of 25 is 5, since $5^2 = 25$ and the negative square root of 25 is -5, since $(-5)^2 = 25$ as well. Other types of roots have appeared on the test (cube root, or $\sqrt[3]{\ }$, is an example), but they tend to be extremely rare.

Note: In the expression $3x^2$, only the x is being squared, not the 3. In other words, $3x^2 = 3(x^2)$. If we wanted to square the 3 as well, we would write $(3x)^2$. (Remember that in the order of operations we raise to a power **before** we multiply, so in $3x^2$ we square x and **then** multiply by 3.)

Rules of Operations with Roots

By convention, the symbol $\sqrt{\ }$ (radical) means the **positive** square root only.

Example: $\sqrt{9} = +3; \ -\sqrt{9} = -3$

Even though there are two different numbers whose square is 9 (both 3 and -3), we say that $\sqrt{9}$ is the positive number 3 only.

When it comes to the four basic arithmetic operations, we treat radicals in much the same way we would treat variables.

Addition and Subtraction: Only like radicals can be added to or subtracted from one another.

Example:

$$\begin{aligned} 2\sqrt{3} + 4\sqrt{2} - \sqrt{2} - 3\sqrt{3} &= \left(4\sqrt{2} - \sqrt{2}\right) + \left(2\sqrt{3} - 3\sqrt{3}\right) \ \left[\text{Note: } \sqrt{2} = 1\sqrt{2}\right] \\ &= 3\sqrt{2} + \left(-\sqrt{3}\right) \\ &= 3\sqrt{2} - \sqrt{3} \end{aligned}$$

Multiplication and Division: To multiply or divide one radical by another, multiply or divide the numbers outside the radical signs, then the numbers inside the radical signs.

Example: $(6\sqrt{3}) \times (2\sqrt{5}) = (6 \times 2) \times (\sqrt{3} \times \sqrt{5}) = 12\sqrt{3 \times 5} = 12\sqrt{15}$

Example: $12\sqrt{15} \div 2\sqrt{5} = (12 \div 2) \times (\sqrt{15} \div \sqrt{5}) = 6\left(\sqrt{\frac{15}{5}}\right) = 6\sqrt{3}$

Example: $\frac{4\sqrt{18}}{2\sqrt{6}} = \left(\frac{4}{2}\right)\left(\frac{\sqrt{18}}{\sqrt{6}}\right) = 2\frac{\sqrt{18}}{\sqrt{6}} = 2\sqrt{3}$

If the number inside the radical is a multiple of a perfect square, the expression can be simplified by factoring out the perfect square.

Example: $\sqrt{72} = \sqrt{36 \times 2} = \sqrt{36} \times \sqrt{2} = 6\sqrt{2}$

Powers of 10

The exponent of a power of 10 tells us how many zeros the number would contain if written out.

Example: $10^6 = 1{,}000{,}000$ (6 zeros) since 10 multiplied by itself six times is equal to 1,000,000.

When multiplying a number by a power of 10, move the decimal point to the right the same number of places as the number of zeros in that power of 10.

Example: $0.029 \times 10^3 = 0.029 \times 1{,}000 = 0.029. = 29$
3 places

When dividing by a power of 10, move the decimal point the corresponding number of places to the left. (Note that dividing by 10^4 is the same as multiplying by 10^{-4}.)

Example: $416.03 \times 10^{-4} = 416.03 \div 10^4 = 0.0416.03 = 0.041603$
4 places

Large numbers or small decimal fractions can be expressed more conveniently using scientific notation. Scientific notation means expressing a number as the product of a decimal between 1 and 10, and a power of 10.

Example: $5{,}600{,}000 = 5.6 \times 10^6$ (5.6 million)

Example: $0.00000079 = 7.9 \times 10^{-7}$

Example: $0.00765 \times 10^7 = 7.65 \times 10^4$

POWERS AND ROOTS EXERCISES

BASIC

For Exercises 1–3, evaluate the expression.

1. 3^6

2. 8^3

3. $\sqrt{2}\sqrt{50}$

4. Write the expression 4^{-3} with a positive exponent.

5. Write 0.0028207 in scientific notation.

6. If $t = 4$, then what is the value of $4t^2 + \frac{1}{2}t^3$?

7. Evaluate $\frac{3^5 \times 9^2}{9^4}$.

8. If $5\sqrt{2x} = 20$, then what is the value of $3x^2$?

9. Simplify $2\sqrt{2} - 3\sqrt{5} + \sqrt{2} + 4\sqrt{5} - 2\left(\sqrt{2} - 6\right)$.

10. Find the value of y for $\sqrt{100 - 4y^2} = 0$.

INTERMEDIATE

For Exercises 11–13, evaluate the expression.

11. $(5 + 3)^2$

12. 6×3^3

13. $\sqrt{8}\sqrt{12}$. Do not evaluate any square roots that would result in a noninteger.

14. If $m = 2$ and $n = -2$, what is the value of $m^2n^3 - (3mn)^2$?

15. If $5^n < 2000$, what is the greatest possible integer value of n?

16. Simplify $12\sqrt{18} \div 3\sqrt{2}$.

17. Find the maximum value of r for $\sqrt{18r - 6r^2}$ to be a real number.

18. What positive number when squared is equal to four times the square of –5?

19. Simplify $(4.825 \times 10^6) \div (3.2 \times 10^8)$. Round to the nearest hundredth.

20. If $k = 3$ and $l = -1$, what is the value of $3l^3k^4$?

ADVANCED

For Exercises 21–23, evaluate the expression.

21. $(4^3)^2$

22. $\frac{2}{3}^2 \times 3^4$

23. $\frac{\sqrt{5}\sqrt{60}}{\sqrt{3}}$

24. Write $10{,}843 \times 10^7$ in scientific notation.

25. True or false? 4.809×10^7 is equivalent to 0.0004809×10^{11}.

26. If $a = -1$ and $b = 3$, what is the value of $\frac{4a^3b^2 - 12a^2b^5}{16(a^3b^2)}$?

27. True or false? Given d, e, and $f \neq 0$, $\frac{d^3ef^5}{2de^3}$ is equal to $\frac{3d^2e^3f^7}{6e^5f^2}$.

28. If x and y are negative odd integers, is the following always, sometimes, or never positive: $4x + 3y - y^x + x^3$?

29. Simplify $\frac{4\sqrt{21} \times 5\sqrt{2}}{10\sqrt{7}}$.

30. Simplify $9^{\frac{1}{2}} \times 4^3 \times 2^{-6}$.

POWERS AND ROOTS EXERCISES ANSWER KEY

BASIC

1. 729

3^6 is the same as 3 multiplied six times: $3 \times 3 \times 3 \times 3 \times 3 \times 3$, or 729.

2. 512

8^3 is the same as $8 \times 8 \times 8$, or 512.

3. 10

$$\sqrt{2}\sqrt{50} = \sqrt{2 \times 50} = \sqrt{100} = 10$$

4. $\mathbf{\frac{1}{4^3}}$

A negative exponent denotes the reciprocal of a base with a positive exponent.

5. $\mathbf{2.8207 \times 10^{-3}}$

0.0028207 can be written in scientific notation by moving the decimal point 3 places to the right, or 2.8207×10^{-3}.

6. 96

To evaluate, substitute 4 for $t: 4\left(4^2\right) + \frac{1}{2}\left(4^3\right) = 4(16) + \frac{1}{2}(64) = 64 + 32 = 96$.

7. 3

Begin by cancelling any common factors in the numerator and denominator. $\frac{3^5 \times \cancel{9^2}}{9^{\cancel{4}^{2}}} = \frac{3^5}{9^2}$.

Then, substitute 3^2 for 9 in the denominator and simplify the exponent: $\frac{3^5}{9^2} = \frac{3^5}{\left(3^2\right)^2} = \frac{3^5}{3^4}$.

Finally, cancel any common factors in the numerator and denominator again: $\frac{3^{\cancel{5}^{1}}}{\cancel{3^4}} = 3$.

8. 192

First, solve $5\sqrt{2x} = 20$ for x. Divide by 5: $\sqrt{2x} = 4$. Square both sides: $2x = 16$. Divide by 2: $x = 8$. Substitute 8 for x in the second equation: $3(8)^2 = 3(64) = 192$.

9. $\mathbf{\sqrt{2} + \sqrt{5} + 12}$

Use the distributive property to multiply. Then use the commutative property to move like square roots together and simplify.

$$2\sqrt{2} - 3\sqrt{5} + \sqrt{2} + 4\sqrt{5} - 2(\sqrt{2} - 6) =$$
$$2\sqrt{2} - 3\sqrt{5} + \sqrt{2} + 4\sqrt{5} - 2\sqrt{2} + 12 =$$
$$2\sqrt{2} + \sqrt{2} - 2\sqrt{2} - 3\sqrt{5} + 4\sqrt{5} + 12 =$$
$$\sqrt{2} + \sqrt{5} + 12$$

10. 5, –5

First, square both sides so you have the expression $100 - 4y^2 = 0$. Then solve for y. $4y^2 = 100$. Divide both sides by 4 and you get $y^2 = 25$. Finally y must be equal to either 5 or –5.

INTERMEDIATE

11. 64

Begin by evaluating the expression inside the parentheses and then find the square: $(8)^2 = 64$.

12. 162

Evaluate the exponent and then multiply: $6 \times 3^3 = 6 \times (3 \times 3 \times 3) = 6(27) = 162$.

13. $4\sqrt{6}$

Begin by expanding each expression in the square roots to find a perfect square: $\sqrt{2 \times 4}\sqrt{3 \times 4} = \sqrt{2} \times \sqrt{4} \times \sqrt{3} \times \sqrt{4}$. Then, take the square roots: $\sqrt{2} \times 2 \times \sqrt{3} \times 2$. Finally, multiply and combine the square roots again: $4\sqrt{2}\sqrt{3} = 4\sqrt{6}$.

14. –176

Substitute the values for the variables, then use PEMDAS to simplify the expression:

$$m^2n^3 - (3mn)^2$$
$$= 2^2 \times -2^3 - (3 \times 2 \times -2)^2$$
$$= 4 \times -8 - (-12)^2$$
$$= -32 - 144 = -176$$

15. 4

Use a guess-and-check method to find the solution: $5^2 = 25$, $5^3 = 125$, $5^4 = 625$. Since 5^5 will be larger than 2,000, the exponent must be an integer less than 5. Therefore, the answer is 4.

16. 12

Begin by writing as a fraction: $\frac{12\sqrt{18}}{3\sqrt{2}}$. Then, use properties of roots to simplify.

$$4\sqrt{\frac{18}{2}} = 4\sqrt{9} = 4 \times 3 = 12$$

17. 3

For the square root to be real, the value inside the radical must be greater than or equal to 0. Find the value that makes $18r - 6r^2 = 0$, as this is the greatest number that will still produce a real root. $18r = 6r^2$. Divide both sides by r: $18 = 6r$. Divide both sides by 6: $3 = r$. The greatest value that will still produce a real root is 3.

18. 10

First, find the square of –5, or 25. Then, find $4 \times 25 = 100$. Find the (positive) square root of 100, which is 10.

19. 0.02

Write the expression as a fraction.

$$\frac{4.825 \times 10^6}{3.2 \times 10^8} = \frac{4.825 \times \cancel{10^6}}{3.2 \times \cancel{10^8}^{\,2}} = \frac{4.825}{3.2 \times 10^2} = \frac{4.825}{320} \approx 0.02$$

20. –243

Substitute the values into the expression: $3(-1)^3(3)^4 = 3(-1)(81) = -3(81) = -243$.

Advanced

21. 4,096

Use the rules of exponents to simplify the expression: $(4^3)^2 = 4^6 = 4{,}096$.

22. 36

Use the rules of exponents to simplify the expression:

$$\left(\frac{2}{3}\right)^2 \times 3^4 = \left(\frac{2^2}{3^2}\right) \times 3^4 = \left(\frac{2^2}{\cancel{3^2}}\right) \times 3^{\cancel{4}^{\,2}} = 2^2 \times 3^2 = 4 \times 9 = 36.$$

23. 10

Simplify the square roots:

$$\sqrt{5}\sqrt{\frac{60}{3}} = \sqrt{5}\sqrt{20} = \sqrt{5 \times 20} = \sqrt{100} = 10.$$

24. 1.0843×10^{11}

To be written in scientific notation, the decimal part can have a digit only in the ones place. So, move the decimal point four places to the left. This means we need to add 4 to 7, the power of 10 of the original number.

25. True

Expand both numbers and then compare: $4.809 \times 10^7 = 48{,}090{,}000.$ and $0.0004809 \times 10^{11} = 48{,}090{,}000$. The two numbers are equivalent.

26. 20.5

First, factor out the numerator and denominator using the rules of exponents and then simplify:
$\frac{4a^3b^2 - 12a^2b^5}{16\left(a^3b^2\right)} = \frac{4a^2b^2\left(a - 3b^3\right)}{4a^2b^2(4a)} = \frac{a - 3b^3}{4a}$. Then substitute the given values and evaluate:

$$\frac{-1 - 3(3)^3}{4(-1)} = \frac{-1 - 3(27)}{-4} = \frac{-1 - 81}{-4} = \frac{-82}{-4} = 20.5.$$

27. True

Simplify both expressions and see if they are equal.

$$\frac{3d^2e^3f^7}{6e^5f^2} = \frac{d^2f^5}{2e^2}$$

$$\frac{d^3ef^5}{2de^3} = \frac{d^2f^5}{2e^2}$$

Because both expressions simplify to the same expression, they are equal.

28. Never

Reason through each term in the expression. Because x is a negative integer, $4x$ will always be a negative integer. The greatest possible value of $4x$ is -4. Because y is negative, $3y$ will always be a negative integer. The greatest possible value of $3y$ is -3. Then, because x is negative, $-y^x$ will be a positive value between 0 and 1, since $-y^x = -\frac{1}{y^{-x}}$. The exponent $-x$ in the denominator will be a positive odd integer, and y raised to a positive odd integer exponent will result in a negative integer in the denominator. Subtracting a negative fraction between -1 and 0 is the same as adding a positive fraction between 0 and 1. Finally, because x is negative, x^3 will always be a negative integer. The greatest possible value of x^3 is -1. Adding these three negative integers and one positive fraction between 0 and 1 will always result in a negative number. It will never be positive.

29. $2\sqrt{6}$

Begin by simplifying the nonroot portions of the expression:

$$\frac{4\sqrt{21} \times 5\sqrt{2}}{10\sqrt{7}} = \frac{20\sqrt{21}\sqrt{2}}{10\sqrt{7}} = \frac{2\sqrt{21}\sqrt{2}}{\sqrt{7}}$$

Then, use the rules of square roots to continue simplifying:

$$\frac{2\sqrt{21}\sqrt{2}}{\sqrt{7}} = \frac{2\sqrt{21\times 2}}{\sqrt{7}} = \frac{2\sqrt{42}}{\sqrt{7}} = 2\sqrt{\frac{42}{7}} = 2\sqrt{6}$$

30. 3

Remove fractional exponents and change negative exponents to positive exponents: $9^{\frac{1}{2}} \times 4^3 \times 2^{-6} = \sqrt{9} \times 4^3 \times \frac{1}{2^6}$. Then, simplify:

$\sqrt{9} \times 4^3 \times \frac{1}{2^6} = 3 \times 64 \times \frac{1}{64} = 3.$

CHAPTER 7

Algebra

UNDERSTANDING ALGEBRA

The use of variables to represent numbers is what differentiates algebra from arithmetic. Calculations in algebra may involve solving for a value of a variable that makes an equation true, or they may involve substituting different values for a variable in an expression. On the GRE, you will see some questions that are strictly algebra based, but you will also see questions that involve the use of algebra along with reasoning, problem solving, and data interpretation skills. For those reasons, algebra is an important area on which to focus your review. You must understand basic equations and how to solve them.

A good place to start is with a review of algebraic terminology to ensure that you understand directions, questions, and explanations as you go along.

Variable: A letter used to represent a quantity whose value is unknown.

Examples: The letters x, y, n, a, b, and c are used frequently to represent variables.

Term: A term is a numerical constant or the product (or quotient) of a numerical constant and one or more variables.

Examples: $3x$, $4x^2$, and $\frac{2a}{c}$

Expression: An algebraic expression is a combination of one or more terms. Terms in an expression are separated by either addition or subtraction signs.

Examples: $3xy$, $4ab - 5cd$, and $x^2 + x - 1$.

Coefficient: In the term $3xy$, the multiplier 3 is called a coefficient. In a simple term such as z, 1 is the coefficient.

Constant: A value that does not change.

Example: In the expression $x + 7$, the number 7 is a constant.

Monomial: A single term, such as $-6x$ or $2a^2$.

Polynomial: The general name for expressions with more than one term.

Binomial: A polynomial with exactly two terms.

Trinomial: A polynomial with exactly three terms.

OPERATIONS WITH ALGEBRAIC EXPRESSIONS

Working efficiently and confidently with algebraic expressions can save you time on the GRE. The operations shown here may be just one important step in solving a problem.

SUBSTITUTION

Substitution is a method used to evaluate an algebraic expression or to express an algebraic expression in terms of other variables.

Example: Evaluate $3x^2 - 4x$ when $x = 2$.

Replace every x in the expression with 2 and then carry out the designated operations.

Remember to follow the order of operations (PEMDAS).

$$\begin{aligned} 3x^2 - 4x &= 3(2)^2 - 4(2) \\ &= 3 \times 4 - 4 \times 2 \\ &= 12 - 8 \\ &= 4 \end{aligned}$$

Example: Express $\frac{a}{b - a}$ in terms of x and y if $a = 2x$ and $b = 3y$.

Here, replace every a with $2x$ and b with $3y$.

$$\frac{a}{b - a} = \frac{2x}{3y - 2x}$$

SYMBOLISM

You are familiar with the operation symbols $+$, $-$, $\times$, and $\div$, but you may also see some unfamiliar symbols in a GRE question. Symbols such as ♦, ❖, or ◘ may be used to describe an operation. These symbols need not confuse you; the question stem in these problems always tells you what a symbol represents. This type of problem may seem odd, but it is really a type of substitution problem.

Example: Let $x \blacklozenge$ be defined by the operation: $x \blacklozenge = \frac{1 - x}{x^2}$, where $x \neq 0$. Evaluate $(-2) \blacklozenge$.

It is helpful to say the meaning of the symbol to yourself as you work the substitution: $x \blacklozenge$ means *the quantity 1 minus the number is then divided by the square of the number.*

For $x = -2$, $x \blacklozenge = \frac{1 - x}{x^2} = \frac{1 - (-2)}{(-2)^2} = \frac{3}{(-2) \times (-2)} = \frac{3}{4}$

Operations with Polynomials

All of the laws of arithmetic operations, such as the commutative, associative, and distributive laws, also apply to polynomials. For polynomials, these laws also make it possible to combine like terms. **Like terms** contain the same variables, and the corresponding variables have the same exponents. For example, $5x^2$ and $-x^2$ are like terms, but $5x^2$ and $4x$ are *not* like terms. The laws for operations with polynomials may be applied one at a time, or they may be combined to simplify expressions.

Commutative law: $2x + 5y = 5y + 2x$ (for addition)

$5a \times 3b = 3b \times 5a$ (for multiplication)

Associative law: $2x - 3x + 5y + 2y$

$= (2x - 3x) + (5y + 2y)$ (for addition)

$= -x + 7y$ Combine like terms.

$(7w \times 2a) \times 3a$

$= 7w \times (2a \times 3a)$ (for multiplication)

$= 7w \times 6a^2$ Simplify.

$= 42wa^2$ Simplify.

Distributive law: $3a(2b - 5c) = (3a \times 2b) - (3a \times 5c)$

$= 6ab - 15ac$ Simplify.

The laws of operations are frequently used in combination to simplify expressions.

Example: $5x(y + 2) - xy + 2x$

$= 5xy + 10x - xy + 2x$ Distributive law

$= 5xy - xy + 10x + 2x$ Commutative law for addition

$= (5xy - xy) + (10x + 2x)$ Associative law for addition

$= 4xy + 12x$ Combine like terms.

The product of two binomials can be found by applying the distributive law twice. Each term in the first binomial is used as a multiplier of the second binomial.

Example: $(x + 5)(x - 2) = x(x - 2) + 5(x - 2)$
$= x^2 - 2x + 5x - 10$
$= x^2 + 3x - 10$

The mnemonic **FOIL** (**F**irst, **O**uter, **I**nner, **L**ast) describes the multiplication of one binomial by another binomial. The **F**irst terms in $(x + 5)(x - 2)$ are both x; the **L**ast terms are 5 and −2. The **O**uter terms in $(x + 5)(x - 2)$ are x and −2; the **I**nner terms are 5 and x.

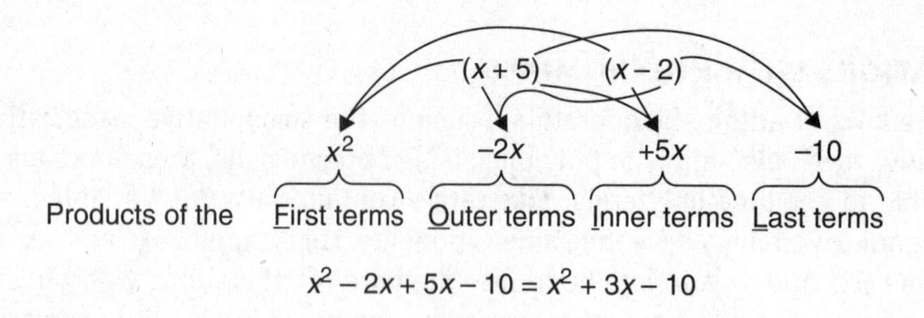

Factoring

Factoring a polynomial means expressing it as a product of two or more simpler expressions. When there is a monomial factor common to every term in the polynomial, it can be factored out using the distributive law. In the example below, $2a$ is the greatest common factor of $2a$ and $6ac$.

Example: $2a + 6ac = 2a(1 + 3c)$

It is helpful to be able to recognize several classic polynomial equations (Classic Quadratics) because they represent polynomials that can be factored. The factoring, in turn, can lead to simplifying expressions.

Difference of two perfect squares: The difference of two squares can be factored into a product: $a^2 - b^2 = (a - b)(a + b)$. In the example below, $9b^2$ and $4c^2$ are the "squares" and the subtraction sign between them indicates a "difference" is taken.

Example: $9b^2 - 4c^2 = (3b - 2c)(3b + 2c)$

Perfect square trinomials: Any polynomial of the form $a^2 + 2ab + b^2$ is equivalent to the square of a binomial. Notice that $(a + b)^2 = a^2 + 2ab + b^2$ (try FOIL). Factoring such a polynomial is just reversing this procedure.

Example: $x^2 + 6x + 9 = x^2 + 2(3)x + 3^2 = (x + 3)(x + 3)$

Polynomials of the form $a^2 - 2ab + b^2$ can also be factored into two identical binomials. Here, though, the binomial is the difference of two terms: $a^2 - 2ab + b^2 = (a - b)(a - b)$.

Example: $w^2 - 10w + 25 = x^2 - 2(5)x + 5^2 = (w - 5)(w - 5)$

In general, polynomials of the form $x^2 + bx + c$ can be factored into two binomials under these conditions:

- The product of the first terms in each binomial equals the first term of the polynomial.
- The product of the last terms of the binomials equals the third term of the polynomial.
- The sum of the remaining products equals the second term of the polynomial.

Example: $x^2 - 8x - 20 = (x - 10)(x + 2)$

The first term in the polynomial is x^2, so each binomial starts with x. The product of -10 and 2 is -20, the last term in the polynomial, and the sum of -10 and 2 is -8, the coefficient of the middle term. All three conditions for factoring are met. Factoring can be thought of as the FOIL method in reverse, so you can check your factoring by using FOIL to obtain the original polynomial.

OPERATIONS WITH ALGEBRAIC EXPRESSIONS EXERCISES

BASIC

1. If $a = -2$ and $c = 4$, what is the value of $4 - 2a + c$?

2. Factor the expression $5x^2 - 5$.

3. Factor the expression $x^2 - 10x + 25$.

4. If $c \square d = 2c + d - 1$, what is $3 \square 10$?

5. Simplify $(2a - b)(2a + b)$.

6. Which of the following are like terms: $6a$, $6b$, $-a$, $3a^2$?

7. Simplify the expression $-4x + 5 - x + 9$.

8. Factor $x^2 - xy + x$.

9. If $b \neq c$, simplify the expression $\frac{b^2 - c^2}{b - c}$.

10. If $n \bigstar p = \frac{p}{2n}$, what is $7 \bigstar 28$?

INTERMEDIATE

11. If $a = -2$, $b = 1$, and $c = 4$, what is the value of $\frac{-2(b + c)}{2c - a}$?

12. In the equation $mx + 5 = y$, m is a constant. If $x = 2$ when $y = 1$, what is the value of x when $y = -1$?

13. What is the coefficient of the a-term in the product of $(a - 7)(a + 3)$?

14. $5c^2 - 2b = c$, what is b in terms of c?

15. Factor $x^2 + x + \frac{1}{4}$.

16. Simplify $xyz\left(\frac{1}{xy} + \frac{1}{yz} + \frac{1}{xz}\right)$.

17. Express $\frac{2a}{b - 3a}$ in terms of x and y if $a = -x$ and $b = 5y$.

18. If $y \neq z$, what is the simplest form of $\frac{xy - zx}{z - y}$?

19. Simplify $(a^2 + b)^2 - (a^2 - b)^2$.

20. If $x = 0.5$ and $y = -2$, what is the value of $10(2x - y)$?

ADVANCED

21. If $m ❖ n$ is defined by the equation $m ❖ n = \frac{m^2 - n + 1}{mn}$ for all nonzero m and n, then what is $3 ❖ 1$?

22. In the equation $y = -2x + b$, b is a constant. If $y = -5$ when $x = -1$, what is the value of y when $x = 0$?

23. Which of the following are like terms: ab, $-bc$, $\frac{1}{2}bc$, ac^2?

24. What is the common monomial factor in the expression $4c^3d - c^2d^2 + 2cd$?

25. If $p \neq q$, simplify the expression $\frac{p^2 - q^2}{-5(q - p)}$.

26. What is the coefficient of the x^2-term in the product of $(x + 1)(x + 2)(x - 1)$?

27. Factor $x^3 - \frac{2}{3}x^2 + \frac{1}{9}x$ completely.

28. If $s ❖ r = \frac{2s^3}{r^2}$, what is $-3 ❖ 6$?

29. If $m = 0.5$ and $n = -0.25$, what is the value of $(m - n)^2$?

30. What is the value of a if $ab + ac = -21$ and $b + c = 3$?

OPERATIONS WITH ALGEBRAIC EXPRESSIONS ANSWER KEY

BASIC

1. 12

In this problem, plug in the values for the given variables.

$$4 - 2a + c = 4 - 2(-2) + 4 = 4 + 4 + 4 = 12$$

Be careful with the signs of the numbers; $4 - 2(-2)$ is the same as $4 + 4$.

2. $5(x - 1)(x + 1)$

Factor the common factor, 5, and then factor the difference of squares.

$$5x^2 - 5 = 5(x^2 - 1) = 5(x - 1)(x + 1)$$

3. $(x - 5)(x - 5)$ or $(x - 5)^2$

This is one of the Classic Quadratics. $x^2 - 10x + 25 = (x - 5)(x - 5)$.

4. 15

The symbol for the operation says to multiply c by 2, add d, and subtract 1.

$$3\square 10 = 2(3) + 10 - 1 = 6 + 10 - 1 = 15$$

5. $4a^2 - b^2$

This is one of the Classic Quadratics, the difference of two squares, in which $(x + y)(x - y) = x^2 - y^2$. If you recognized this as the difference of two squares, you could have skipped the whole FOIL process. However, using FOIL to simplify:

$$(2a - b)(2a + b) = (2a)(2a) + (2a)b - (2a)b - (b)(b) = 4a^2 - b^2$$

6. $6a, -a$

Only these two terms have the same variable with the same exponent.

7. $-5x + 14$

Combine the x-terms and then combine the constants:

$$-4x + 5 - x + 9 = -5x + 14$$

8. $x(x - y + 1)$

The common factor is x. Use the distributive property to factor out x.

9. $b + c$

Learning to recognize the Classic Quadratics really pays off here. The numerator is a difference of squares, so write the two binomials in the numerator. Then cancel $(b - c)$ in both the numerator and the denominator.

$$\frac{b^2 - c^2}{b - c} = \frac{\cancel{(b - c)}(b + c)}{\cancel{b - c}} = b + c$$

10. 2

The symbol for the operation says to divide p by 2 times n. Substitute for the variables and calculate:

$$\frac{p}{2n} = \frac{28}{2(7)} = \frac{28}{14} = 2$$

Intermediate

11. –1

Substitute for the variables. Evaluate the numerator and the denominator and then write the fraction in its simplest form.

$$\frac{-2(b+c)}{2c-a} = \frac{-2(1+4)}{2(4)-(-2)} = \frac{-2(5)}{8+2} = \frac{-10}{10} = -1$$

12. 3

Substitute the given values for x and y; solve for m.

$$\begin{aligned} mx + 5 &= y \\ m(2) + 5 &= 1 \\ 2m + 5 - 5 &= 1 - 5 \\ 2m &= -4 \\ m &= \frac{-4}{2} \\ m &= -2 \end{aligned}$$

Now that you know the value of m, you can solve for x when $y = -1$:

$$\begin{aligned} -2x + 5 &= y \\ -2x + 5 &= -1 \\ -2x &= -6 \\ x &= 3 \end{aligned}$$

13. –4

Multiply using FOIL: $(a - 7)(a + 3) = a^2 + 3a - 7a - 21 = a^2 + -4a - 21$. -4 is the coefficient before the a-term.

14. $\frac{5c^2 - c}{2}$

Isolate the variable b in the equation $5c^2 - 2b = c$.

$$\begin{aligned} 5c^2 - 2b &= c \\ 5c^2 - 2b + 2b - c &= c - c + 2b \\ 5c^2 - c &= 2b \\ \frac{5c^2 - c}{2} &= b \end{aligned}$$

15. $\left(x + \frac{1}{2}\right)\left(x + \frac{1}{2}\right)$

This is a Classic Quadratic. The last term in $x^2 + x + \frac{1}{4}$ is the product $\frac{1}{2} \times \frac{1}{2} = \frac{1}{4}$. The middle term is $(2)\frac{1}{2}x = x$.

16. $z + x + y$

Distribute the term *xyz* to each fraction inside the parentheses. Then simplify each term.

$$xyz\left(\frac{1}{xy} + \frac{1}{yz} + \frac{1}{xz}\right) = \frac{xyz}{xy} + \frac{xyz}{yz} + \frac{xyz}{xz} = z + x + y$$

17. $\frac{-2x}{5y + 3x}$

Substitute for a and b in the expression: $\frac{2a}{b - 3a} = \frac{2(-x)}{5y - 3(-x)} = \frac{-2x}{5y + 3x}$.

18. $-x$

$$\frac{xy - zx}{z - y} = \frac{x(y - z)}{z - y} = \frac{-x\cancel{(z - y)}}{\cancel{z - y}} = -x$$

When two binomials are the opposite of each other, such as $(y - z)$ and $(z - y)$ here, factor out -1 from one of the binomials to simplify the expression.

19. $4a^2b$

Use FOIL to multiply each Classic Quadratic expression and then combine terms.

$$\begin{aligned}(a^2 + b)^2 - (a^2 - b)^2 &= (a^4 + 2a^2b + b^2) - (a^4 - 2a^2b + b^2)\\ &= a^4 + 2a^2b + b^2 - a^4 + 2a^2b - b^2\\ &= 4a^2b\end{aligned}$$

20. 30

Substitute the given values; watch signs as you evaluate inside the parentheses first.

$$10(2x - y) = 10[2(0.5) - (-2)] = 10[1 - (-2)] = 10(1 + 2) = 30$$

ADVANCED

21. 3

$$3 ❖ 1 = \frac{3^2 - 1 + 1}{3 \times 1} = \frac{9 - 1 + 1}{3} = \frac{9}{3} = 3$$

22. −7

Substitute the given values for x and y in the equation $y = -2x + b$ and solve for b:

$$y = -2x + b;\ -5 = -2(-1) + b;\ -5 = 2 + b;\ b = -7$$

So, $y = -2x - 7$.

Substitute 0 for x:

$$y = -2(0) - 7$$
$$y = 0 - 7$$
$$y = -7$$

23. $-bc, \frac{1}{2}bc$

Only these terms have the same variables raised to the same power.

24. cd

Every term contains the common monomial factor cd:

$$4c^3d - c^2d^2 + 2cd = cd(4c^2 - cd + 2)$$

25. $\frac{p+q}{5}$

Factor the difference of squares in the numerator; then factor -1 in the denominator. Simplify as shown.

$$\frac{p^2 - q^2}{-5(q-p)} = \frac{(p-q)(p+q)}{5(-1)(q-p)} = \frac{\cancel{(p-q)}(p+q)}{5\cancel{(p-q)}} = \frac{p+q}{5}$$

26. 2

Use FOIL twice to simplify the expression. Start with $(x+1)$ and $(x-1)$ so that you'll end up with the difference of two squares.

$$(x+1)(x+2)(x-1) = (x+2)(x+1)(x-1) = (x+2)(x^2-1) = x^3 + 2x^2 - x - 2$$

The coefficient of the x^2 term is 2.

27. $x\left(x - \frac{1}{3}\right)\left(x - \frac{1}{3}\right)$

First factor out the x, which leaves you with a Classic Quadratic, a perfect square trinomial. Then factor the perfect square:

$$x^3 - \frac{2}{3}x^2 + \frac{1}{9}x = x\left(x^2 - \frac{2}{3}x + \frac{1}{9}\right) = x\left(x - \frac{1}{3}\right)\left(x - \frac{1}{3}\right)$$

28. $-\frac{3}{2}$

$$\text{If } s ❖ r = \frac{2s^3}{r^2},\ -3 ❖ 6 = \frac{2(-3)^3}{6^2} = \frac{2(-27)}{36} = -\frac{27}{18} = -\frac{3}{2}.$$

29. 0.5625

If $m = 0.5$ and $n = -0.25$, $(m-n)^2 = [0.5 - (-0.25)]^2 = (0.5 + 0.25)^2 = (0.75)^2 = 0.5625$.

30. −7

If $ab + ac = -21$, $a(b+c) = -21$. If $b + c = 3$, then $a(3) = -21$ and $a = -7$.

RULES OF EXPONENTS

The rules of exponents for working with numbers also apply to algebraic expressions. These rules are used to simplify expressions or to make two expressions look more alike. In the expression x^a, the **base** is x and the **exponent** is a. An expression such as x^a is sometimes referred to as a **power.** Consider the rules of exponents, where the bases x and y are nonzero real numbers and the exponents a and b are integers. Note that these restrictions apply to all the rules shown in the table.

Statement of Rule	Meaning and Examples
$x^{-a} = \frac{1}{x^a}$	x^{-a} and x^a are reciprocals of each other. *Examples:* $2^{-3} = \frac{1}{2^3}$, $4^2 = \frac{1}{4^{-2}}$, and $x^{-3} = \frac{1}{x^3}$
$(x^a)(x^b) = x^{a+b}$	When the bases are the same in a product, add the exponents. *Examples:* $(5^3)(5^1) = 5^4 = 625$ and $(x^2)(x^3) = x^5$
$\frac{x^a}{x^b} = x^{a-b}$	When the bases are the same in a quotient, subtract the exponents. *Examples:* $\frac{3^3}{3^{-1}} = 3^{3-(-1)} = 3^4 = 81$ and $\frac{m^2}{m^3} = m^{2-3} = m^{-1} = \frac{1}{m}$
$x^0 = 1$	Any nonzero quantity to the zero power equals 1. However, 0^0 is not defined. *Examples:* $7^0 = 1$, $(-6)^0 = 1$, $(2n)^0 = 1$
$(xy)^a = (x^a)(y^a)$	A product to a power can be written as individual factors to the same power. *Examples:* $(2^2)(5^2) = (10)^2 = 100$ and $(2mn)^3 = (2^3)(m^3)(n^3) = 8m^3n^3$
$\left(\frac{x}{y}\right)^a = \frac{x^a}{y^a}$	A quotient to a power can be written as a numerator and a denominator to the same power. *Examples:* $\left(\frac{4}{5}\right)^3 = \frac{4^3}{5^3}$ and $\left(\frac{3m}{n}\right)^{-2} = \frac{(3m)^{-2}}{n^{-2}} = \frac{n^2}{(3m)^2} = \frac{n^2}{9m^2}$
$(x^a)^b = x^{ab}$	To raise a power to another power, multiply the exponents. *Examples:* $((-3)^2)^3 = (-3)^{2 \times 3} = (-3)^6$ and $(m^4)^2 = m^8$

It is a good idea to become familiar with the algebraic form of the exponent rules as well as the wording of the rules. That way, you will not be tempted to simplify expressions incorrectly.

RULES OF EXPONENTS EXERCISES

BASIC

Use the rules of exponents to simplify the following. Note: None of the variables are equal to zero.

1. $2^3 \times 2^2$
2. $(3 \times 2)^2$
3. $(rs)^4$
4. $(5b)^0$
5. $\left(\frac{c}{d}\right)^6$
6. $a^2 \times a^8$
7. $(t^4)^3$
8. $(-1)^0$
9. $(28)^1$
10. $\frac{b^7}{b^6}$

INTERMEDIATE

Use the rules of exponents to simplify the following.

11. $(-3)^2 \times (-3)^{-1}$
12. $\frac{k^3 \times k^4}{k^2}$
13. $6^3 \times 6^{-2}$
14. $\frac{a^{-1}}{a^5}$
15. $(2a)^4$
16. $(ab^2)^3$
17. $(3b \times 4b^2)^1$
18. $\frac{5^1 - 2^2}{3 \times 3^0}$
19. $(3b^0)^4$
20. $j^3k^2\, j^4k$

ADVANCED

Use the rules of exponents to simplify the following.

21. $\left(\frac{5a}{2b}\right)^2$
22. $\left(\frac{-3}{4st}\right)^3$
23. $\frac{(2b)^{-2}}{(3c)^{-1}}$
24. $\left(\frac{5a}{b}\right)^2 \times \frac{ab}{3}$
25. $\left(\frac{2x}{5}\right)^2 \times \frac{25x}{4}$
26. $\frac{4s^2t}{tv} \times (2v)^3$
27. $\left(\frac{3x}{4}\right)^2 + \left(\frac{y^2}{2}\right)^4$
28. $\frac{x^4y^{-2}}{x^{-3}y^4}$
29. $\frac{r^0s^4}{t} \div \left(\frac{3s}{t}\right)^2$
30. $\left(\frac{-3}{x^2}\right)^3 \times (6xy)^0 \times \left(\frac{x^5}{9}\right)^1$

RULES OF EXPONENTS ANSWER KEY

Basic

1. 32

The powers 2^3 and 2^2 have the same base, 2. To multiply, add the exponents.

$$2^3 \times 2^2 = 2^{3+2} = 2^5 = 32$$

2. 36

PEMDAS [or the Order of Operations] says to do parentheses first, then exponents:

$$(3 \times 2)^2 = 6^2 = 36$$

3. $r^4 s^4$

Apply the exponent 4 to each factor inside the parentheses. (Remember, *rs* means $r \times s$.)

$$(rs)^4 = r^4 s^4$$

4. 1

Any number or expression raised to the 0 power is 1.

$$(5b)^0 = 1$$

5. $\frac{c^6}{d^6}$

To raise a fraction to the sixth power, raise the numerator and denominator each to the sixth power.

$$\left(\frac{c}{d}\right)^6 = \frac{c^6}{d^6}$$

6. a^{10}

The expressions a^2 and a^8 have the same base (a). To multiply the expressions, add the exponents.

$$a^2 \times a^8 = a^{2+8} = a^{10}$$

7. t^{12}

To raise a power to an exponent, multiply the exponents.

$$(t^4)^3 = t^{4 \times 3} = t^{12}$$

8. 1

Any number or expression raised to the 0 power is 1.

$$(-1)^0 = 1$$

9. 28

Any number or expression to the first power is equal to that number or expression.

$$(28)^1 = 28$$

10. *b*

The expressions in the numerator and the denominator have the same base (b). To divide, subtract the exponent in the denominator from the exponent in the numerator. Keep the same base.

$$\frac{b^7}{b^6} = b^{7-6} = b^1 = b$$

Intermediate

11. −3

The powers being multiplied have the same base, −3. To multiply, add the exponents. Then simplify the power.

$$(-3)^2 \times (-3)^{-1} = (-3)^{2+(-1)} = (-3)^1 = -3$$

12. k^5

First, simplify the numerator using the rule for multiplying powers with the same base. Then use the rule for dividing powers with the same base.

$$\frac{k^3 \times k^4}{k^2} = \frac{k^{3+4}}{k^2} = \frac{k^7}{k^2} = k^{7-2} = k^5$$

13. 6

The powers 6^3 and 6^{-2} have the same base. To multiply the powers, add the exponents.

$$6^3 \times 6^{-2} = 6^{3+(-2)} = 6^1 = 6$$

14. $\frac{1}{a^6}$ or a^{-6}

To divide, subtract the exponents.

$$\frac{a^{-1}}{a^5} = a^{-1-5} = a^{-6} = \frac{1}{a^6}$$

15. $16a^4$

The expression $2a$ is the product of 2 and a. To raise a product to a power, raise each factor to the power.

$$(2a)^4 = 2^4 \times a^4 = 16a^4$$

16. a^3b^6

The expression ab^2 is the product of a and b^2. To raise a product to a power, raise each factor to the power.

$$(ab^2)^3 = a^3 \times (b^2)^3 = a^3b^{2\times3} = a^3b^6$$

17. $12b^3$

Simplify inside parentheses first. Remember that the variable *b* has an implied power of 1.

$$(3b \times 4b^2)^1 = (3 \times 4 \times b \times b^2)^1 = (12b^{1+2})^1 = (12b^3)^1$$

Any expression raised to the power of 1 is equal to that expression.

$$(12b^3)^1 = 12b^3$$

18. $\frac{1}{3}$

Apply the rules for powers of 0 and 1.

$$\frac{5^1 - 2^2}{3 \times 3^0} = \frac{5 - 2^2}{3 \times 1}$$

Simplify the remaining power.

$$\frac{5 - 2^2}{3 \times 1} = \frac{5 - 4}{3 \times 1}$$

Simplify the numerator. Simplify the denominator.

$$\frac{5 - 4}{3 \times 1} = \frac{1}{3}$$

19. 81

The exponent 0 applies only to the variable *b* and NOT to the coefficient 3.

Simplify inside parentheses. Then, apply the exponent 4.

$$(3b^0)^4 = (3 \times 1)^4 = 3^4 = 81$$

20. j^7k^3

Use properties of multiplication to change the order of the factors and group like bases together. Remember that any number or variable without an exponent has an implied exponent of 1. To multiply powers with the same base, add the exponents.

$$j^3\,k^2\,j^4\,k = j^3\,j^4\,k^2\,k^1 = j^{3+4}\,k^{2+1} = j^7\,k^3$$

Advanced

21. $\frac{25a^2}{4b^2}$

Raise the numerator and the denominator to the second power. Then, apply the exponent 2 to each factor inside parentheses.

$$\left(\frac{5a}{2b}\right)^2 = \frac{(5a)^2}{(2b)^2} = \frac{5^2a^2}{2^2b^2} = \frac{25a^2}{4b^2}$$

22. $\dfrac{-27}{64s^3t^3}$

Raise the numerator and the denominator to the third power. Then, apply the exponent 3.

$$\left(\frac{-3}{4st}\right)^3 = \frac{(-3)^3}{(4st)^3} = \frac{-3\times-3\times-3}{4^3s^3t^3} = \frac{-27}{64s^3t^3}$$

23. $\dfrac{3c}{4b^2}$

To apply a negative exponent, which means the reciprocal of a positive one, move the base to the other "level" of the fraction and make the exponent positive. Then, simplify using the rules of exponents.

$$\frac{(2b)^{-2}}{(3c)^{-1}} = \frac{(3c)^1}{(2b)^2} = \frac{3c}{2^2b^2} = \frac{3c}{4b^2}$$

24. $\dfrac{25a^3}{3b}$

Apply the exponent 2:

$$\left(\frac{5a}{b}\right)^2 \times \frac{ab}{3} = \frac{(5a)^2}{b^2} \times \frac{ab}{3} = \frac{5^2a^2}{b^2} \times \frac{ab}{3} = \frac{25a^2}{b^2} \times \frac{ab}{3}$$

Multiply fractions by multiplying across the numerator and across the denominator.

$$\frac{25a^2}{b^2} \times \frac{ab}{3} = \frac{(25a^2)(a^1b^1)}{(b^2)(3)} = \frac{25a^{2+1}b^1}{3b^2} = \frac{25a^3b^1}{3b^2}$$

Simplify further by dividing powers of the same base, b.

$$\frac{25a^3b^1}{3b^2} = \frac{25a^3b^{1-2}}{3} = \frac{25a^3b^{-1}}{3} = \frac{25a^3}{3b}$$

25. x^3

Apply the exponent 2. Then, divide out common factors before multiplying across.

$$\left(\frac{2x}{5}\right)^2 \times \frac{25x}{4} = \frac{\cancel{4}x^2}{\cancel{25}} \times \frac{\cancel{25}x}{\cancel{4}} = \frac{x^2 \times x}{1} = \frac{x^{2+1}}{1} = x^3$$

26. $32s^2v^2$

Apply the exponent 3 to $(2v)$. Then, divide out common factors before multiplying across.

$$\frac{4s^2t}{tv} \times (2v)^3 = \frac{4s^2t}{tv} \times \frac{8v^3}{1} = \frac{32s^2\cancel{t^1}v^3}{\cancel{t^1}v^1} = 32s^2v^{3-1} = 32s^2v^2$$

27. $\dfrac{9x^2 + y^8}{16}$

Raising each expression inside parentheses to the given power results in a common denominator of 16. To add fractions with the same denominator, add the numerators.

$$\left(\frac{3x}{4}\right)^2 + \left(\frac{y^2}{2}\right)^4 = \frac{(3x)^2}{4^2} + \frac{(y^2)^4}{2^4} = \frac{9x^2}{16} + \frac{y^8}{16} = \frac{9x^2 + y^8}{16}$$

28. $\dfrac{x^7}{y^6}$

To apply a negative exponent that appears in the numerator, move the base into the denominator and use a positive exponent. To apply a negative exponent that appears in the denominator, move the base into the numerator and use a positive exponent.

$$\frac{x^4y^{-2}}{x^{-3}y^4} = \frac{x^4x^3}{y^2y^4} = \frac{x^{4+3}}{y^{2+4}} = \frac{x^7}{y^6}$$

29. $\dfrac{s^2t}{9}$

Apply the exponent 2 to the expression $\frac{3s}{t}$. Then, divide fractions by multiplying the first fraction by the reciprocal of the second fraction.

$$\frac{r^0s^4}{t} \div \left(\frac{3s}{t}\right)^2 = \frac{s^4}{t} \div \frac{9s^2}{t^2} = \frac{s^4}{t} \times \frac{t^2}{9s^2} = \frac{s^{4-2}t^{2-1}}{9} = \frac{s^2t}{9}$$

30. $\dfrac{-3}{x}$

Apply the rules for exponents 0 and 1.

$$\left(\frac{-3}{x^2}\right)^3 \times (6xy)^0 \times \left(\frac{x^5}{9}\right)^1 = \left(\frac{-3}{x^2}\right)^3 \times 1 \times \frac{x^5}{9} = \left(\frac{-3}{x^2}\right)^3 \times \frac{x^5}{9}$$

Apply the exponent 3.

$$\left(\frac{-3}{x^2}\right)^3 \times \frac{x^5}{9} = \frac{(-3)^3}{(x^2)^3} \times \frac{x^5}{9} = \frac{-27}{x^{2\times 3}} \times \frac{x^5}{9} = \frac{-27}{x^6} \times \frac{x^5}{9}$$

Multiply across.

$$\frac{-27}{x^6} \times \frac{x^5}{9} = \frac{-27x^5}{9x^6}$$

Simplify.

$$\frac{\overset{-3}{\cancel{-27}}\, x^5}{\underset{1}{\cancel{9}}\, x^6} = \frac{-3x^5}{x^6} = -3x^{5-6} = -3x^{-1} = \frac{-3}{x}$$

SOLVING LINEAR EQUATIONS

An **equation** is an algebraic sentence that says that two expressions are equal to each other. The two expressions consist of numbers, variables, and arithmetic operations to be performed on these numbers and variables. To **solve** for a variable, you can manipulate the equation until you have isolated that variable on one side of the equal sign, leaving any numbers or other variables on the other side. Of course, you must be careful to manipulate the equation only in accordance with the equality postulate: whenever you perform an operation on one side of the equation, you must perform the same operation on the other side. Otherwise, the two sides of the equation will no longer be equal.

Linear Equations with One Variable

A **linear** or first-degree equation is an equation in which all the variables are raised to the first power (there are no squares or cubes). In order to solve such an equation, we'll perform operations on both sides of the equation in order to get the variable we're solving for all alone on one side. The operations that can be performed without upsetting the balance of the equation are addition and subtraction, and multiplication or division by a number other than 0. Typically, at each step in the process, you'll need to use the reverse of the operation that's being applied to the variable in order to isolate the variable.

The equation $n + 6 = -10$ is a linear equation with one variable, in which 6 is added to the variable to get -10. Use the reverse of addition to isolate n; subtract 6 from both sides of the equation:

$$\begin{aligned} n + 6 &= -10 \\ n + 6 - 6 &= -10 - 6 \\ n &= -16 \end{aligned}$$

Alternate version of example above:

$$\begin{array}{r} n + 6 = -10 \\ -6 \quad\;\; -6 \\ \hline n = -16 \end{array}$$

Here's an example of a linear equation in which the variable appears on both sides of the equation.

Example: If $4x - 7 = 2x + 5$, what is x?

1. Get all the terms with the variable on one side of the equation. Combine like terms.

$$4x - 7 = 2x + 5$$
$$4x - 2x - 7 = 2x - 2x + 5$$
$$2x - 7 = 5$$

2. Get all constant terms on the other side of the equation.

$$2x - 7 + 7 = 5 + 7$$
$$2x = 12$$

3. Isolate the variable by dividing both sides by its coefficient.

$$\frac{2x}{2} = \frac{12}{2}$$
$$x = 6$$

You can easily check your work when solving this kind of equation. The answer represents the value of the variable that makes the equation true. Therefore, to check that it's correct, substitute the value found for the variable into the original equation. If the equation holds true, you've found the correct answer. In the given example, the answer was $x = 6$. Replacing x with 6 in the original equation gives:

$$4x - 7 = 2x + 5$$
$$4(6) - 7 = 2(6) + 5$$
$$24 - 7 = 12 + 5$$
$$17 = 17 \checkmark$$

Substituting 6 for x gives a statement that is true: 17 equals 17. So the answer is correct.

Equations with fractional coefficients can be solved using the same approach—isolate the variable on one side and the constants on the other—but first it is best to multiply to get rid of the fraction format. This will give you an equivalent equation to solve, but you will not have to deal with the fractions. Let's see how to solve such a problem.

Example: If $\frac{x - 2}{3} + \frac{x - 4}{10} = \frac{x}{2}$, what is x?

1. Multiply both sides of the equation by the lowest common denominator (LCD). Here the LCD is 30.

$$30\left(\frac{x - 2}{3}\right) + 30\left(\frac{x - 4}{10}\right) = 30\left(\frac{x}{2}\right)$$
$$10(x - 2) + 3(x - 4) = 15x$$

2. Clear parentheses using the distributive property and combine like terms.

$$10x - 20 + 3x - 12 = 15x$$
$$13x - 32 = 15x$$

3. Isolate the variable. Again, combine like terms.

$$-32 = 15x - 13x$$
$$-32 = 2x$$

4. Divide both sides by the coefficient of the variable.

$$x = \frac{-32}{2} = -16$$

Literal Equations

If a problem involves more than one variable, we cannot find a specific value for a variable; we can only solve for one variable in terms of the others. To do this, try to get the desired variable alone on one side and all the other variables on the other side.

Example: In the formula $V = \frac{PN}{R + NT}$, solve for N in terms of P, R, T, and V.

1. Clear denominators by cross multiplying.

$$V = \frac{PN}{R + NT}$$

$$V(R + NT) = PN$$

2. Remove parentheses by distributing.

$$VR + VNT = PN$$

3. Put all terms containing N on one side and all other terms on the other side.

$$VNT - PN = -VR$$

4. Factor out the common factor N.

$$N(VT - P) = -VR$$

5. Divide by the coefficient of N to get N alone.

$$N = \frac{-VR}{VT - P}$$

Note: You can reduce the number of negative signs in the fraction by multiplying *both* the numerator and the denominator by –1.

$$N = \frac{VR}{P - VT}$$

Rearranging the terms of an equation to isolate one variable is referred to as *solving a literal equation.* You could also solve the same equation for a different variable.

Simultaneous Equations

Earlier, you solved an equation for one variable and were able to find a numerical value for that variable. In the previous example, you were not able to find a numerical value for N because the equation contained variables other than just N. In general, if you want to find numerical values for all your variables, you will need as many distinct equations as you have variables. Let's say, for example, that you have one equation with two variables: $x - y = 7$. There are an infinite number of solution sets to this equation: for example, $x = 8$ and $y = 1$ (since $8 - 1 = 7$), or $x = 9$ and $y = 2$ (since $9 - 2 = 7$), and so on.

If you are given two different equations with the same two variables, you can solve the equations simultaneously to obtain a unique solution set. Isolate the variable in one equation, then plug that expression into the other equation. This method, called *substitution*, works well when the coefficient of one variable is 1.

Example: Find the values of m and n if $2m + 5n = 7$ and $m + 4n = 2$.

1. Isolate the variable m in the second equation.

$$m + 4n = 2$$
$$m = -4n + 2$$

2. Plug the expression for m into the first equation and solve for n.

$$2m + 5n = 7$$
$$2(-4n + 2) + 5n = 7$$
$$-8n + 4 + 5n = 7$$
$$-3n + 4 = 7$$
$$-3n = 3$$
$$n = -1$$

3. Plug the value for n into the second equation to find the value of m.

$$m + 4n = 2$$
$$m + 4(-1) = 2$$
$$m = 6$$

Be careful to identify the results correctly. The values that make the simultaneous equations true are $m = 6$ and $n = -1$.

It is also possible to solve two equations with two variables using the *combination* method. Let's take a look at how that works.

For the simultaneous equations $8x + 4y = 140$ and $x + 2y = 55$, you cannot eliminate a variable by simply adding or subtracting the equations as they are given. The coefficients of the respective variables are not opposites of each other. But if you multiply both sides of one equation by -2 to make one set of coefficients opposites, then you can solve the system.

Example: Find the values of x and y if $8x + 4y = 140$ and $x + 2y = 55$.

1. Write one equation under the other.

$$8x + 4y = 140$$
$$x + 2y = 55$$

2. Multiply the second equation by -2 to create opposite coefficients for y.

$$8x + 4y = 140$$
$$-2(x + 2y) = -2(55)$$

3. Combine the equations by adding the second one to the first.

$$8x + 4y = 140$$
$$+ (-2x - 4y = -110)$$
$$6x = 30$$

4. Solve for x.

$$x = 5$$

5. To find the value of y, substitute for x in either equation.

$$x + 2y = 55$$
$$5 + 2y = 55$$
$$2y = 50$$
$$y = 25$$

The values that make the simultaneous equations true are $x = 5$ and $y = 25$.

SOLVING LINEAR EQUATIONS EXERCISES

Basic

Solve each of the following equations for the variable.

1. $2r = 10$
2. $t + 4 = 16$
3. $k - 7 = 4$
4. $\frac{w}{6} = 9$
5. $3x = -12$
6. $2m - 7 = 5$
7. $\frac{y}{2} + 3 = 9$
8. Isolate T: $30R = 4T + B$

Solve each of the following systems of equations for *x* and *y*.

9. $3x = 9$
   $2x + 4y = 22$
10. $x + y = 10$
    $x - y = 2$

Intermediate

Solve each of the following equations for the variable.

11. $5x + 3 = 4x - 6$
12. $3b = 6b - 2b + 4$
13. $96 - 2b = -18b$
14. $3(x + 2) = 5x + 1$
15. $\frac{2x}{2} + \frac{x}{4} = 10$
16. $\frac{2b}{5} - 6 = 10$
17. Isolate b: $A = \frac{bh}{2}$
18. Isolate t: $A = P(1 + rt)$

Solve each of the following systems of equations for *x* and *y*.

19. $y = 3x$
    $x + y = 8$
20. $y - 2x = -5$
    $y - x = -3$

Advanced

Solve each of the following equations for the variable.

21. $3(x + 2) = 14 - 2(3 - 2x)$
22. $5(6 - 3b) = 3b + 3$
23. $6w - 4 - 3w = 8 - 12 - 4w$
24. $\frac{r}{6} - \frac{3r}{5} = \frac{1}{2}$
25. $\frac{2}{3}j - \frac{1}{2} = \frac{1}{6}j + \frac{11}{2}$
26. Isolate F: $C = \frac{5}{9}(F - 32)$
27. Isolate b: $A = \frac{1}{2}(a + b)h$

Solve each of the following systems of equations for both variables.

28. $x - 5y = 2$
    $2x + y = 4$
29. $2c = 14 + 4g$
    $3g = 2 - c$
30. $4x + 3y = 10$
    $3x + 5y = 13$

SOLVING LINEAR EQUATIONS ANSWER KEY

Basic

1. 5

To solve for r, divide both sides by 2.

$$2r = 10$$
$$\frac{2r}{2} = \frac{10}{2}$$
$$r = 5$$

2. 12

To solve for t, subtract 4 from both sides.

$$t + 4 = 16$$
$$t + 4 - 4 = 16 - 4$$
$$t = 12$$

3. 11

To solve for k, add 7 to both sides.

$$k - 7 = 4$$
$$k - 7 + 7 = 4 + 7$$
$$k = 11$$

4. 54

To solve for w, multiply both sides by 6.

$$\frac{w}{6} = 9$$
$$\frac{w}{6} \times 6 = 9 \times 6$$
$$w = 54$$

5. −4

To solve for x, divide both sides by 3.

$$3x = -12$$
$$\frac{3x}{3} = \frac{-12}{3}$$
$$x = -4$$

6. 6

To solve for m, first isolate the term $2m$ by adding 7 to both sides. Then, isolate m by dividing both sides by 2.

$$2m - 7 = 5$$
$$2m - 7 + 7 = 5 + 7$$
$$2m = 12$$
$$\frac{2m}{2} = \frac{12}{2}$$
$$m = 6$$

7. 12

To solve for y, first isolate the term $\frac{y}{2}$ by subtracting 3 from both sides. Then, isolate y by multiplying both sides by 2.

$$\frac{y}{2} + 3 = 9$$
$$\frac{y}{2} + 3 - 3 = 9 - 3$$
$$\frac{y}{2} = 6$$
$$\frac{y}{2} \times 2 = 6 \times 2$$
$$y = 12$$

8. $T = \frac{30R - B}{4}$

To solve for T, first isolate the term $4T$ on the right side by subtracting B from both sides. Then, isolate T by dividing both sides by 4.

$$30R = 4T + B$$
$$30R - B = 4T + B - B$$
$$30R - B = 4T$$
$$\frac{30R - B}{4} = \frac{4T}{4}$$
$$\frac{30R - B}{4} = T$$

9. $x = 3$ and $y = 4$

The equation $3x = 9$ has only one variable and can therefore be solved independently. Divide both sides by 3.

$$3x = 9$$
$$\frac{3x}{3} = \frac{9}{3}$$
$$x = 3$$

Now use $x = 3$ to solve $2x + 4y = 22$. Substitute 3 for x and solve for y.

$$2x + 4y = 22$$
$$2(3) + 4y = 22$$
$$6 + 4y = 22$$
$$6 - 6 + 4y = 22 - 6$$
$$4y = 16$$
$$\frac{4y}{4} = \frac{16}{4}$$
$$y = 4$$

10. $x = 6$ and $y = 4$

In the given equations, the y terms have opposite coefficients. So, combine the equations.

$$x + y = 10$$
$$+(x - y = 2)$$
$$\overline{\quad 2x = 12 \quad}$$
$$\frac{2x}{2} = \frac{12}{2}$$
$$x = 6$$

Now use $x = 6$ to solve for y. Choose either equation and substitute 6 for x to solve for y. We'll choose the first one, but we'd get the same solution with the second.

$$x + y = 10$$
$$6 + y = 10$$
$$y = 4$$

Intermediate

11. $x = -9$

Solve by moving all x terms to the left side and all constant terms to the right side.

$$5x + 3 = 4x - 6$$
$$5x - 4x + 3 = 4x - 4x - 6$$
$$x + 3 = -6$$
$$x + 3 - 3 = -6 - 3$$
$$x = -9$$

12. $b = -4$

Combine like terms on the right side. Then, move all b terms to the left side.

$$3b = 6b - 2b + 4$$
$$3b = 4b + 4$$
$$3b - 4b = 4b - 4b + 4$$
$$-b = 4$$
$$\frac{-b}{-1} = \frac{4}{-1}$$
$$b = -4$$

13. $b = -6$

Collect all b terms on the right side and all constant terms on the left side. Then, solve for b by dividing to eliminate the coefficient -16.

$$96 - 2b = -18b$$
$$96 - 2b + 2b = -18b + 2b$$
$$96 = -16b$$
$$\frac{96}{-16} = \frac{-16b}{-16}$$
$$-6 = b$$

14. $x = \frac{5}{2}$

First, simplify the left side using the distributive property. Then, collect x terms on one side and constants on the other side.

$$\begin{aligned} 3(x+2) &= 5x+1 \\ 3x+6 &= 5x+1 \\ 3x-5x+6 &= 5x-5x+1 \\ -2x+6 &= 1 \\ -2x+6-6 &= 1-6 \\ -2x &= -5 \\ \frac{-2x}{-2} &= \frac{-5}{-2} \\ x &= \frac{5}{2} \end{aligned}$$

15. $x = 8$

Eliminate fractions by multiplying the entire equation by the lowest common multiple of 2 and 4, which is 4.

$$\begin{aligned} \frac{2x}{2}+\frac{x}{4} &= 10 \\ 4\left(\frac{2x}{2}+\frac{x}{4}\right) &= (4)10 \\ 2(2x)+x &= 40 \\ 4x+x &= 40 \\ 5x &= 40 \\ \frac{5x}{5} &= \frac{40}{5} \\ x &= 8 \end{aligned}$$

16. $b = 40$

Isolate the term $\frac{2b}{5}$ on the left side by adding 6 to both sides. Eliminate the fraction by multiplying both sides by 5. Then, divide by 2 to isolate the variable.

$$\begin{aligned} \frac{2b}{5}-6 &= 10 \\ \frac{2b}{5}-6+6 &= 10+6 \\ \frac{2b}{5} &= 16 \\ \frac{5}{2}\times\frac{2b}{5} &= \frac{5}{2}\times 16 \\ b &= 40 \end{aligned}$$

17. $b = \frac{2A}{h}$

Multiply both sides by 2. Then, divide both sides by h.

$$\begin{aligned} A &= \frac{bh}{2} \\ 2\times A &= \frac{bh}{2}\times 2 \\ 2A &= bh \\ \frac{2A}{h} &= b \end{aligned}$$

18. $t = \frac{\frac{A}{P}-1}{r}$ **or** $t = \frac{A-P}{Pr}$

Distribute the P. Then, put all terms without t on one side and all terms with t on the other. Divide to isolate t.

$$\begin{aligned} A &= P(1+rt) \\ A &= P+Prt \\ A-P &= Prt \\ \frac{A-P}{Pr} &= t \end{aligned}$$

19. $x = 2$ **and** $y = 6$

The equation $y = 3x$ tells us that y and $3x$ are interchangeable. So, substitute $3x$ for y in the other equation.

$$\begin{aligned} x + y &= 8 \\ x + 3x &= 8 \\ 4x &= 8 \\ \frac{4x}{4} &= \frac{8}{4} \\ x &= 2 \end{aligned}$$

If $x = 2$ and $y = 3x$, then $y = 3 \times 2 = 6$.

So, $x = 2$ and $y = 6$.

20. $x = 2$ and $y = -1$

In each equation, y has the coefficient 1. Multiply both sides of the second equation by -1, then combine equations.

$$\begin{aligned} y - 2x = -5 \quad &\rightarrow \quad y - 2x = -5 \\ y - x = -3 \quad &\rightarrow \quad \underline{-y + x = 3} \\ & \qquad\quad -x = -2 \\ & \qquad\quad \frac{-x}{-1} = \frac{-2}{-1} \\ & \qquad\quad x = 2 \end{aligned}$$

Now use $x = 2$ to solve for y.

$$\begin{aligned} y - 2x &= -5 \\ y - 2(2) &= -5 \\ y - 4 &= -5 \\ y - 4 + 4 &= -5 + 4 \\ y &= -1 \end{aligned}$$

ADVANCED

21. $x = -2$

First, simplify each side by distributing and combining like terms. Then, collect x terms on the left side and constants on the right side.

$$\begin{aligned} 3(x + 2) &= 14 - 2(3 - 2x) \\ 3x + 6 &= 14 - 6 + 4x \\ 3x + 6 &= 8 + 4x \\ 3x + 6 - 6 &= 8 + 4x - 6 \\ 3x &= 2 + 4x \\ 3x - 4x &= 2 + 4x - 4x \\ -x &= 2 \\ \frac{-x}{-1} &= \frac{2}{-1} \\ x &= -2 \end{aligned}$$

22. $b = \frac{3}{2}$

Simplify the left side by distributing. Then, collect b terms on one side and constant terms on the other side. Remember to express the final answer in its simplest form.

$$\begin{aligned} 5(6 - 3b) &= 3b + 3 \\ 30 - 15b &= 3b + 3 \\ 30 - 30 - 15b &= 3b + 3 - 30 \\ -15b &= 3b - 27 \\ -15b - 3b &= 3b - 3b - 27 \\ -18b &= -27 \\ \frac{-18b}{-18} &= \frac{-27}{-18} \\ b &= \frac{-27}{-18} \\ b &= \frac{3}{2} \end{aligned}$$

23. $\mathbf{w = 0}$

Combine like terms on each side of the equation first. Notice that the term −4 now appears on each side of the equation. When the same term appears on both sides of an equation, you can remove it from each side without upsetting the balance of the equation.

$$\begin{aligned} 6w - 4 - 3w &= 8 - 12 - 4w \\ 3w - 4 &= -4 - 4w \\ 3w &= -4w \\ 7w &= 0 \\ \frac{7w}{7} &= \frac{0}{7} \\ w &= 0 \end{aligned}$$

24. $\mathbf{r = -\frac{15}{13}}$

Eliminate fractions by multiplying the entire equation by the least common multiple of 6, 5, and 2.

$$\begin{aligned} 30 \times \left(\frac{r}{6} - \frac{3r}{5}\right) &= \left(\frac{1}{2}\right) \times 30 \\ 5r - 6(3r) &= 15 \\ 5r - 18r &= 15 \\ -13r &= 15 \\ \frac{-13r}{-13} &= \frac{15}{-13} \\ r &= -\frac{15}{13} \end{aligned}$$

25. $\mathbf{j = 12}$

Eliminate fractions by multiplying each side of the equation by 6. Then, collect j terms on one side of the equation and constants on the other side.

$$\begin{aligned} 6 \times \left(\frac{2}{3}j - \frac{1}{2}\right) &= 6 \times \left(\frac{1}{6}j + \frac{11}{2}\right) \\ 4j - 3 &= j + 33 \\ 4j - j - 3 &= j - j + 33 \\ 3j - 3 &= 33 \\ 3j - 3 + 3 &= 33 + 3 \\ 3j &= 36 \\ \frac{3j}{3} &= \frac{36}{3} \\ j &= 12 \end{aligned}$$

26. $\mathbf{F = \frac{9}{5}C + 32}$

Isolate the expression $F - 32$ by multiplying both sides by the reciprocal of $\frac{5}{9}$, which is $\frac{9}{5}$. Then, isolate F by adding 32 to both sides.

$$\begin{aligned} \frac{9}{5} \times C &= \frac{9}{5} \times \frac{5}{9}(F - 32) \\ \frac{9}{5}C &= F - 32 \\ \frac{9}{5}C + 32 &= F - 32 + 32 \\ \frac{9}{5}C + 32 &= F \end{aligned}$$

27. $b = \frac{2A}{h} - a$ **or** $b = \frac{2A - ah}{h}$

Multiply both sides by 2, and divide both sides by h to isolate the expression $(a + b)$. Then, isolate b.

$$
\begin{aligned}
A &= \frac{1}{2}(a + b)h \\
2 \times A &= 2 \times \frac{1}{2}(a + b)h \\
2A &= (a + b)h \\
\frac{2A}{h} &= \frac{(a + b)h}{h} \\
\frac{2A}{h} &= a + b \\
\frac{2A}{h} - a &= a - a + b \\
\frac{2A}{h} - a &= b
\end{aligned}
$$

28. $x = 2$ **and** $y = 0$

Create opposite coefficients on x by multiplying the first equation by -2.

$$
\begin{aligned}
(-2)x - 5y = 2(-2) &\rightarrow -2x + 10y = -4 \\
2x + y = 4 &\rightarrow \underline{2x + y = 4} \\
& \qquad 11y = 0 \\
& \qquad \frac{11y}{11} = \frac{0}{11} \\
& \qquad y = 0
\end{aligned}
$$

Now use $y = 0$ to solve for x.

$$
\begin{aligned}
2x + y &= 4 \\
2x + 0 &= 4 \\
2x &= 4 \\
\frac{2x}{2} &= \frac{4}{2} \\
x &= 2
\end{aligned}
$$

29. $c = 5$ **and** $g = -1$

Solve the first equation for c.

$$
\begin{aligned}
2c &= 14 + 4g \\
\frac{2c}{2} &= \frac{14 + 4g}{2} \\
c &= 7 + 2g
\end{aligned}
$$

Now use $c = 7 + 2g$ to solve for g. Substitute $7 + 2g$ for c in the second equation.

$$
\begin{aligned}
3g &= 2 - c \\
3g &= 2 - (7 + 2g) \\
3g &= 2 - 7 - 2g \\
3g &= -5 - 2g \\
3g + 2g &= -5 - 2g + 2g \\
5g &= -5 \\
\frac{5g}{5} &= \frac{-5}{5} \\
g &= -1
\end{aligned}
$$

Now use $g = -1$ to solve for c.

$$
\begin{aligned}
c &= 7 + 2g \\
c &= 7 + 2(-1) \\
c &= 7 + (-2) \\
c &= 5
\end{aligned}
$$

30. $x = 1$ **and** $y = 2$

Multiply to create opposite coefficients on *x*. Then, combine the equations.

$$3(4x + 3y) = 3(10) \rightarrow 12x + 9y = 30$$
$$-4(3x + 5y) = -4(13) \rightarrow -12x - 20y = -52$$
$$-11y = -22$$
$$\frac{-11y}{-11} = \frac{-22}{-11}$$
$$y = 2$$

Now use $y = 2$ to solve for *x*. Substitute 2 for *y* in one of the equations.

$$4x + 3y = 10$$
$$4x + 3(2) = 10$$
$$4x + 6 = 10$$
$$4x + 6 - 6 = 10 - 6$$
$$4x = 4$$
$$\frac{4x}{4} = \frac{4}{4}$$
$$x = 1$$

QUADRATIC EQUATIONS

If you set the polynomial $ax^2 + bx + c$ equal to zero, where a, b, and c are constants and $a \neq 0$, there is a special name for it. It is called a quadratic equation. You can find the value(s) for x that make the equation true.

Example: $x^2 - 3x + 2 = 0$

To find the solutions, also called roots of the equation, start by factoring whenever possible. You can factor $x^2 - 3x + 2$ into $(x - 2)(x - 1)$, making the quadratic equation:

$$(x - 2)(x - 1) = 0$$

Now you have a product of two binomials that is equal to 0. The only time a product of two factors is equal to 0 is when *at least* one of the factors is equal to 0. If the product of $(x - 2)$ and $(x - 1)$ is equal to 0, that means either the first binomial equals 0 or the second binomial equals 0.

To find the roots, set each binomial equal to 0. That gives you:

$$(x - 2) = 0 \text{ or } (x - 1) = 0$$

Solving for x, you get $x = 2$ or $x = 1$. As a check, plug in each value into the original equation.

Check:

Let $x = 2$

$$\begin{aligned} x^2 - 3x + 2 &= 0 \\ (2)^2 - 3(2) + 2 &\stackrel{?}{=} 0 \\ 4 - 6 + 2 &\stackrel{?}{=} 0 \\ -2 + 2 &\stackrel{?}{=} 0 \\ 0 &= 0 \checkmark \end{aligned}$$

Let $x = 1$

$$\begin{aligned} x^2 - 3x + 2 &= 0 \\ (1)^2 - 3(1) + 2 &\stackrel{?}{=} 0 \\ 1 - 3 + 2 &\stackrel{?}{=} 0 \\ -2 + 2 &\stackrel{?}{=} 0 \\ 0 &= 0 \checkmark \end{aligned}$$

Both values, $x = 2$ and $x = 1$, are solutions of the quadratic equation $x^2 - 3x + 2 = 0$.

Here is another example of a quadratic that can be factored to determine its solutions. Always make sure the equation is set equal to zero before you start to factor.

Example: Find the solutions of $3x^2 - 5x + 2 = 0$.

1. Factor the equation into two binomials.

$$\begin{aligned} 3x^2 - 5x + 2 &= 0 \\ (3x - 2)(x - 1) &= 0 \end{aligned}$$

2. Set each factor equal to zero and solve for x.

$$\begin{aligned} 3x - 2 &= 0 \quad & x - 1 &= 0 \\ 3x &= 2 & x &= 1 \\ x &= \frac{2}{3} \end{aligned}$$

3. Write the solutions.

$$x = \frac{2}{3} \text{ or } x = 1$$

A quadratic equation that can be factored into the square of a binomial will have only one solution. For example, the equation $x^2 - 14x + 49 = 0$ can be factored like this: $(x - 7)(x - 7) = 0$. Its only solution is $x = 7$.

The solutions to a quadratic equation in the form $ax^2 + bx + c = 0$ can also be found using the **quadratic formula**. Provided a, b, and c are real numbers and $a \neq 0$, then:

$$x = \frac{-b \pm \sqrt{b^2 - 4ac}}{2a}$$

Example: Find the solutions of $2x^2 + 9x + 9 = 0$.

Identify a, b, and c: $a = 2$, $b = 9$, $c = 9$.

$$\begin{aligned} x &= \frac{-b \pm \sqrt{b^2 - 4ac}}{2a} \\ &= \frac{-9 \pm \sqrt{9^2 - 4(2)(9)}}{2(2)} \\ & \frac{-9 \pm \sqrt{9}}{4} \\ &= \frac{-9 \pm 3}{4} \\ &= -3 \text{ or } -\frac{3}{2} \end{aligned}$$

The symbol $\pm$ indicates there are two possible solutions, one found with addition and the other with subtraction, as you saw in this example. If the value under the radical, $b^2 - 4ac$, is negative, the two solutions are complex numbers. This is a situation you will not see on the GRE. If the value of $b^2 - 4ac$ equals 0, there is just one solution to the equation.

QUADRATIC EQUATIONS EXERCISES

Basic

Find the value(s) for the variables.

1. $x^2 = 9$
2. $y^2 = 121$
3. $(x - 4)(x + 5) = 0$
4. $x(x - 2) = 0$
5. $(x - 9)^2 = 0$
6. $(b - 2)(b - 5) = 0$
7. $a^2 + 5a + 6 = 0$
8. $t^2 - 2t - 8 = 0$
9. $m^2 - 10m + 16 = 0$
10. $p^2 + p - 8 = 0$

Intermediate

Solve the quadratic equations below.

11. $y^2 - 7y = 0$
12. $d^2 - 20d + 100 = 0$
13. $3b^2 - 5b - 2 = 0$
14. $x^2 - 5x = 6$
15. $8j^2 - 24j - 32 = 0$
16. $4n^2 + 5 = 9n$
17. $n^2 - 11n + 1 = 0$
18. $x^2 + 3x + 1 = 0$
19. $3g^2 + 9g - 6 = 0$
20. $h^2 = 2h + 8$

Advanced

Solve the quadratic equations below.

21. $6w^2 - w - 15 = 0$
22. $4m^2 - 25 = 0$
23. $3x^2 + 4x = x^2 + 5x + 3$
24. $p^2 - 20p + 96 = 0$
25. $2n^2 - n - 1 = 0$
26. $2n^2 - 3n + 1 = 0$
27. $\frac{1}{6}x^2 - \frac{1}{2}x + \frac{1}{6} = 0$
28. $5x^2 - 35x - 55 = 0$
29. $-5h^2 - 5h = 1$
30. $-3b^2 = -10b - 8$

QUADRATIC EQUATIONS ANSWER KEY

BASIC

1. 3, –3

Think: what number, squared, is equal to 9? Since $3^2 = 9$ and $(-3)^2 = 9$, the equation has two solutions.

2. 11, –11

Solve by taking the square root of both sides. y can be either 11 or –11.

3. 4, –5

The expression $(x - 4)(x + 5)$ equals 0 when one or both factors are equal to 0. Set each factor equal to 0 and solve.

$$x - 4 = 0 \qquad x + 5 = 0$$
$$x = 4 \qquad x = -5$$

4. 0, 2

Set each factor equal to 0 and solve.

$$x = 0 \qquad x - 2 = 0$$
$$x = 2$$

5. 9

The equation $(x - 9)^2 = 0$ is true when $x - 9 = 0$.

$$x - 9 = 0$$
$$x = 9$$

6. 2, 5

The equation is true when $b - 2 = 0$ and when $b - 5 = 0$.

$$b - 2 = 0 \qquad b - 5 = 0$$
$$b = 2 \qquad b = 5$$

7. –2, –3

Factor. Then, set each factor equal to 0.

$$a^2 + 5a + 6 = 0$$
$$(a + 2)(a + 3) = 0$$

Set each factor equal to 0 and solve.

$$a + 2 = 0 \qquad a + 3 = 0$$
$$a = -2 \qquad a = -3$$

8. 4, –2

Factor. Then, set each factor equal to 0.

$$t^2 - 2t - 8 = 0$$
$$(t - 4)(t + 2) = 0$$

Set each factor equal to 0 and solve.

$$t - 4 = 0 \qquad t + 2 = 0$$
$$t = 4 \qquad t = -2$$

9. 8, 2

Though the expression on the left-hand side of this equation can be factored, you can also use the quadratic formula with $a = 1$, $b = -10$, and $c = 16$.

$$\frac{-b \pm \sqrt{b^2 - 4ac}}{2a} = \frac{-(-10) \pm \sqrt{(-10)^2 - 4(1)(16)}}{2(1)}$$
$$= \frac{10 \pm \sqrt{100 - (64)}}{2}$$
$$= \frac{10 \pm \sqrt{36}}{2}$$
$$= \frac{10 \pm 6}{2}$$

Now separate the two solutions.

$$\frac{10 + 6}{2} = \frac{16}{2} = 8 \qquad \frac{10 - 6}{2} = \frac{4}{2} = 2$$

If you use factoring, $m^2 - 10m + 16 = (m - 8)(m - 2) = 0$. Setting each factor equal to 0 and solving for m, you get $m = 8$ and $m = 2$.

10. $\frac{-1 + \sqrt{33}}{2}$, $\frac{-1 - \sqrt{33}}{2}$

Use the quadratic formula with $a = 1$, $b = 1$, and $c = -8$.

$$\frac{-b \pm \sqrt{b^2 - 4ac}}{2a} = \frac{-1 \pm \sqrt{1^2 - 4(1)(-8)}}{2(1)}$$
$$= \frac{-1 \pm \sqrt{1 - (-32)}}{2}$$
$$= \frac{-1 \pm \sqrt{33}}{2}$$

INTERMEDIATE

11. 0, 7

Factor the left side. Then, find the value of y that makes each factor equal to 0.

$$y^2 - 7y = 0$$
$$y(y - 7) = 0$$

Set each factor equal to 0 and solve.

$$y = 0 \qquad y - 7 = 0$$
$$y = 7$$

12. 10

Factor.

$$d^2 - 20d + 100 = 0$$
$$(d - 10)(d - 10) = 0$$

The two factors are identical, so set one equal to 0 and solve.

$$d - 10 = 0$$
$$d = 10$$

13. $-\frac{1}{3}$**, 2**

Factor.

$$3b^2 - 5b - 2 = 0$$
$$(3b + 1)(b - 2) = 0$$

Set each factor equal to 0 and solve.

$$3b + 1 = 0 \qquad b - 2 = 0$$
$$3b = -1 \qquad b = 2$$
$$b = -\frac{1}{3}$$

14. 6, –1

Rearrange the terms so that the equation takes the form $ax^2 + bx + c = 0$. Then, factor to solve.

$$x^2 - 5x = 6$$
$$x^2 - 5x - 6 = 0$$
$$(x - 6)(x + 1) = 0$$

Set each factor equal to 0 and solve.

$$x - 6 = 0 \qquad x + 1 = 0$$
$$x = 6 \qquad x = -1$$

15. 4, −1

First, factor the common factor 8 from each term.

$$8j^2 - 24j - 32 = 0$$
$$8(j^2 - 3j - 4) = 0$$

Continue factoring the quadratic expression.

$$8(j - 4)(j + 1) = 0$$

Set each factor equal to 0 and solve. (The factor 8 is never equal to 0.)

$$j - 4 = 0 \qquad j + 1 = 0$$
$$j = 4 \qquad j = -1$$

16. $\frac{5}{4}$, 1

Rearrange the terms so that the equation takes the form $ax^2 + bx + c = 0$. Then, factor to solve.

$$4n^2 + 5 = 9n$$
$$4n^2 - 9n + 5 = 0$$
$$(4n - 5)(n - 1) = 0$$

Set each factor equal to 0 and solve.

$$4n - 5 = 0$$
$$4n = 5 \qquad n - 1 = 0$$
$$n = \frac{5}{4} \qquad n = 1$$

17. $\frac{11 + \sqrt{117}}{2}, \frac{11 - \sqrt{117}}{2}$

Use the quadratic formula with $a = 1$, $b = -11$, and $c = 1$.

$$\frac{-b \pm \sqrt{b^2 - 4ac}}{2a} = \frac{-(-11) \pm \sqrt{(-11)^2 - 4(1)(1)}}{2(1)}$$
$$= \frac{11 \pm \sqrt{121 - 4}}{2}$$
$$= \frac{11 \pm \sqrt{117}}{2}$$

18. $\frac{-3+\sqrt{5}}{2}, \frac{-3-\sqrt{5}}{2}$

Solve using the quadratic formula with $a = 1$, $b = 3$, and $c = 1$.

$$\begin{aligned}\frac{-b \pm \sqrt{b^2 - 4ac}}{2a} &= \frac{-3 \pm \sqrt{3^2 - 4(1)(1)}}{2(1)} \\ &= \frac{-3 \pm \sqrt{9-4}}{2} \\ &= \frac{-3 \pm \sqrt{5}}{2}\end{aligned}$$

19. $\frac{-3+\sqrt{17}}{2}, \frac{-3-\sqrt{17}}{2}$

Solve using the quadratic formula with $a = 3$, $b = 9$, and $c = -6$.

$$\begin{aligned}3g^2 + 9g - 6 &= 0 \\ \frac{-b \pm \sqrt{b^2 - 4ac}}{2a} &= \frac{-9 \pm \sqrt{9^2 - 4(3)(-6)}}{2(3)} \\ &= \frac{-9 \pm \sqrt{81+72}}{6} \\ &= \frac{-9 \pm \sqrt{153}}{6} = \frac{-9 \pm \sqrt{9 \times 17}}{6} = \frac{-9 \pm 3\sqrt{17}}{6} = \frac{-3 \pm \sqrt{17}}{2}\end{aligned}$$

20. 4, –2

Rearrange the terms so that the equation takes the form $ax^2 + bx + c = 0$.

$$\begin{aligned}h^2 &= 2h + 8 \\ h^2 - 2h - 8 &= 0\end{aligned}$$

Factor.

$$(h-4)(h+2) = 0$$

Set each factor equal to 0 and solve.

$$\begin{aligned}h - 4 &= 0 \qquad & h + 2 &= 0 \\ h &= 4 & h &= -2\end{aligned}$$

Advanced

21. $-\frac{3}{2}, \frac{5}{3}$

Factor.

$$6w^2 - w - 15 = 0$$
$$(2w + 3)(3w - 5) = 0$$

Set each factor equal to 0 and solve.

$$2w + 3 = 0 \qquad 3w - 5 = 0$$
$$2w = -3 \qquad 3w = 5$$
$$w = -\frac{3}{2} \qquad w = \frac{5}{3}$$

22. $-\frac{5}{2}, \frac{5}{2}$

This is a Classic Quadratic: the difference of two squares.

$$4m^2 - 25 = 0$$
$$(2m + 5)(2m - 5) = 0$$

Set each factor equal to 0 and solve.

$$2m + 5 = 0 \qquad 2m - 5 = 0$$
$$2m = -5 \qquad 2m = 5$$
$$m = -\frac{5}{2} \qquad m = \frac{5}{2}$$

23. $\frac{3}{2}, -1$

Get the equation into the standard form. Then, factor.

$$3x^2 + 4x = x^2 + 5x + 3$$
$$2x^2 - x - 3 = 0$$
$$(2x - 3)(x + 1) = 0$$

Set each factor equal to 0 and solve.

$$2x - 3 = 0 \qquad x + 1 = 0$$
$$2x = 3 \qquad x = -1$$
$$x = \frac{3}{2}$$

24. 12, 8

Factor.

$$\begin{aligned} p^2 - 20p + 96 &= 0 \\ (p - 12)(p - 8) &= 0 \end{aligned}$$

Set each factor equal to 0 and solve.

$$\begin{aligned} p - 12 &= 0 & \qquad p - 8 &= 0 \\ p &= 12 & p &= 8 \end{aligned}$$

25. $-\frac{1}{2}$**, 1**

Factor.

$$\begin{aligned} 2n^2 - n - 1 &= 0 \\ (2n + 1)(n - 1) &= 0 \end{aligned}$$

Set each factor equal to 0 and solve.

$$2n + 1 = 0$$
$$2n = -1$$
$$n = \frac{-1}{2} \qquad n - 1 = 0$$
$$n = 1$$

26. 1, $\frac{1}{2}$

Use the quadratic formula with $a = 2$, $b = -3$, and $c = 1$.

$$\begin{aligned} \frac{-b \pm \sqrt{b^2 - 4ac}}{2a} &= \frac{-(-3) \pm \sqrt{(-3)^2 - 4(2)(1)}}{2(2)} \\ &= \frac{3 \pm \sqrt{9 - 8}}{4} \\ &= \frac{3 \pm \sqrt{1}}{4} \\ &= \frac{3 \pm 1}{4} \end{aligned}$$

Now separate the two solutions.

$$\frac{3 + 1}{4} = \frac{4}{4} = 1 \qquad \frac{3 - 1}{4} = \frac{2}{4} = \frac{1}{2}$$

You can also solve this problem using factoring. $2n^2 - 3n + 1 = (2n - 1)(n - 1) = 0$. Set each factor equal to 0 and solve for n: $n = 1$ and $n = \frac{1}{2}$.

27. $\dfrac{3+\sqrt{5}}{2}, \dfrac{3-\sqrt{5}}{2}$

Clear fractions by multiplying both sides of the equation by 6.

$$\frac{1}{6}x^2 - \frac{1}{2}x + \frac{1}{6} = 0 \rightarrow x^2 - 3x + 1 = 0$$

Use the quadratic formula with $a = 1$, $b = -3$, and $c = 1$.

$$\begin{aligned}\frac{-b \pm \sqrt{b^2 - 4ac}}{2a} &= \frac{-(-3) \pm \sqrt{(-3)^2 - 4(1)(1)}}{2(1)} \\ &= \frac{3 \pm \sqrt{9-4}}{2} \\ &= \frac{3 \pm \sqrt{5}}{2}\end{aligned}$$

28. $\dfrac{7+\sqrt{93}}{2}, \dfrac{7-\sqrt{93}}{2}$

Make a, b, and c easier to work with by dividing both sides of the equation by 5.

$$5x^2 - 35x - 55 = 0 \rightarrow x^2 - 7x - 11 = 0$$

Use the quadratic formula with $a = 1$, $b = -7$, and $c = -11$.

$$\begin{aligned}\frac{-b \pm \sqrt{b^2 - 4ac}}{2a} &= \frac{-(-7) \pm \sqrt{(-7)^2 - 4(1)(-11)}}{2(1)} \\ &= \frac{7 \pm \sqrt{49 + 44}}{2} \\ &= \frac{7 \pm \sqrt{93}}{2}\end{aligned}$$

29. $\dfrac{-5+\sqrt{5}}{10}, \dfrac{-5-\sqrt{5}}{10}$

Get the equation into the standard form.

$$\begin{aligned}-5h^2 - 5h &= 1 \\ -5h^2 - 5h - 1 &= 0 \\ 5h^2 + 5h + 1 &= 0\end{aligned}$$

Use the quadratic formula with $a = 5$, $b = 5$, and $c = 1$.

$$\frac{-b \pm \sqrt{b^2 - 4ac}}{2a} = \frac{-5 \pm \sqrt{5^2 - 4(5)(1)}}{2(5)}$$
$$= \frac{-5 \pm \sqrt{25 - 20}}{10}$$
$$= \frac{-5 \pm \sqrt{5}}{10}$$

30. $\mathbf{b = -\frac{2}{3}, b = 4}$

Rearrange the equation into the form $ax^2 + bx + c = 0$.

$$-3b^2 + 10b + 8 = 0$$

Factor.

$$3b^2 - 10b - 8 = 0$$
$$(3b + 2)(b - 4) = 0$$

Set each factor equal to 0 and solve.

$$3b + 2 = 0$$
$$3b = -2 \qquad b - 4 = 0$$
$$b = -\frac{2}{3} \qquad b = 4$$

SOLVING INEQUALITIES

Inequalities may be written with the symbols shown here:

$<$ less than	$\leq$ less than or equal to
$>$ greater than	$\geq$ greater than or equal to

Examples: $x > 4$ means all numbers greater than 4.

$x < 0$ means all numbers less than zero (the negative numbers).

$x \geq -2$ means x can be -2 or any number greater than -2.

$x \leq \frac{1}{2}$ means x can be $\frac{1}{2}$ or any number less than $\frac{1}{2}$.

A range of values is often expressed on a number line. Two ranges are shown below.

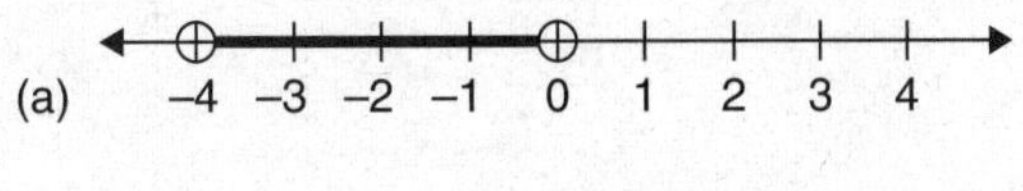

(a) Represents the set of all numbers between –4 and 0 excluding the endpoints –4 and 0, or $-4 < x < 0$.

(b) Represents the set of all numbers greater than –1, up to and including 3, or $-1 < x \leq 3$.

To solve inequalities, use the same methods used in solving equations, with one exception.

If the inequality is multiplied or divided by a negative number, the direction of the inequality is reversed. For instance, if both sides of the inequality $-3x < 2$ are multiplied by -1, the result is $3x > -2$.

Example: Solve for x and represent the solution set on a number line.

$3 - \frac{x}{4} \geq 2$

1. Multiply both sides by 4. $\quad 12 - x \geq 8$
2. Subtract 12 from both sides. $\quad -x \geq -4$
3. Divide both sides by -1 and reverse the direction of the inequality symbol. $\quad x \leq 4$

Note: the solution set to an inequality is not a single number, but a range of possible values. Here the values include 4 and all numbers less than 4. These numbers are located to the left of 4 on the number line.

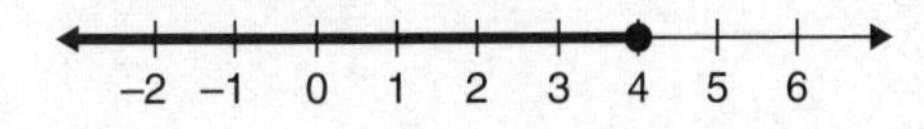

SOLVING INEQUALITIES EXERCISES

Basic

Solve.

1. $x - 3 > 7$
2. $5 \geq x + 3$
3. $x + 5 < 3$
4. $-6x > 18$
5. $9 \leq \frac{x}{5}$
6. $2x + 1 \leq 7$
7. $\frac{1}{2}x + 3 > -2$
8. $-3x > 6$
9. $-8 < -2x$
10. $-x + 3 \geq 5$

Intermediate

11. $2x + 3 > 4x - 9$
12. $\frac{x + 3}{2} \geq 7$
13. $-2(x - 4) < -8$
14. $2x - 3x \leq 6$
15. $6x - 3 > 5x + 2x + 1$
16. $-x + 3 \geq 5$
17. $\frac{3x}{-2} \geq -6$
18. $0.5x + 3 > 7$
19. $-\frac{1}{3}x > 5 - 2x$
20. $2(x + 10) \geq 30$

Advanced

21. $3x + \frac{5}{8} > 2x + \frac{5}{6}$
22. $5(x - 3) + 7 < 2x + 7$
23. $-\frac{3}{4}x \geq -\frac{5}{8}$
24. $5x - 2(x + 3) \geq 12$
25. $\frac{7}{10}x > -\frac{4}{15}$
26. $-5 \geq \frac{10(x - 2)}{-4}$
27. $\frac{1}{4}(8 - 12d) < \frac{2}{5}(10d + 15)$
28. $4 - 3(3 - n) \leq 3(2 - 5n)$
29. $2x - 3(x - 4) > 4 - 2(x - 7)$
30. $\frac{3}{2}x - \frac{3}{4}x < \frac{7}{4}x - 1$

SOLVING INEQUALITIES ANSWER KEY

BASIC

1. $x > 10$

Add 3 to both sides of the inequality.

$$\begin{aligned} x - 3 &> 7 \\ x - 3 + 3 &> 7 + 3 \\ x &> 10 \end{aligned}$$

2. $x \leq 2$

Subtract 3 from both sides of the inequality. Remember that the inequalities $2 \geq x$ and $x \leq 2$ are equivalent.

$$\begin{aligned} 5 &\geq x + 3 \\ 5 - 3 &\geq x + 3 - 3 \\ 2 &\geq x \end{aligned}$$

3. $x < -2$

Subtract 5 from both sides of the inequality.

$$\begin{aligned} x + 5 &< 3 \\ x + 5 - 5 &< 3 - 5 \\ x &< -2 \end{aligned}$$

4. $x < -3$

Divide both sides of the inequality by -6. Remember that you must reverse the inequality symbol (from $>$ to $<$) when dividing by a negative number.

$$\begin{aligned} -6x &> 18 \\ \frac{-6x}{-6} &< \frac{18}{-6} \\ x &< -3 \end{aligned}$$

5. $x \geq 45$

Multiply both sides of the inequality by 5.

$$\begin{aligned} 9 &\leq \frac{x}{5} \\ 5 \times 9 &\leq 5 \times \frac{x}{5} \\ 45 &\leq x \end{aligned}$$

6. $x \leq 3$

First, subtract 1 from both sides of the inequality. Then, divide both sides by 2.

$$\begin{aligned} 2x + 1 &\leq 7 \\ 2x + 1 - 1 &\leq 7 - 1 \\ 2x &\leq 6 \\ \frac{2x}{2} &\leq \frac{6}{2} \\ x &\leq 3 \end{aligned}$$

7. $x > -10$

First, subtract 3 from both sides of the equation. Then, multiply both sides by 2 to eliminate the fraction.

$$\begin{aligned} \frac{1}{2}x + 3 &> -2 \\ \frac{1}{2}x + 3 - 3 &> -2 - 3 \\ \frac{1}{2}x &> -5 \\ 2 \times \frac{1}{2}x &> 2 \times -5 \\ x &> -10 \end{aligned}$$

8. $x < -2$

Divide both sides of the inequality by -3. Remember to reverse the direction of the inequality when you divide by a negative number.

$$\begin{aligned} -3x &> 6 \\ \frac{-3x}{-3} &< \frac{6}{-3} \\ x &< -2 \end{aligned}$$

9. $x < 4$

Divide both sides of the inequality by -2. Remember to reverse the direction of the inequality when you divide by a negative number.

$$\begin{aligned} -8 &< -2x \\ \frac{-8}{-2} &> \frac{-2x}{-2} \\ 4 &> x \end{aligned}$$

10. $x \leq -2$

First, subtract 3 from both sides of the inequality. Then, multiply both sides by –1. Reverse the direction of the inequality when you multiply by a negative number.

$$\begin{aligned} -x + 3 &\geq 5 \\ -x + 3 - 3 &\geq 5 - 3 \\ -x &\geq 2 \\ -1 \times (-x) &\leq -1 \times 2 \\ x &\leq -2 \end{aligned}$$

Intermediate

11. $x < 6$

Isolate all the x terms on one side of the inequality and all constant terms on the other side.

$$\begin{aligned} 2x + 3 &> 4x - 9 \\ 12 &> 2x \\ 6 &> x \end{aligned}$$

12. $x \geq 11$

First, isolate $x + 3$ by multiplying both sides of the inequality by 2. Then, isolate x by subtracting 3 from both sides.

$$\begin{aligned} \frac{x + 3}{2} &\geq 7 \\ x + 3 &\geq 14 \\ x &\geq 11 \end{aligned}$$

13. $x > 8$

First, simplify the left side of the inequality by distributing –2. Then, isolate x. Remember to reverse the direction of the inequality when you divide by a negative number.

$$\begin{aligned} -2(x - 4) &< -8 \\ -2x + 8 &< -8 \\ -2x &< -16 \\ \frac{-2x}{-2} &> \frac{-16}{-2} \\ x &> 8 \end{aligned}$$

14. $x \geq -6$

First, simplify the left side of the inequality by combining like terms. Then, solve for x by dividing both sides by –1. Remember to reverse the direction of the inequality when you divide by a negative number.

$$\begin{aligned} 2x - 3x &\leq 6 \\ -x &\leq 6 \\ \frac{-x}{-1} &\geq \frac{6}{-1} \\ x &\geq -6 \end{aligned}$$

15. $x < -4$

Simplify the right side by combining like terms. Then, isolate x using inverse operations.

$$\begin{aligned} 6x - 3 &> 5x + 2x + 1 \\ 6x - 3 &> 7x + 1 \\ -4 &> x \end{aligned}$$

16. $x \leq -2$

Isolate the x term and then divide by –1. Remember to reverse the direction of the inequality when you divide by a negative number.

$$\begin{aligned} -x + 3 &\geq 5 \\ -x &\geq 2 \\ \frac{-x}{-1} &\leq \frac{2}{-1} \\ x &\leq -2 \end{aligned}$$

17. $x \leq 4$

First isolate $3x$. Then isolate x. Remember to reverse the direction of the inequality when you multiply by a negative number.

$$\begin{aligned} \frac{3x}{-2} &\geq -6 \\ -2 \times \left(\frac{3x}{-2}\right) &\leq -2 \times (-6) \\ 3x &\leq 12 \\ x &\leq 4 \end{aligned}$$

18. $\boldsymbol{x > 8}$

One way to avoid calculations involving a decimal is to replace the decimal 0.5 with the equivalent fraction $\frac{1}{2}$. First, isolate the x term; then, multiply by 2 to solve for x.

$$\begin{aligned} 0.5x + 3 &> 7 \\ \frac{1}{2}x + 3 &> 7 \\ \frac{1}{2}x &> 4 \\ 2 \times \frac{1}{2}x &> 2 \times 4 \\ x &> 8 \end{aligned}$$

19. $\boldsymbol{x > 3}$

You can avoid having to add fractions by eliminating them. Multiply both sides of the inequality by 3. Then isolate x.

$$\begin{aligned} 3 \times \left(-\frac{1}{3}x\right) &> 3 \times (5 - 2x) \\ -x &> 15 - 6x \\ 5x &> 15 \\ x &> 3 \end{aligned}$$

20. $\boldsymbol{x \geq 5}$

Simplify the left side of the inequality by distributing 2 to each term inside the parentheses. Then isolate x.

$$\begin{aligned} 2(x + 10) &\geq 30 \\ 2x + 20 &\geq 30 \\ 2x &\geq 10 \\ x &\geq 5 \end{aligned}$$

ADVANCED

21. $\boldsymbol{x > \frac{5}{24}}$

Eliminate fractions by multiplying both sides of the inequality by the least common multiple of 8 and 6, 24.

$$\begin{aligned} 3x + \frac{5}{8} &> 2x + \frac{5}{6} \\ 24 \times \left(3x + \frac{5}{8}\right) &> 24 \times \left(2x + \frac{5}{6}\right) \\ 72x + 15 &> 48x + 20 \\ 24x &> 5 \\ x &> \frac{5}{24} \end{aligned}$$

22. $\boldsymbol{x < 5}$

Simplify the left side before attempting to isolate x.

$$\begin{aligned} 5(x - 3) + 7 &< 2x + 7 \\ 5x - 15 + 7 &< 2x + 7 \\ 5x - 8 &< 2x + 7 \\ 3x &< 15 \\ x &< 5 \end{aligned}$$

23. $\boldsymbol{x \leq \frac{5}{6}}$

Eliminate fractions by multiplying both sides of the inequality by 8. Remember to reverse the direction of the inequality when you divide by a negative number.

$$\begin{aligned} 8 \times \left(-\frac{3}{4}x\right) &\geq 8 \times \left(-\frac{5}{8}\right) \\ -6x &\geq -5 \\ \frac{-6x}{-6} &\geq \frac{-5}{-6} \\ x &\leq \frac{5}{6} \end{aligned}$$

24. $\boldsymbol{x \geq 6}$

Simplify the left side before isolating x.

$$\begin{aligned} 5x - 2(x + 3) &\geq 12 \\ 5x - 2x - 6 &\geq 12 \\ 3x - 6 &\geq 12 \\ 3x &\geq 18 \\ x &\geq 6 \end{aligned}$$

25. $x > \frac{-8}{21}$

Multiply both sides by the least common multiple of 10 and 15, 30, to eliminate the fractions.

$$\begin{aligned} 30 \times \left(\frac{7}{10}x\right) &> 30 \times \left(-\frac{4}{15}\right) \\ 21x &> -8 \\ x &> \frac{-8}{21} \end{aligned}$$

26. $x \geq 4$

Begin by multiplying both sides by −4. Remember to reverse the direction of the inequality when you multiply by a negative number. Once the fraction is eliminated, isolate *x*.

$$\begin{aligned} -5 &\geq \frac{10(x-2)}{-4} \\ -4 \times (-5) &\leq -4 \times \left(\frac{10(x-2)}{-4}\right) \\ 20 &\leq 10(x-2) \\ 20 &\leq 10x - 20 \\ 40 &\leq 10x \\ 4 &\leq x \end{aligned}$$

27. $-\frac{4}{7} < d$

Eliminate fractions by multiplying both sides of the inequality by the least common multiple of 4 and 5, 20. Then simplify each side before isolating *x*.

$$\begin{aligned} \frac{1}{4}(8-12d) &< \frac{2}{5}(10d+15) \\ 20 \times \frac{1}{4}(8-12d) &< 20 \times \frac{2}{5}(10d+15) \\ 5(8-12d) &< 8(10d+15) \\ 40 - 60d &< 80d + 120 \\ -80 &< 140d \\ \frac{-80}{140} &< \frac{140d}{140} \\ -\frac{80}{140} &< d \\ -\frac{4}{7} &< d \end{aligned}$$

28. $n \leq \frac{11}{18}$

Simplify each side of the inequality before isolating *n*.

$$\begin{aligned} 4 - 3(3-n) &\leq 3(2-5n) \\ 4 - 9 + 3n &\leq 6 - 15n \\ -5 + 3n &\leq 6 - 15n \\ 18n &\leq 11 \\ n &\leq \frac{11}{18} \end{aligned}$$

29. $x > 6$

Simplify each side of the inequality before isolating *x*.

$$\begin{aligned} 2x - 3(x-4) &> 4 - 2(x-7) \\ 2x - 3x + 12 &> 4 - 2x + 14 \\ -x + 12 &> 18 - 2x \\ x &> 6 \end{aligned}$$

30. $x > 1$

Eliminate fractions by multiplying each term in the inequality by 4. Remember to reverse the direction of the inequality when you divide by a negative number.

$$\begin{aligned} \frac{3}{2}x - \frac{3}{4}x &< \frac{7}{4}x - 1 \\ 4\times\left(\frac{3}{2}x - \frac{3}{4}x\right) &< 4 \times \left(\frac{7}{4}x - 1\right) \\ 6x - 3x &< 7x - 4 \\ 3x &< 7x - 4 \\ -4x &< -4 \\ \frac{-4x}{-4} &> \frac{-4}{-4} \\ x &> 1 \end{aligned}$$

TRANSLATION INTO ALGEBRA

Identifying information about a situation given in words and translating the words into algebra is often the first step in solving a problem. This involves defining a variable to represent an unknown value and then writing an expression, equation, or inequality.

The Translation Table below lists some common English words and phrases and the corresponding algebraic symbols.

Translation Table

Equals, is, was, will, be, has, costs, adds up to, is the same as	=
Times, of, multiplied by, product of, twice, double, half, triple	• or ×
Divided by, per, out of, each, ratio of __ to __	÷
Plus, added to, sum, combined, and, more than, total	+
Minus, subtracted from, less than, decreased by, difference between	−
What, how much, how many, a number	x, n, etc.

TRANSLATING TO EXPRESSIONS

To solve some word problems, you will have to translate the information given into algebra. This means representing the unknown value with a variable and then writing an expression. Here are some examples:

Examples: Raul's age is 1 more than twice his sister's age:

Let s represent the sister's age. Then Raul's age is given by $2s + 1$.

The cube of a number is multiplied by 2 and then 1 is subtracted from the product.

Let n represent the number. Then $2n^3 - 1$ represents the result of the operations on n.

A number of chairs, c, is arranged for a banquet so that 10 of the chairs are placed at the head table and the remaining chairs are divided equally among 20 other dining tables.

Then $\frac{c - 10}{20}$ represents the number of chairs at each table other than the head table.

TRANSLATING TO EQUATIONS

For Problem Solving questions, you may have to write an equation to solve the problem. The equation may be unique to the problem, or it may be a formula you are familiar with.

Examples: Company Z spent $\frac{2}{5}$ of its revenues one year on its payroll and $\frac{1}{6}$ of what was left on research. What were the year's original revenues for Company Z if there was $500,000 left after its payroll and research expenditures?

Let R represent the original revenues.

After $\frac{2}{5}R$ was spent on payroll, there was $\frac{3}{5}R$ left for other uses, including research.

The amount spent on research is given by $\frac{1}{6} \times \frac{3}{5}R = \frac{1}{\cancel{6}_2} \times \frac{\cancel{3}^1}{5}R = \frac{1}{10}R$.

Write the equation and solve for R.

$$\begin{aligned} R - \frac{2}{5}R - \frac{1}{10}R &= 500{,}000 \\ R\left(1 - \frac{2}{5} - \frac{1}{10}\right) &= 500{,}000 && \text{Factor } R \text{ from each term.} \\ R\left(\frac{10}{10} - \frac{4}{10} - \frac{1}{10}\right) &= 500{,}000 && \text{Write fractions with a common denominator.} \\ R\left(\frac{5}{10}\right) &= 500{,}000 && \text{Simplify fractions.} \\ R &= 2 \times 500{,}000 \\ R &= 1{,}000{,}000 \end{aligned}$$

The original revenue for Company Z was $1,000,000.

The distance formula $d = rt$, where d represents distance, r represents the rate of speed, and t represents time, is frequently the basis for questions on the GRE. To use the formula, the rate and the time must be given in the same units of time, for example, in *miles per hour* and *hours*. Consider this question:

Two cyclists finish a bike course at average speeds of 22 mph and 26 mph, respectively. If it took the first cyclist 38 minutes to finish the course, approximately how long did it take the second cyclist?

The rate for each cyclist is given in mph, so write the time for each one as part of an hour: 38 minutes $= \frac{38}{60}$ hour, and the unknown time for the second cyclist can be written as $\frac{m}{60}$ hour. Each cyclist rode the same distance, so you can equate the products of rate and time for the two cyclists.

Write the equation and solve for m.

$$\begin{aligned} \text{Cyclist 1's distance} &= \text{Cyclist 2's distance} \\ r_1 t_1 &= r_2 t_2 \\ 22\left(\frac{38}{60}\right) &= 26\left(\frac{m}{60}\right) \\ 22(38) &= 26m \\ m &= \frac{22 \times 38}{26} \approx 32 \end{aligned}$$

It took Cyclist 2 about 32 minutes to finish the course.

Translating to Inequalities

Some word problems are solved with an inequality rather than an equation. Key words such as *at least*, *no more than*, or *at most* are clues that an inequality is appropriate.

Example: It costs a manufacturer \$25 to produce a certain phone. Assume that 250 phones are produced, none are defective, and all will be sold to customers. What must the selling price of the phones be if a profit (revenue less production cost) of at least \$2,000 is needed on the order?

Let p represent the selling price of one phone. Then the total profit is $250(p - 25)$.

Write the inequality and solve for p.

$$\begin{aligned} 250(p - 25) &\geq 2{,}000 \\ 250p - 6{,}250 &\geq 2{,}000 \\ 250p &\geq 8{,}250 \\ p &\geq 33 \end{aligned}$$

The selling price must be at least \$33 per phone to guarantee a profit of \$2,000.

Simple and Compound Interest

Some questions on the GRE involve information about earning interest on an investment or paying interest on a loan. The interest may be **simple interest**, which is computed only on the **principal**—the initial amount invested. Simple interest is usually applied to investments with a time period less than one year. The formula for the computation of simple interest is

$$I = Prt$$

where I is the interest, P is the principal, r is the annual rate expressed as a decimal, and t is the time expressed in years.

To compute the amount, A, of an investment, P, at the end of t years, at annual rate, r, use the formula $A = P(1 + rt)$.

Example: If $3,000 is invested at a simple interest rate of 4%, what is the value of the investment after 9 months?

1. Identify what is given.

$$P = 3{,}000 \qquad r = 4\%$$
$$t = 9 \text{ months} = \frac{3}{4} \text{ year}$$

2. Write the equation and solve for A.

$$\begin{aligned} A &= P(1 + rt) \\ &= 3{,}000\left(1 + 0.04\left(\frac{3}{4}\right)\right) \\ &= 3{,}000(1 + 0.03) \\ &= 3{,}000 \times 1.03 \\ &= 3{,}090 \end{aligned}$$

The value of the investment after 9 months is $3,090.

In the case of compound interest, the interest is computed on the principal as well as any interest earned. To compute the amount of an investment, A, involving compound interest, use this formula: $A = P\left(1 + \frac{r}{C}\right)^{tC}$, where C is the number of times compounded annually, r is the annual rate expressed as a decimal, and t is the time in years.

Example: If $6,000 is invested at 5% annual interest, compounded semiannually, what is the balance after 3 years?

1. Identify what is given.

$$P = 6{,}000 \quad r = 5\%$$
$$t = 3 \text{ years} \quad C = 2 \text{ (semiannually)}$$

2. Write the equation and solve for A.

$$\begin{aligned} A &= P\left(1 + \frac{r}{C}\right)^{tC} \\ &= 6{,}000\left(1 + \frac{0.05}{2}\right)^{3\times 2} \\ &= 6{,}000(1.025)^{6} \\ &\approx 6{,}958.16 \end{aligned}$$

The value of the investment after 3 years is $6,958.16.

TRANSLATION INTO ALGEBRA EXERCISES

Basic

Find an algebraic expression or equation to represent each of the following.

1. Javier is 2 years older than three times his sister's age, s. How old is Javier?
2. The product of 5 and one-half of n.
3. 8 more than k is the same as 2.
4. The sum of the cube of 7 and the value of 1.
5. The quotient of h divided by -11.
6. b is at most the sum of 2 and -1.

Solve.

7. The difference between a number and 4.5 is -1. What is the number?
8. The sum of 18 and a number is at most 10. What could the number be?
9. The quotient of a number divided by -15 is less than 4. What could the number be?
10. Two numbers have a sum of 51. One number is 3 more than the other. What is the greater number?

Intermediate

Find an algebraic expression or equation to represent each of the following.

11. The sum of 4 and b is multiplied by -3, and 0.5 is subtracted from the result.
12. 9 is at least $2\frac{1}{2}$ less than x.
13. The product of 4 less than x and 6 more than x is 20.

Solve.

14. Carla drives for 1.5 hours at 55 miles per hour, then stops for lunch. Then she drives for 2.5 hours at 60 miles per hour. How far does she travel in total?

15. Two consecutive even integers have a sum of 70. What is the lesser integer?

16. The length of a rectangle is 3 less than twice its width. If the perimeter is 36, what are the dimensions of the rectangle?

17. Two numbers have the ratio 4:5. The sum of the numbers is 81. What are the numbers?

18. If $10,800 is invested at a simple interest rate of 4%, what is the value of the investment after 18 months?

19. If $4,500 is invested at a simple interest rate of 6%, what is the value of the investment after 10 months?

20. Two consecutive integers have a sum of –35. What are the integers?

ADVANCED

21. Sixty people attended a concert. Children's tickets sold for $8, and adult tickets sold for $12. If $624 was collected in ticket money, how many children and how many adults attended the concert?

22. Two cars started from the same point and traveled on a straight course in opposite directions for exactly 3 hours, at which time they were 300 miles apart. If one car traveled, on average, 10 miles per hour faster than the other car, what was the average speed of each car for the 3-hour trip?

23. For a given two-digit positive integer, the ones digit is 1 less than the tens digit. The sum of the digits is 15. Find the integer.

24. If the ratio of $3x$ to $5y$ is 4:7, what is the ratio of x to y?

25. At a snack stand, hot dogs cost $3.50, and hamburgers cost $5. If the snack stand sold 27 snacks and had total sales of $118.50, how many hot dogs were sold? How many hamburgers?

26. The sum of x and $\frac{1}{2}$ of y is 10. The value of x minus y is -5. What are the numbers represented by x and y?

27. If $9,500 is invested at 4.5% annual interest, compounded quarterly, what is the balance after 2 years?

28. The length of a rectangle is 5 less than 5 times its width. If the area is 60 square units, what is the length of the rectangle?

29. If $2,300 is invested at $5\frac{1}{2}\%$ annual interest, compounded semiannually, what is the balance after 5 years?

30. Hector invested a total of $6,000. Part of the money was invested in a money market account that paid 9% simple annual interest, and the remainder of the money was invested in a fund that paid 7% simple annual interest. If the interest earned at the end of the first year from these investments was $490, how much did Hector invest at 9%, and how much was invested at 7%?

TRANSLATION INTO ALGEBRA ANSWER KEY

BASIC

1. $J = 2 + 3s$

Let J represent Javier's age. Three times s is $3s$, and 2 years older means plus 2. Another correct expression is $J = 3s + 2$.

2. $5\left(\frac{1}{2}n\right)$

"Product" means to multiply. Half of n means $\frac{1}{2}$ times n. A multiplication sign can be used instead of parentheses. Another correct expression is $\frac{5n}{2}$.

3. $8 + k = 2$

"8 more than k" means 8 plus k, and "is" translates to equals. Another correct equation is $k + 8 = 2$.

4. $7^3 + 1$

The "cube" of a number means that number raised to the third power. "Sum" indicates addition.

5. $\frac{h}{-11}$

"Quotient" indicates division. Another correct expression is $h \div -11$.

6. $b \leq 2 + (-1)$

The phrase "at most" means "no more than," which translates to the $\leq$ symbol.

7. $n = 3.5$

"Difference" indicates subtraction, so write the subtraction equation $n - 4.5 = -1$. Then solve by adding 4.5 to both sides.

$$\begin{aligned} n - 4.5 &= -1 \\ n - 4.5 + 4.5 &= -1 + 4.5 \\ n &= 3.5 \end{aligned}$$

8. $x \leq -8$

The phrase "at most" means "no more than." Write and solve $18 + x \leq 10$. Subtract 18 from both sides. The solution includes all numbers less than or equal to -8.

$$\begin{aligned} 18 + x &\leq 10 \\ 18 - 18 + x &\leq 10 - 18 \\ x &\leq -8 \end{aligned}$$

9. $x > -60$

Write the inequality $\frac{x}{-15} < 4$. Solve by multiplying both sides by -15. Whenever you multiply or divide both sides of an inequality by a negative, reverse the symbol.

$$\frac{x}{-15} < 4$$
$$-15 \times \frac{x}{-15} > 4(-15)$$
$$x > -60$$

10. 27

Let $x =$ the lesser number, and let $x + 3 =$ the greater number. Write the equation $x + x + 3 = 51$. Combine like terms on the left side, and then solve the equation.

$$x + x + 3 = 51$$
$$2x + 3 = 51$$
$$2x = 48$$
$$x = 24$$

Since $x = 24$, the lesser number is 24 and the greater number is $x + 3 = 24 + 3 = 27$. The question asked for the greater number, so the answer is 27.

INTERMEDIATE

11. $-3(4 + b) - 0.5$

To indicate that a sum is multiplied by a number, use parentheses. If you wrote $-3 \times 4 + b - 0.5$, the multiplication would only apply to 4, and not b.

12. $9 \geq x - 2\frac{1}{2}$

The phrase "at least" indicates an inequality that uses the symbol $\geq$. Note that another way to write this inequality would be $x - 2\frac{1}{2} \leq 9$.

13. $(x - 4)(x + 6) = 20$

Since the factors are binomials, use parentheses to group them. No multiplication sign is needed. Note that "4 less than x" means $x - 4$, not $4 - x$.

14. 232.5 miles

To find distance, use $d = rt$. Add the distances of the two different parts of the trip to find the total.

$$d = r_1t_1 + r_2t_2$$
$$d = 55(1.5) + 60(2.5)$$
$$d = 82.5 + 150$$
$$d = 232.5$$

15. 34

Let $n =$ the lesser integer. The next even integer would be 2 more than n, so the greater integer is $n + 2$.

$$\begin{aligned} n + n + 2 &= 70 \\ 2n + 2 &= 70 \\ 2n &= 68 \\ n &= 34 \end{aligned}$$

If $n = 34$, then $n + 2 = 36$. Check: $34 + 36 = 70$.

The question asked for the lesser integer, so the answer is 34.

16. 7 and 11

Always make a sketch for polygon problems. Draw a rectangle and label the sides. Since the length is compared to the width, let $w =$ width and $2w - 3 =$ length.

w

2w – 3

To find the perimeter of a rectangle, find the sum of all four sides.

$$\begin{aligned} (w) + (2w - 3) + (w) + (2w - 3) &= 36 \\ 6w - 6 &= 36 \\ 6w &= 42 \\ w &= 7 \end{aligned}$$

The width is 7, so the length is $2w - 3 = 2(7) - 3 = 11$.

To check the answer, find the perimeter using the dimensions you found: $7 + 11 + 7 + 11 = 36$, so the answer is correct. The problem asks for the dimensions, so both numbers are needed to answer the question.

17. 36 and 45

Think about numbers that have a ratio of 4:5. Some ratios include $\frac{4}{5}, \frac{8}{10}, \frac{12}{15}$, and $\frac{40}{50}$. Notice that both 4 and 5 are always multiplied by the same number to find the numbers in an equivalent ratio. Think of 4 and 5 both being multiplied by x. Write the equation $4x + 5x = 81$. Combine like terms and use division to find that $x = 9$. The question asks for the two numbers in the ratio. Since $x = 9$, the numbers are $4x = 36$ and $5x = 45$.

18. $11,448

Use the simple interest formula. Since $t =$ the number of years, use $\frac{18}{12}$ for t.

$$\begin{aligned} A &= P(1 + rt) \\ A &= 10{,}800\left(1 + 0.04 \times \frac{18}{12}\right) \\ A &= 10{,}800(1 + 0.04 \times 1.5) \\ A &= 10{,}800(1 + 0.06) \\ A &= 10{,}800(1.06) \\ A &= 11{,}448 \end{aligned}$$

19. $4,725

Use the simple interest formula. Since $t =$ the number of years, use $\frac{10}{12}$ for t.

$$\begin{aligned} A &= P(1 + rt) \\ A &= 4{,}500\left(1 + 0.06 \times \frac{10}{12}\right) \\ A &= 4{,}500\left(1 + \frac{0.6}{12}\right) \\ A &= 4{,}500(1+0.05) \\ A &= 4{,}500(1.05) \\ A &= 4{,}725 \end{aligned}$$

20. –18 and –17

"Consecutive" means in a row. For example, consecutive integers include 4, 5, 6, 7, etc. Each is one more than the previous integer. Let $n =$ the first integer and $n + 1 =$ the greater integer. The sum of these two expressions is –35. Solve $n + n + 1 = -35$.

$$\begin{aligned} n + n + 1 &= -35 \\ 2n + 1 &= -35 \\ 2n &= -36 \\ n &= -18 \end{aligned}$$

The lesser integer is –18, so the greater integer is $n + 1 = -18 + 1 = -17$.

Advanced

21. 24 children, 36 adults

Write and solve a system of equations. Let $c =$ the number of children's tickets and $a =$ the number of adult tickets. Write one equation for the number of tickets sold and another equation for the money earned.

$$\begin{aligned} c + a &= 60 \\ 8c + 12a &= 624 \end{aligned}$$

Then solve the system using substitution. Subtract a from both sides of the first equation to find that $c = 60 - a$. Substitute $60 - a$ for c into the second equation.

$$\begin{aligned} 8(60 - a) + 12a &= 624 \\ 480 - 8a + 12a &= 624 \\ 480 + 4a &= 624 \\ 4a &= 144 \\ a &= 36 \end{aligned}$$

This means that there were 36 adult tickets sold. Substitute this into the first equation to find the number of children's tickets sold.

$$\begin{aligned} c + a &= 60 \\ c + 36 &= 60 \\ c &= 24 \end{aligned}$$

So, 36 adult tickets and 24 children's tickets were sold.

22. 45 miles per hour and 55 miles per hour

Think of two partial distances being added together to form one long distance of 300 miles. Let $x =$ the rate of one car and $x + 10 =$ the rate of the faster car. Use the formula distance = rate × time, or $d = rt$.

$$\begin{aligned} d_1 + d_2 &= 300 \\ r_1t_1 + r_2t_2 &= 300 \\ x(3) + (10 + x)3 &= 300 \\ 3x + 30 + 3x &= 300 \\ 6x + 30 &= 300 \\ 6x &= 270 \\ x &= 45 \end{aligned}$$

This means that the slower car was driving 45 miles per hour and $x + 10 = 55$ is the rate of the other car. The question asks for the rate of both cars, so the answer is 45 miles per hour and 55 miles per hour.

23. 87

Write and solve a system of equations. Let x = the tens digit and y = the ones digit. Since the sum of the digits is 15, write the equation $x + y = 15$. Since the ones digit is 1 less than the tens digit, think "ones digit = tens digit minus 1", and write the equation $y = x - 1$. Solve to find both digits of the number. Substitute $x - 1$ for y in the equation $x + y = 15$:

$$\begin{aligned} x + x - 1 &= 15 \\ 2x - 1 &= 15 \\ 2x &= 16 \\ x &= 8 \end{aligned}$$

The tens digit is 8, and the ones digit is $15 - 8 = 7$. Therefore, the number is 87.

24. $\frac{20}{21}$

Write a proportion to show the equivalent ratios. Then multiply both sides of the equation by $\frac{5}{3}$ to isolate $\frac{x}{y}$. Express the answer as a ratio.

$$\begin{aligned} \frac{3x}{5y} &= \frac{4}{7} \\ \frac{5}{3} \times \frac{3x}{5y} &= \frac{4}{7} \times \frac{5}{3} \\ \frac{x}{y} &= \frac{20}{21} \end{aligned}$$

So, the answer is $\frac{20}{21}$.

25. 11 hot dogs and 16 hamburgers

Write and solve a system of equations. Let h = the number of hot dogs sold and b = the number of hamburgers sold. Write one equation for the number of snacks sold and another equation for the money earned.

$$\begin{aligned} h + b &= 27 \\ 3.5h + 5b &= 118.5 \end{aligned}$$

Solve the first equation for b: $b = 27 - h$.

Substitute $27 - h$ for b in the second equation.

$$\begin{aligned} 3.5h + 5(27 - h) &= 118.5 \\ 3.5h + 135 - 5h &= 118.5 \\ -1.5h + 135 &= 118.5 \\ -1.5h &= -16.5 \\ h &= 11 \end{aligned}$$

Substitute 11 for h in the first equation to find b:

$$\begin{aligned} 11 + b &= 27 \\ b &= 16 \end{aligned}$$

There were 11 hot dogs sold and 16 hamburgers sold.

26. $x = 5$ and $y = 10$

Write out the system of equations.

$$\begin{aligned} x + \frac{1}{2}y &= 10 \\ x - y &= -5 \end{aligned}$$

To eliminate the x terms, combine the equations by subtracting the second equation from the first.

$$\begin{aligned} x + \frac{1}{2}y &= 10 \\ -(x - y &= -5) \\ \frac{1}{2}y - (-y) &= 10 - (-5) \\ \frac{3}{2}y &= 15 \\ y &= 15 \times \frac{2}{3} \\ y &= 10 \end{aligned}$$

Now, substitute 10 for y in either equation to find x. Since the second equation is easier to work with, use that one:

$$\begin{aligned} x - y &= -5 \\ x - (10) &= -5 \\ x &= 5 \end{aligned}$$

The answer is $x = 5$ and $y = 10$.

27. $10,389.20

Use the compound interest formula. Express the rate 4.5% as 0.045.

$$\begin{aligned} A &= P\left(1 + \frac{r}{C}\right)^{tC} \\ A &= 9{,}500\left(1 + \frac{0.045}{4}\right)^{2 \times 4} \\ A &= 9{,}500(1 + 0.01125)^{8} \\ A &= 9{,}500(1.01125)^{8} \\ A &\approx 9{,}500(1.0936) \\ A &\approx 10{,}389.20 \end{aligned}$$

28. $l = 15$

Draw and label a diagram. Since the length is compared to the width, let w = the width.

w

$5w - 5$

Since the area of a rectangle equals *length* × *width*, write an expression with the factors being multiplied and set it equal to the area, which is 60.

$$\begin{aligned} w(5w - 5) &= 60 \\ 5w^2 - 5w &= 60 \\ 5w^2 - 5w - 60 &= 0 \\ 5(w^2 - w - 12) &= 0 \\ 5(w + 3)(w - 4) &= 0 \\ w + 3 &= 0;\ w - 4 = 0 \\ w &= -3;\ w = 4 \end{aligned}$$

Since the rectangle cannot have a negative width, use the solution $w = 4$. Then the length is $5w - 5 = 5(4) - 5 = 15$.

29. $3,016.80

Use the compound interest formula.

$$\begin{aligned} A &= P\left(1 + \frac{r}{C}\right)^{tC} \\ A &= 2{,}300\left(1 + \frac{0.055}{2}\right)^{5\times 2} \\ A &= 2{,}300(1 + 0.0275)^{10} \\ A &= 2{,}300(1.0275)^{10} \\ A &\approx 2{,}300(1.31165) \\ A &\approx 3{,}016.80 \end{aligned}$$

30. \$3,500 in the 9% account and \$2,500 in the 7% account

Let x = the amount invested in the 9% account. Therefore, 6,000 – x represents the amount invested in the 7% account. The amounts invested must add up to 6,000. To find each amount of interest, use $I = Prt$. (Note that this formula gives the amount of interest, not the total amount of investment plus interest.) The two amounts of interest add up to 490.

$$\begin{aligned} 0.09(x) + 0.07(6{,}000 - x) &= 490 \\ 0.09x + 420 - 0.07x &= 490 \\ 0.02x + 420 &= 490 \\ 0.02x &= 70 \\ x &= 3{,}500 \end{aligned}$$

Hector invested \$3,500 in the 9% account. Subtract this amount from 6,000 to find the amount invested in the 7% account: 6,000 – 3,500 = 2,500. Hector invested \$2,500 in the 7% account.

FUNCTIONS

Classic function notation problems may appear on the test. An algebraic expression of only one variable may be defined as a **function**, usually *f* or *g*, of that variable.

Example: What is the value of the function $f(x) = x^2 - 1$ when $x = 1$?

In the function $f(x) = x^2 - 1$, if x is 1, then $f(1) = 1^2 - 1 = 0$. In other words, when the **input** to the function is 1, the **output** is 0. The set of input numbers for a function is the **domain** and the set of output values is the **range** of the function. Every input value of a function has exactly one output value. However, more than one input value may have the same output value. Consider the function $f(x) = x^2 - 1$ again. If $x = -1$, $f(-1) = (-1)^2 - 1 = 1 - 1 = 0$. So for $f(x) = x^2 - 1$, $f(1) = f(-1) = 0$.

Restricted Domain of a Function

A function may be defined for all real numbers or it may be defined only for a subset of the real numbers.

Example: $g(x) = -x^2 + 2$, where $-3 \leq x \leq 3$

The domain of the function $g(x)$ is restricted to values of x between -3 and 3 inclusive. On the GRE you may be asked to find the minimum value of a function or the maximum value of a function. The maximum value of $g(x) = -x^2 + 2$ occurs when $x = 0$; $g(0) = -x^2 + 2 = 0 + 2 = 2$. On the graph, this is represented by the point (0, 2). To find the minimum value of $g(x) = -x^2 + 2$, where $-3 \leq x \leq 3$, find the value when $x = -3$ or $x = 3$.

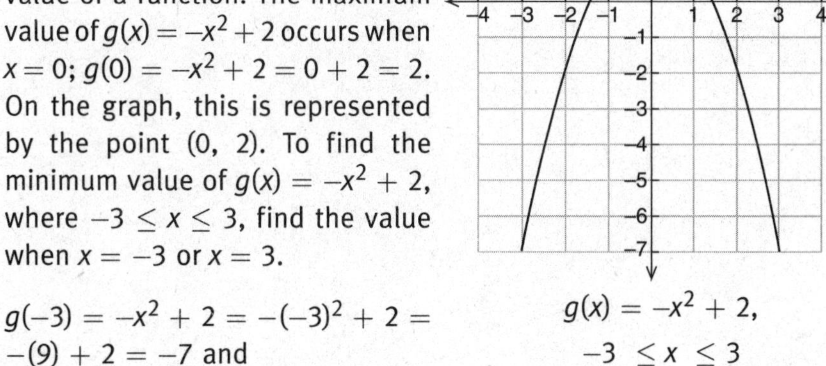

$g(x) = -x^2 + 2$,
$-3 \leq x \leq 3$

$g(-3) = -x^2 + 2 = -(-3)^2 + 2 = -(9) + 2 = -7$ and

$g(3) = -x^2 + 2 = -(3)^2 + 2 = -(9) + 2 = -7$

On the GRE, a minimum (or maximum) value problem can be solved in one of two ways. Either plug the answer choices into the function and find which gives you the least (or greatest) value, or use what you know about number operations. In the case $g(x) = -x^2 + 2$, the function g will be at a maximum when $-x^2$ is as small as possible. This leads you to consider the case where $x = 0$. Conversely, the minimum value occurs when $-x^2$ is as large as possible, considering the restricted domain. This occurs when $x = -3$ or $x = 3$. That value is -7.

The domain of a function may also be restricted to avoid having a zero in the denominator of a fraction or to avoid taking the square root of a negative number.

Examples: $f(x) = \frac{2x - 5}{x - 3}$, where $x \neq 3$ — Restrict domain of f to avoid $x - 3 = 0$.

$h(x) = \sqrt{x + 10}$, where $x \geq -10$ — Restrict the domain of h to avoid $x + 10 < 0$.

ABSOLUTE VALUE FUNCTION

The function $g(x) = |x|$ is called the **absolute value function**. The domain of g is all real numbers—unless it is restricted in a certain problem—and represents the distance between any number x and 0 on the number line. Because the absolute value function represents a distance, its output is always a positive number or zero.

FUNCTIONS EXERCISES

BASIC

1. What is the domain of this function? {(–3, 4), (–2, 5), (–1, 6), (0, 7), (1, 8)}

2. What is the range of this function? {(–3, 4), (–2, 5), (–1, 6), (0, 7), (1, 8)}

3. Determine if this table represents a function.

x	y
8	4
6	3
4	2
2	1
0	0

4. Does this set of ordered pairs represent a function? {(8, 5), (4, 6), (2, 3), (4, 9), (1, 7)}

5. Write the equation $y = 2x + 7$ using function notation.

6. If $f(x) = 4x + 1$, what is the value of $f(-3)$?

7. If $f(x) = -2(x - 4)$, what is the value of $f(-7)$?

8. What is the range of the function $f(x) = 6 + x$ if the domain is $\{-4, 0, 4\}$?

9. If $g(x) = x^2 + x$, what is the value of $g(9)$?

10. For what value should the domain be restricted for the function $f(x) = \frac{x + 2}{x - 2}$?

INTERMEDIATE

11. What is the minimum value for the function $f(x) = x^2 - 5$?

12. What is the maximum value for the function $g(x) = -2x^2 - 1$?

13. What is the minimum value for the function $f(x) = -x^2 + x + 4$ where $-6 \leq x \leq 6$?

14. For what values should the domain be restricted for the function $g(x) = \sqrt{x + 8}$?

15. Does the graph below represent a function? Why or why not?

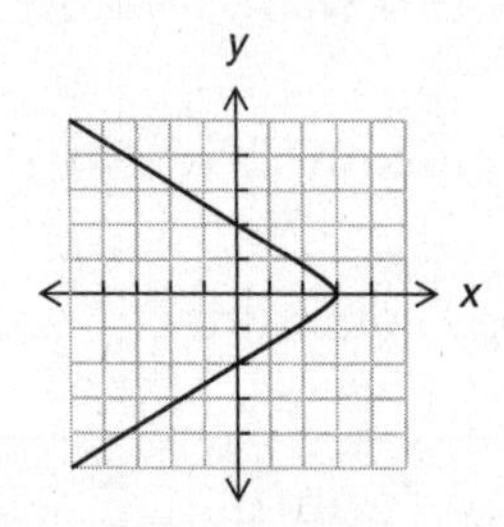

16. Express the relationship between x and $f(x)$ algebraically; x and $f(x)$ have a linear relationship.

x	4	5	6	−4	−5	−6
$f(x)$	25	31	37	−23	−29	−35

17. Determine the range for this function table.

$g(x) = |x| + 2$

x	$g(x)$
−8.8	
−6.6	
−4.4	
−2.2	
0	

18. What is the range of the function shown in the graph?

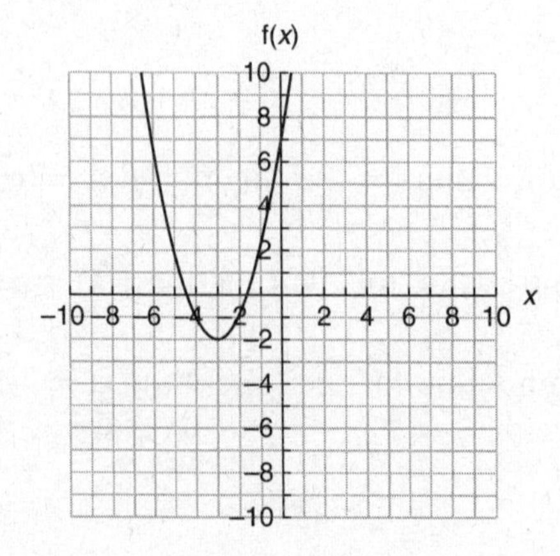

19. What is the domain for the function shown in the graph?

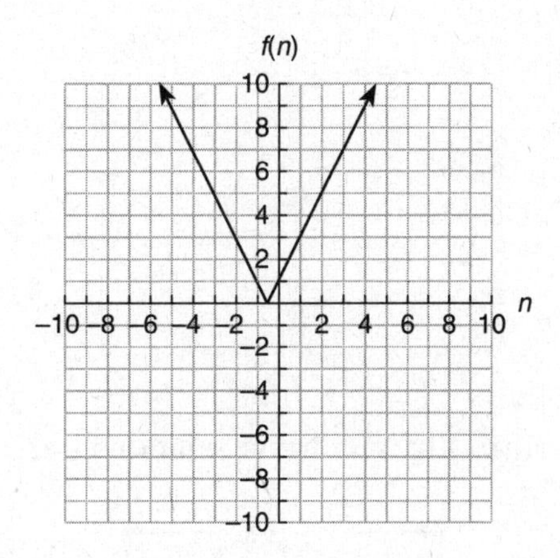

20. If $f(x) = \frac{1}{2}x + 8$, what value of x makes $f(x) = 0$?

ADVANCED

21. For what values should the domain be restricted for this function?

$$f(n) = \frac{n - 6}{n^2 - 6n}$$

22. What is the domain of the function shown below?

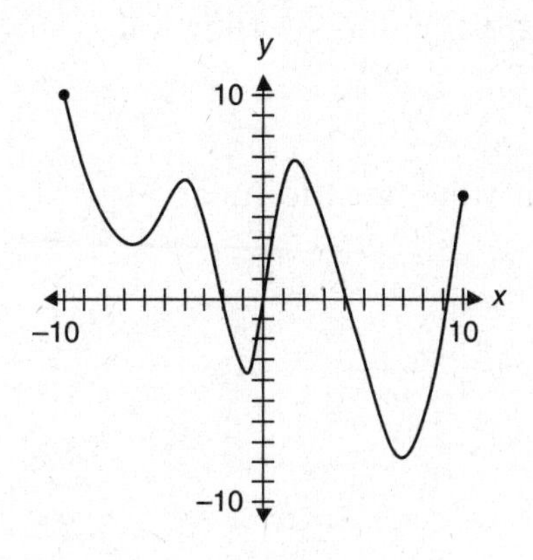

23. What is $h(-4)$ if $h(x) = h^3 + 3h^2 + h$?

24. If $g(a) = a^2 - 1$ and $f(a) = a + 4$, what is $g(f(-4))$?

25. What is the domain for this function?

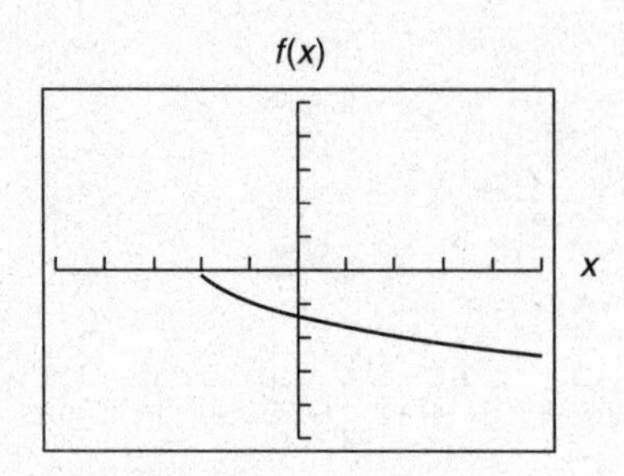

26. What equation describes this quadratic relationship?

x	$f(x)$
3	10
6	37
9	82
12	145

27. If $f(x) = \dfrac{2x + 6}{4}$ and $g(x) = 2x - 1$, what is $f(g(x))$?

28. What is the minimum value for the function $f(x) = -x^2 - x + 1$ where $-6 \leq x \leq 6$?

29. What is the maximum value for the function $g(x) = -2x^2 + 5$ where $-5 < x < 5$?

30. What is the minimum value for the function $|x| + 3 = g(x)$?

FUNCTIONS ANSWER KEY

BASIC

1. {–3, –2, –1, 0, 1}

The domain of a function is the set of *x*-values in a set of ordered pairs. In the set {(–3, 4), (–2, 5), (–1, 6), (0, 7), (1, 8)} the *x*-values (domain) are listed first in each pair.

2. {4, 5, 6, 7, 8}

The range of a function is the set of *y*-values in a set of ordered pairs. In the set {(–3, 4), (–2, 5), (–1, 6), (0, 7), (1, 8)} the *y*-values (range) are listed second in each pair.

3. Yes, this table represents a function.

Each element of the domain is paired with exactly one element of the range.

4. No, this set of ordered pairs does not represent a function.

The *x*-value of 4 is paired with two different *y*-values; therefore, it is not a function.

5. $f(x) = 2x + 7$

In a function, x represents the domain, and $f(x)$ represents the range.

6. –11

Replace x with -3 in $f(x) = 4x + 1$ to find the value of $f(-3)$:

$$f(-3) = 4(-3) + 1$$
$$f(-3) = -12 + 1$$
$$f(-3) = -11$$

7. 22

Replace x with -7 in $f(x) = -2(x - 4)$ to find the value of $f(-7)$:

$$f(-7) = -2(-7 - 4)$$
$$f(-7) = -2(-11)$$
$$f(-7) = 22$$

8. {2, 6, 10}

Replace x with each value of the domain and solve for the range.

Replace x with -4 in $f(x) = 6 + x$. $\quad f(x) = 6 + -4 = 2$

Replace x with 0 in $f(x) = 6 + x$. $\quad f(x) = 6 + 0 = 6$

Replace x with 4 in $f(x) = 6 + x$. $\quad f(x) = 6 + 4 = 10$

9. 90

Replace x with 9 in $g(x) = x^2 + x$ to find the value of $g(9)$.

$$g(9) = 9^2 + 9$$
$$g(9) = 81 + 9$$
$$g(9) = 90$$

10. $x = 2$

Restrict the domain of x to avoid $x - 2 = 0$; 0 is not permissible as a divisor.

INTERMEDIATE

11. -5

The minimum value for the function $f(x) = x^2 - 5$ is found by replacing x with 0; $0^2 - 5 = 0 - 5 = -5$. Because the value of any number squared (besides 0) is positive, the value of x^2 is always positive; replacing x with 0 gives the minimum of the function.

12. -1

The maximum value for the function $g(x) = -2x^2 - 1$ is found by replacing x with 0; $-2(0)^2 - 1 = 0 - 1 = -1$. Because the coefficient of x^2 is negative, replacing x with 0 gives the maximum of the function.

13. -38

The minimum value for the function $f(x) = -x^2 + x + 4$ where $-6 \leq x \leq 6$ is found by replacing x with -6; $f(-6) = -(-6)^2 + (-6) + 4 = -36 - 6 + 4 = -38$. Replacing x with any other value of the domain results in a value greater than -38.

14. All values less than -8; $x < -8$

Restrict the domain of x to avoid $x + 8 < 0$; the square root of a negative number is nonreal.

15. Yes

We're used to thinking of functions as $y = f(x)$; that is, given a particular x-value, we can find a single value for y. There are multiple (indeed, infinite) points on this graph at which two different y-values exist for the same x-value. Therefore, this graph does not represent a function in the sense that you might be used to. However, it is a function expressed as $x = f(y)$: for a given y value, there is a single x value. Therefore, this is a function.

16. $f(x) = 6x + 1$

Because you are told that this is a linear relationship, try to express the relationship in the slope-intercept form of a line, $f(x) = y = mx + b$, where m is the slope and b is the y-intercept. First, find the slope using the equation $m = \frac{\text{change in } y}{\text{change in } x} = \frac{31 - 25}{5 - 4} = \frac{6}{1} = 6$.

Next, plug in values of m, x, and y into the slope-intercept form to find b.

$$\begin{aligned} y &= mx + b \\ 25 &= (6)(4) + b \\ 25 &= 24 + b \\ b &= 1 \end{aligned}$$

Finally, use function notation to express the relationship between x and $f(x)$:

$$f(x) = 6x + 1$$

Plug in the values of x to make sure the values of $f(x)$ correspond with your chosen rule.

17. {10.8, 8.6, 6.4, 4.2, 2}

Replace x in the function with each value of the domain to find the range.

18. $f(x) \geq -2$

To determine the range of the function, look at all the values of $f(x)$. The values are greater than or equal to -2, so the range is $f(x) \geq -2$.

19. The domain of the function is the set of all real numbers.

The domain represents all the values for n. Since there are no constraints on what n can be, the domain is the set of all real numbers.

20. $x = -16$

Replace $f(x)$ with 0 and then solve for x.

$$\begin{aligned} \frac{1}{2}x + 8 &= 0 \\ \frac{1}{2}x &= -8 \\ x &= -16 \end{aligned}$$

Advanced

21. 6 and 0

Factor the denominator: $n^2 - 6n = n(n - 6)$; the value of n cannot be 6 or 0, as a denominator of 0 is not permissible.

22. $-10 \leq x \leq 10$

The x-values on the graph range from -10 to 10, so the domain includes all of the values between these numbers.

23. –20

Replace h with –4 in $h(x) = h^3 + 3h^2 + h$ to find the value of $h(-4)$.

$$h(-4) = (-4)^3 + 3(-4)^2 + (-4)$$
$$h(-4) = -64 + 3(16) + (-4)$$
$$h(-4) = -64 + 48 - 4$$
$$h(-4) = -20$$

24. –1

Replace a in $f(a)$ with –4: $f(-4) = -4 + 4 = 0$. Then replace a in $g(a)$ with the value of $f(-4)$, or 0: $g(f(-4)) = g(0) = (0)^2 - 1 = -1$.

25. $x \geq -2$

Notice the values on the x-axis. For this function, x begins at –2 and increases, so the domain of this function is any value greater than or equal to –2.

26. $f(x) = x^2 + 1$

Because you are told that this is a quadratic relationship, try to express the relationship in the general form of a quadratic equation, $f(x) = y = ax^2 + bx + c$. Before setting up a complicated series of three equations to solve for a, b, and c, look at the values of x and $f(x)$ that are given. You know that $f(x)$ is related to x^2, so square the given values of x and see how they relate to $f(x)$.

x	x^2	$f(x)$	$f(x) - x^2$
3	9	10	1
6	36	37	1
9	81	82	1
12	144	145	1

Since $f(x) - x^2$ is a constant for all values of x and $f(x)$ given, $f(x) - x^2 = 1 = c$, so b must equal 0. Finally, use function notation to express the relationship between x and $f(x)$:

$$f(x) = x^2 + 1$$

Plug in the values of x to make sure the values of $f(x)$ correspond with your chosen rule.

27. $x + 1$

To find $f(g(x))$, replace the variable in $f(x)$ with $g(x)$, or $2x - 1$.

$$f(x) = \frac{2x + 6}{4}$$

$$f(g(x)) = \frac{2(g(x)) + 6}{4} = \frac{2(2x - 1) + 6}{4} = \frac{4x - 2 + 6}{4} = \frac{4x + 4}{4} = x + 1$$

28. -41

To determine the minimum value of the function $f(x) = -x^2 - x + 1$ where $-6 \leq x \leq 6$, find the value of $x = 6$, which yields the least value of $f(x)$ in the domain.

$$-(6)^2 - 6 + 1 = -36 - 6 + 1 = -41$$

29. 5

The maximum value for the function $g(x) = -2x^2 + 5$ where $-5 < x < 5$ is found when $x = 0$.

$$-2(0)^2 + 5 = 0 + 5 = 5$$

30. 3

The minimum for this function occurs at the least possible value for the absolute value of x, which is $x = 0$.

$$|0| + 3 = 3$$

COORDINATE GEOMETRY

Questions on the GRE sometimes refer to lines or curves graphed on a coordinate plane. An ***xy*-coordinate plane** is formed by two number lines at right angles to each other, intersecting at their zero points. The intersecting lines form four quadrants, named by the Roman numerals, I, II, III, and IV. The horizontal number line is the ***x*-axis**, the vertical number line is the ***y*-axis** and the point of intersection is the **origin** (*O*). Every point on the plane is named by its *x*- and *y*-coordinates, (*x*, *y*).

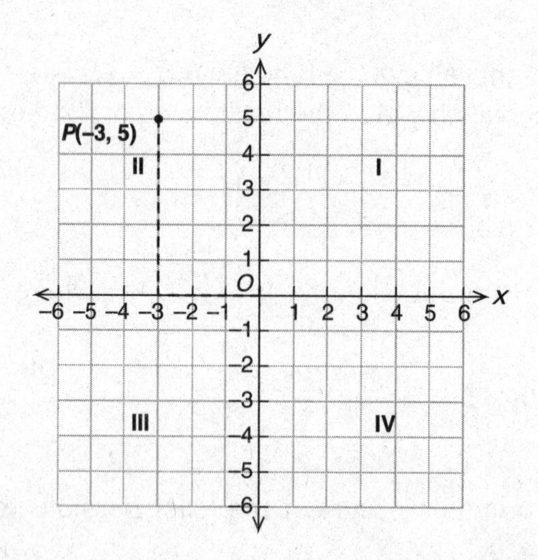

To graph point *P*, whose coordinates are (–3, 5), start at the origin, go left 3 units or spaces, and then go up 5 units or spaces. To reflect point *P*:

- across the *x*-axis, change the sign of the *y*-coordinate.
- across the *y*-axis, change the sign of the *x*-coordinate.
- about the origin, change the sign of both the *x*- and *y*-coordinate.

The reflections of point *P* are shown on the following graph.

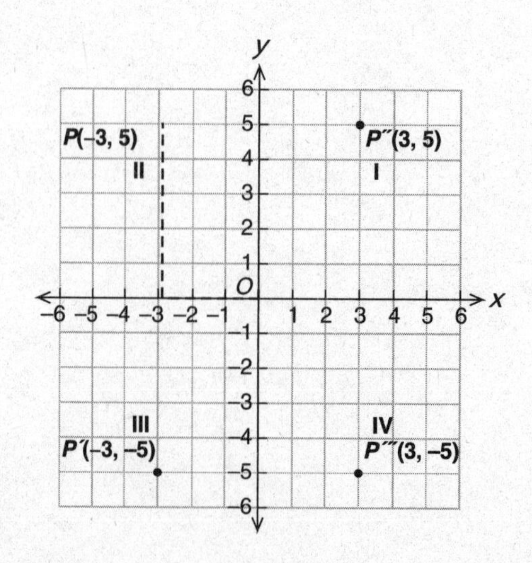

Distance between Points

There are several ways to find the distance between two points in the *xy*-plane, depending on the location of the points. If two points lie on the same horizontal or vertical grid line, just count the units or spaces between them. The distance from C to A below is $5 + 3 = 8$ units or spaces. Likewise, the distance from A to B is 4 units or spaces.

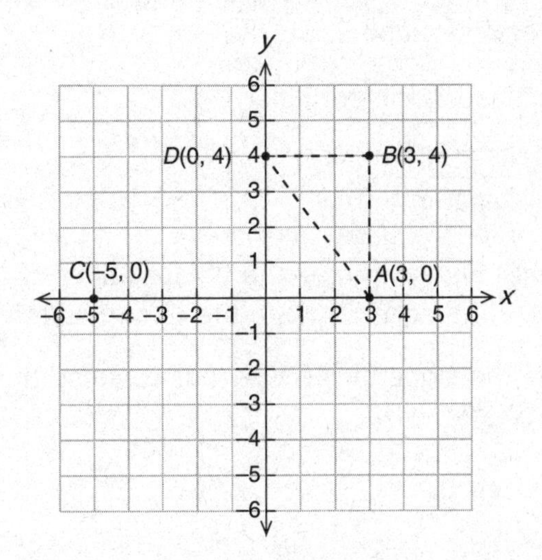

To find the distance from A to D, consider the line segment AD as the hypotenuse of a right triangle, whose legs are segments AB and BD. Because triangle ABD has legs of 3 and 4, it is a 3:4:5 triangle. So the distance from A to D is the same as the length of the hypotenuse of the triangle, which is 5. You will see special triangles such as this frequently on the GRE. For more discussion of special right triangles, see chapter 8.

The distance between two points in the *xy*-plane can also be found using the Pythagorean theorem. To find the distance between the points C and B in the figure shown, use the fact that the length of CA is 8 and the length of AB is 4. Plug these numbers into the Pythagorean theorem:

$$\begin{aligned} CB^2 &= CA^2 + AB^2 \\ CB^2 &= 8^2 + 4^2 \\ CB^2 &= 64 + 16 \\ CB^2 &= 80 \\ CB &= \sqrt{80} \approx 9 \end{aligned}$$

The distance between the points C and B is about 9 units or spaces.

Slope of a Line

The graph of an equation in the variables x and y is the set of all points (x, y) whose coordinates satisfy the equation. The graph of a linear equation in the form $y = mx + b$ is a straight line in the *xy*-plane with slope m and y-intercept b. This form of the equation of a line is known as **slope-intercept form**.

The ***y*-intercepts** of a graph are the y-values of the points where the graph crosses the y-axis. The graph of a linear function has only one y-intercept. The ***x*-intercepts** of a graph are the x-values of the points where the graph crosses the x-axis. The graph of a linear function has at most one x-intercept.

The **slope** of a line tells you how steeply that line goes up or down. If a line gets higher as you move to the right, it has a positive slope. If it goes down as you move to the right, it has a negative slope.

To find the slope of a line, use this formula:

$$\text{Slope} = \frac{\text{rise}}{\text{run}} = \frac{\text{change in } y}{\text{change in } x}$$

Rise means the difference between the y-coordinate values of any two points on the line, and *run* means the difference between the x-coordinate values.

Example: What is the slope of the line that contains the points (1, 2) and (4, −5)?

$$\text{Slope} = \frac{-5-2}{4-1} = \frac{-7}{3} = -\frac{7}{3}$$

To determine the slope of a line from an equation, put the equation into the slope-intercept form.

Example: What is the slope of the line represented by the equation $3x + 2y = 4$?

$$3x + 2y = 4$$
$$2y = -3x + 4$$
$$y = -\frac{3}{2}x + 2, \text{ so } m \text{ is } -\frac{3}{2}$$

The larger the absolute value of the slope, the steeper the slope. The slope of any horizontal line is 0 because the *rise* (the numerator of the slope formula) is 0. The slope of any vertical line is undefined because the *run*, (the denominator of the slope formula) is 0.

Parallel lines have the same slope. **Perpendicular** lines have slopes that are negative reciprocals of each other. For example, all lines parallel to $y = -\frac{3}{2}x + 2$ have a slope of $-\frac{3}{2}$, and all lines perpendicular to $y = -\frac{3}{2}x + 2$ have a slope of $\frac{2}{3}$.

GRAPHING SYSTEMS OF EQUATIONS AND INEQUALITIES

A graph can be used to show the solutions of systems of equations or systems of inequalities. Let's look at the graph of a system of two equations first.

Example: What is the solution of the system of linear equations shown below?

$$y = -x + 8$$
$$y = 4x - 7$$

The graphs of the two equations are shown below. The point of intersection, (3, 5), is the solution to the system.

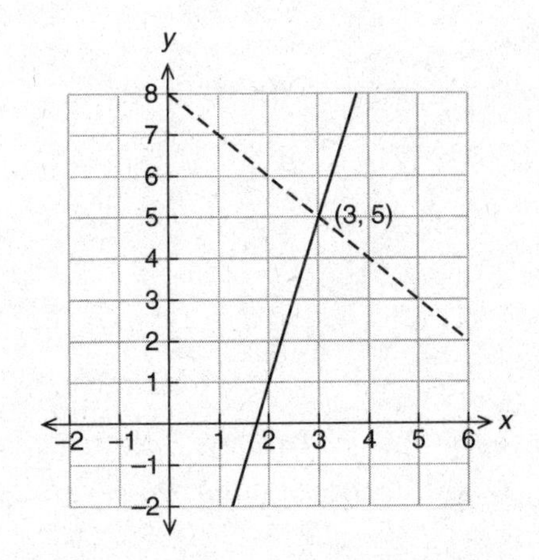

Example: What is the solution of the system of inequalities shown?

$$y \leq x + 2$$
$$y \geq -x$$

The graphs of the two inequalities are shown. The solutions to the inequality $y \leq x + 2$ are all the points on and below the line $y = x + 2$. The solutions to the inequality $y \geq -x$ are all the points on and above the line $y = -x$. The solution to the system includes all the points in the region that is double-shaded.

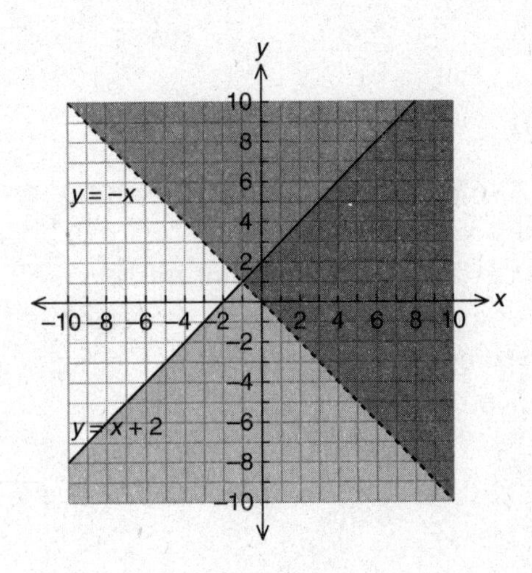

SYMMETRY

A point whose coordinates are (a, b) can be reflected across the line $y = x$ in the coordinate plane. The coordinates of the reflected point will be (b, a). To say it another way, the points (a, b) and (b, a) are symmetric about the line $y = x$.

The line $y = x$ is graphed below. Consider the line $y = 4x + 2$. The equation of the reflection of the line $y = 4x + 2$ is found by interchanging y and x in the equation and then solving for y:

$$\begin{aligned} x &= 4y + 2 \\ -4y &= -x + 2 \\ y &= \frac{-1}{-4}x - \frac{2}{4} \\ y &= \frac{1}{4}x - \frac{1}{2} \end{aligned}$$

Both lines are shown on the graph that follows. Notice that when point $A(1, 6)$ is reflected across the line $y = x$, the coordinates are reversed and you have $A'(6, 1)$.

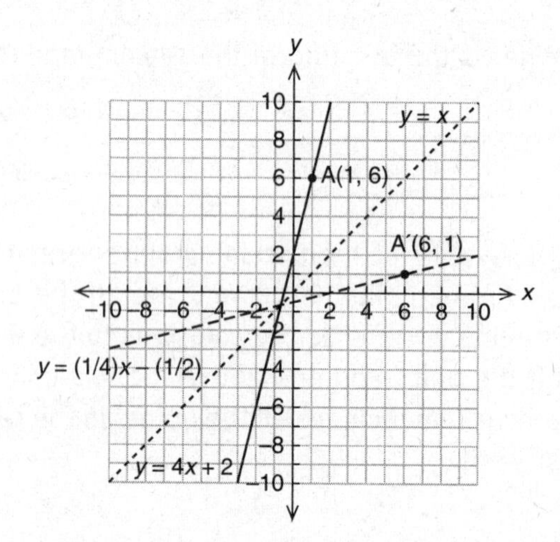

PARABOLAS

The graph of a quadratic equation of the form $ax^2 + bx + c$, where a, b, and c are constants and $a \neq 0$, is a parabola. When a is positive, the parabola opens upward and the **vertex** is the minimum point on its graph. Conversely, when a is negative, the parabola opens downward and the **vertex** is the maximum point on its graph. A vertical line through the vertex is a line of symmetry of the parabola. An example of each case is shown.

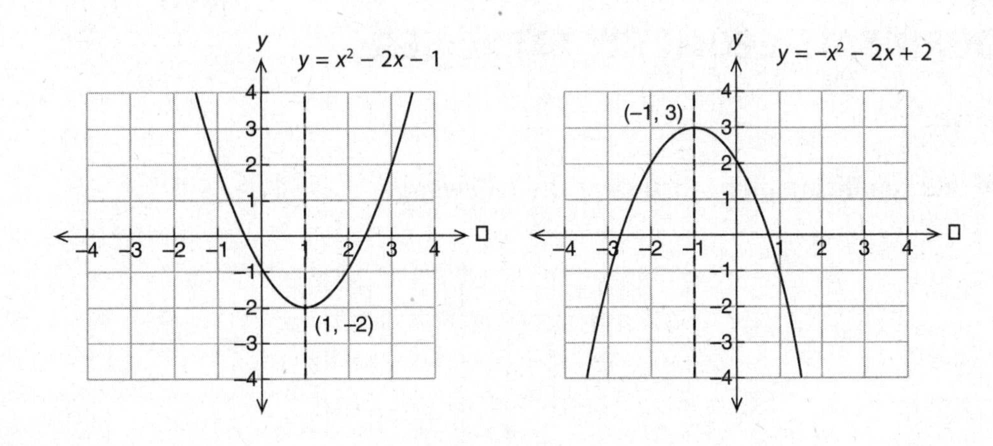

Circles

The graph of $x^2 + y^2 = 64$ is a circle centered at the origin with radius 8. It is the larger circle on the following graph. The smaller circle on the graph is centered at (–2, –2) with radius 3. The equation of the smaller circle is $(x + 2)^2 + (y + 2)^2 = 9$.

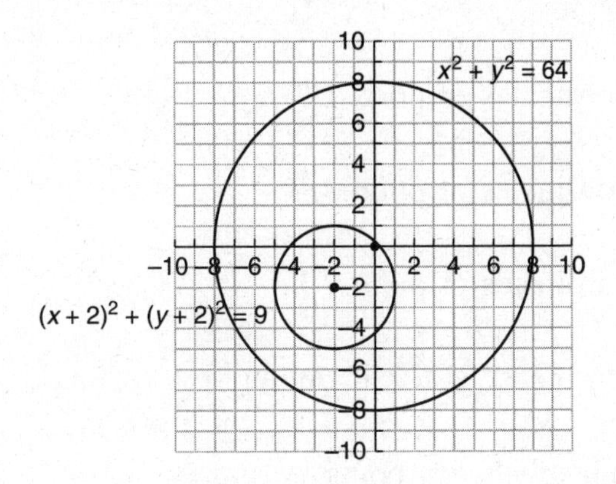

COORDINATE GEOMETRY EXERCISES

BASIC

Use this coordinate plane to answer the following.

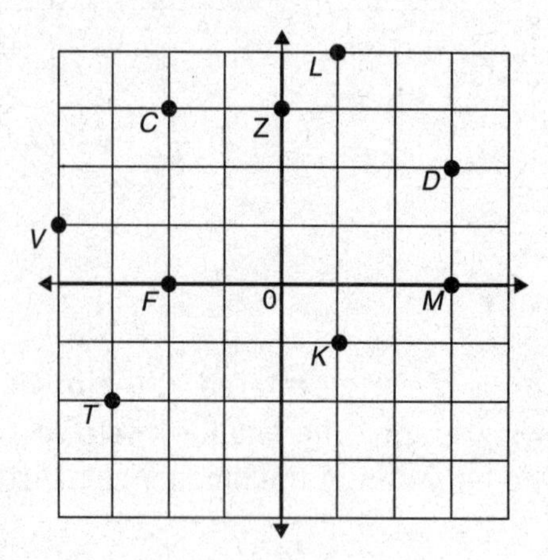

1. What are the coordinates of point *C*?

2. What are the coordinates of point *M*?

3. What are the coordinates of point *T*?

4. What quadrant do points *V* and *C* lie in?

5. What is the distance between points *K* and *L*?

6. What point lies at (0, 3)?

7. What are the coordinates of point *D* if it is reflected over the *x*-axis?

8. What are the coordinates of point *V* if it is reflected over the *y*-axis?

9. What are the coordinates of point *K* if it is reflected about the origin?

10. What is the approximate distance between point *C* and point *M*?

INTERMEDIATE

11. What is the slope of the line that contains the points (4, 8) and (–1, –2)?

12. What is the slope of the line represented by the equation $y = 8x + 3$?

13. What is the slope of the line represented by the equation $4x + 3y = 12$?

14. What is the slope of a line that is parallel to $y = -2x - 4$?

15. What is the slope of a line that is perpendicular to $y = \frac{1}{3}x - 7$?

16. What is the slope of the line that contains the points (–4, 5) and (2, 5)?

17. What is the slope of the line that contains the points (6, 3) and (6, –2)?

18. What is the slope of the line that is represented by the equation $3y = -\frac{1}{2}x$?

19. What is the length of $\overline{CD}$?

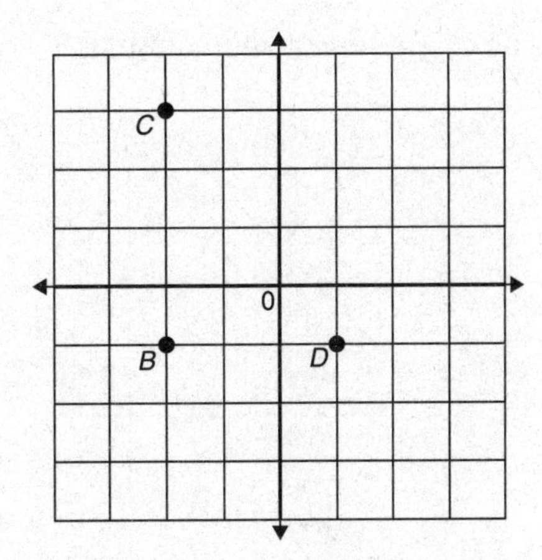

20. What is the solution to the system shown on the coordinate plane?

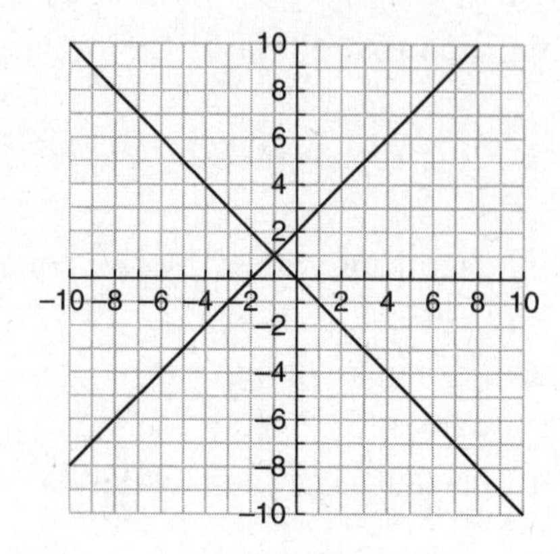

ADVANCED

21. What is the slope of a line parallel to a line that contains the points (–3, 7) and (–3, –2)?

22. What is the slope of a line perpendicular to a line that contains the points (5, −1) and (−2, −1)?

23. Will these two lines have a point of intersection?

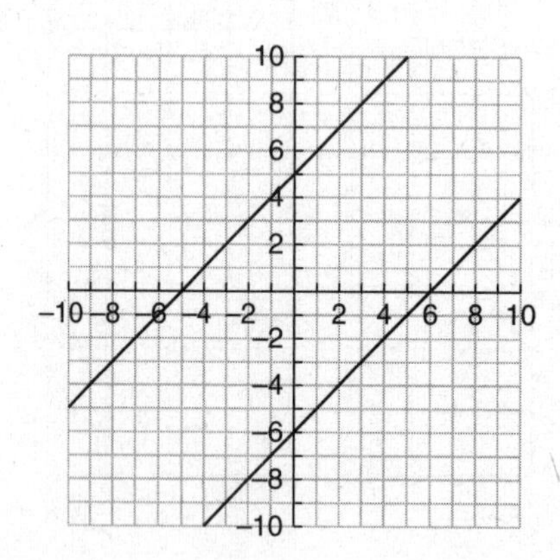

24. What is the solution to this system of inequalities?

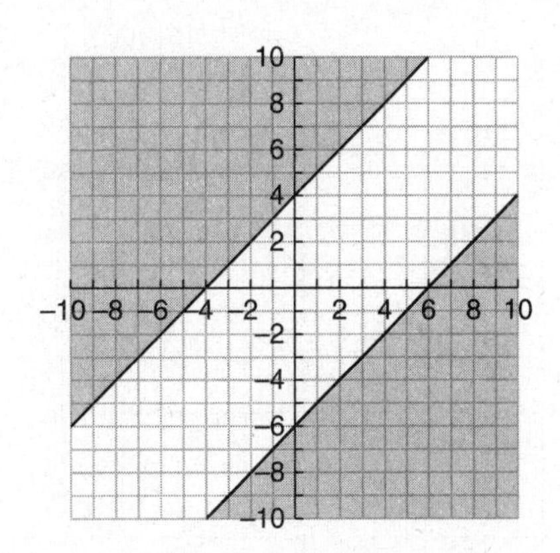

25. What is the slope of a line that is perpendicular to $2x - 5y = 20$?

26. What is the x-intercept of the line represented by $y = 4x - 8$?

27. Determine if the point $(-2, 12)$ lies on the graph of $y = -2.5x + 7$.

28. Which line is steeper?

$y = -3x + 4$

$4x - 3y = 9$

29. Is the origin part of the solution for this system of inequalities?

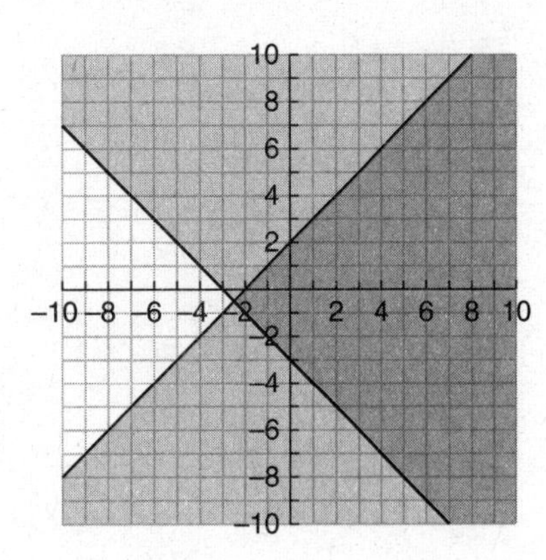

30. What transformation occurs if point C is reflected over the x-axis and then over the y-axis?

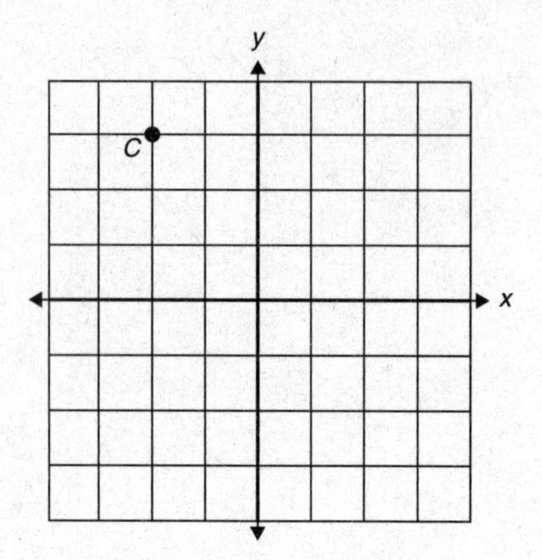

COORDINATE GEOMETRY ANSWER KEY

Basic

1. (–2, 3)

To find the location of point *C*, start at the origin and count two spaces to the left and then count up three spaces.

2. (3, 0)

To find the location of point *M*, start at the origin and count three spaces to the right. When a point is located on the *x*-axis, it will have 0 as a *y*-coordinate.

3. (–3, –2)

To find the location of point *T*, start at the origin and count three spaces to the left and then count down two spaces.

4. Quadrant II

The location of an ordered pair with a negative *x*-coordinate and a positive *y*-coordinate is in Quadrant II.

5. 5 units

Points *K* and *L* are on the same vertical line, so count the number of spaces between points *K* and *L* to determine the distance.

6. Point *Z*

From the origin, count three units up. Point *Z* is located at (0, 3). Any point that lies on the *y*-axis will have 0 as the *x*-coordinate.

7. (3, –2)

The original location of point *D* is (3, 2). To reflect a point over the *x*-axis, change the sign of the *y*-coordinate.

8. (4, 1)

The original location of point *V* is (–4, 1). To reflect a point over the *y*-axis, change the sign of the *x*-coordinate.

9. (–1, 1)

The original location of point *K* is (1, –1). To reflect a point about the origin, change the sign of both the *x*- and *y*-coordinates.

10. A little less than 6

Points *C*, *F*, and *M* form a right triangle. To find the distance from *C* to *M*, use the Pythagorean theorem.

$$CM^2 = CF^2 + FM^2$$
$$CM^2 = 3^2 + 5^2$$
$$CM^2 = 9 + 25$$
$$CM^2 = 34$$
$$CM = \sqrt{34} \approx \text{a little less than } 6$$

INTERMEDIATE

11. 2

Substitute the coordinates of each point in the slope formula:

$$\text{slope} = \frac{\text{change in } y}{\text{change in } x} = \frac{y_2 - y_1}{x_2 - x_1}.$$

$$\text{slope} = \frac{-2 - 8}{-1 - 4} = \frac{-10}{-5} = 2$$

12. 8

This equation is written in slope-intercept form, $y = mx + b$, so the slope is the coefficient of the x-term.

13. $m = -\frac{4}{3}$

To find the slope m of the line $4x + 3y = 12$, rewrite it in slope-intercept form $y = mx + b$.

$$4x + 3y = 12$$
$$3y = -4x + 12$$
$$y = -\frac{4}{3}x + \frac{12}{3}$$
$$y = -\frac{4}{3}x + 4$$

14. −2

The slope of the line $y = -2x - 4$ is -2. Parallel lines have the same slope, so the slope of any line parallel to $y = -2x - 4$ is -2.

15. −3

The slope of the line $y = \frac{1}{3}x - 7$ is $\frac{1}{3}$. To find the slope, of a line that is perpendicular, find the negative reciprocal of the slope. The negative reciprocal of $\frac{1}{3}$ is -3.

16. 0

Substitute the coordinates of each point in the slope formula:

$$\text{slope} = \frac{5 - 5}{4 - (-2)} = \frac{0}{6} = 0$$

17. Undefined

Substitute the coordinates of each point in the slope formula:

$$\text{slope} = \frac{-2-3}{6-6} = \frac{-5}{0}$$

You cannot divide by 0, so the slope is undefined. This line is vertical.

18. $-\frac{1}{6}$

To find the slope of the line $3y = -\frac{1}{2}x$, it needs to be in slope-intercept form.

$$\begin{aligned} 3y &= -\frac{1}{2}x \\ y &= \frac{-\frac{1}{2}x}{3} \\ y &= -\frac{1}{2}x\left(\frac{1}{3}\right) \\ y &= -\frac{1}{6}x \end{aligned}$$

19. 5

Lines *BD* and *BC* are the legs of a right triangle and line *CD* is the hypotenuse. The hypotenuse of a right triangle with leg lengths of 3 and 4 (lines *BD* and *BC* respectively) is 5. (This is the common 3:4:5 Pythagorean triple.)

20. (–1, 1), or $x = -1$, $y = 1$

The point of intersection of the two lines is the solution to the system.

ADVANCED

21. Undefined

Substitute the coordinates of each point in the slope formula:

$$\text{slope} = \frac{-2-7}{-3-(-3)} = \frac{-9}{0}$$

You cannot divide by 0, so the slope of this line is undefined. The slopes of parallel lines are equal, so the slope of a parallel line is also undefined.

22. Undefined

Substitute the coordinates of each point in the slope formula:

$$\text{slope} = \frac{-1-(-1)}{-2-5} = \frac{0}{-7} = 0$$

The given line has a slope of 0, which means it is horizontal, so a perpendicular line would be vertical. The slope of a vertical line is undefined.

23. No

The lines shown on this graph are parallel. This can be verified by determining that the slope of both lines equals 1. Parallel lines, by definition, do not intersect.

24. No solution

The two regions representing the inequalities do not have any points of intersection. Since no points are in the shaded regions of both inequalities, there is no solution to this system.

25. $-\frac{5}{2}$

First, find the slope of the line represented by $2x - 5y = 20$ by writing it in slope-intercept form:

$$\begin{aligned} 2x - 5y &= 20 \\ -5y &= -2x + 20 \\ y &= \frac{2}{5}x - \frac{20}{5} \\ y &= \frac{2}{5}x - 4 \end{aligned}$$

The slope of this line is $\frac{2}{5}$. So the slope of a line perpendicular to this line would be the negative reciprocal, or a slope of $-\frac{5}{2}$.

26. 2

The x-intercept(s) of a graph are the x-value(s) of the point(s) where the graph crosses the x-axis. This means the y-coordinate is 0. Replace y with 0 in the equation and solve for x.

$$\begin{aligned} y &= 4x - 8 \\ 0 &= 4x - 8 \\ 8 &= 4x \\ 2 &= x \end{aligned}$$

The x-intercept is 2.

27. Yes

If the point (–2, 12) lies on the graph of $y = -2.5x + 7$, then replacing x and y with the ordered pair will make a true equation.

$$\begin{aligned} y &= -2.5x + 7 \\ 12 &= -2.5(-2) + 7 \\ 12 &= 5 + 7 \\ 12 &= 12 \end{aligned}$$

28. $\mathbf{y = -3x + 4}$

The equation $4x - 3y = 9$ needs to be written in slope-intercept form so that the slopes can be compared.

$$4x - 3y = 9$$
$$-3y = -4x + 9$$
$$y = \frac{4}{3}x - 3$$

The slope of $y = -3x + 4$ is -3 and the slope of $4x - 3y = 9$ is $\frac{4}{3}$. The absolute value of -3 is greater than the absolute value of $\frac{4}{3}$. Therefore, the slope of $y = -3x + 4$ is steeper.

29. Yes

The point (0, 0) is the origin and is included in the section that shows shading for both inequalities.

30. A reflection about the origin

The original location of point C is $(-2, 3)$. To reflect a point over the x-axis, change the sign of the y-coordinate. The first reflection has point C at $(-2, -3)$. To reflect a point over the y-axis, change the sign of the x-coordinate. The new location of point C is $(2, -3)$. The original ordered pair was $(-2, 3)$. The sign of both coordinates changed, which indicates a reflection about the origin.

GRAPHS OF FUNCTIONS

To graph a function in the *xy*-plane, use the *x*-axis for the input and the *y*-axis for the output. You can represent every input value, *x*, and its corresponding output value, *y*, as an ordered pair (x, y). The output of a function may also be referred to as $f(x)$, so you can write $y = f(x)$.

Two functions are graphed here. The first is a linear function $f(x) = 2x - 1$ and the second is a quadratic function $g(x) = x^2 - 1$.

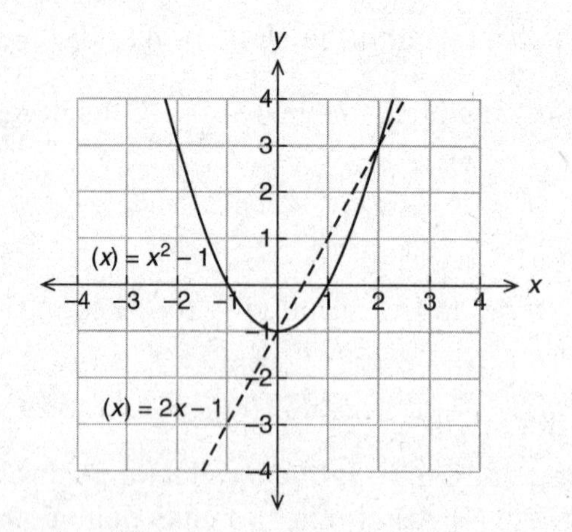

The graphs of the functions intersect at two points. These are the points for which $f(x) = g(x)$. To find the points using algebra, set $f(x) = g(x)$ and solve for *x*.

$$\begin{aligned} x^2 - 1 &= 2x - 1 \\ x^2 - 2x &= -1 + 1 \\ x^2 - 2x &= 0 \\ x(x - 2) &= 0 \\ x &= 0 \text{ or } 2 \end{aligned}$$

Use either function to find the corresponding *y*-value for each point.

For example, if $x = 0$, $f(0) = 2x - 1 = 2(0) - 1 = -1$ and if $x = 2$, $f(2) = 2x - 1 = 2(2) - 1 = 3$. The points of intersection are $(0, -1)$ and $(2, 3)$.

PIECEWISE FUNCTION

A **piecewise function** is defined by more than one equation, where each equation applies to a different part of the domain of the function. The absolute value function is an example of a piecewise function. Two "pieces" are needed in the definition of the function; these are $y = -x$ for all *x* less than 0 and $y = x$ for all $x \geq 0$. A piecewise function is usually described as shown for $f(x) = |x|$:

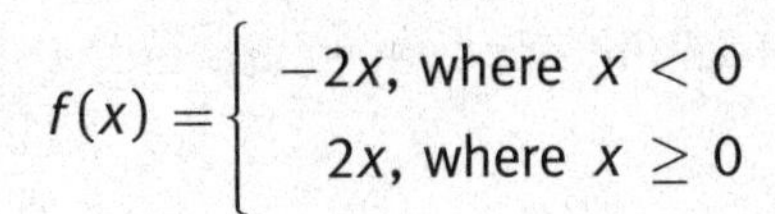

$$f(x) = \begin{cases} -2x, \text{ where } x < 0 \\ 2x, \text{ where } x \geq 0 \end{cases}$$

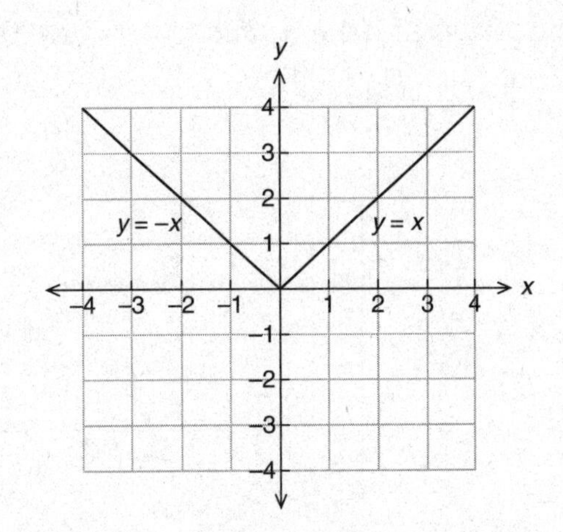

The graph of a function may also be shifted horizontally or vertically c units, where c is a positive number. For example, to shift the graph of $f(x) = |x|$, follow these rules:

- The graph of $f(x) + c$ is the graph of $f(x)$ shifted **upward** c units or spaces.
- The graph of $f(x) - c$ is the graph of $f(x)$ shifted **downward** c units or spaces.
- The graph of $f(x + c)$ is the graph of $f(x)$ shifted **to the left** c units or spaces.
- The graph of $f(x - c)$ is the graph of $f(x)$ shifted **to the right** c units or spaces.

Notice the placement of the constant in each description. To verify the direction of a shift of a function, plot several points of the original function and several points of the shifted function.

Consider the graph of $f(x) = x^2$ for $x > 0$ and the graph of $g(x) = \sqrt{x}$ for $x > 0$ shown. Because the domain is restricted for this graph, only part of the parabola is shown for $f(x) = x^2$.

Notice that these graphs are symmetric about the line $y = x$.

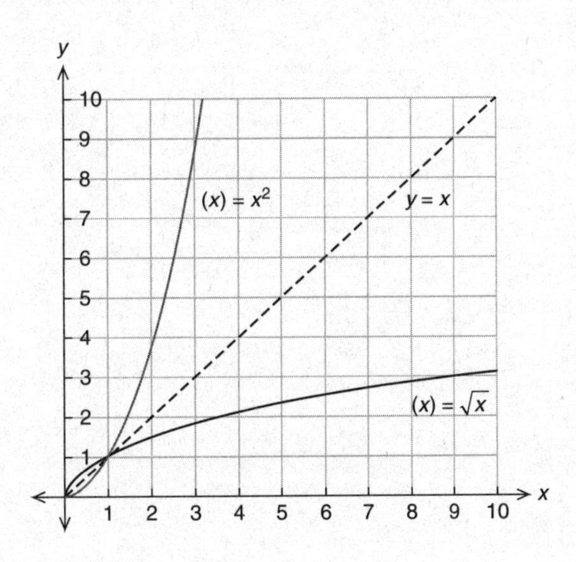

To shift the graphs of $f(x) = x^2$ and $g(x) = \sqrt{x}$ three units or spaces upward, graph $f(x) = x^2 + 3$ and $g(x) = \sqrt{x} + 3$.

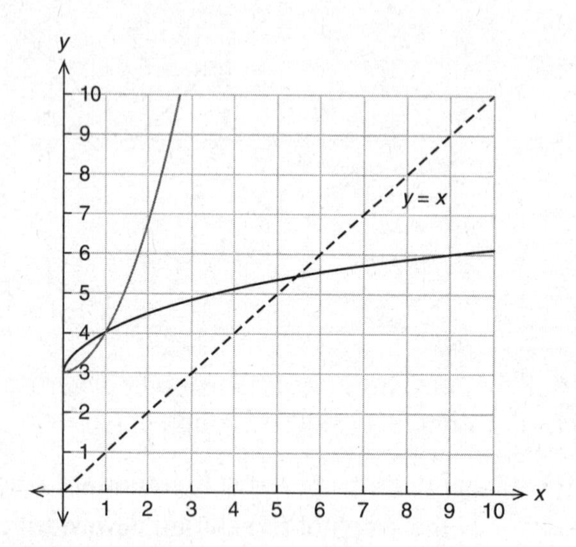

Notice that the graphs still have the same position **relative** to each other, but they are no longer symmetric across the line $y = x$.

The graph of a function may also be vertically stretched away or compressed toward the *x*-axis by a factor of *c*, where *c* is a positive number. That is, you can stretch the graph by making the slope larger and steeper, or you can compress the graph by making the slope smaller and less steep. To change the vertical shape of the graph of $f(x)$, follow these rules:

- The graph of $c \times f(x)$ is the graph of $f(x)$ stretched away from the *x*-axis by a factor of *c*.
- The graph of $\frac{1}{c} \times f(x)$ is the graph of $f(x)$ compressed toward the *x*-axis by a factor of *c*.

GRAPHS OF FUNCTIONS EXERCISES

BASIC

1. If the point $(2, y)$ lies on the graph of $f(x) = x^2 + 3$, what is y?

2. If the point $(x, 0)$ lies on the graph of $f(x) = 3x - 12$, what is x?

3. Complete the following ordered pairs for the function $f(x) = x^2 - 2x + 1$.

   $(0, \_\_\_)\ (-1, \_\_\_)\ (4, \_\_\_)$

4. Describe the graph of $y = 7$.

5. Describe the graph of $x = -3$

6. The graphs of $f(x) = 3x$ and $g(x) = 2x + 4$ intersect at the point (x, y). What is the value of x at the point of intersection?

7. The graphs of $f(x) = x^2$ and $g(x) = 5x - 6$ intersect at two points. Find the x-coordinate of each point of intersection.

8. The graph of $f(x) = x^2$ is shown. How is the graph of the function $g(x) = x^2 - 4$ related to the graph of $f(x)$?

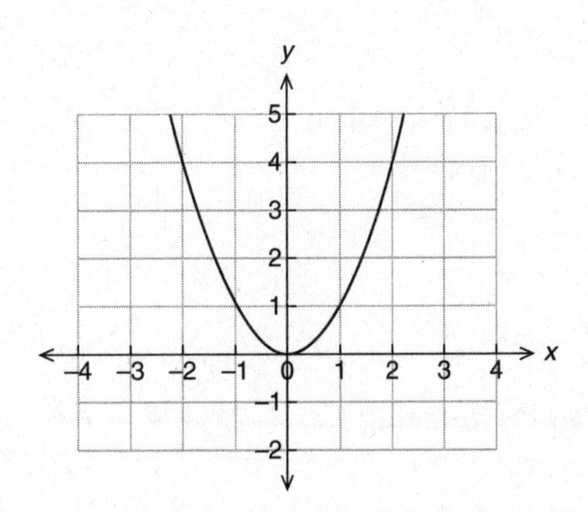

9. The graph of $h(x) = |x|$ is shown. How is the graph of the function $k(x) = |x - 5|$ related to the graph of $h(x)$?

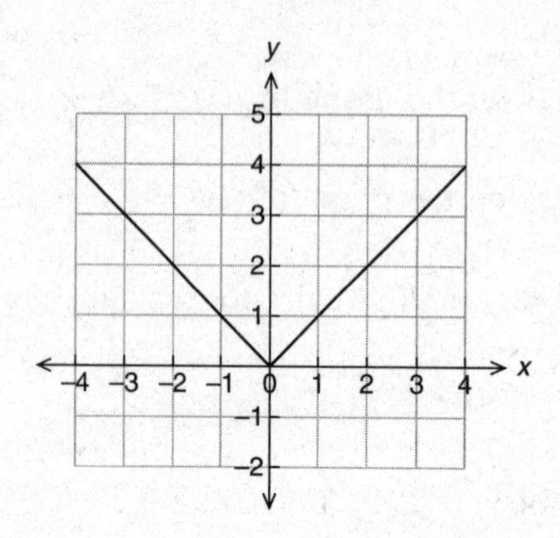

10. Sketch the graph of the piecewise function $f(x)$ on the coordinate plane provided.

$$f(x) = \begin{cases} -2x, \text{ where } x < 0 \\ 2x, \text{ where } x \geq 0 \end{cases}$$

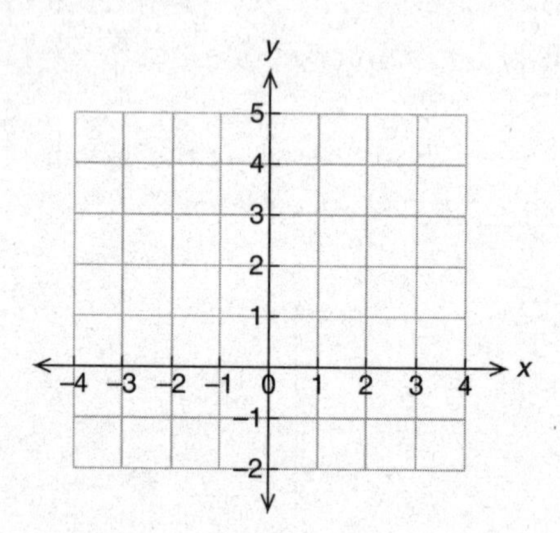

INTERMEDIATE

For Exercises 11–14, use the function $f(x) = 2x + 4$.

11. Complete the following ordered pairs for $f(x)$.

(−1, ___) (0, ___) (___, 0) (___, 6)

12. On the coordinate plane provided, graph the ordered pairs found in Question 11. Describe the graph of $f(x)$.

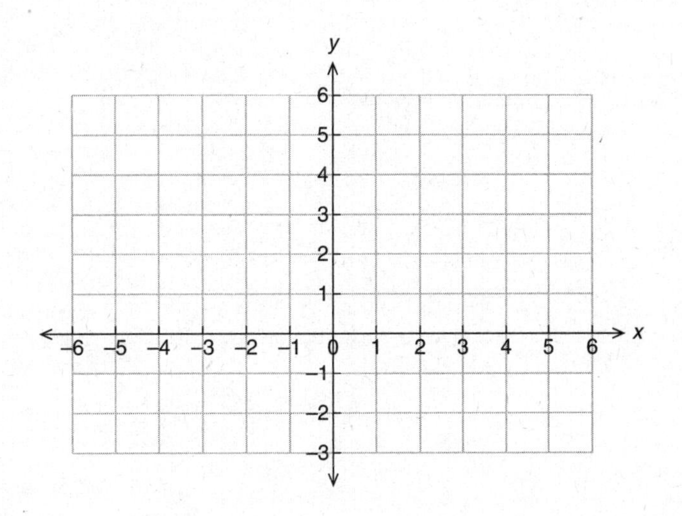

13. Suppose $g(x) = 3x - 2$. At what point(s) do $f(x)$ and $g(x)$ intersect?

14. Suppose $h(x) = x^2 + 1$. At what point(s) do $f(x)$ and $h(x)$ intersect?

15. Describe the relationship between the graphs of $f(x) = 3x^2$ and $g(x) = 3(x - 1)^2$.

16. Describe the relationship between the graphs of $f(x) = x^2$ and $g(x) = \frac{1}{2}x^2$.

17. The graph of $f(x) = 2|x|$ is shown. On the coordinate plane provided, sketch the graph of $g(x) = 2|x + 4|$.

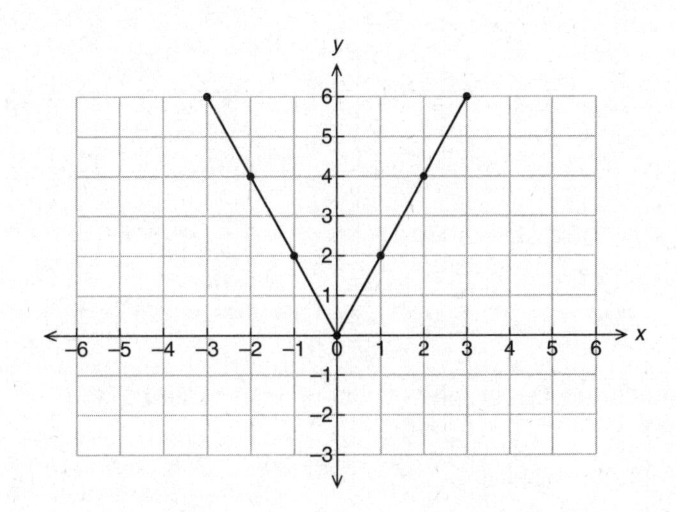

18. The graph of $f(x) = 2|x|$ is shown. On the coordinate plane provided, sketch the graph of $g(x) = 2|x| - 2$.

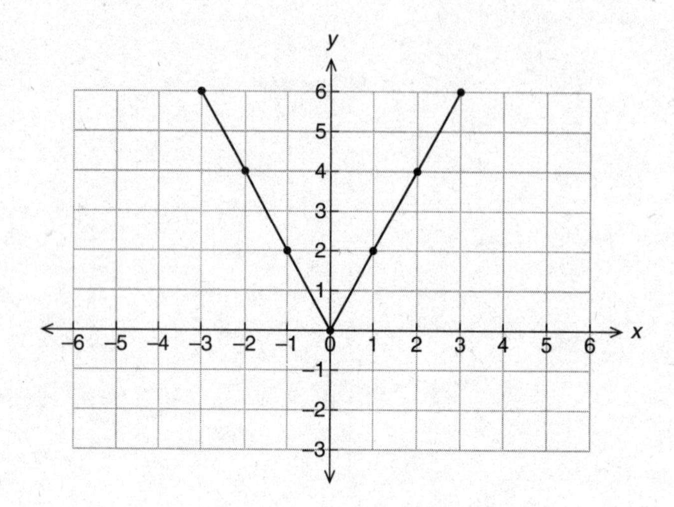

19. Sketch the graph of the piecewise function $f(x)$ on the coordinate plane provided.

$$f(x) = \begin{cases} -x, \text{ where } x < 0 \\ 3x, \text{ where } x \geq 0 \end{cases}$$

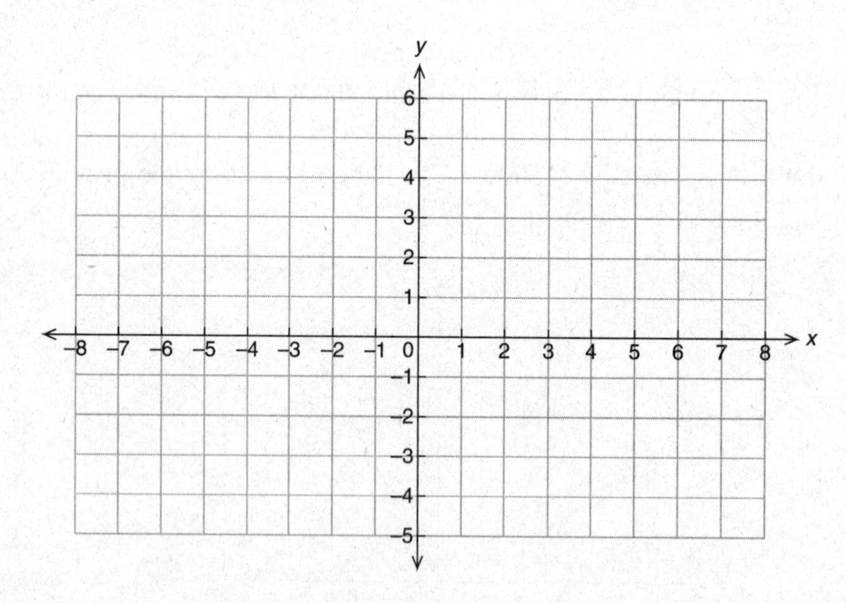

20. At what point(s) do the graphs of $f(x) = x^2 - 4$ and $g(x) = x^2 + 2x$ intersect?

ADVANCED

For Exercises 21–24, use the graph of $f(x) = -x - 2$ shown.

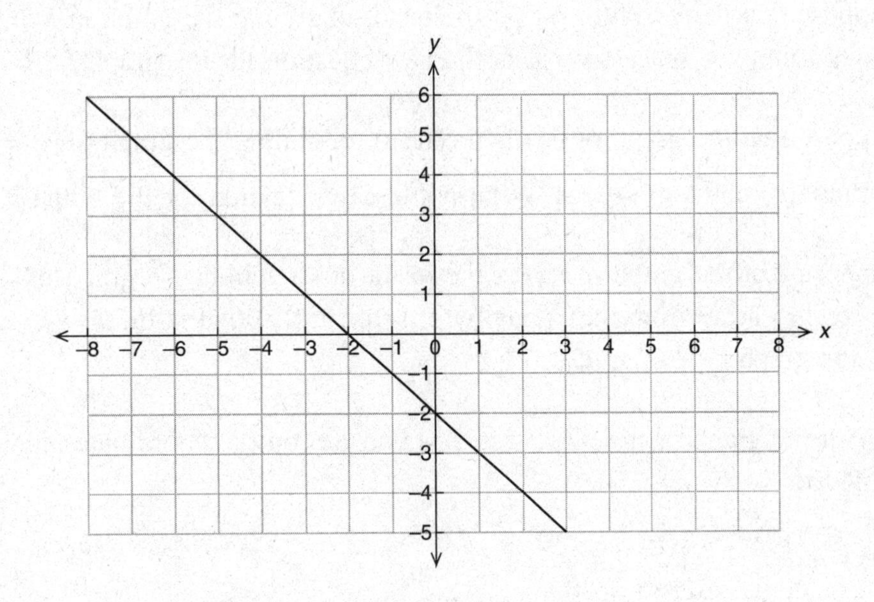

21. If $g(x) = f(x) + 4$, what is the y-intercept of $g(x)$?

22. If $h(x) = f(x - 4)$, what is the x-intercept of $h(x)$?

23. On the following coordinate plane, sketch the graph of $j(x) = x^2$ to show that the graphs of $f(x)$ and $j(x)$ do not intersect.

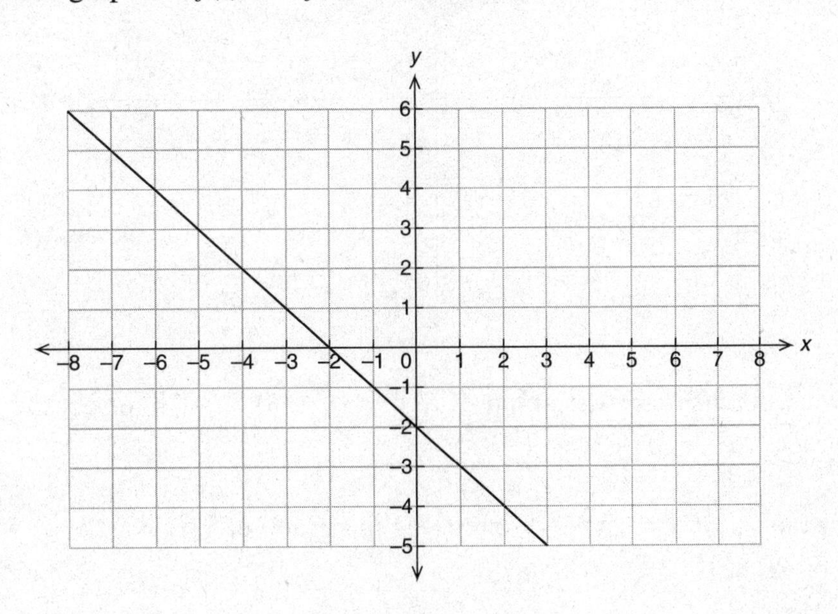

24. Show algebraically that the graph of $f(x)$ does not intersect the graph of $j(x) = x^2$.

25. Suppose that the graph of $f(x)$ is the result of sliding the graph of $y = 2x^2$ down 3 units or spaces. What is the new equation for the graph $f(x)$?

26. Suppose that the graph of $f(x)$ is the result of sliding the graph of $y = \frac{1}{2}x$ to the left 7 units or spaces. What is the new equation for the graph $f(x)$?

27. Suppose that the graph of $f(x)$ is the result of stretching the graph of $y = x + 5$ away from the x-axis by a factor of 2. What is the new equation for the graph $f(x)$?

28. Sketch the graph of the piecewise function $f(x)$ on the coordinate plane provided.

$$f(x) = \begin{cases} x^2, \text{where } x < 0 \\ x, \text{where } x \geq 0 \end{cases}$$

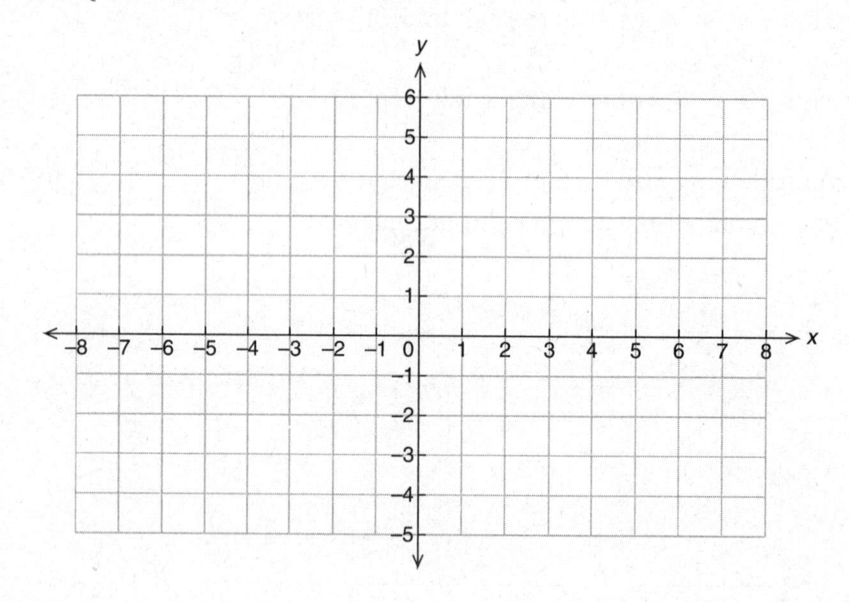

29. At what point(s) does the graph of $f(x) = x^2 + 3x + 11$ intersect the graph of $g(x) = 2x^2 - 3x + 4$?

30. The graph of $f(x) = 2x^2 - 5$ intersects the graph of $g(x) = x^2 + a$ when $x = 3$. What is a?

GRAPHS OF FUNCTIONS ANSWER KEY

Basic

1. 7

The function $f(x) = x^2 + 3$ can be written $y = x^2 + 3$. To find y when x is 2, substitute 2 for x and solve.

$$\begin{aligned} y &= x^2 + 3 \\ y &= (2)^2 + 3 \\ y &= 4 + 3 \\ y &= 7 \end{aligned}$$

2. 4

The function $f(x) = 3x - 12$ can be written $y = 3x - 12$. To find x when $y = 0$, substitute 0 for y and solve.

$$\begin{aligned} y &= 3x - 12 \\ 0 &= 3x - 12 \\ 12 &= 3x \\ 4 &= x \end{aligned}$$

3. (0, 1) (–1, 4) (4, 9)

Write the function $f(x) = x^2 - 2x + 1$ in the form $y = x^2 - 2x + 1$. Complete each ordered pair by substituting the given value of x, then solving for y.

when $x = 0$	when $x = -1$	when $x = 4$
$y = x^2 - 2x + 1$	$y = x^2 - 2x + 1$	$y = x^2 - 2x + 1$
$y = 0^2 - 2 \times 0 + 1$	$y = (-1)^2 - 2 \times (-1) + 1$	$y = 4^2 - 2 \times 4 + 1$
$y = 0 - 0 + 1$	$y = 1 + 2 + 1$	$y = 16 - 8 + 1$
$y = 1$	$y = 4$	$y = 9$
(0, 1)	(–1, 4)	(4, 9)

4. **The graph of $y = 7$ is a horizontal line that lies 7 units or spaces above the x-axis. Each point on the line $y = 7$ has y-coordinate 7.**

5. **The graph of $x = -3$ is a vertical line that lies 3 units to the left of the y-axis. Each point on the line $x = -3$ has x-coordinate -3.**

6. 4

To find the *x*-coordinate of the point of intersection, set the function rules equal and solve for *x*.

$$f(x) = g(x)$$
$$3x = 2x + 4$$
$$x = 4$$

7. $x = 2, x = 3$

To find the *x*-coordinate of each point of intersection, set the function rules equal and solve for *x*.

$$f(x) = g(x)$$
$$x^2 = 5x - 6$$
$$x^2 - 5x + 6 = 0$$
$$(x - 2)(x - 3) = 0$$

$$x - 2 = 0 \quad \text{or} \quad x - 3 = 0$$
$$x = 2 \qquad\qquad x = 3$$

8. The graph of $g(x) = x^2 - 4$ is a translation (shift) of the graph of $f(x) = x^2$ four units or spaces downward.

9. The graph of $k(x) = |x - 5|$ is a translation (shift) of the graph of $h(x) = |x|$ five units or spaces to the right.

10. The correct graph is shown. Be sure your graph comes to a point at (0, 0) and passes through the points (1, 2) and (2, 4) on the right side and the points (–1, 2) and (–2, 4) on the left side.

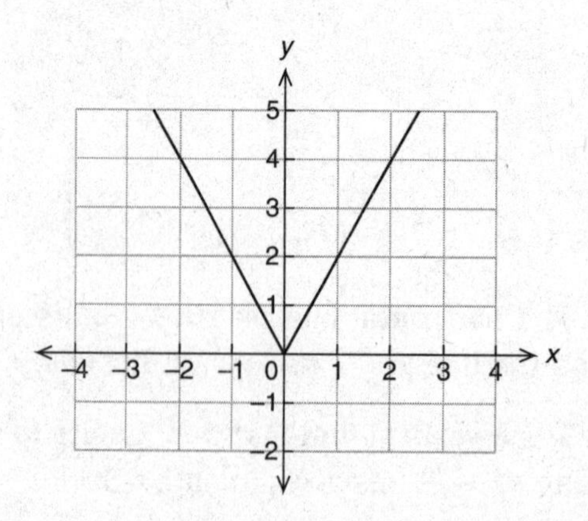

INTERMEDIATE

11. (–1, 2) (0, 4) (–2, 0) (1, 6)

Write the function $f(x) = 2x + 4$ in the form $y = 2x + 4$. Complete each ordered pair by substituting the given coordinate, then solving for the missing coordinate.

when $x = -1$	when $x = 0$	when $y = 0$	when $y = 6$
$y = 2x + 4$	$y = 2x + 4$	$y = 2x + 4$	$y = 2x + 4$
$y = 2(-1) + 4$	$y = 2(0) + 4$	$0 = 2x + 4$	$6 = 2x + 4$
$y = -2 + 4$	$y = 0 + 4$	$-4 = 2x$	$2 = 2x$
$y = 2$	$y = 4$	$-2 = x$	$1 = x$
(–1, 2)	(0, 4)	(–2, 0)	(1, 6)

12. The graph of $f(x)$ is a line with x-intercept at (–2, 0) and y-intercept at (0, 4) and a slope of 2.

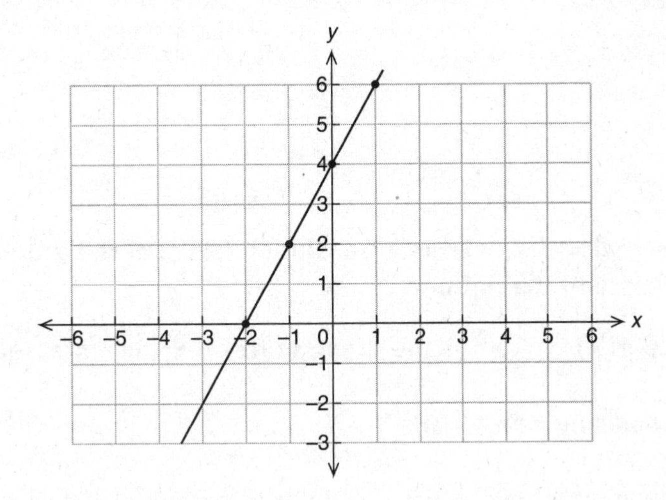

13. (6, 16)

To find the x-coordinate(s) of the point(s) of intersection, set the functions equal and solve for x.

$$f(x) = g(x)$$
$$2x + 4 = 3x - 2$$
$$6 = x$$

To find y when $x = 6$, substitute 6 for x in one of the functions.

$$y = 2x + 4$$
$$y = 2(6) + 4$$
$$y = 12 + 4$$
$$y = 16$$

When $x = 6$, $y = 16$. So, the graphs intersect at the point (6, 16).

14. (3, 10), (–1, 2)

To find the *x*-coordinate(s) of the point(s) of intersection, set the functions equal and solve for *x*.

$$f(x) = h(x)$$
$$2x + 4 = x^2 + 1$$
$$0 = x^2 - 2x - 3$$
$$0 = (x - 3)(x + 1)$$

$x - 3 = 0$, $x = 3$ or $x + 1 = 0$, $x = -1$

Now find the *y*-coordinate for each point of intersection.

when $x = 3$	when $x = -1$
$y = 2x + 4$	$y = 2x + 4$
$y = 2(3) + 4$	$y = 2(-1) + 4$
$y = 6 + 4$	$y = -2 + 4$
$y = 10$	$y = 2$
(3, 10)	(–1, 2)

15. The graph of $g(x) = 3(x - 1)^2$ is a translation (shift) of the graph of $f(x) = 3x^2$ one unit or space to the right.

16. The graph of $g(x) = \frac{1}{2}x^2$ is the graph of $f(x) = x^2$, compressed vertically toward the *x*-axis by a factor of $\frac{1}{2}$.

17. To sketch the graph of $g(x) = 2|x + 4|$, shift the graph of $f(x) = 2|x|$ to the left four units or spaces. The correct graph of $g(x) = 2|x + 4|$ is shown.

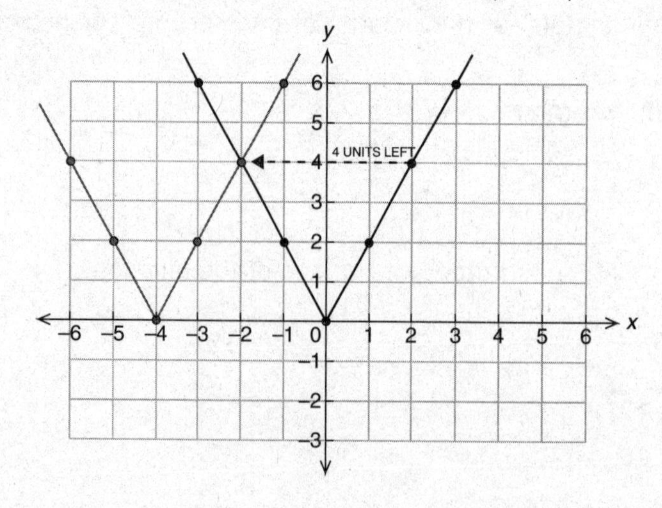

18. To graph $g(x) = 2|x| - 2$, shift the graph of $f(x) = 2|x|$ down 2 units or spaces. The correct graph of $g(x) = 2|x| - 2$ is shown.

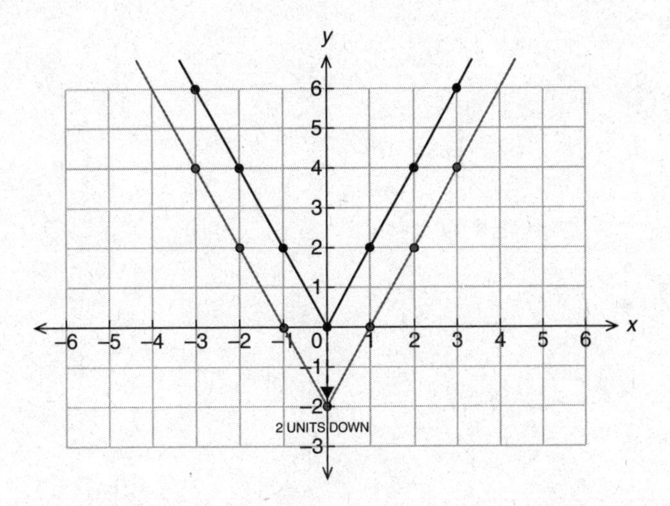

19. The graph of $f(x)$ has two parts. When $x < 0$, sketch the graph of the line $y = -x$. When $x \geq 0$, sketch the graph of the line $y = 3x$. The correct graph of $f(x)$ is shown.

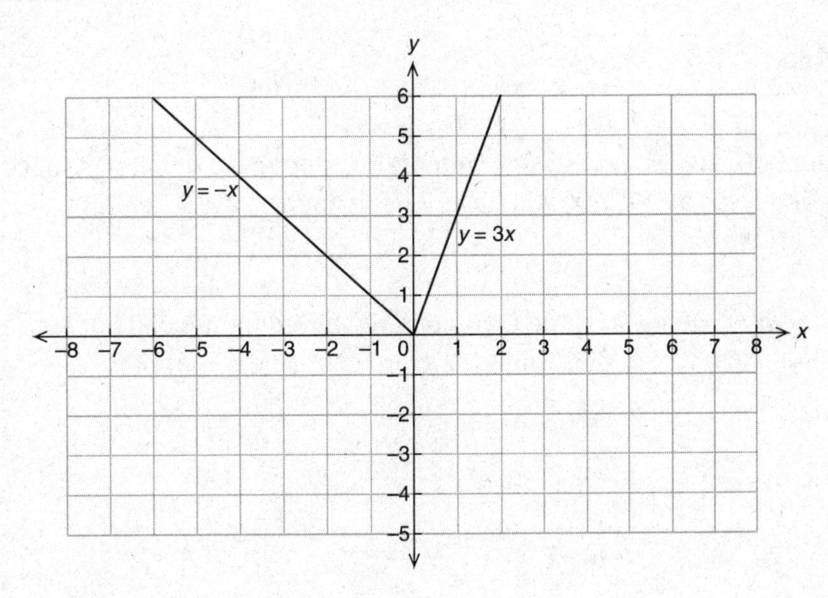

20. (–2, 0)

To find the x-coordinate(s) of the point(s) of intersection, set the functions equal and solve for x.

$$\begin{aligned} f(x) &= g(x) \\ x^2 - 4 &= x^2 + 2x \\ \cancel{x^2} - 4 &= \cancel{x^2} + 2x \\ -4 &= 2x \\ -2 &= x \end{aligned}$$

When $x = -2$,

$$\begin{aligned} y &= x^2 - 4 \\ y &= (-2)^2 - 4 \\ y &= 4 - 4 \\ y &= 0 \end{aligned}$$

Advanced

21. 2

The graph of $g(x) = f(x) + 4$ is the graph of $f(x)$, shifted up 4 units or spaces. Since $f(x)$ has y-intercept at -2, $g(x)$ will have y-intercept at $-2 + 4$, or 2.

22. 2

The graph of $h(x) = f(x - 4)$ is the graph of $f(x)$, shifted right 4 units or spaces. Since $f(x)$ has x-intercept at -2, $h(x)$ will have x-intercept at $-2 + 4$, or 2.

23. The correct graph is shown. The graph of $j(x) = x^2$ is a parabola that opens upward with vertex at the origin. The graphs of $f(x)$ and $j(x)$ do not intersect.

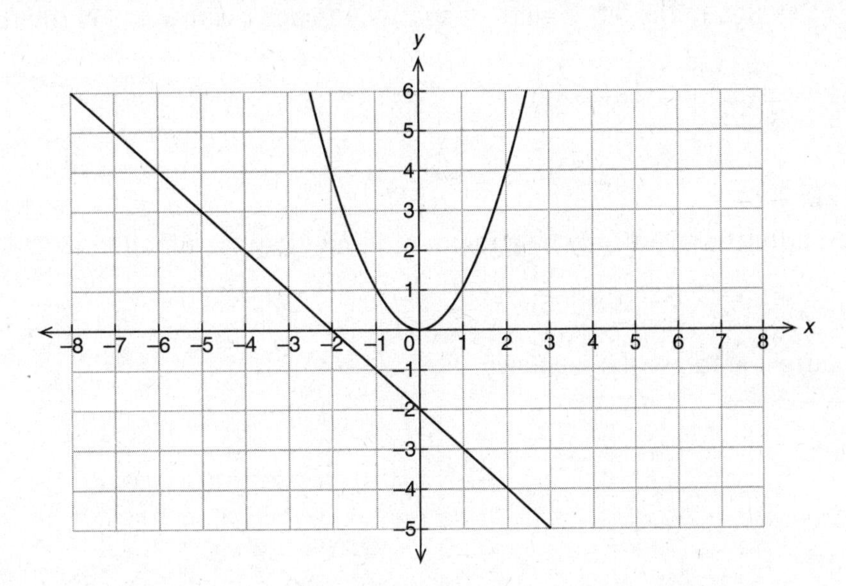

24. To show that the graph of $f(x)$ does not intersect the graph of $j(x) = x^2$, set the functions equal and show that the equation has no solutions.

$$\begin{aligned} f(x) &= g(x) \\ -x - 2 &= x^2 \\ 0 &= x^2 + x + 2 \\ x &= \frac{-1 \pm \sqrt{1^2 - 4(1)(2)}}{2(1)} \\ x &= \frac{-1 \pm \sqrt{1 - 8}}{2} \\ x &= \frac{-1 \pm \sqrt{-7}}{2} \end{aligned}$$

The square root of a negative number is not defined. Therefore, the equation has no solution in the real numbers. Since the equation has no solution, the graphs have no points of intersection.

25. $\mathbf{y = 2x^2 - 3}$

To shift a function downward 3 units or spaces, subtract 3 from the function.

$$y = 2x^2 \rightarrow y = (2x^2) - 3$$

26. $\mathbf{y = \frac{1}{2}(x + 7)}$

To shift a function to the left 7 units or spaces, replace x with $x + 7$ in the function.

$$y = \frac{1}{2}x \rightarrow y = \frac{1}{2}(x + 7)$$

27. $\mathbf{y = 2(x + 5)}$

To stretch a function vertically by a factor of 2, multiply the function by 2.

$$y = x + 5 \rightarrow y = 2(x + 5)$$

28. The correct graph of $f(x)$ is shown.

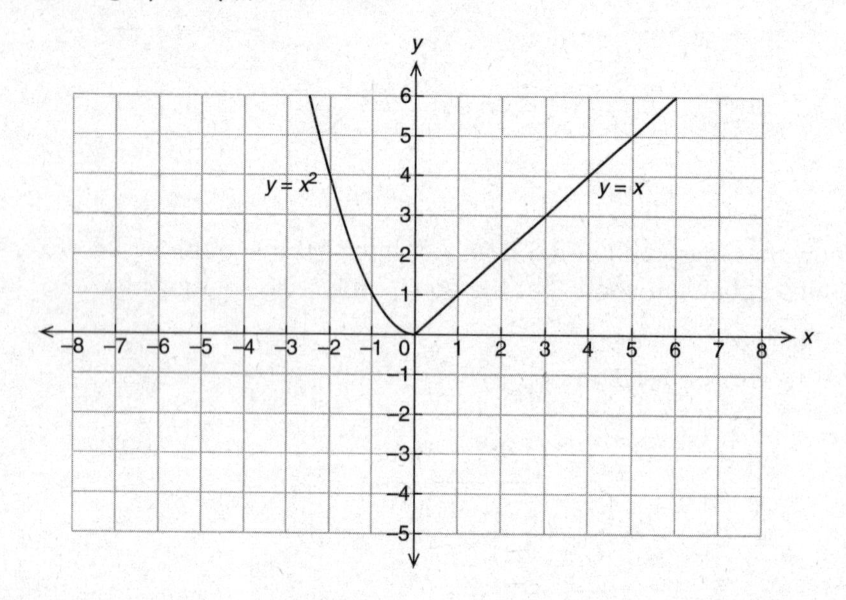

29. (7, 81), (–1, 9)

Set the function rules equal and solve.

$$\begin{aligned} f(x) &= g(x) \\ x^2 + 3x + 11 &= 2x^2 - 3x + 4 \\ 0 &= x^2 - 6x - 7 \\ 0 &= (x - 7)(x + 1) \end{aligned}$$

$$\begin{aligned} x - 7 &= 0 \\ x &= 7 \end{aligned} \quad \text{or} \quad \begin{aligned} x + 1 &= 0 \\ x &= -1 \end{aligned}$$

Now find the y-coordinate for each point of intersection.

when $x = 7$	when $x = -1$
$y = x^2 + 3x + 11$	$y = x^2 + 3x + 11$
$y = 7^2 + 3 \times 7 + 11$	$y = (-1)^2 + 3 \times (-1) + 11$
$y = 49 + 21 + 11$	$y = 1 - 3 + 11$
$y = 81$	$y = 9$
(7, 81)	(–1, 9)

30. 4

If the graphs of $f(x)$ and $g(x)$ intersect when $x = 3$, then the function rules must be equal when $x = 3$. Set the function rules equal. Then substitute 3 for x and solve to find a.

$$\begin{aligned} f(x) &= g(x) \\ 2x^2 - 5 &= x^2 + a \\ 2(3)^2 - 5 &= 3^2 + a \\ 2(9) - 5 &= 9 + a \\ 18 - 5 &= 9 + a \\ 13 &= 9 + a \\ 4 &= a \end{aligned}$$

CHAPTER 8

Geometry

UNDERSTANDING GEOMETRY

Geometry topics include lines and angles, triangles—including isosceles, equilateral, and special right triangles—polygons, circles, multiple figures, three-dimensional figures (uniform solids), area, perimeter, and volume. You do not need to know how to do geometry proofs for the test.

The geometry tested on the GRE is basic. There are only a few fundamental definitions and formulas you need to know. The GRE emphasizes new ways of applying these elementary rules.

Diagrams

Pay a lot of attention to diagrams. There can be a lot of information "hidden" in a diagram. If a diagram of an equilateral triangle gives you the length of one side, for instance, it actually gives the length of all sides. Similarly, if you are given the measure of one of the angles formed by the intersection of two lines, you can easily find the measure of all four angles. In fact, many geometry questions specifically test your ability to determine what additional information is implied by the information you are given in the diagram.

The diagrams provide basic information such as what kind of figure you are dealing with (is it a triangle? a quadrilateral?), the order of the points on lines, etc. However, the figures on the GRE are not drawn to scale unless otherwise stated. Because they are not drawn to scale, you can't rely on how figures are drawn to conclude anything about the size of the figure. If a line looks straight in the diagram, you can assume it is straight. But you must be careful when using the diagram to judge relative lengths, angles, sizes, etc., since these may not be drawn accurately. A square that looks twice as big as another is not necessarily twice as big. If an angle looks like a right angle, you cannot assume it is one, unless it is marked as such. If one side of a triangle looks longer than another, you cannot assume it is unless some other information tells you that it is. If a figure looks like a square, you don't know it is a square; you only know it is a quadrilateral. This is especially important to bear in mind in the Quantitative Comparison section, where the diagram may lead you to believe that one column is greater, but logic will prove that you need more information.

You can also, on occasion, use the diagram to your advantage by looking at the question logically. For instance:

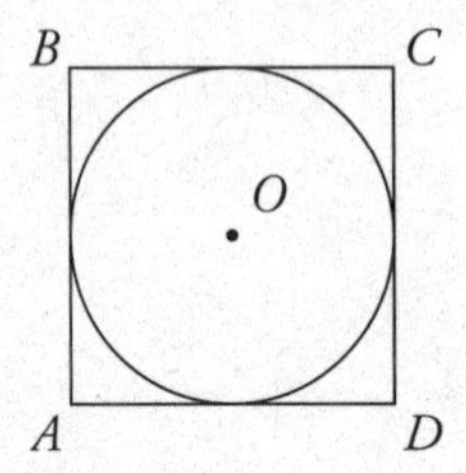

In the figure shown, the circle with center O has area 4π. What is the area of square $ABCD$?

(A) 4
(B) 2π
(C) 12
(D) 16
(E) 8π

We know from the question stem that we have a square and a circle, and we can see from the diagram that the circle is inscribed in the square—that is, it touches the square on all four sides. Whatever the area of the circle, we can see that the square's area must be bigger; otherwise the circle wouldn't fit inside it. So the right answer must be larger than the area of the circle, or larger than 4π. Now we can approximate 4π to a little more than 12. Since the correct answer is larger than 12, it must be either choice (D) or (E). And since you would expect to see π in the area of the circle, as it is here, that means π will not be in the area of the surrounding square. (The correct answer is 16.)

This example highlights an important point: π appears very often on geometry problems, so you should have some idea of its value. It is approximately equal to 3.14, but for most purposes you only need remember that it is slightly greater than 3.

LINES AND ANGLES

A line is a one-dimensional, geometrical abstraction—infinitely long with no width. It is not physically possible to draw a line; any physical line would have a finite length and some width, no matter how long and thin we tried to make it. Two points determine a straight line; given any two points, there is exactly one straight line that passes through them.

Lines: A **line segment** is a section of a straight line, of finite length, with two endpoints. A line segment is named by its endpoints, as in segment *AB*. The **midpoint** is the point that divides a line segment into two equal parts.

6

A *M* *B*

Example: In the figure shown, *A* and *B* are the endpoints of the line segment *AB* and *M* is the midpoint ($AM = MB$). What is the length of *AB*?

Since *AM* is 6, *MB* is also 6, and so *AB* is $6 + 6$, or 12.

Two lines are **parallel** if they lie in the same plane and never intersect each other regardless of how far they are extended. If line ℓ_1 is parallel to line ℓ_2, we write $\ell_1 \parallel \ell_2$.

Angles: An **angle** is formed by two lines or line segments intersecting at a point. The point of intersection is called the **vertex** of the angle. Angles are measured in degrees (°).

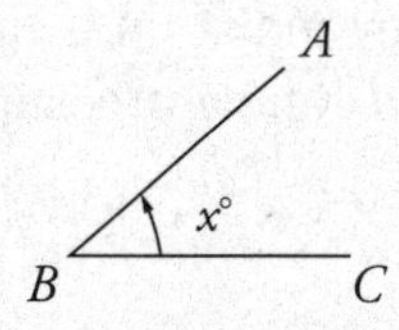

Angle *x*, $\angle ABC$, and $\angle B$ all denote the same angle shown in the diagram.

An **acute angle** is an angle whose degree measure is between 0° and 90°. A **right angle** is an angle whose degree measure is exactly 90°. An **obtuse angle** is an angle whose degree measure is between 90° and 180°. A **straight angle** is an angle whose degree measure is exactly 180°.

acute ($x < 90$) right ($y = 90$) obtuse ($90 < z < 180$) straight ($w = 180$)

The sum of the measures of the angles on one side of a straight line is 180°.

straight ($x + y + z = 180$)

The sum of the measures of the angles around a point is 360°.

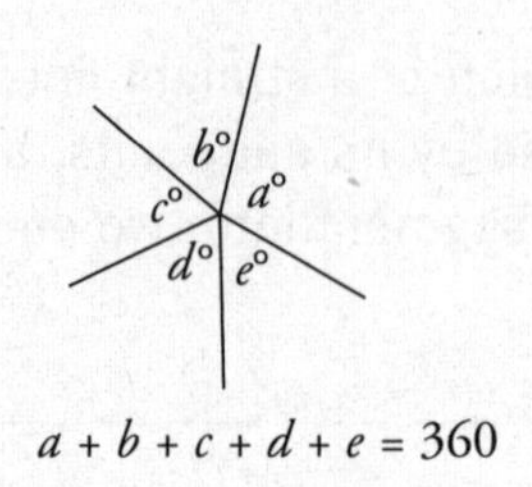

Two lines are **perpendicular** if they intersect at a 90° angle. The shortest distance from a point to a line is the line segment drawn from the point to the line such that it is perpendicular to the line. If line ℓ_1 is perpendicular to line ℓ_2, we write $\ell_1 \perp \ell_2$. If $\ell_1 \perp \ell_2$ and $\ell_2 \perp \ell_3$, then $\ell_1 \parallel \ell_3$.

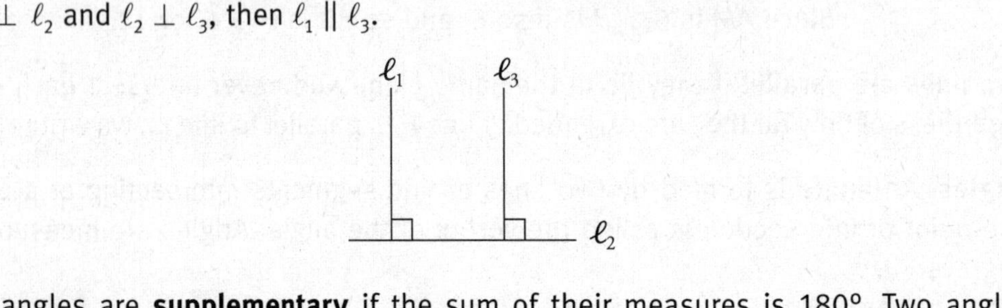

Two angles are **supplementary** if the sum of their measures is 180°. Two angles are **complementary** if together they make up a right angle (i.e., if the sum of their measures is 90°). Angles *c* and *d* below are supplementary; angles *a* and *b* are complementary.

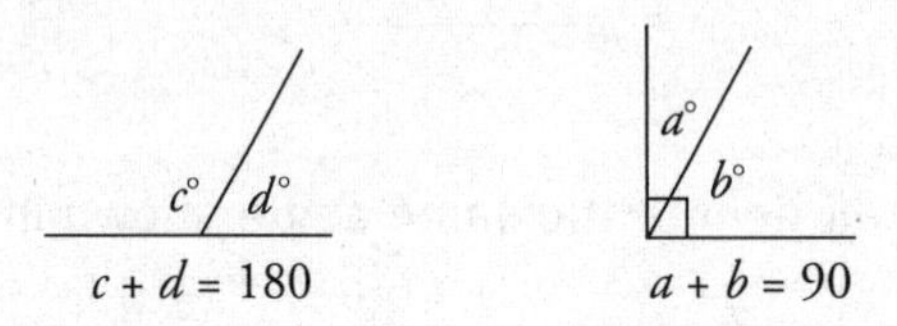

A line or line segment bisects an angle if it splits the angle into two smaller, equal angles. If line segment *BD* below bisects $\angle ABC$, then $\angle ABD$ has the same measure as $\angle DBC$. The two smaller angles are each half the size of $\angle ABC$.

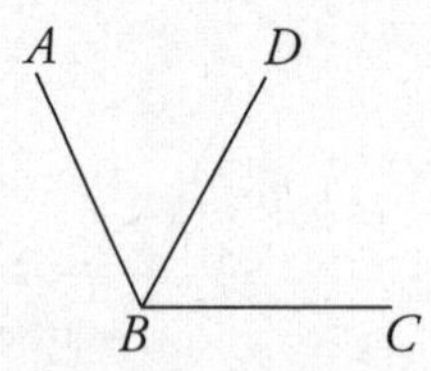

Vertical angles are a pair of opposite angles formed by two intersecting line segments. At the point of intersection, two pairs of vertical angles are formed. Angles *a* and *c* below are vertical angles, as are *b* and *d*.

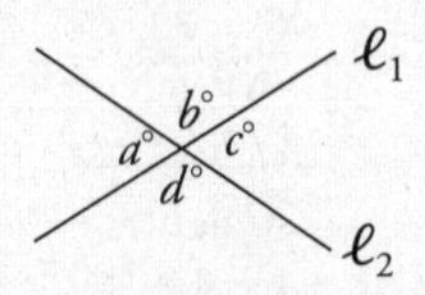

The two angles in a pair of vertical angles have the same degree measure. In the diagram above, $a = c$ and $b = d$. In addition, since ℓ_1 and ℓ_2 are straight lines,

$$a + b = c + d = a + d = b + c = 180°$$

In other words, each angle is supplementary to each of its two adjacent angles.

If two parallel lines intersect with a third line (called a *transversal*), each of the parallel lines will intersect the third line at the same angle. In the figure below, $a = e$. Since a and e are equal, and $c = a$ and $e = g$ (vertical angles), we know that $a = c = e = g$. Similarly, $b = d = f = h$.

If $\ell_1 \parallel \ell_2$, then
$a = c = e = g$ and
$b = d = f = h$.

In other words, when two parallel lines intersect with a third line, all acute angles formed are equal, all obtuse angles formed are equal, and any acute angle is supplementary to any obtuse angle.

Points in the Coordinate Plane

Questions on the GRE sometimes refer to lines or curves graphed on a coordinate plane. An ***xy*-coordinate plane** is formed by two number lines at right angles to each other, intersecting at their zero points. The intersecting lines divide the plane into four quadrants, named counterclockwise by the Roman numerals I, II, III, and IV. The horizontal number line is the ***x*-axis**, the vertical number line is the ***y*-axis,** and the point of intersection is the **origin**. Every point on the plane is named by its x- and y-coordinates, (x, y).

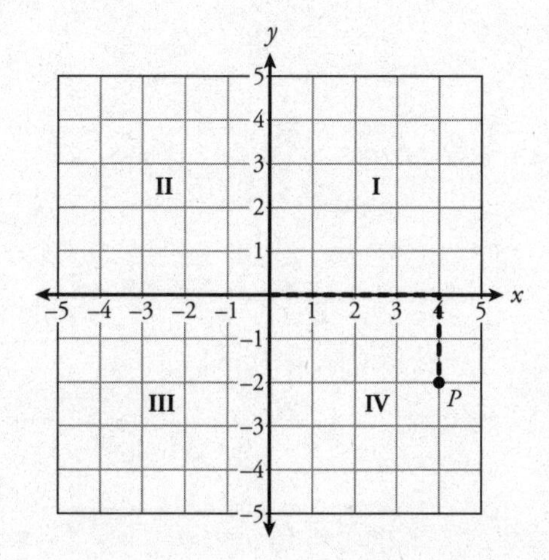

To graph point P, whose coordinates are $(4, -2)$, start at the origin, go right 4 units or spaces and then go down 2 units or spaces.

SLOPE

The slope of a line tells you how steeply that line goes up or down from a horizontal position. If a line gets higher as you move to the right, it has a positive slope. If it goes down as you move to the right, it has a negative slope.

To find the slope of a line, use the following formula:

$$\textbf{Slope} = \frac{\textbf{rise}}{\textbf{run}} = \frac{\textbf{change in } y}{\textbf{change in } x}$$

Rise means the difference between the y-coordinate values of the two points on the line, and *run* means the difference between the x-coordinate values.

The slope-intercept equation of a line is $y = mx + b$, where m is the slope and b is the y-intercept. The y-intercept is the value of y where the line crosses the y-axis, in other words, the value of y when $x = 0$.

Example: What is the slope of the line represented by the equation $y = 2x - 5$?

The equation is in slope-intercept form, so the slope is 2 because 2 is the coefficient of x.

Example: What is the slope of the line that contains the points (1, 2) and (4, −5)?

$$Slope = \frac{-5-2}{4-1} = \frac{-7}{3} = -\frac{7}{3}$$

To determine the slope of a line from an equation, put the equation into the slope-intercept form.

Example: What is the slope of the line represented by the equation $3x + 2y = 4$?

$$3x + 2y = 4$$
$$2y = -3x + 4$$
$$y = -\frac{3}{2}x + 2, \text{ so } m \text{ is } -\frac{3}{2}.$$

LINES AND ANGLES EXERCISES

BASIC

In Exercises 1–4, classify each angle x as acute, right, obtuse, or straight.

1. $x = 180°$

2. $x = 90°$

3. $x = 25°$

4. $x = 130°$

5. What is the value of x in the figure shown?

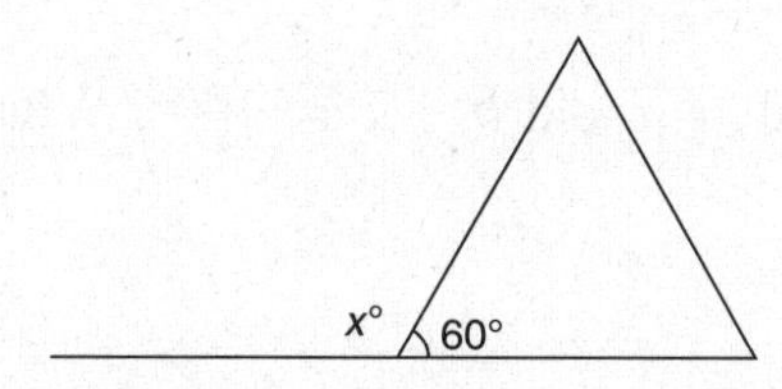

Use the figure shown for Exercises 6–8.

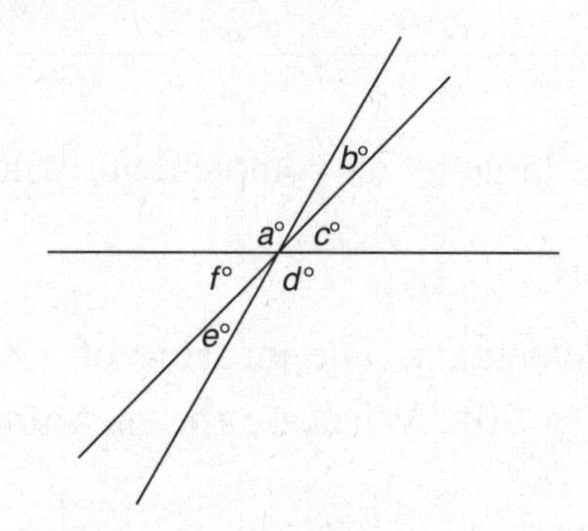

6. If $a = 60°$ and $c = 45°$, what is the value of b?

7. What is the value of $a + f + b$?

8. What is the value of $a + b + c + d + e + f$?

9. If angle x has a measure of 90°, what is the measure of its supplement?

10. If angle x has a measure of 36°, what is the measure of its complement?

Intermediate

11. The four angles around a point measure y, $2y$, 35°, and 55°, respectively. What is the value of y?

12. In the figure shown, ℓ_1, ℓ_2, and ℓ_3 are parallel. What is the value of $a + 2c + e$?

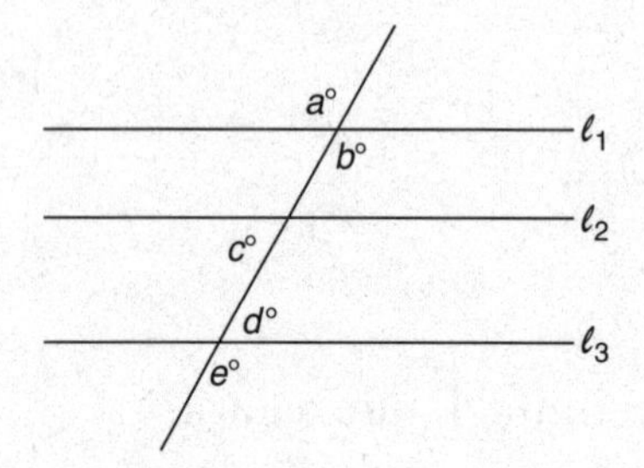

13. In the figure shown, $a = d$ and $b + c = 100°$. What is the value of a?

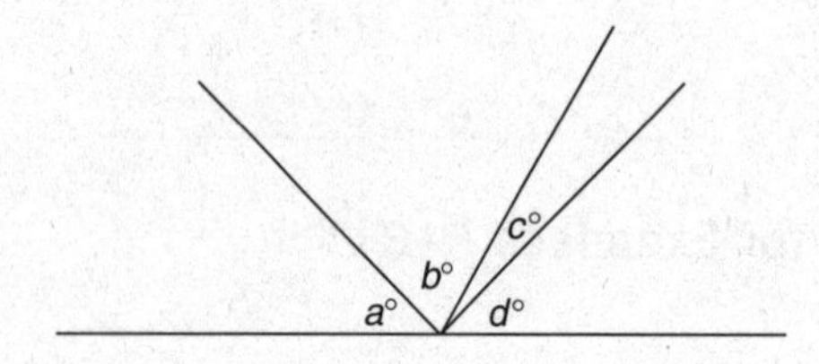

14. If an angle is twice as large as its complement, what is the measure of the larger angle?

15. $\angle A$ and $\angle B$ are supplementary. The measure of $\angle A$ is $(3x + 50)°$. The measure of $\angle B$ is $(7x + 90)°$. What are the measures of $\angle A$ and $\angle B$?

16. In the figure shown, M is the midpoint of segment AB. What is the length of AB?

$4x + 9$ $6x + 5$
A M B

17. What is the slope of the line whose equation is $3x + 5y + 4 = 0$?

18. $\angle A$ and $\angle B$ are vertical angles. The measure of $\angle A$ is $(4x - 35)°$. The measure of $\angle B$ is $(x + 10)°$. What is the measure of $\angle A$?

19. Point X is between points A and B on a line. $AB = 45$, $AX = 2x$, and $XB = 2x + 5$. What is the value of x?

20. In the figure shown, ℓ_1 and ℓ_2 are parallel. What is the value of x?

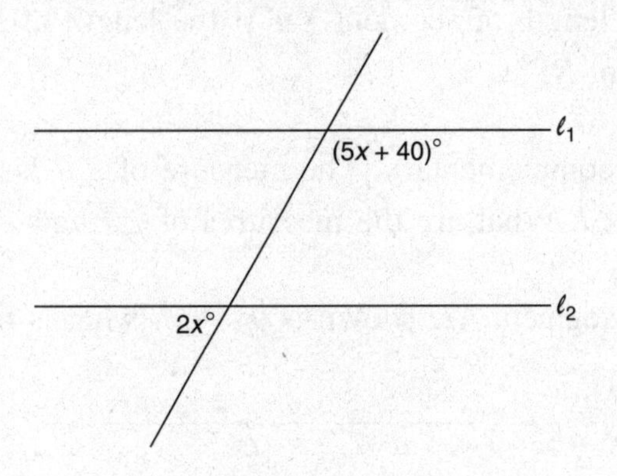

ADVANCED

Use the figure shown for Exercises 21–22.

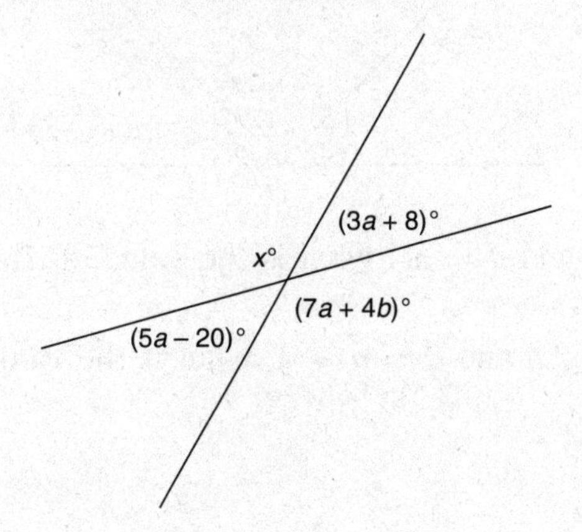

21. What are the values of a and b?

22. What is the measure of $\angle x$?

23. If the ratio of $x:y:z$ in the figure shown is 1:2:3, what are the values of x, y, and z?

$y°$

$x°$ $z°$

straight

24. $\angle A$ and $\angle B$ are supplementary. The measure of $\frac{1}{2}\angle A$ equals the measure of $2\angle B$. What are the measures of $\angle A$ and $\angle B$?

25. The measure of an angle is 48 more than the measure of its complement. What is the measure of the angle?

26. Point X lies on a line segment RS. The length of segment RX is 2 more than half of the length of segment XS. If the length of XS is 10, what is the length of segment RS?

27. $\angle J$ and $\angle K$ are complementary. The measure of $\angle K$ is 15 more than twice the measure of $\angle J$. What are the measures of $\angle J$ and $\angle K$?

28. If the length of segment AD shown is $9x + 7$, what is the length of AD?

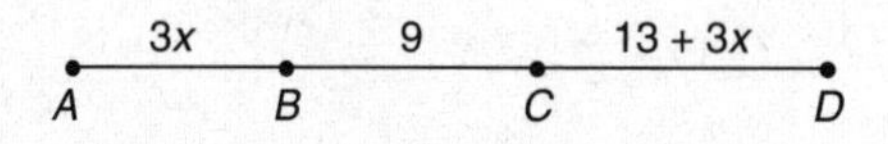

Use the figure shown for Exercises 29–30.

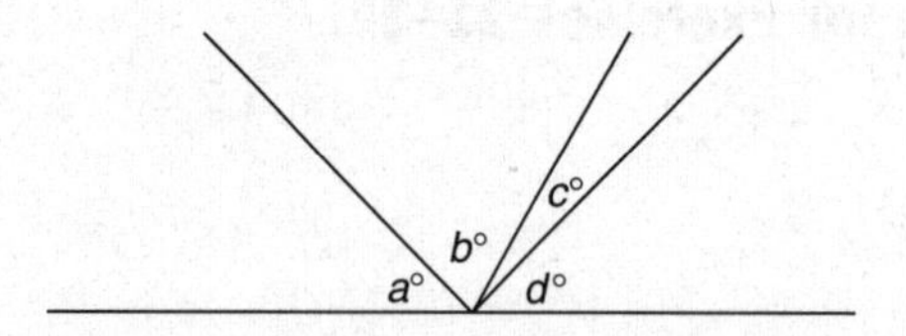

29. If $b = \frac{a}{2}$, $c = \frac{b}{2}$, and $d = 3c$, what is the value of d?

30. If $a = b - 5$, $b = 2c$, and $d = a + 1$, what is the value of c?

LINES AND ANGLES ANSWER KEY

BASIC

1. Straight

A straight angle is 180°.

2. Right

A right angle is 90°.

3. Acute

An acute angle is between 0° and 90°.

4. Obtuse

An obtuse angle is between 90° and 180°.

5. 120°

Angle x is supplementary to the 60° angle. $x + 60° = 180°$ so $x = 120°$.

6. 75°

$a + b + c = 180°$ because they are all angles on the same side of a line. $60° + b + 45° = 180°$. $105° + b = 180°$ so $b = 75°$.

7. 180°

$a + f + b = 180°$ because the three angles together form a straight line.

8. 360°

$a + b + c + d + e + f = 360°$ because the six angles together are all possible angles around a single point.

9. 90°

Supplementary angles sum to 180°. Because $90° + x = 180°$, $x = 90°$.

10. 54°

Complementary angles sum to 90°. Because $36° + x = 90°$, $x = 54°$.

INTERMEDIATE

11. 90°

The angles around a point sum to 360. Therefore, $y + 2y + 35° + 55° = 360°$; $3y + 90° = 360°$; $3y = 270°$, so $y = 90°$.

12. 360°

When two parallel lines intersect with a third line, any acute angle is supplementary to any obtuse angle. Angle a is obtuse, angle c is acute, and angle e is obtuse. $a + c = 180°$ and $c + e = 180°$. $a + c + c + e = 180° + 180° = 360°$.

13. 40°

$a + b + c + d = 180°$ because the four angles together compose a straight line. We know $b + c = 100°$ and $a = d$, so $a + 100° + a = 180°$. Thus, $2a + 100° = 180°$; $2a = 80°$, so $a = 40°$.

14. 60°

Let $2x$ = the larger angle and x = the complement of that angle. Then $2x + x = 90°$ because the angles are complementary. Thus, $3x = 90°$; $x = 30°$, so $2x = 60°$.

15. $\angle A = 62°$ and $\angle B = 118°$

The sum of the measures of angles A and B must be 180° because they are supplementary. Therefore, $3x + 50 + 7x + 90 = 180$.

$$\begin{aligned} 10x + 140 &= 180 \\ 10x &= 40 \\ x &= 4 \end{aligned}$$

The measure of $\angle A = (3x + 50)° = (3(4) + 50)° = 12° + 50° = 62°$.
The measure of $\angle B = (7x + 90)° = (7(4) + 90)° = 28° + 90° = 118°$.

16. 34

$4x + 9 = 6x + 5$ because M is the midpoint. Thus, $4x + 4 = 6x$; $4 = 2x$, so $2 = x$. Substitute 2 for x and add the length of the segments together to find the total length:

$$\begin{aligned} &4x + 9 + 6x + 5 \\ &= 4(2) + 9 + 6(2) + 5 \\ &= 8 + 9 + 12 + 5 \\ &= 34 \end{aligned}$$

17. $-\frac{3}{5}$

Put the equation $3x + 5y + 4 = 0$ in slope-intercept form: $3x + 5y + 4 = 0$; $5y = -3x - 4$. Divide both sides by 5: $y = -\frac{3}{5}x - \frac{4}{5}$. The slope m is the coefficient of x, or $-\frac{3}{5}$.

18. 25°

$\angle A = \angle B$ because vertical angles are equal. Therefore, $4x - 35 = x + 10$; $3x = 45$; $x = 15$. Substitute the value of x to find the measure of $\angle A$: $4x - 35 = 4(15) - 35 = 60 - 35 = 25$.

19. 10

$AX + XB = AB$ because X is between A and B. Therefore, $2x + 2x + 5 = 45$; $4x + 5 = 45$; $4x = 40$; $x = 10$.

20. 20

Angle $2x$ and angle $5x + 40$ are supplementary because when two parallel lines intersect with a third line, any acute angle is supplementary to any obtuse angle. So, $2x + 5x + 40 = 180$; $7x + 40 = 180$; $7x = 140$; $x = 20$.

Advanced

21. $a = 14$, $b = 8$

Vertical angles are equal, so $5a - 20 = 3a + 8$; $2a = 28$; $a = 14$. Angle $(3a + 8)$ and angle $(7a + 4b)$ are supplementary because together they form a straight line. Therefore, $3a + 8 + 7a + 4b = 180$. Substitute 14 for a: $3(14) + 8 + 7(14) + 4b = 180$; $42 + 8 + 98 + 4b = 180$; $148 + 4b = 180$; $4b = 32$; $b = 8$.

22. 130°

Vertical angles are equal, so the measure of $\angle x = 7a + 4b$. Substitute the values of $a = 14$ and $b = 8$ to solve: $\angle x = 7a + 4b = 7(14) + 4(8) = 98 + 32 = 130$.

23. $x = 30°$, $y = 60°$, $z = 90°$

The ratio of $x{:}y{:}z$ is 1:2:3, so let $x = 1a$, $y = 2a$, and $z = 3a$. x, y, and z together compose a straight line. $x + y + z = 180$; $1a + 2a + 3a = 180$; $6a = 180$; $a = 30$. Then $x = 1a = 1(30) = 30$; $y = 2a = 2(30) = 60$; $z = 3a = 3(30) = 90$.

24. $\angle A = 144°$, $\angle B = 36°$

Because $\frac{1}{2}\angle A = 2\angle B$, $\angle A = 4\angle B$. $\angle A$ and $\angle B$ are supplementary. Therefore, $\angle A + \angle B = 180$. Substituting, $4\angle B + \angle B = 180$; $5\angle B = 180$; $\angle B = 36$. $\angle A + \angle B = 180$; $\angle A + 36 = 180$; $\angle A = 144$.

25. 69°

Let $x =$ the angle. Its complement equals $x - 48°$. $x + x - 48° = 90°$; $2x - 48° = 90°$; $2x = 138°$; $x = 69°$.

26. 17

If point X lies on segment RS, then $RX + XS = RS$. If the length of segment RX is 2 more than half of the length of segment XS, then $RX = \frac{1}{2}XS + 2$. Because $XS = 10$, $RX = \frac{1}{2}(10) + 2 = 5 + 2 = 7$. $RX + XS = RS$, so $7 + 10 = 17 = RS$.

27. $\angle J = 25°$, $\angle K = 65°$

$\angle K = 2\angle J + 15$. $\angle J$ and $\angle K$ are complementary, so $\angle J + \angle K = 90$. Substituting, $\angle J + (2\angle J + 15) = 90$; $3\angle J + 15 = 90$; $3\angle J = 75$; $\angle J = 25$. So, $\angle K = 2\angle J + 15 = 2(25) + 15 = 50 + 15 = 65$.

28. 52

$$\begin{aligned} AD &= AB + BC + CD \\ 9x + 7 &= 3x + 9 + (13 + 3x) \\ 9x + 7 &= 6x + 22 \\ 3x + 7 &= 22 \\ 3x &= 15 \\ x &= 5 \\ AD &= 9x + 7 = 9(5) + 7 = 45 + 7 = 52 \end{aligned}$$

29. 54°

Express a, b, and d in terms of c.

$$c = \frac{b}{2}, \text{ so } b = 2c. \; b = \frac{a}{2}, \text{ so } a = 2b = 2(2c) = 4c. \; d = 3c.$$

Because the four angles together compose a straight line, $a + b + c + d = 180$. Substituting,

$$\begin{aligned} 4c + 2c + c + 3c &= 180^\circ \\ 10c &= 180^\circ \\ c &= 18^\circ \\ d &= 3c = 3(18^\circ) = 54^\circ \end{aligned}$$

30. $c = 27°$

Express a, b, and d in terms of c.

$$\begin{aligned} &b = 2c \text{ so } a = 2c - 5 \\ &d = 2c - 5 + 1 = 2c - 4 \end{aligned}$$

Because the four angles together compose a straight line, $a + b + c + d = 180$. Substituting,

$$\begin{aligned} 2c - 5 + 2c + c + 2c - 4 &= 180^\circ \\ 7c - 9 &= 180^\circ \\ 7c &= 189^\circ \\ c &= 27^\circ \end{aligned}$$

TRIANGLES AND PYTHAGOREAN THEOREM

General Triangles

A **triangle** is a closed figure with three angles and three straight sides.

The sum of the **interior angles** of any triangle is 180 degrees.

Each interior angle is supplementary to an adjacent **exterior angle**. The degree measure of an exterior angle is equal to the sum of the measures of the two nonadjacent (remote) interior angles, or 180° minus the measure of the adjacent interior angle.

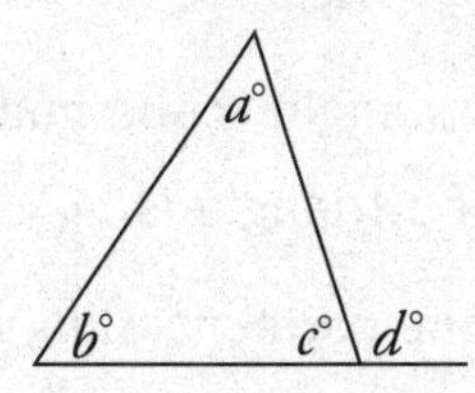

The **height** of a triangle is the perpendicular distance from a vertex to the side opposite the vertex. The height, or altitude, can fall inside the triangle, outside the triangle, or on one of the sides.

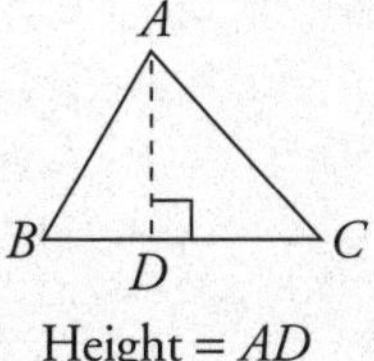

Height = AD

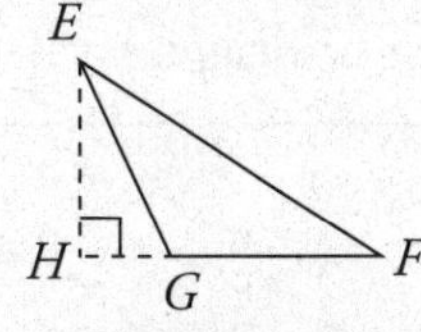

Height = EH

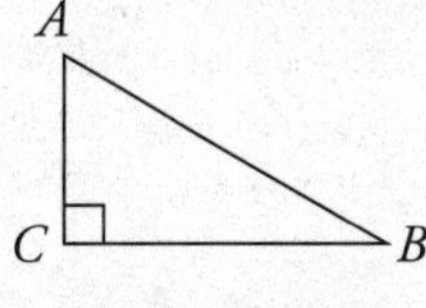

Height = AC

Sides and angles: The length of any side of a triangle is less than the sum of the lengths of the other two sides, and it is greater than the positive difference of the lengths of the other two sides.

$$b + c > a > b - c$$
$$a + b > c > a - b$$
$$a + c > b > a - c$$

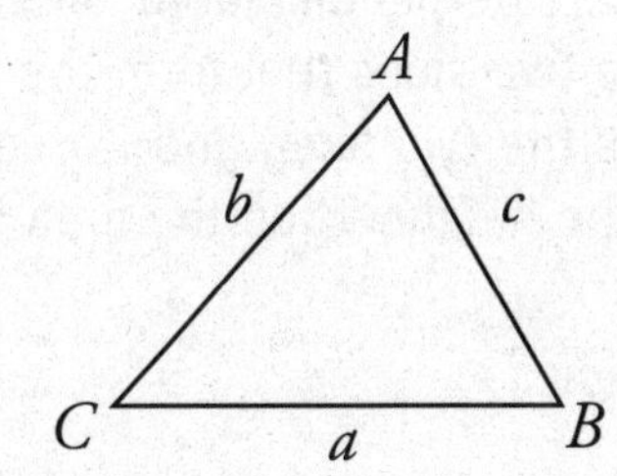

If the lengths of two sides of a triangle are unequal, the **greater angle** lies **opposite the longer side** and vice versa. In the figure shown, if $\angle A > \angle B > \angle C$, then $a > b > c$.

Example: In the diagram shown, what is an inequality that shows the relationship between the sides of the triangle?

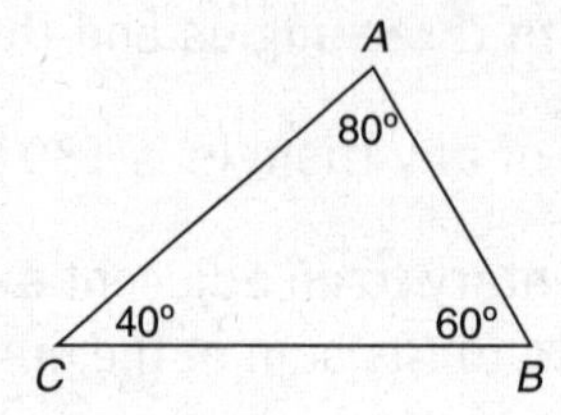

Because $\angle A > \angle B > \angle C$, $BC > AC > AB$.

Area of a triangle: The area of a triangle is the number of square units inside the triangle. A formula for the area of a triangle is $A = \frac{1}{2}bh$, where b is the length of the base of the triangle and h is the corresponding height.

Example: In the diagram shown, the base has length 4 and the height has length 3. What is the area of the triangle?

> The area of a triangle is $\frac{1}{2}$ base $\times$ height.

$$A = \frac{1}{2}bh$$
$$= \frac{1}{2} \times 4 \times 3 = 6$$

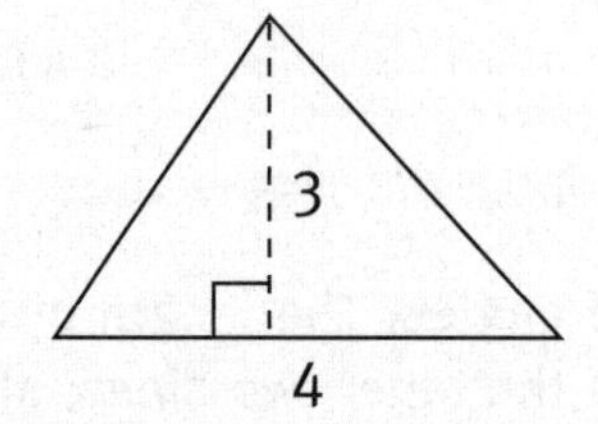

Remember that the height is perpendicular to the base. Therefore, when two sides of a triangle are perpendicular to each other, the area is easy to find. In a right triangle, we call the two sides that form the 90° angle the **legs**. The easiest base and height are always the two legs, and it doesn't matter which is called the base and which is called the height. Then the area is one-half the product of the legs, or

$$A = \frac{1}{2}bh$$
$$A = \frac{1}{2}ab$$

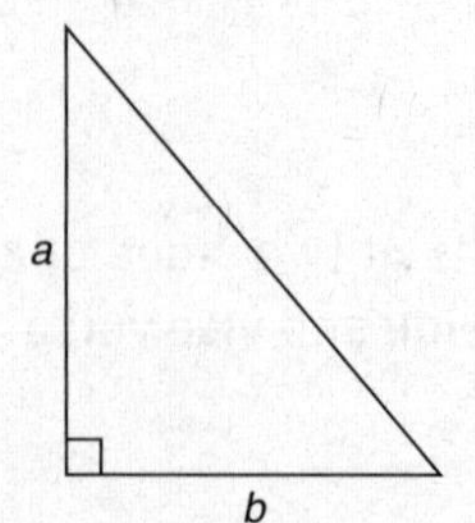

Example: What is the area of a right triangle with legs 6 and 8?

$$A = \frac{1}{2}\ell_1 \times \ell_2$$
$$= \frac{1}{2} \times 6 \times 8 = 24$$

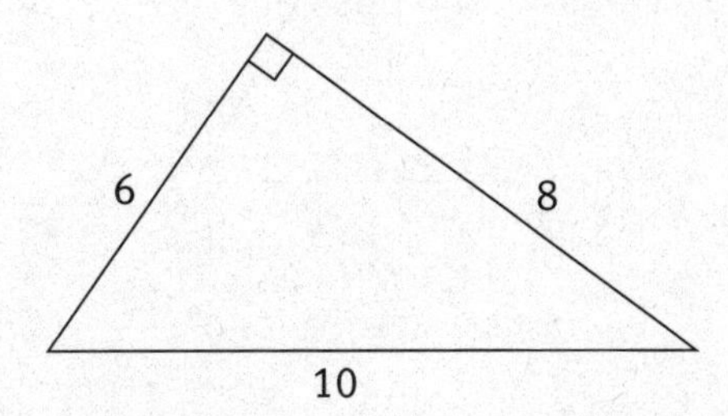

Perimeter of a triangle: The **perimeter** of a triangle is the distance around the triangle. In other words, the perimeter is equal to the sum of the lengths of the sides.

Example: In the triangle shown, the sides are of length 5, 6, and 8. What is the perimeter?

The perimeter is the sum of the sides: $5 + 6 + 8$, or 19.

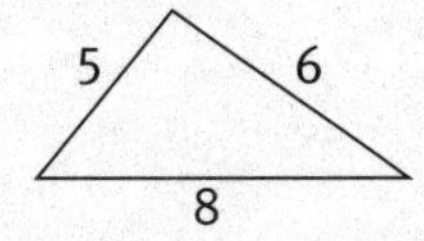

Isosceles triangles: An **isosceles triangle** is a triangle that has two sides of equal length. The two equal sides are called **legs**, and the third side is called the **base**.

Since the two legs have the same length, the two angles opposite the legs must have the same measure. In the figure shown, $PQ = PR$, and $\angle R = \angle Q$.

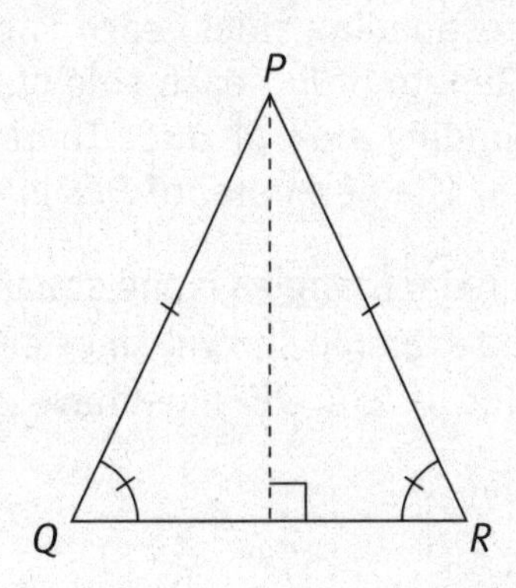

Equilateral triangles: An **equilateral triangle** has three sides of equal length and three 60° angles.

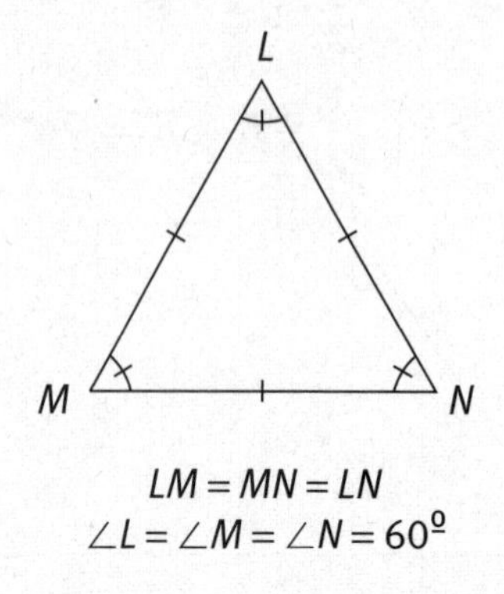

$LM = MN = LN$
$\angle L = \angle M = \angle N = 60^{\circ}$

Similar triangles: Triangles are **similar** if they have the same shape—if corresponding angles have the same measure. For instance, any two triangles whose angles measure 30°, 60°, and 90° are similar. In similar triangles, corresponding sides are in the same ratio. Triangles are **congruent** if corresponding angles have the same measure and corresponding sides have the same length.

Example: What is the perimeter of ΔDEF below?

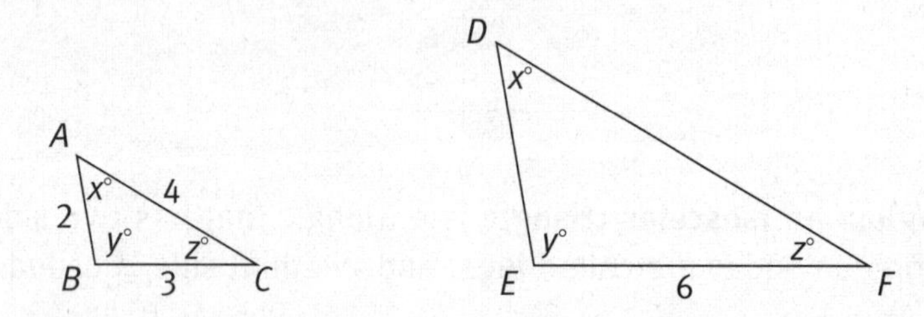

Each triangle has an $x°$ angle, a $y°$ angle, and a $z°$ angle; therefore, they are similar, and corresponding sides are in the same ratio. BC and EF are corresponding sides; each is opposite the $x°$ angle. Since EF is twice the length of BC, each side of DEF will be twice the length of the corresponding side of ABC. Therefore $DE = 2(AB)$ or 4, and $DF = 2(AC)$ or 8. The perimeter of DEF is $4 + 6 + 8 = 18$.

The ratio of the areas of two similar triangles is the square of the ratio of corresponding lengths. For instance, in the example shown, since each side of DEF is 2 times the length of the corresponding side of ABC, DEF must have 2^2 or 4 times the area of ABC.

$$\frac{\text{Area } \Delta DEF}{\text{Area } \Delta ABC} = \left(\frac{DE}{AB}\right)^2 = \left(\frac{2}{1}\right)^2 = 4$$

Right Triangles and the Pythagorean Theorem

A right triangle has one interior angle of 90°. The longest side (which lies opposite the right angle, the largest angle of a right triangle) is called the **hypotenuse**. As mentioned before, the other two sides are called the **legs**.

Pythagorean Theorem

The Pythagorean theorem holds for all right triangles and states that the square of the hypotenuse is equal to the sum of the squares of the legs.

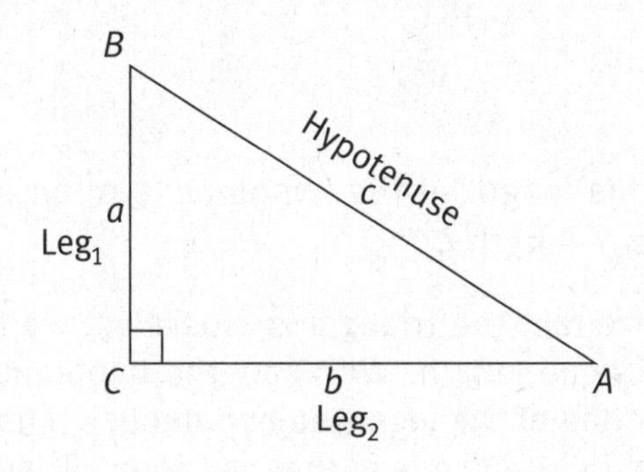

$$(\text{Leg}_1)^2 + (\text{Leg}_2)^2 = (\text{Hypotenuse})^2$$

or

$$a^2 + b^2 = c^2$$

Some sets of integers happen to satisfy the Pythagorean theorem. These sets of integers are commonly referred to as "Pythagorean triples." One very common set that you might remember is 3, 4, and 5. Since $3^2 + 4^2 = 5^2$, you can have a right triangle with legs of lengths 3 and 4 and hypotenuse of length 5. This is probably the most common kind of right triangle on the GRE. You should be familiar with the numbers so that whenever you see a right triangle with legs of 3 and 4, you will immediately know the hypotenuse must have length 5. In addition, any multiple of these lengths makes another Pythagorean triple; for instance, $6^2 + 8^2 = 10^2$, so 6, 8, and 10 also make a right triangle. One other triple that is seen frequently is 5, 12, and 13. The Pythagorean theorem is very useful; whenever you're given the lengths of two sides of a right triangle, you can find the length of the third side with the Pythagorean theorem.

Example: What is the length of the hypotenuse of a right triangle with legs of length 9 and 10?

Use the Pythagorean theorem: the square of the length of the hypotenuse equals the sum of the squares of the lengths of the legs. Here the legs are 9 and 10, so we have

$$\begin{aligned}\text{Hypotenuse}^2 &= 9^2 + 10^2\\ &= 81 + 100\\ &= 181\\ \text{Hypotenuse} &= \sqrt{181}\end{aligned}$$

Example: What is the length of the hypotenuse of an isosceles right triangle with legs of length 4?

Since we're told the triangle is isosceles, we know two of the sides have the same length. We know the hypotenuse can't be the same length as one of the legs (the hypotenuse must be the longest side), so it must be the two legs that are equal. Therefore, in this example, the two legs have length 4, and we can use the Pythagorean theorem to find the hypotenuse.

$$\begin{aligned}\text{Hypotenuse}^2 &= 4^2 + 4^2\\ &= 16 + 16\\ &= 32\\ \text{Hypotenuse} &= \sqrt{32} = 4\sqrt{2}\end{aligned}$$

An isosceles right triangle is a special right triangle. The ratio of its sides is $1:1:\sqrt{2}$.

You can always use the Pythagorean theorem to find the lengths of the sides in a right triangle. There are two frequently-tested special right triangles, though, that always have the same ratios. They are:

$1:1:\sqrt{2}$ (for isosceles 45°–45°–90° right triangles)

$1:\sqrt{3}:2$ (for 30°–60°–90° right triangles)

Thus, special right triangles include those with the following side ratios:

- 3:4:5
- 5:12:13
- $1:1:\sqrt{2}$
- $1:\sqrt{3}:2$

Example: What is the hypotenuse of a right triangle if the legs are 5 and 5?

This is a multiple of a $1:1:\sqrt{2}$ triangle, so the sides are $5:5:5\sqrt{2}$. The hypotenuse is $5\sqrt{2}$.

Example: What is the hypotenuse of a right triangle if the legs are 12 and 16?

This is a multiple of a 3:4:5, so the sides are 12:16:20. The hypotenuse is 20.

Example: The hypotenuse of a right triangle is 4 and one leg is 2. What is the other leg?

This is a multiple of $1:\sqrt{3}:2$, so the sides are $2:2\sqrt{3}:4$. The other leg is $2\sqrt{3}$.

If you don't remember the special right triangles, you can still use the Pythagorean theorem to calculate the length of a side, as we did in examples on the previous page.

TRIANGLES AND PYTHAGOREAN THEOREM EXERCISES

BASIC

1. The angles of a triangle are 50°, 60°, and $x°$. What is the value of x?

2. What is the area of a triangle with a height of 8 and a base of 10?

3. What is the perimeter of a triangle with sides of lengths 6, 7, and 9?

4. What is the perimeter of a right triangle with legs of lengths 3 and 4?

5. What is the area of a right triangle with legs of lengths 12 and 16?

6. Angle x is an exterior angle of a triangle with remote interior angles 50° and 80°. What is the value of x in degrees?

7. The measure of an angle across from one of the legs of an isosceles triangle is 40°. What are the measures of the other two angles?

8. For two similar triangles, the ratio of their corresponding sides is 2:3. What is the ratio of their perimeters?

9. What is the ratio of the areas of two congruent triangles?

10. The legs of a right triangle are of length 1 and $\sqrt{3}$. What is the hypotenuse?

INTERMEDIATE

11. The sides of a triangle are of length 5, 8, and c. What is the range of the possible values of c?

12. The hypotenuse of an isosceles right triangle is 16. What is the area of the triangle?

13. What is the length of a diagonal of the rectangle shown?

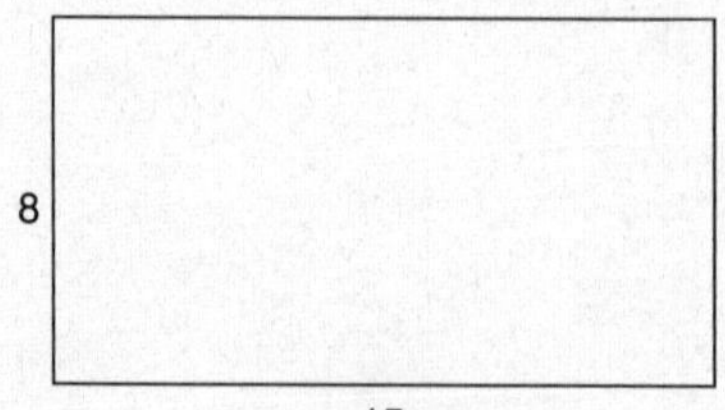

14. A square has a diagonal of length 5. What is the length of a side of the square?

15. A triangle has sides with lengths 4, 6, and 9. A larger similar triangle's shortest side has length 12. What is the perimeter of the larger triangle?

16. Triangle ABC is congruent to triangle DEF, $AC = 8$, $AB = 10$, and $BC = 15$. What is EF?

17. For similar triangles, the ratio of their corresponding sides is 2:3. What is the ratio of their areas?

18. The area of a triangle with a base of length 6 is equal to 12. What is the height of the triangle?

19. Triangle ABC is similar to triangle DEF, $AC = 9$, $AB = 6$, $DF = 18$, $DE = 12$, and $BC = 12$. What is EF?

20. What is the value of x in the triangle shown?

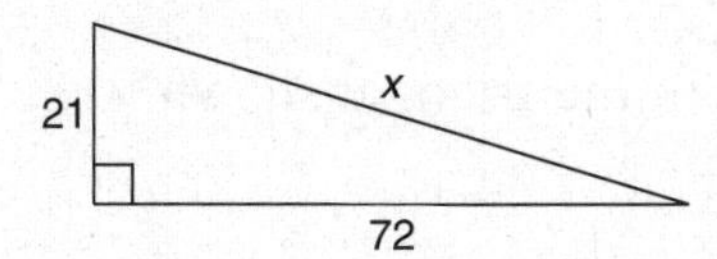

ADVANCED

21. What is the area of the triangle shown?

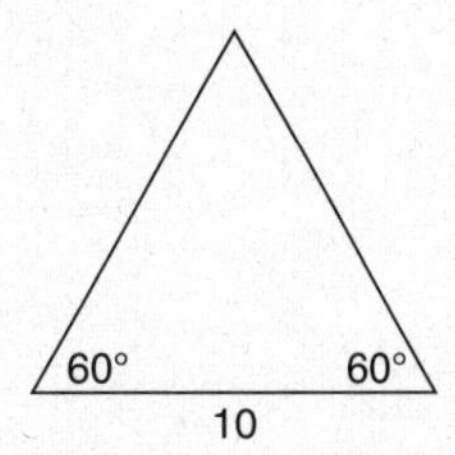

22. The sides of a triangle are of length 4, 8, and $4\sqrt{3}$. What is the area of the triangle?

23. What is the area of the triangle shown?

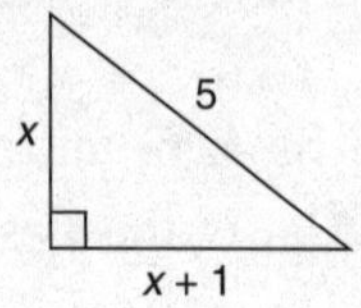

24. A 25-foot ladder rests against the side of a house. If the base of the ladder is 7 feet away from the house, how high off the ground is the top of the ladder?

25. A rectangle inscribed in a circle has a length of 8 and a width of 6. What is the area of the circle in terms of π?

26. What is the area of a right triangle with hypotenuse 61 and one leg 11?

27. The lengths of the legs of a right triangle are $3x$ and $x + 1$. The hypotenuse is $3x + 1$. What is the value of x ?

28. What is the side length of an equilateral triangle with an altitude of 6?

29. What is the value of x in the triangle shown?

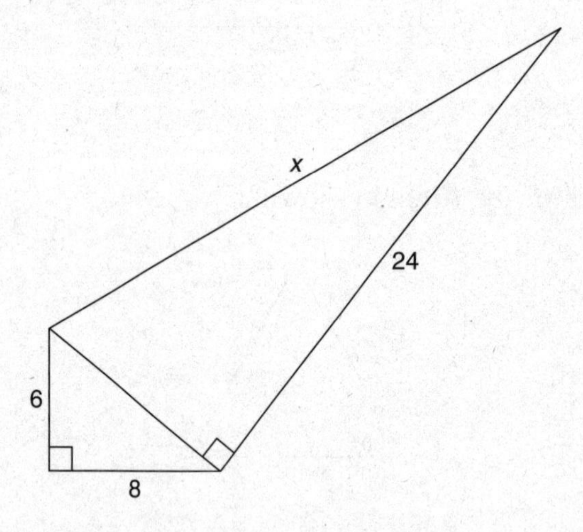

30. What is the value of x in the triangle shown?

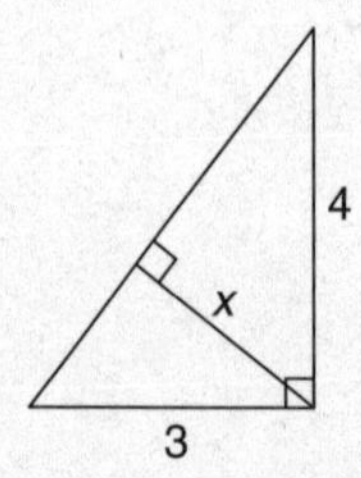

TRIANGLES AND PYTHAGOREAN THEOREM ANSWER KEY

BASIC

1. 70°

The sum of the interior angles of a triangle is 180°. Therefore, $50° + 60° + x = 180°$; $110° + x = 180°$; $x = 70°$.

2. 40

The formula for the area of a triangle is $A = \frac{1}{2}bh$. $A = \frac{1}{2}(10)(8) = 40$.

3. 22

The perimeter is equal to the sum of the lengths of the sides. $P = 6 + 7 + 9 = 22$.

4. 12

A right triangle with legs 3 and 4 has a hypotenuse of 5. It is a special right triangle. $P = 3 + 4 + 5 = 12$.

5. 96

The legs of a right triangle are the base and height of the triangle.

$$A = \frac{1}{2}bh.\ A = \frac{1}{2}(12)(16) = 96.$$

6. 130°

An exterior angle of a triangle is equal to the sum of the remote interior angles.

Thus, $50° + 80° = 130°$.

7. 40°, 100°

Base angles of an isosceles triangle are equal, so the other base angle is 40°. The sum of all the angles is 180°. Solve to find the measure of the remaining angle x: $40° + 40° + x = 180°$; $80° + x = 180°$; $x = 100°$.

8. 2:3

The ratio of the perimeters of two similar triangles equals the ratio of the corresponding sides, 2:3.

9. 1:1

Corresponding sides of congruent triangles are equal, so the areas will be the same; they will have a ratio of 1:1.

10. 2

If the legs of a right triangle are 1 and $\sqrt{3}$, the right triangle is a 30°-60°-90° right triangle and the hypotenuse is 2. You can also use the Pythagorean theorem: $c^2 = 1^2 + (\sqrt{3})^2 = 1 + 3 = 4$. $c = \pm 2$. Since c is a side of a triangle, it must be positive, and $c = 2$.

INTERMEDIATE

11. $3 < c < 13$

The third side of a triangle is greater than the difference of the two sides but less than the sum of the two sides.

$$a - b < c < a + b$$
$$8 - 5 < c < 8 + 5$$
$$3 < c < 13$$

12. 64

The sides of an isosceles right triangle are in the ratio $1:1:\sqrt{2}$. So, divide the hypotenuse by $\sqrt{2}$ to get the length of a leg: $\frac{16}{\sqrt{2}} = \frac{16\sqrt{2}}{\sqrt{2} \times \sqrt{2}} = \frac{16\sqrt{2}}{2} = 8\sqrt{2}$. The legs of a right triangle are the base and height of the triangle. $A = \frac{1}{2}bh$. $A = \frac{1}{2}(8\sqrt{2})(8\sqrt{2}) = 64$.

13. 17

The diagonal of a rectangle is the hypotenuse of a right triangle whose legs are the length and the width of the rectangle. $c^2 = 8^2 + 15^2 = 64 + 225 = 289$; $c = 17$. 8:15:17 is a Pythagorean triple.

14. $\frac{5\sqrt{2}}{2}$

The diagonal of a square is the hypotenuse of an isosceles right triangle whose legs are sides of the square. The sides of an isosceles right triangle are in the ratio $1:1:\sqrt{2}$. So, divide the hypotenuse by $\sqrt{2}$ to get the length of a leg: $\frac{5}{\sqrt{2}} = \frac{5\sqrt{2}}{\sqrt{2} \times \sqrt{2}} = \frac{5\sqrt{2}}{2}$.

15. 57

The triangles are similar. The shortest side of the smaller triangle is 4. The shortest side of the larger triangle is 12. The ratio of corresponding sides is 4:12 or 1:3. So the sides of the larger triangle are $4(3) = 12$, $6(3) = 18$, and $9(3) = 27$. $P = 12 + 18 + 27 = 57$.

16. 15

Triangle *ABC* is congruent to triangle *DEF*. The corresponding sides are *AB* and *DE*, *AC* and *DF*, and *BC* and *EF*. Because the two triangles are congruent, the sides are in the ratio 1:1. So, if $BC = 15$, $EF = 15$.

17. 4:9

The ratio of the areas of two similar triangles equals the square of the ratio of the corresponding sides: $2^2:3^2 = 4:9$.

18. 4

$$A = \frac{1}{2}bh.\ 12 = \frac{1}{2}(6)(h);\ 12 = 3h;\ 4 = h.$$

19. 24

Triangle *ABC* is similar to triangle *DEF*. The corresponding sides are *AB* and *DE*, *AC* and *DF*, and *BC* and *EF*. $\frac{AB}{DE} = \frac{AC}{DF} = \frac{BC}{EF}$. $\frac{6}{12} = \frac{9}{18} = \frac{12}{EF}$. The scale factor is 1:2. So $EF = 24$.

20. 75

Use the Pythagorean theorem. $x^2 = 21^2 + 72^2 = 441 + 5{,}184 = 5{,}625$. $x = 75$. This is a multiple of the Pythagorean triple 7:24:25.

Advanced

21. $25\sqrt{3}$

If two angles of a triangle are each 60°, the third angle must also be 60°. So the triangle is equilateral with side length 10. Draw an altitude. It is the side opposite the 60° angle in a 30°-60°-90° triangle. Its length is one-half the hypotenuse times the square root of 3.

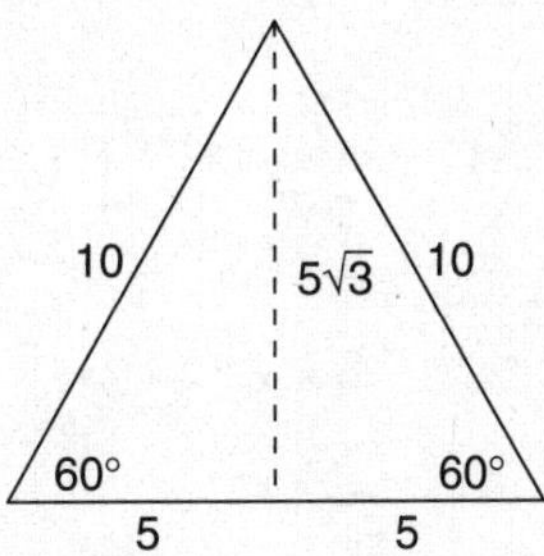

$A = \frac{1}{2}bh.\ A = \frac{1}{2}(10)(5\sqrt{3}) = 25\sqrt{3}.$

22. $8\sqrt{3}$

The ratio of the sides is $4:4\sqrt{3}:8$. So, the triangle is a 30°-60°-90° triangle with legs 4 and $4\sqrt{3}$. The area of a right triangle is one-half the product of the legs: $A = \frac{1}{2}bh$. Thus, $A = \frac{1}{2}(4)(4\sqrt{3}) = 8\sqrt{3}$.

23. 6

The triangle is a special right triangle. It is a 3:4:5 triangle.

The area of a right triangle is one-half the product of the legs: $A = \frac{1}{2}bh$. $A = \frac{1}{2}(3)(4) = 6$.

24. 24 feet

The ladder, the ground, and the house form a right triangle with hypotenuse 25 and leg 7.

If you notice that this is a 7:24:25 Pythagorean triple, you're done. Alternatively, use the Pythagorean theorem: $25^2 = 7^2 + x^2$; $625 = 49 + x^2$; $576 = x^2$; $x = 24$.

25. 25π

The diagonal of a rectangle inscribed in a circle is the diameter of the circle. The diagonal of the rectangle is equal to the hypotenuse of a right triangle with legs 6 and 8. Therefore, the hypotenuse is 10 because it is a multiple of a 3:4:5 right triangle. The area of a circle is $A = \pi r^2$. The diameter of a circle is $2r$, so $r = 5$. The area of the circle is $A = \pi r^2 = \pi(5)^2 = 25\pi$.

26. 330

You can use the Pythagorean theorem to find the other leg. Thus, $61^2 = 11^2 + x^2$; $3{,}721 = 121 + x^2$; $3{,}600 = x^2$; $x = 60$. The area of a right triangle is one-half the product of the legs: $A = \frac{1}{2}bh$. $A = \frac{1}{2}(60)(11) = 330$.

27. 4

Use the Pythagorean theorem.

$$\begin{aligned}(3x + 1)^2 &= (3x)^2 + (x + 1)^2 \\ 9x^2 + 6x + 1 &= 9x^2 + x^2 + 2x + 1 \\ 6x &= x^2 + 2x \\ 4x &= x^2 \\ 4 &= x\end{aligned}$$

28. $4\sqrt{3}$

The altitude of this equilateral triangle divides the triangle into two 30°-60°-90° triangles with side lengths x, 6, and $2x$, where $2x$ is the hypotenuse of the 30°-60°-90° triangles and side length of the equilateral triangle. You can use the Pythagorean theorem, but since 6 is the longer leg of the 30°-60°-90° triangle and a 30°-60°-90° triangle has sides of ratio $1:\sqrt{3}:2$, you can say that $6 = x\sqrt{3}$. From here,

$$x = \frac{6}{\sqrt{3}}$$

$$x = \frac{6\sqrt{3}}{\sqrt{3}\sqrt{3}} = \frac{6\sqrt{3}}{3} = 2\sqrt{3}$$

However, the question asks for the length of the side of the equilateral triangle, which is $2x$, so we need to double the value of x.

$$2 \times 2\sqrt{3} = 4\sqrt{3}$$

The answer is $4\sqrt{3}$.

29. 26

The right triangle with legs 6 and 8 has hypotenuse 10 because it is a double 3:4:5 triangle. The other triangle has legs 10 and 24, which are doubles of 5 and 12. A 5:12:13 right triangle has hypotenuse 13. So, the hypotenuse of a right triangle with legs of length 10 and 24 is $2(13) = 26$.

30. 2.4

If the legs of a right triangle are of length 3 and 4, the hypotenuse is 5. We also know the area of the triangle is one-half times the product of the lengths of the two legs. Therefore,

$$A = \frac{1}{2}(3)(4) = 6$$

The line segment marked by x can also be used as the height when the hypotenuse is used as the base, since the hypotenuse and the line marked by x are perpendicular. Since the area of the triangle is 6,

$$6 = \frac{1}{2}(5)(x)$$

$$6 = 2.5x$$

$$\frac{6}{2.5} = x$$

$$2.4 = x$$

Therefore, x is equal to 2.4.

POLYGONS

A **polygon** is a closed figure whose sides are straight line segments.

The **perimeter** of a polygon is the sum of the lengths of the sides.

A **vertex** of a polygon is the point where two adjacent sides meet.

A **diagonal** of a polygon is a line segment connecting two nonadjacent vertices.

A **regular polygon** has sides of equal length and interior angles of equal measure.

The number of sides determines the specific name of the polygon. A **triangle** has three sides, a **quadrilateral** has four sides, a **pentagon** has five sides, and a **hexagon** has six sides. Triangles and quadrilaterals are by far the most important polygons on the GRE.

Interior and exterior angles: A polygon can be divided into triangles by drawing diagonals from a given vertex to all other nonadjacent vertices. For instance, the pentagon shown can be divided into 3 triangles. Since the sum of the interior angles of each triangle is 180°, the sum of the interior angles of a pentagon is $3 \times 180° = 540°$.

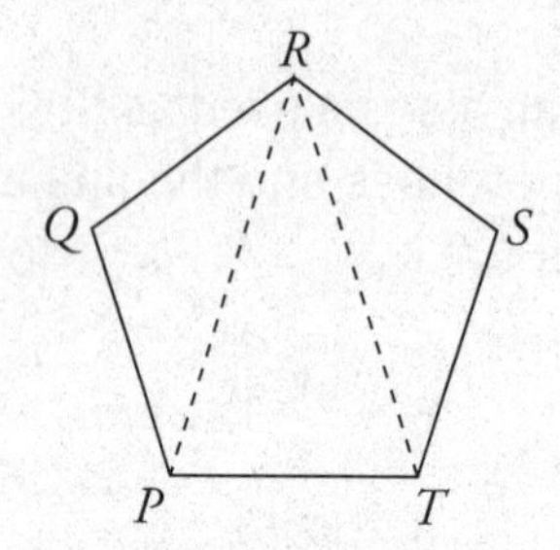

Example: What is the measure of one interior angle of the regular hexagon shown?

Find the sum of the interior angles and divide by the number of interior angles, or 6. (Since all angles are equal, each of them is equal to one-sixth of the sum.)

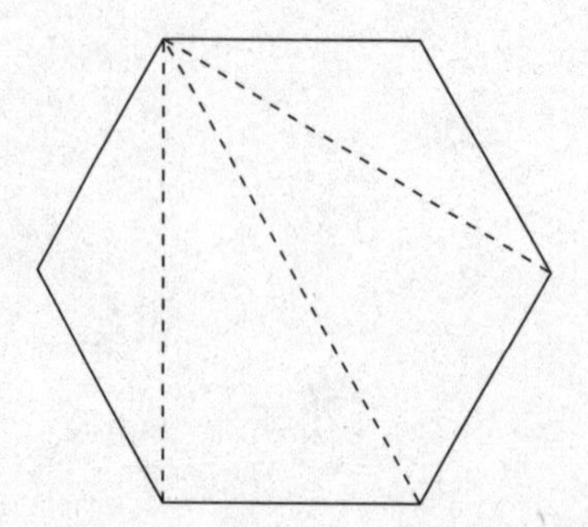

Since we can draw 4 triangles in a 6-sided figure, the sum of the interior angles will be $4 \times 180°$, or $720°$. Therefore, each of the six interior angles has measure $\frac{720}{6}$, or 120 degrees.

In general, the sum of the interior angles of a polygon equals $180°(n - 2)$, where n is the number of sides in the polygon.

QUADRILATERALS

The most important quadrilaterals to know for the GRE are the rectangle and square. Any quadrilateral could show up on the test, but concentrate on the most important figures and principles as they will yield you the most points.

Quadrilateral: A four-sided polygon. The sum of its four interior angles is 360°.

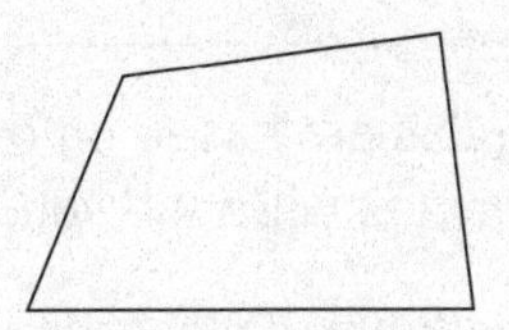

Parallelogram: A quadrilateral with two pairs of parallel sides.

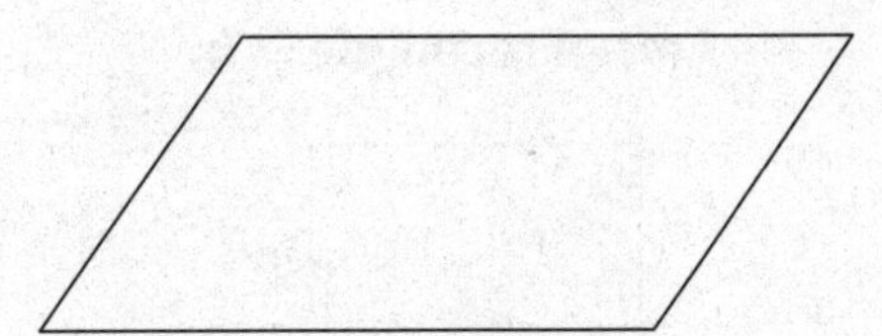

Rectangle: A parallelogram with four equal angles, each a right angle.

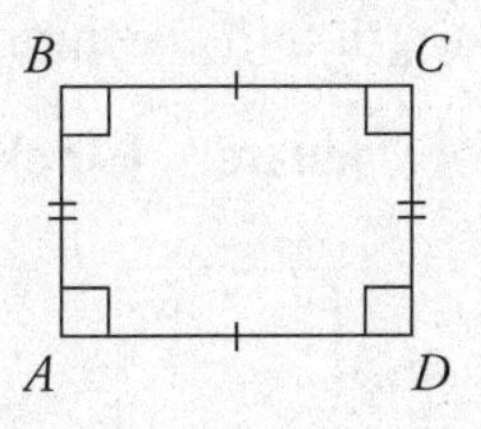

$AB = CD \qquad AD = CB$

The opposite sides of a rectangle are equal in length. Also, the diagonals of a rectangle have equal length.

Square: A rectangle with four equal sides.

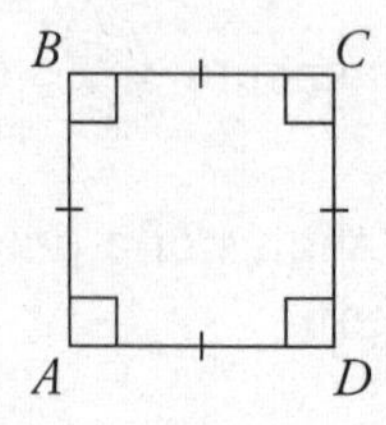

$AB = BC = CD = DA$

Trapezoid: A quadrilateral having only two parallel sides.

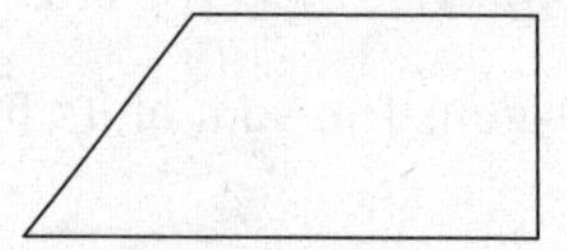

Areas of quadrilaterals: All formulas are based on common sense, observation, and deductions. Memorizing the formulas will save you time.

For the case of a rectangle, we multiply the lengths of any two adjacent sides, called the length and width, or:

Area of rectangle = lw

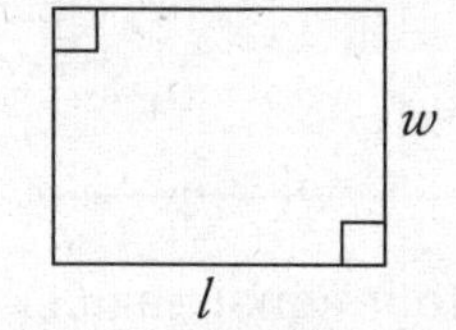

For the case of a square, since length and width are equal, we say:

Area of a square = (side)^2 = s^2

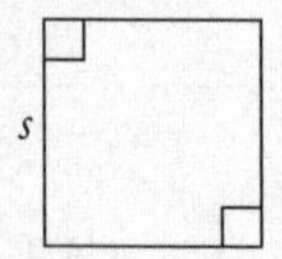

Area of a parallelogram = bh

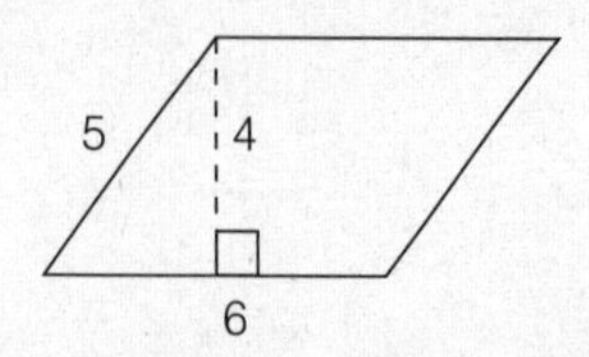

Area $= 6 \times 4 = 24$

Area of a trapezoid = (average of parallel sides)(height)

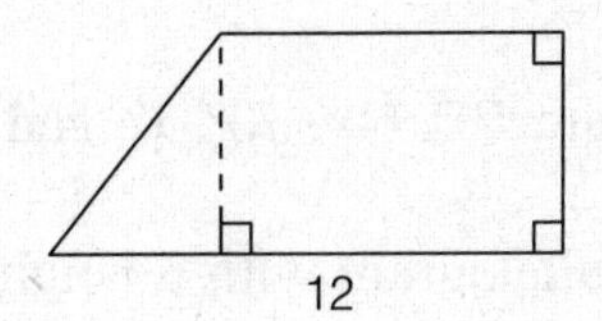

$$\text{Area} = \left(\frac{8 + 12}{2}\right) \times 5 = 50$$

The areas of other figures can usually be found using the methods we'll discuss later in the Multiple Figures section.

POLYGONS EXERCISES

BASIC

1. The angles of a pentagon are 50°, 60°, 70°, x°, and $2x$°. What is the value of x?

2. What is the area of a parallelogram with a height of 6 and a base of 8?

3. What is the perimeter of a quadrilateral with sides 6, 7, 8, and 9?

4. What is the sum of the interior angles of a heptagon, or seven-sided polygon?

5. What is the measure of one interior angle of a regular octagon?

6. What is the measure of an exterior angle of a regular pentagon?

7. What is the perimeter of a square with a side of length 6?

8. If a triangle has two sides of length 4 and 6, what is the range of possible values for the length of the third side x?

9. What is the perimeter of a regular hexagon with a side of length 8?

10. What is the area of a rectangle with a length of 6 and a width of 4?

INTERMEDIATE

11. The length of a rectangle is $2x - 1$. The width is $3x + 5$. The perimeter is 38. What is the value of x?

12. The following figure shows three squares and a triangle. What is the area of square A?

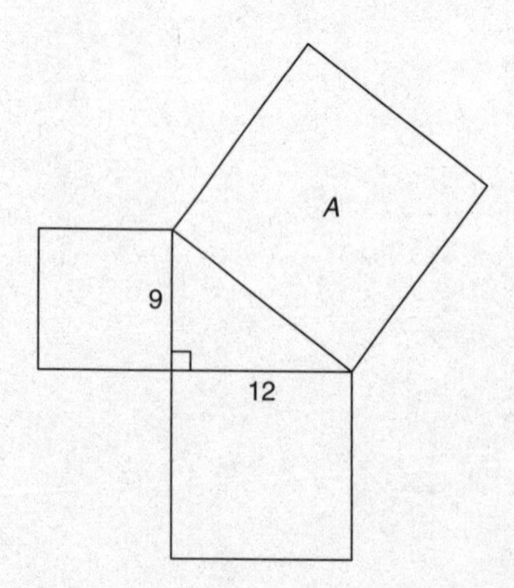

13. What is the area of the rectangle shown?

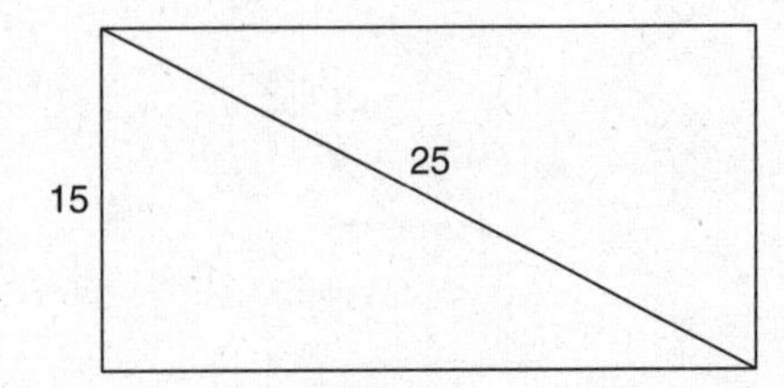

14. The area of a rectangle is 588 square feet, and its length is three times its width. What are the measures of the length and the width?

15. Regular pentagon *ABCDE* has a side length of 28. Regular pentagon *FGHIJ* has a perimeter of 60 and an area of about 248. What is the ratio of the area of *ABCDE* to the area of *FGHIJ*?

16. What is the area of the quadrilateral shown?

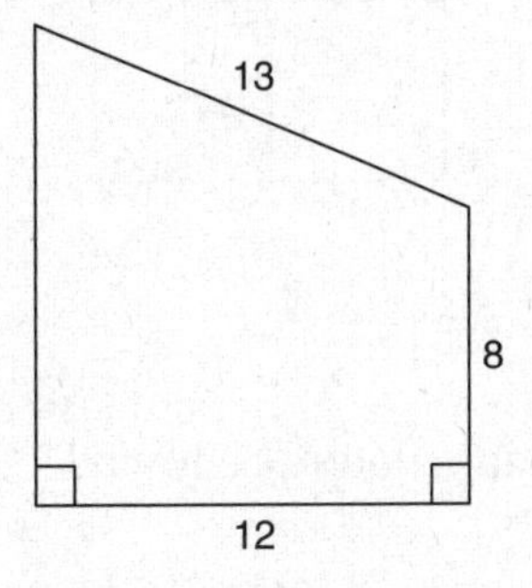

17. In similar hexagons, the ratio of the areas is 16:25. What is the ratio of their corresponding sides?

18. What is the perimeter of the square shown?

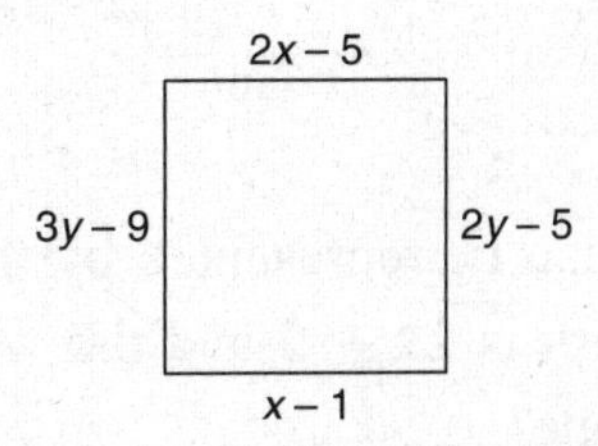

19. What are the values of x and y in the figure shown?

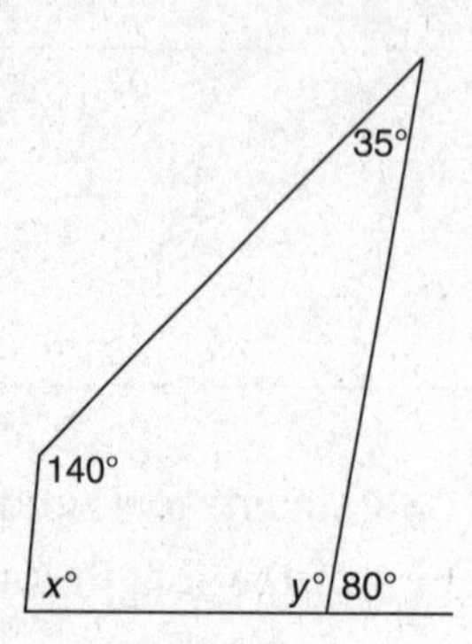

20. Four congruent squares are joined together in the figure shown. What is the ratio of the nonshaded area to the shaded area?

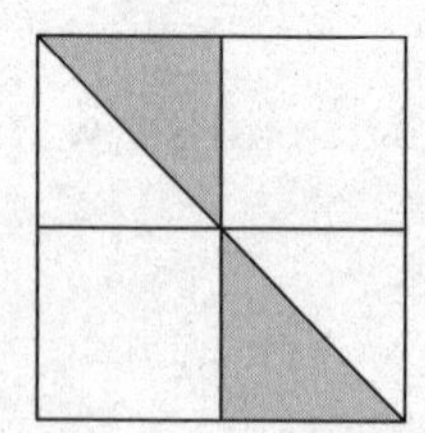

ADVANCED

21. What is the area of the quadrilateral shown?

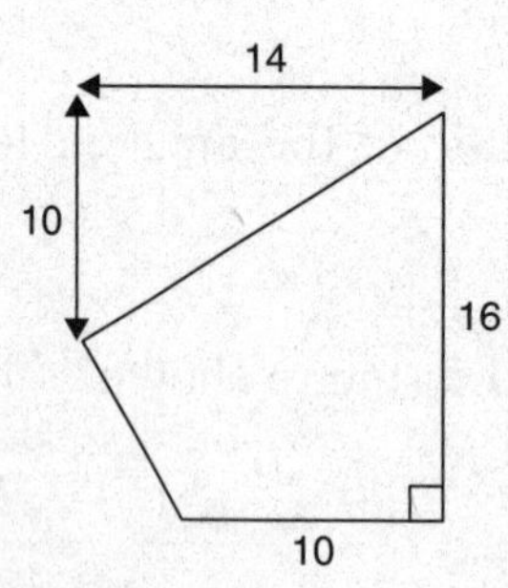

22. The area of a rectangle can be represented by the expression $2x^2 + 9x + 10$. The length is $2x + 5$ and the width is 6. What is the value of the area of the rectangle?

23. A rectangle with an area of 18 has a length that is 3 more than its width. What are the length and width of the rectangle?

24. A ceiling has a length of 30 feet and a width of 30 feet. The ceiling is going to be tiled with 24-inch by 36-inch tiles. How many tiles are needed to cover the ceiling?

25. What is the area of quadrilateral *ABCD* shown?

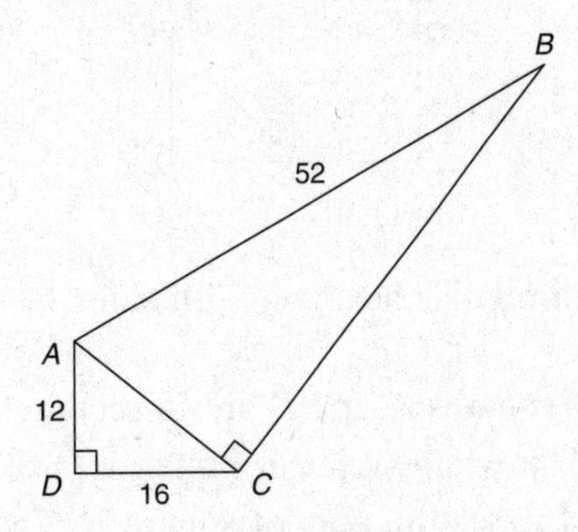

26. In the figure shown, *ABCD* is a rectangle, $AB = 8$, and $BC = 6$. *R*, *S*, *T*, and *Q* are midpoints of the sides of *ABCD*. What is the perimeter of *RSTQ*?

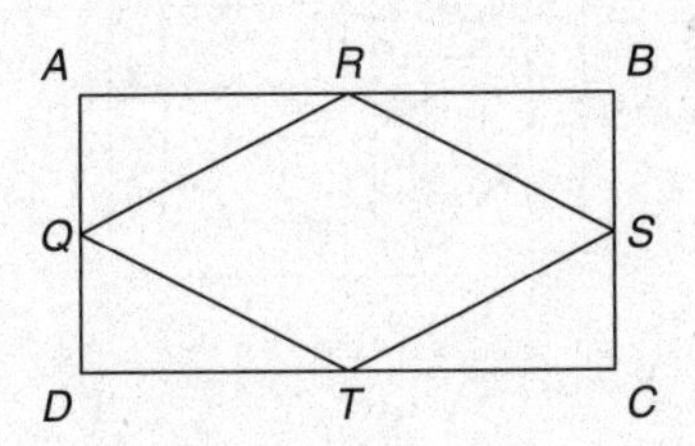

27. In the figure shown, the small square divides the diagonal of the large square into three equal parts. What is the ratio of the area of the large square to the area of the small square?

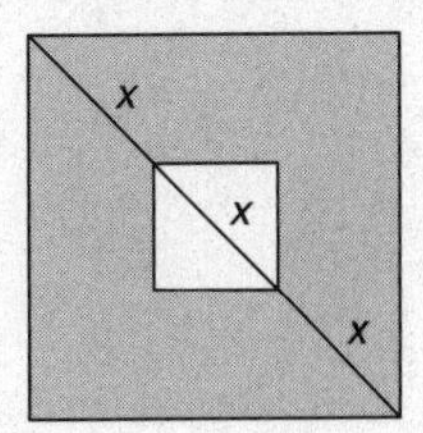

28. In the figure shown, *ABCD* is a square with a side length of 16. *R*, *S*, *T*, and *Q* are midpoints of the sides of *ABCD*. What is the area of *RSTQ* ?

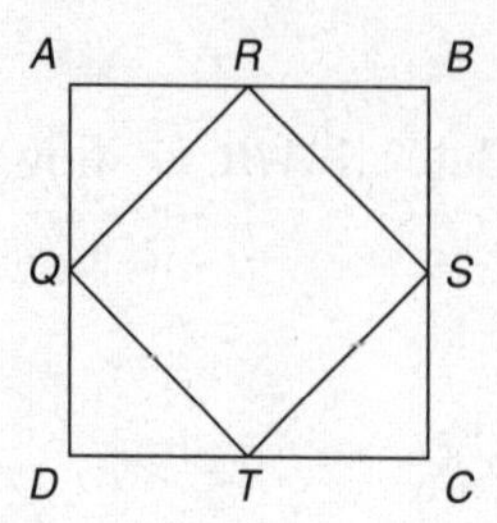

29. What is the area of a regular hexagon with sides of length 6?

30. In the figure shown, squares *A* and *B* are in rectangle *R*. The area of rectangle *R* is 216. The width (vertical dimension) of rectangle *R* is 12. The area of square *A* is 4 times the area of square *B*. The sum of the areas of squares *A* and *B* is 80. What is the value of *x* ?

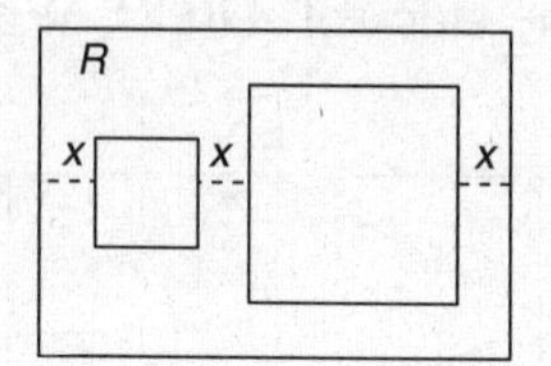

POLYGONS EXERCISES ANSWER KEY

Basic

1. 120

The sum of the interior angles of a pentagon is $(5 - 2)180 = 3(180) = 540$. Thus, $50 + 60 + 70 + x + 2x = 540$; $3x + 180 = 540$; $3x = 360$; $x = 120$.

2. 48

The formula for the area of a parallelogram is $A = bh$. $A = (6)(8) = 48$.

3. 30

The perimeter is equal to the sum of the lengths of the sides. $P = 6 + 7 + 8 + 9 = 30$.

4. 900

The sum of the interior angles of a seven-sided polygon is $(7 - 2)180 = 5(180) = 900$.

5. 135

An octagon is an eight-sided figure. The sum of the interior angles of an octagon is $(8 - 2)180 = 6(180) = 1{,}080$. Divide by 8 to get the measure of one interior angle: $1{,}080 \div 8 = 135$.

6. 72

The sum of the interior angles of a pentagon is $(5 - 2)180 = 3(180) = 540$. Divide by 5 to get the measure of one interior angle: $540 \div 5 = 108$. An exterior angle is supplementary to an interior angle: $x + 108 = 180$; $x = 72$.

7. 24

The perimeter of a square is $P = 4s$. $P = 4(6) = 24$.

8. $2 < x < 10$

For any triangle, the length of any side must be greater than the difference and less than the sum of the lengths of the other two sides. Since the two sides given here are 4 and 6, the third side of length x can be expressed as follows:

$$6 - 4 < x < 6 + 4$$
$$2 < x < 10$$

The length of the third side must be between 2 and 10.

9. 48

The perimeter of a regular hexagon is $P = 6s$. $P = 6(8) = 48$.

10. 24

The area of a rectangle is $A = lw$. $A = (6)(4) = 24$.

Intermediate

11. 3

The perimeter is the sum of all the sides. Therefore, $(2x - 1) + (3x + 5) + (2x - 1) + (3x + 5) = 38$; $10x + 8 = 38$; $10x = 30$; $x = 3$.

12. 225

The triangle is a 9:12:15 triangle, a multiple of a 3:4:5 triangle. The triangle's hypotenuse, 15, is also the side length of the square. $A = s^2 = 15^2 = 225$.

13. 300

The triangle is a 15:20:25 triangle, a multiple of a 3:4:5 triangle. The length of the rectangle is 20, the width is 15. $A = lw = (20)(15) = 300$.

14. Width = 14, Length = 42

Let w = the width. The length equals $3w$. $A = (3w)(w) = 3w^2$; $3w^2 = 588$; $w^2 = 196$; $w = 14$. Thus, $3w = 3(14) = 42$.

15. 49:9

The regular pentagons are similar. The length of one of the sides of *ABCDE* is 28. The perimeter of *FGHIJ* is 60. Since there are 5 sides, the length of one side can be found by dividing the perimeter, in this case 60, by the number of sides, in this case, 5. Each side has a length of 12. The ratio of the areas of two similar polygons equals the square of the ratios of the corresponding sides. The ratio of the sides is 28:12 or 7:3. The ratio of the areas is $7^2:3^2$ or 49:9.

16. 126

Divide the quadrilateral into a right triangle and a rectangle.

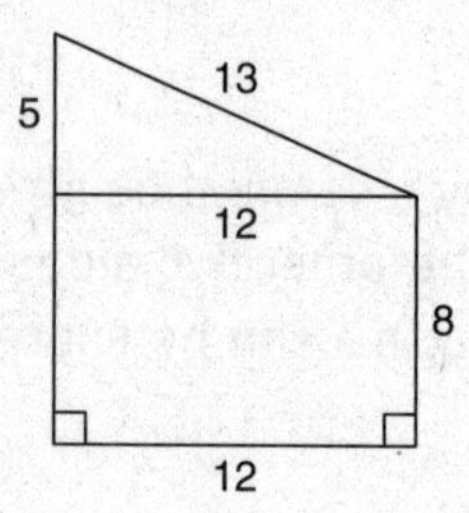

The triangle is a 5:12:13 triangle. For the triangle, $A = \frac{1}{2}bh$. $A = \frac{1}{2}(12)(5) = 30$.

For the rectangle, $A = lw$. $A = (12)(8) = 96$. Add the areas: $30 + 96 = 126$.

17. 4:5

The ratio of the areas of two similar polygons equals the square of the ratio of the corresponding sides. In this case, the ratio of the areas, 16:25, is the same as $4^2:5^2$. So, the ratio of the corresponding sides is 4:5.

18. 12

All sides of a square are equal. In this case, $2x - 5 = x - 1$; $2x = x + 4$; $x = 4$. One of the sides of the square is $x - 1 = 4 - 1 = 3$, so the other three sides also have length 3. The perimeter is $P = 4s = 4(3) = 12$.

19. $x = 85$, $y = 100$

The 80° angle and y are supplementary. Therefore, $y + 80 = 180$; $y = 100$. The sum of the interior angles of a quadrilateral is 360. $x + y + 140 + 35 = 360$; $x + 100 + 140 + 35 = 360$; $x + 275 = 360$; $x = 85$.

20. 3:1

If diagonals are drawn in all the squares, 8 congruent triangles are formed, 2 of which are shaded. There are 6 nonshaded triangles. The ratio is 6:2 or 3:1.

Advanced

21. 142

Divide the quadrilateral into two right triangles and a rectangle.

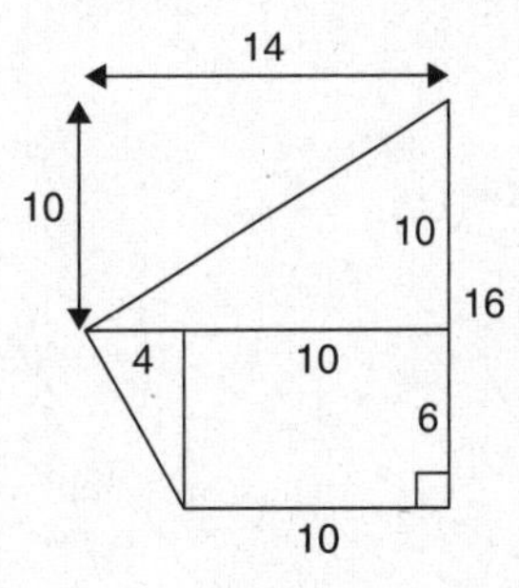

Area of small triangle: $A = \frac{1}{2}bh$. $A = \frac{1}{2}(4)(6) = 12$.

Area of large triangle: $A = \frac{1}{2}bh$. $A = \frac{1}{2}(14)(10) = 70$.

Area of rectangle: $A = lw$. $A = (10)(6) = 60$.

Add the areas: $A = 12 + 70 + 60 = 142$.

22. 78

To find the width in terms of x, divide the area by the length in terms of x:

$$\frac{2x^2 + 9x + 10}{2x + 5} = \frac{(2x + 5)(x + 2)}{(2x + 5)} = x + 2$$

So, $x + 2 = 6$ and $x = 4$. The area is then $2x^2 + 9x + 10 = 2(4)^2 + 9(4) + 10 = 32 + 36 + 10 = 78$. Alternately, length $= 2x + 5 = 2(4) + 5 = 13$, and width $= 6$. $A = lw$. $A = (13)(6) = 78$.

23. Length = 6, Width = 3

Let $w =$ the width. The length equals $w + 3$. Then $18 = (w)(w + 3)$; $18 = w^2 + 3w$; $0 = w^2 + 3w - 18$; $0 = (w + 6)(w - 3)$; $0 = w + 6$, or $0 = w - 3$. So, $w = -6$, or $w = 3$. Width is a distance and cannot be negative, so the width is 3. The length is $w + 3 = 3 + 3 = 6$.

24. 150

The area of the ceiling is $A = (30)(30) = 900$ square feet. The tiles are 24 inches, which is 2 feet, by 36 inches, which is 3 feet. The area of a tile is $A = (2)(3) = 6$ square feet. Divide the area of the ceiling by the area of a tile to get the number of tiles needed.

$900 \div 6 = 150$

25. 576

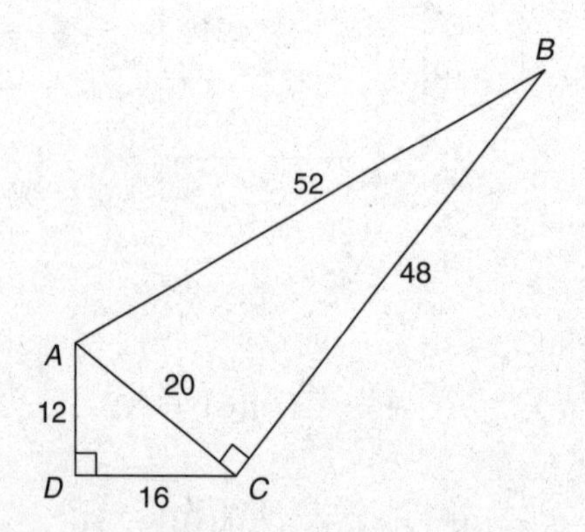

The triangles are multiples of a 3:4:5 right triangle and a 5:12:13 right triangle. The hypotenuse of the triangle with legs 12 and 16 is 20. The missing leg in the other triangle is 48.

Area of small triangle: $A = \frac{1}{2}bh$. $A = \frac{1}{2}(12)(16) = 96$.

Area of large triangle: $A = \frac{1}{2}bh$. $A = \frac{1}{2}(20)(48) = 480$.

$A = 96 + 480 = 576$.

26. 20

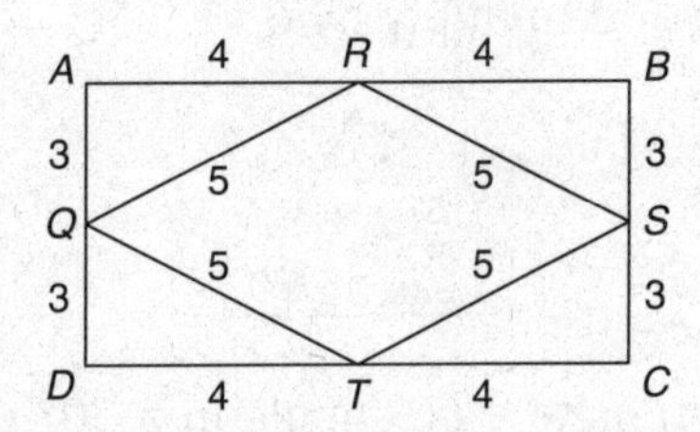

$AR = RB = DT = TC = 4$

$AQ = QD = BS = SC = 3$

The triangles formed are congruent 3:4:5 right triangles. The perimeter of *RSTQ* is 20.

27. 9:1

We can use Picking Numbers for this question. Let a side of the large square be 6. The diagonal of the large square is $6\sqrt{2}$, because a diagonal divides a square into two congruent isosceles right triangles. In addition, $6\sqrt{2} = 3x$. Therefore, $2\sqrt{2} = x =$ diagonal of small square. So, a side of the small square is 2. The ratio of the sides of the large square to the small square is 6:2 or 3:1. The ratio of their areas is $3^2{:}1^3$, or 9:1.

28. 128

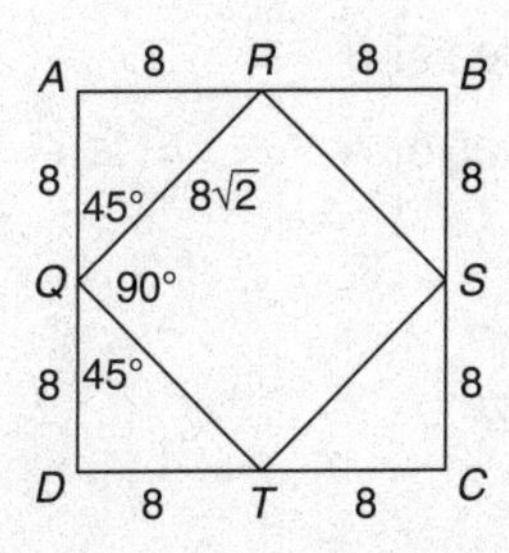

The triangles formed are all isosceles right triangles with legs of length 8. *RSTQ* is a square because all the angles are 90° as shown in the figure above, and its sides are the hypotenuses of four identical right triangles; each side measures $8\sqrt{2}$, as shown. $A = s^2$. $A = (8\sqrt{2})^2 = 128$.

29. $54\sqrt{3}$

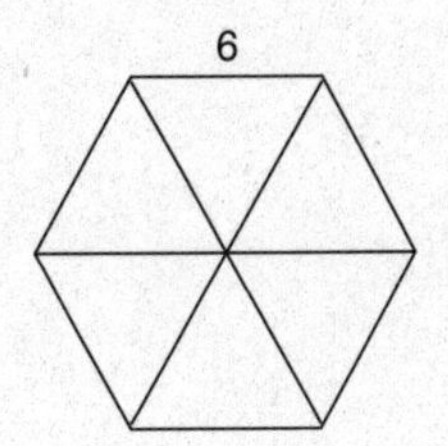

A regular hexagon can be divided into 6 congruent equilateral triangles. Find the area of one of the triangles and multiply by 6.

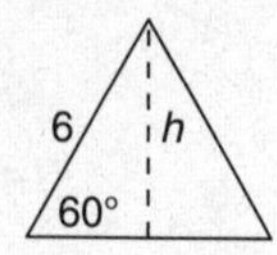

The altitude h is the side opposite a 60° angle in a 30°-60°-90° triangle; h is equal to one-half the hypotenuse times the square root of 3, or $3\sqrt{3}$. The area of one of the triangles, then, is:

$$A = \frac{1}{2}(6)(3\sqrt{3}) = 9\sqrt{3}$$

The area of the hexagon is $A = 6(9\sqrt{3}) = 54\sqrt{3}$.

30. 2

$$\begin{aligned} \text{Area of } A + \text{Area of } B &= 80 \\ \text{Area of } A &= 4(\text{Area of } B) \end{aligned}$$

$$\begin{aligned} 4(\text{Area of } B) + \text{Area of } B &= 80 \\ 5(\text{Area of } B) &= 80 \\ \text{Area of } B &= 16 \\ \text{Area of } A &= 4(16) = 64 \end{aligned}$$

$$\begin{aligned} \text{Area of } R &= lw \\ 216 &= 12l \\ 18 &= l \end{aligned}$$

The length of the rectangle is 18. A side of square A is $\sqrt{64} = 8$, and a side of square B is $\sqrt{16} = 4$.

$$\begin{aligned} 18 &= x + 4 + x + 8 + x \\ 18 &= 3x + 12 \\ 6 &= 3x \\ 2 &= x \end{aligned}$$

CIRCLES

A **circle** is labeled by its center point: circle O means the circle with center point O.

Diameter: A line segment, generally denoted by the variable d, that connects two points on the circle and passes through the center of the circle.

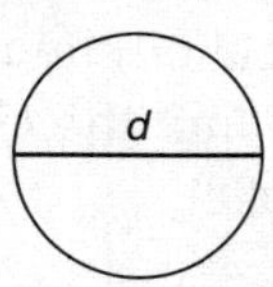

Radius: A line segment, generally denoted by the variable r, from the center of the circle to any point on the circle. The radius of a circle is one-half the length of the diameter.

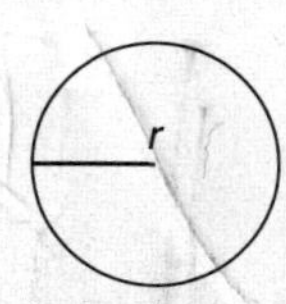

Chord: A line segment joining two points on the circle. Segment c is a chord.

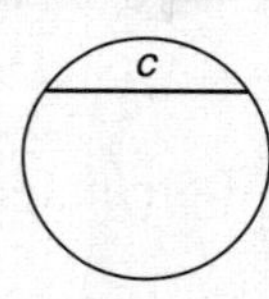

Central angle: An angle formed by two radii. Angle x is a central angle.

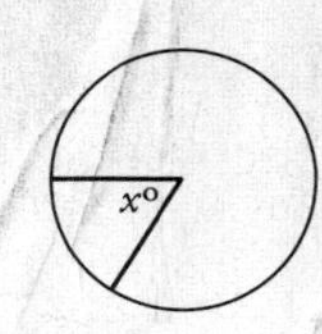

Tangent: A line that touches only one point on the circumference of the circle. A line drawn tangent to a circle is perpendicular to the radius at the point of tangency. Line t is tangent to circle O at point T.

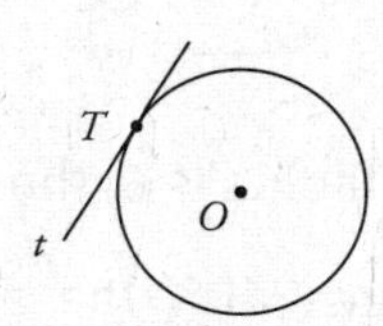

Circumference and arc length: The distance around a circle is called the circumference.

The number π (pi) is the ratio of a circle's circumference C to its diameter d.

$$\pi = \frac{C}{d}$$

$$C = \pi d$$

or

$$C = 2\pi r$$

The value of π is a never-ending decimal that starts with 3.1415926 but is usually approximated 3.14. For the GRE, it is usually sufficient to remember that π is a little more than 3.

An **arc** is a portion of the circumference of a circle. In the figure shown, *AB* is an arc of the circle, with the same degree measure as central angle *AOB*. The shorter distance between *A* and *B* along the circle is called the **minor arc**; the longer distance *AXB* is the **major arc**. An arc that is exactly half the circumference of the circle is called a **semicircle** (in other words, half a circle).

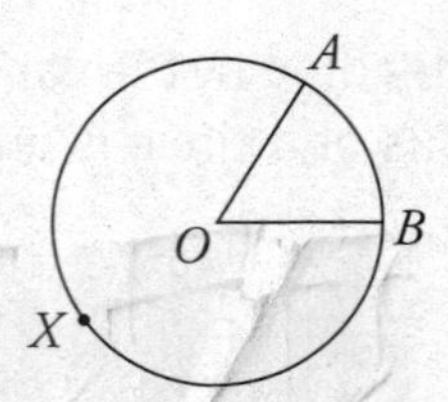

The length of an arc is the same fraction of a circle's circumference as its degree measure is of the degree measure of the circle (360°). For an arc with a central angle measuring *n* degrees:

$$\text{Arc length} = \left(\frac{n}{360}\right)(\text{circumference})$$

$$= \frac{n}{360} \times 2\pi r$$

Example: What is the length of arc *ABC* of the circle with center *O* shown?

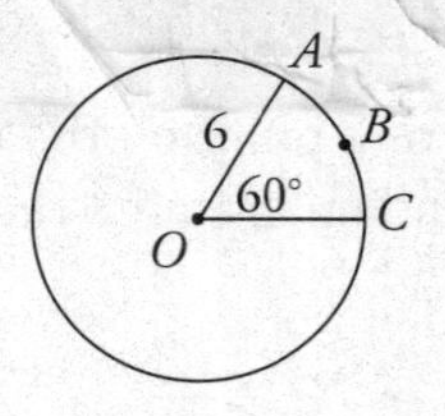

Since $C = 2\pi r$, if the radius is 6, the circumference is $2 \times \pi \times 6 = 12\pi$.

Since $\angle AOC$ measures 60°, the arc is $\frac{60°}{360°}$, or one-sixth, of the circumference.

Therefore, the length of the arc is one-sixth of 12π, which is $\frac{12\pi}{6}$ or 2π.

Area of a circle: The area of a circle is given by the formula

$$\boxed{A = \pi r^2}$$

A **sector** is a portion of the circle that is bounded by two radii and an arc. In the circle shown with center *O*, *OAB* is a sector. To determine the area of a sector of a circle, use the same method we used to find the length of an arc. Determine what fraction of 360° is in the degree measure of the central angle of the sector, then multiply that fraction by the area of the circle. In a sector whose central angle measures *n* degrees:

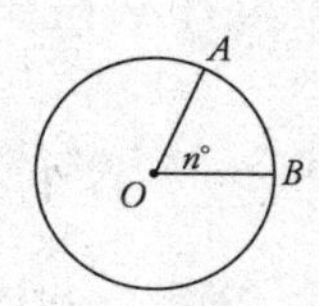

$$\text{Area of sector} = \left(\frac{n}{360}\right) \times (\text{Area of circle})$$
$$= \left(\frac{n}{360}\right) \times \pi r^2$$

Example: What is the area of sector *AOC* in the circle with center *O* shown?

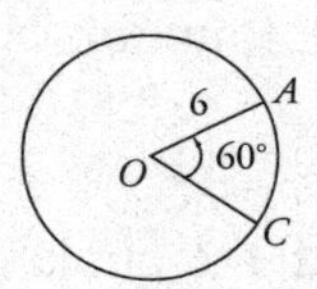

Since $\angle AOC$ measures 60°, a 60° "slice" is $\frac{60°}{360°}$, or one-sixth, of the circle.

So the sector has area $\frac{1}{6} \times \pi r^2 = \frac{1}{6} \times 36\pi = 6\pi$.

CIRCLES EXERCISES

BASIC

1. The radius of a circle is 4 centimeters. What is the diameter of the circle?

2. The diameter of a circle is 7 inches. What is the circumference of the circle?

3. The radius of a circle is 6 meters. What is the circumference of the circle?

4. The circumference of a circle is 10π centimeters. What is the length of the diameter of the circle?

5. The radius of a circle is 5 inches. What is the area of the circle?

6. The diameter of a circle is 12 feet. What is the area of the circle?

7. The area of a circle is 64π square centimeters. What is the radius?

8. The distance around a circular fountain is 38π meters. What is the radius of the fountain?

9. If the radius of a circle is tripled, the circle's area is multiplied by what amount?

10. Find the measure of the smaller angle made by the hands of a clock at 12:30.

INTERMEDIATE

For Exercises 11–13, use the following information:

The circle with center O shown has radius 4. Find the following:

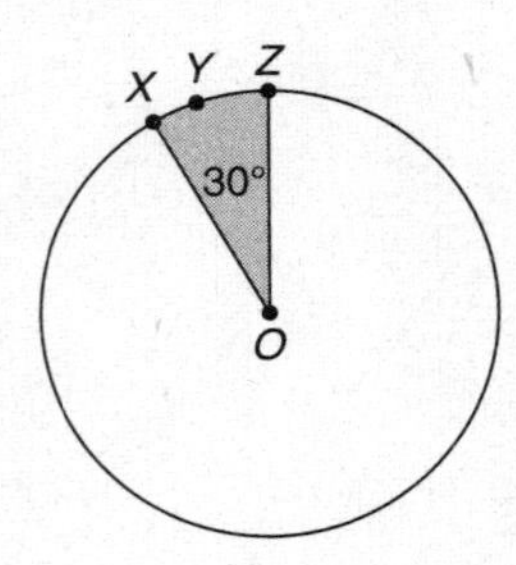

11. Circumference of the circle

12. Length of minor arc XYZ

13. Area of the shaded region

For Exercises 14–16, use the following information:

The figure shows two concentric circles, each with center *C*. Given that the larger circle has radius 14 and the smaller circle has radius 6, find the following:

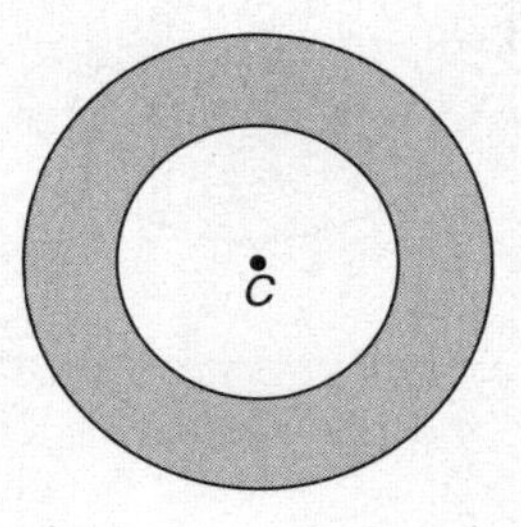

14. Circumference of the larger circle

15. Area of the smaller circle

16. Area of the shaded region

17. If the radius of circle *C* is 12, then what is the area of sector *CED*?

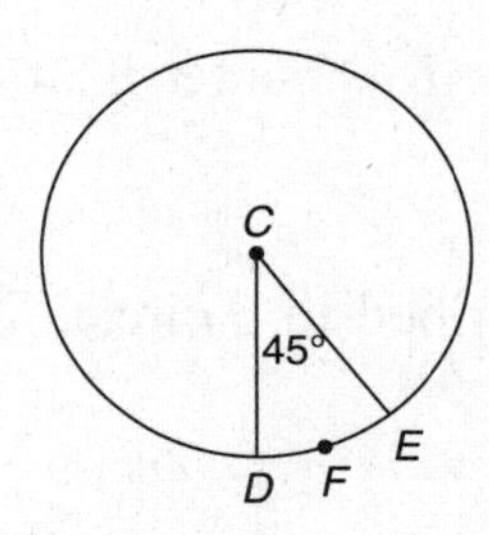

18. If the radius of the circle with center *C* is 6 and the measure of angle *ACB* is 120, what is the length of the minor arc from *A* to *B*?

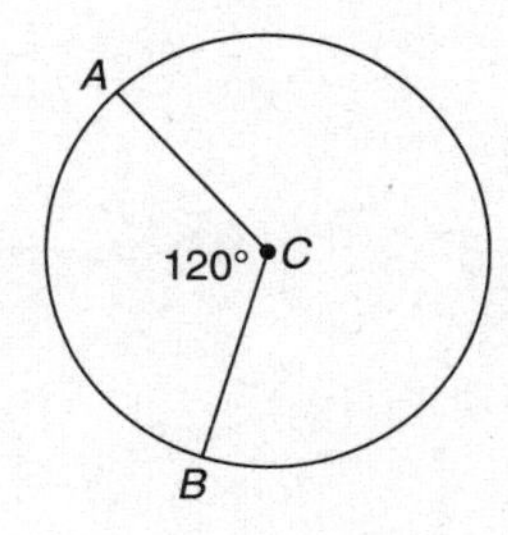

19. What is the number of degrees that the hour hand of a clock moves between noon and 2:30 in the afternoon of the same day?

20. Rectangle $RSTU$ has a perimeter of 42. The half circle with diameter RS has an area of 8π. What is the area of the unshaded part of the figure?

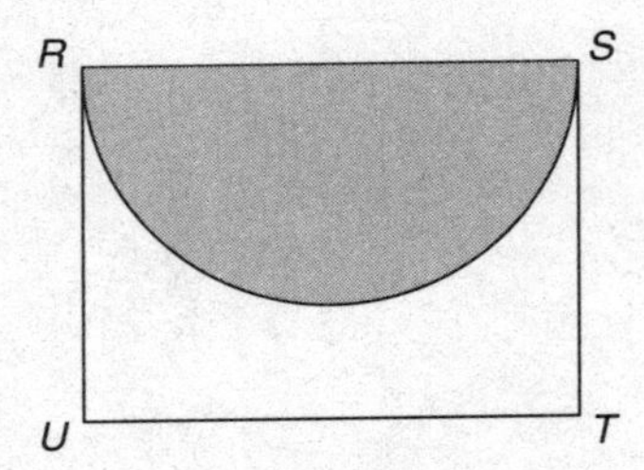

Advanced

21. A circular image is to be enlarged. The new radius will be 25 percent larger than the original. What is the ratio of the area of the current image to the area of the new image?

22. The diameter of a circle is formed by the line joining points (5, –2) and (–5, 3). Find the circumference of the circle.

23. What is the area of a circle if its center is at (3, 0) and the circle passes through (–1, –3)?

24. A 5 by 12 rectangle is inscribed in a circle. What is the circumference of the circle?

25. In the figure shown, if the radius of circle P is 3 times the radius of circle A, $\angle BAC = \angle QPR$, and the shaded area of circle A is 3π square units, then what is the area of the shaded part of circle P?

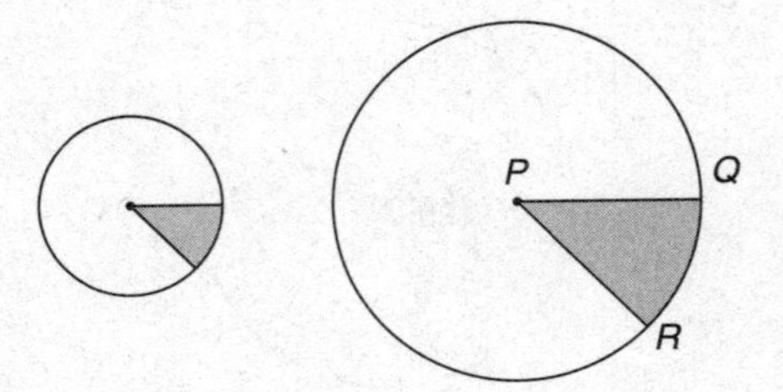

26. An 8-by-6 rectangle is inscribed in a circle. What is the circumference of the circle?

27. In the figure, the square has two sides that are tangent to the circle. If the area of the circle is $16x^2\pi$, what is the area of the square, in terms of x?

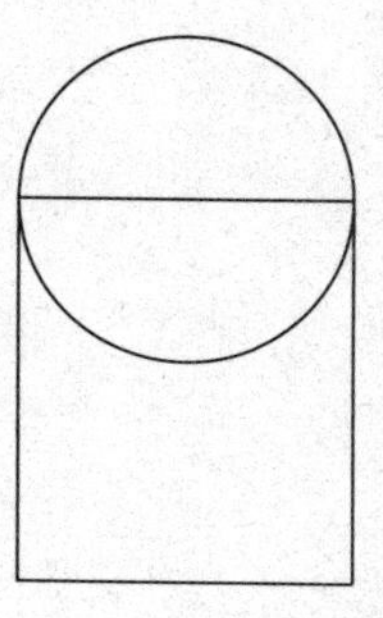

28. Two congruent, adjacent circles are cut out of a 16-by-8 rectangle. The circles have the maximum diameter possible. What is the area of the paper remaining after the circles have been cut out?

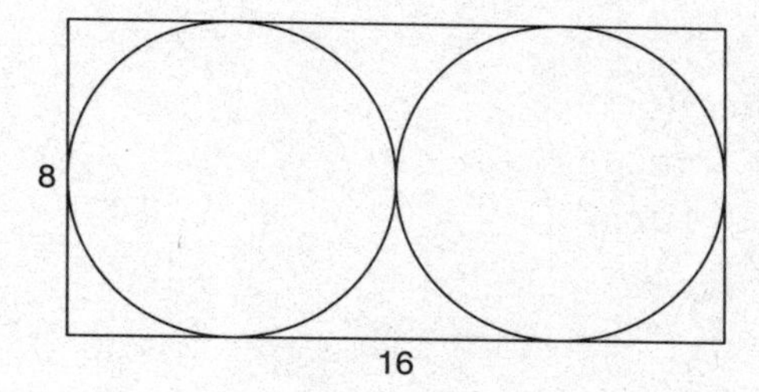

29. Points C and D are equidistant from the line l, and line segment CD intersects line l at a 45-degree angle. If the shortest straight-line distance between point C and line l is 5 units, what is the area of a circle drawn with its center on line l and line segment CD as its diameter?

30. A wheel has a diameter of x centimeters, and a second wheel has a diameter of y centimeters. The first wheel covers a distance of d meters in 200 revolutions. In terms of x and y, how many revolutions does the second wheel make in covering d meters?

CIRCLES EXERCISES ANSWER KEY

BASIC

1. 8cm

$$d = 2r$$
$$d = 2 \times 4$$
$$d = 8\,\text{cm}$$

2. 7π or about 21.98in.

$$C = \pi d$$
$$C = 7\pi \approx 3.14 \times 7$$
$$C = 21.98\,\text{inches}$$

3. 12πm or 37.68m

$$C = 2\pi r$$
$$C = 2(\pi)(6) = 12\pi \approx 12(3.14)$$
$$C = 37.68\,\text{m}$$

4. 10cm

$$C = \pi d$$
$$10\pi = \pi d$$
$$d = 10\,\text{cm}$$

5. 25πin.^{2} or 78.5in.^{2}

$$A = \pi r^2$$
$$A = 5^2(\pi) = 25\pi \approx 3.14 \times 25$$
$$A \approx 78.5\,\text{in.}^2$$

6. 36πft^2

First, find the radius.

$$r = \frac{1}{2}d$$
$$r = \frac{1}{2}(12)$$
$$r = 6$$

Now find the area.

$$A = \pi r^2$$
$$A = \pi \times 6^2$$
$$A = 36\pi\,\text{ft}^2$$

7. 8

$$
\begin{aligned}
A &= \pi r^2 \\
64\pi &= \pi r^2 \\
\frac{64\pi}{\pi} &= \frac{\pi r^2}{\pi} \\
64 &= r^2 \\
\sqrt{64} &= \sqrt{r^2} \\
r &= 8\,\text{cm}
\end{aligned}
$$

8. 19

The distance around the circular fountain is the circumference of a circle.

$$
\begin{aligned}
38\pi &= 2\pi r \\
\frac{38\pi}{2\pi} &= \frac{2\pi r}{2\pi} \\
r &= 19\,\text{m}
\end{aligned}
$$

9. 9

Area is $\pi \times r^2$, so area would be $\pi(3r)^2 = \pi \times 3^2 \times r^2 = 9\pi r^2$ for triple the radius. The area is 9 times bigger than it was before the tripling.

10. 165°

Each hour represents $\frac{1}{12}$ of a rotation, or $\frac{1}{12}$ of the degrees of a circle, which has 360 degrees. Therefore the degree measure between each number on the clock is:

$$360° \times \frac{1}{12} = 30°$$

The time 12:30 will put the hour hand between the positions of 12 o'clock and 1 o'clock, half of the distance between them. Half of 30 degrees equals 15 degrees. So the hour hand will be 15 degrees away from the "12" (for 12 o'clock). The minute hand will be at "6" (for 6 o'clock) at 12:30, so you almost have a straight line, except that the hour hand is rotated 15 degrees from the vertical, shrinking the angle between the hands. So, find the angle measurement: $180° - 15° = 165°$.

INTERMEDIATE

11. 8π

$$C = 2\pi r$$
$$C = 2\pi \times 4 = 8\pi$$

12. $\frac{2\pi}{3}$

$$\frac{\text{measure of angle}}{360} = \frac{\text{length of arc}}{\text{circumference}}$$
$$\frac{30}{360} = \frac{\text{length of arc}}{8\pi}$$
$$\frac{1}{12} = \frac{l}{8\pi}$$
$$8\pi = 12l$$
$$\frac{8\pi}{12} = l$$
$$\frac{2\pi}{3} = l$$

13. $\frac{4\pi}{3}$

Area of a circle: $A = \pi r^2 = \pi(4^2) = 16\pi$

Shaded region: $\frac{30°}{360°} \times \pi(4^2) = \frac{1}{12} \times 16\pi = \frac{4\pi}{3}$

14. 28π

$$C = 2\pi r$$
$$C = 2 \times \pi \times 14$$
$$C = 28\pi$$

15. 36π

$$A = \pi r^2 = \pi(6^2) = 36\pi$$

16. 160π

Area of larger circle: $A = \pi r^2 = \pi(14^2) = 196\pi$

Area of shaded region = area of larger circle – area of smaller circle.

$196\pi - 36\pi = 160\pi$

17. 18π

$$\begin{aligned}\text{Area of sector} &= \frac{n^\circ}{360^\circ} \times \pi r^2\\ &= \frac{45^\circ}{360^\circ} \times \pi(12)^2\\ &= \frac{1}{8} \times 144\pi\\ &= \frac{144}{8}\pi\\ &= 18\pi\end{aligned}$$

18. 4π

$C = 2\pi r$, so the circumference $= 12\pi$

The length of the arc is $\frac{120}{360}$, or $\frac{1}{3}$ of the circumference.

$$\frac{1}{3} \times 12\pi = 4\pi$$

19. 75

The hour hand starts at 12 and moves until it is halfway between 2 and 3. The angle covered between each hour on the clock is $\frac{360}{12} = 30$. The hour hand has covered 2.5 of these divisions, so $30 \times 2.5 = 75$.

20. 104 – 8π

The total perimeter of the unshaded part is made up of three sides of the rectangle and the perimeter of the half circle.

The area of a half circle $= \frac{1}{2}\pi r^2$.

$8\pi = \frac{1}{2}\pi r^2$; therefore, $r = 4$, and the diameter of the semicircle is 8. This is also the length of the rectangle. Since the perimeter is 42, calculate the width of the rectangle:

$$\begin{aligned}P &= 2l + 2w\\ 42 &= 2(8) + 2(w)\\ 42 &= 16 + 2w\\ 2w &= 26\\ w &= 13\end{aligned}$$

Find the area of the rectangle:

$$\begin{aligned}A &= lw\\ A &= (8)(13) = 104\end{aligned}$$

The unshaded area is the area of the rectangle minus the area of the semicircle. So the area is $104 - 8\pi$.

Advanced

21. 16:25

Picking Numbers is a great strategy to use here. If the diameter of the old image is 4, the radius of the new image will be 5, since 25% of 4 = 1. The area of the old image is 16π, and the area of the new image is 25π.

The areas of the images are in the ratio $16\pi:25\pi$, or 16:25.

22. $5\pi\sqrt{5}$

If you draw the distance from (5, −2) to (−5, 3) on a coordinate plane, you are also drawing a right triangle, where the leg that runs parallel to the *x*-axis has a length of 10, and the leg that runs parallel to the *y*-axis has a length of 5. Use the Pythagorean theorem to find the diameter, represented by *c*:

$$(10)^2 + (5)^2 = c^2$$

$$100 + 25 = c^2$$

$$125 = c^2, \text{so } c = \sqrt{125} = \sqrt{25}\sqrt{5} = 5\sqrt{5}$$

The diameter is $5\sqrt{5}$, so the circumference is $5\pi\sqrt{5}$.

23. 25π

If you draw the distance from (3, 0) to (−1, −3) on a coordinate plane, you are also drawing a right triangle, where the leg that runs parallel to the *x*-axis has a length of 4 and the leg that runs parallel to the *y*-axis has a length of 3. The hypotenuse of the triangle is 5 because this is a 3:4:5 right triangle, and it also represents the radius of the circle. Alternatively, use the Pythagorean theorem to find the radius, here represented by *c*:

$$(3)^2 + (4)^2 = c^2$$

$$9 + 16 = c^2$$

$$25 = c^2, \text{ so } c = 5$$

The radius is 5, so the area is $5^2\pi = 25\pi$.

24. 13π

The diagonal of the rectangle is the diameter of the circle. The diagonal is the hypotenuse of a 5:12:13 triangle and is therefore 13. Circumference $= \pi d = 13\pi$.

25. 27π

If the radii of the two circles are in the ratio 1:3, then the areas of the circles will be in the ratio $(1)^2:(3)^2$, which is 1:9. So the areas of the shaded parts are also in the ratio 1:9. Since $\angle BAC$ is congruent to $\angle QPR$, if the shaded area of the smaller circle is 3π, then the shaded area of the larger circle will be 27π.

26. 10π

The diagonal of the rectangle inscribed in a circle is the same as the diameter of the circle. The diagonal is the hypotenuse of a 6:8:10 triangle, a multiple of a 3:4:5 triangle, and is therefore 10. Circumference $= \pi d = 10\pi$.

27. $64x^2$

Note that figures on the GRE are not necessarily drawn to scale. The quadrilateral in the figure looks like a rectangle, not a square, but the question stem calls it a square, so that's how you have to treat it. If the area of the circle is $16x^2\pi$, the radius will be the square root of $16x^2$ (or $4x$). The diameter will be $8x$. The diameter is also the side of the square. Area of the square is $(8x)^2 = 64x^2$.

28. $128 - 32\pi$

The area remaining is the area of the rectangle – (2 × area of a circle). The area of the rectangle is found by multiplying the length times the width.

$$A = lw$$
$$A = (16)(8) = 128$$

For the circles, the diameter of the circle is the same as the width of the rectangle. The radius is one-half of the diameter. Therefore the radius for these circles is 4, and the area of one circle is $\pi(4)^2 = 16\pi$. The sum of the areas of the two circles is $16\pi + 16\pi = 32\pi$.

The area of the paper remaining is $128 - 32\pi$.

29. 50π

If O is the point at the center of the circle that lies on line l, and CD is the diameter of that circle, then point O must be the point of intersection between line segment CD and line l. Since points C and D are equidistant from line l, $CO = DO =$ radius of circle O.

The shortest straight-line distance from point C to line l is specified in the question as 5 units. Since line segment CD intersects line l at a 45-degree angle, a 45°-45°-90° right triangle is formed by the following three line segments: line segment CO, the line segment that represents the shortest straight-line distance between point C and line l, and line l itself. Since the shortest straight-line distance between point C and line l (a leg of the right triangle) is 5 units, the hypotenuse of that triangle, which is line segment CO, is equal to $5\sqrt{2}$. Since $CO = DO =$ radius of circle O, the area of the circle with a radius of $5\sqrt{2}$ is equal to πr^2, which here is $\pi\left(5\sqrt{2}\right)^2$, or $\pi(5)^2\left(\sqrt{2}\right)^2 = \pi(25)(2) = 50\pi$.

30. $200\frac{x}{y}$

Total distance covered by the first wheel $= d$, which is equivalent to 200 × circumference. Circumference $= x\pi$.

Circumference of the second wheel $= y\pi$.

Revolutions covered by the second wheel $= r$.

Total distance covered by the second wheel $= d =$ total distance covered by the first wheel.

Therefore, $ry\pi = 200x\pi$, and $r = \frac{200x}{y} = 200\frac{x}{y}$.

MULTIPLE FIGURES

You can expect to see some problems on the GRE that involve several different types of figures. They test your understanding of various geometrical concepts and relationships, not just your ability to memorize a few formulas. For instance, the hypotenuse of a right triangle may be the side of a neighboring rectangle or the diameter of a circumscribed circle. Keep looking for the relationships between the different figures until you find one that leads you to the correct answer.

One common kind of multiple-figures question involves irregularly shaped regions formed by two or more overlapping figures, often with one region shaded. When you are asked to find the area of such a region, any or all of the following methods may work:

(1) Break up that shaded area into smaller pieces. Find the area of each piece using an appropriate formula. Add those areas together.

(2) Find the area of the whole figure and the area of the unshaded region, then subtract the unshaded area from the total area.

Example: Rectangle *ABCD* has an area of 72 and is composed of 8 equal squares. Find the area of the shaded region.

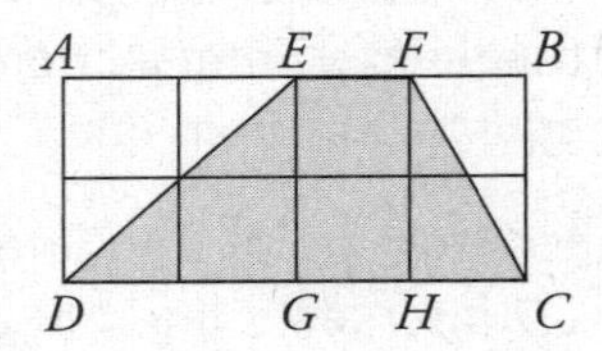

For this problem, you can use either of the two approaches described above or a third approach. First, divide 8 into 72 to get the area of each square, which is 9. Since the area of a square equals its side squared, each side of the small squares must have length 3. Now you have a choice of methods.

(1) You can break up the trapezoid into right triangle *DEG*, rectangle *EFHG*, and right triangle *FHC*.

The area of triangle *DEG* is $\frac{1}{2} \times 6 \times 6$, or 18. The area of rectangle *EFHG* is 3×6, or 18. The area of triangle *FHC* is $\frac{1}{2} \times 6 \times 3$, or 9. The total area is $18 + 18 + 9$, or 45.

(2) The area of the whole rectangle *ABCD* is 72. The area of unshaded triangle *AED* is $\frac{1}{2} \times 6 \times 6$, or 18. The area of unshaded triangle *FBC* is $\frac{1}{2} \times 6 \times 3$, or 9. Therefore, the total unshaded area is $18 + 9 = 27$. The area of the shaded region is the area of the rectangle minus the unshaded area, or $72 - 27 = 45$.

(3) Count shaded squares. There are 8 equal squares in total. One is completely blank, two are half shaded (which means that those two blocks equal one shaded block), and two make a rectangle that is half shaded, again making for two blocks equaling one shaded one. Each block has an area of 9. Multiply 9 by 5, the equivalent number of shaded blocks, to get a shaded area of 45.

Inscribed and circumscribed figures: A polygon is **inscribed** in a circle if all the vertices of the polygon lie on the circle. A polygon is **circumscribed** about a circle if all the sides of the polygon are tangent to the circle.

When a rectangle is inscribed in a circle, a diagonal of the rectangle is a diameter of the circle. In the figure, *DB* is the diameter of circle *O* and a diagonal of *ABCD*.

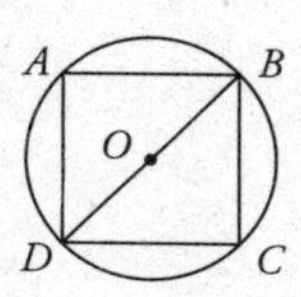

When a square is circumscribed about a circle, a diagonal of the rectangle passes through a diameter of the circle. Moreover, a side of the square is equal in length to the length of the circle's diameter.

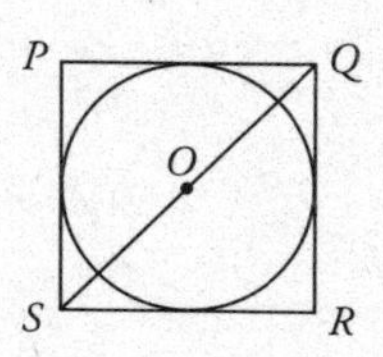

Square *ABCD* is inscribed in circle *O*.

(We can also say that circle *O* is circumscribed about square *ABCD*.)

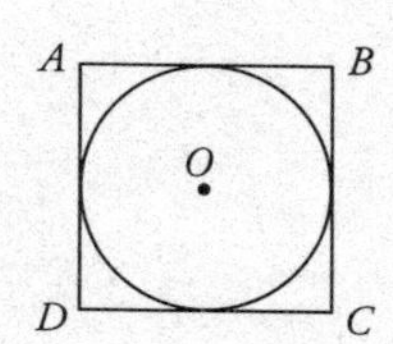

Square *PQRS* is circumscribed about circle *O*.

(We can also say that circle *O* is inscribed in square *PQRS*.)

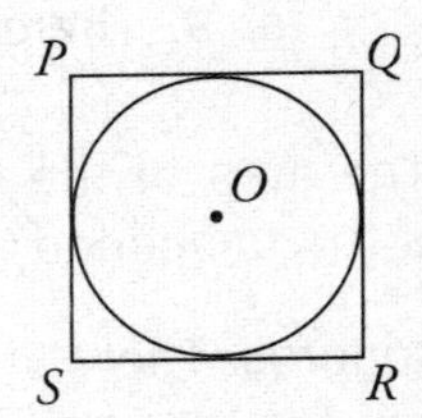

A triangle inscribed in a semicircle such that one side of the triangle coincides with the diameter of the semicircle is a right triangle.

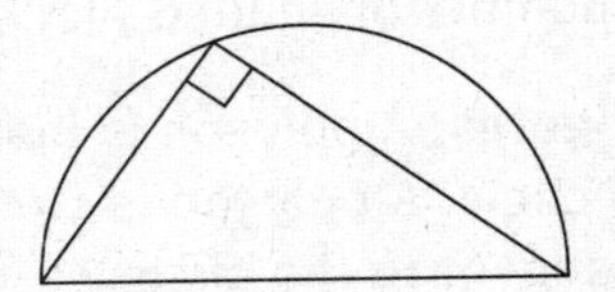

MULTIPLE FIGURES EXERCISES

BASIC

1. In the figure shown, the square is inscribed in a circle, and the area of the circle is 9π. What is the length of the diagonal of the square?

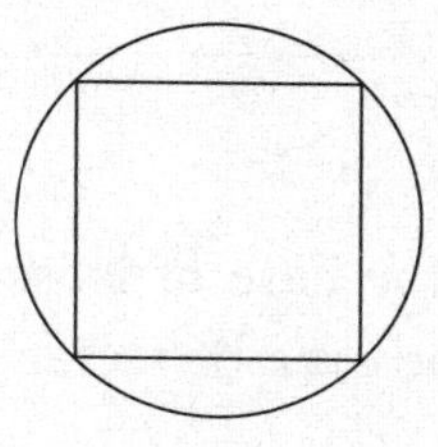

2. In the figure shown, the circle is inscribed in the square. The area of the circle is 100π. What is the area of the square?

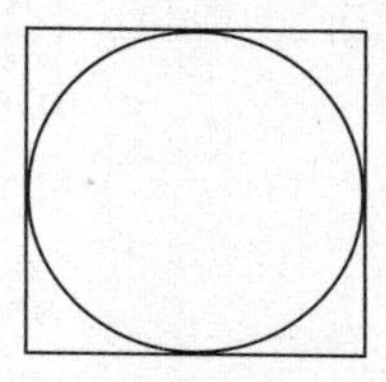

3. In the figure shown, the square is inscribed in a circle, and each side of the square is 10. What is the area of the circle?

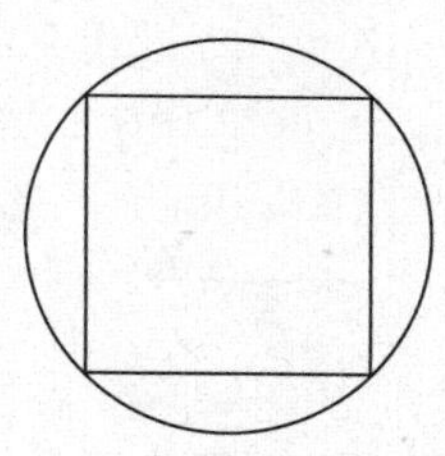

4. In the figure shown, the circle is inscribed in a square, and each side of the square is 4. What is the circumference of the circle?

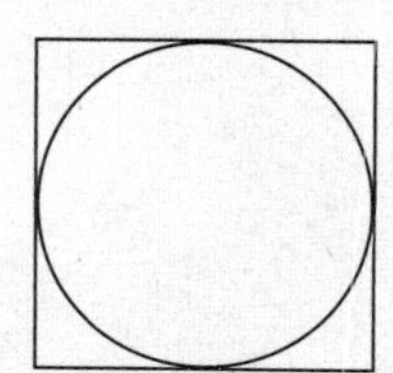

5. In the figure shown, the rectangle is inscribed in the circle. The length of the diagonal of the rectangle is 12. What is the circumference of the circle?

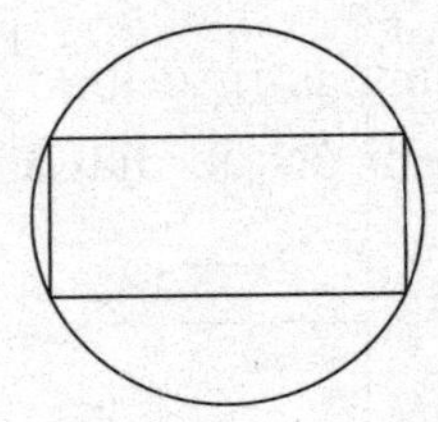

6. In the figure shown, AB is the base of isosceles triangle ABR and a side of rectangle $ABCD$. What is the area of rectangle $ABCD$?

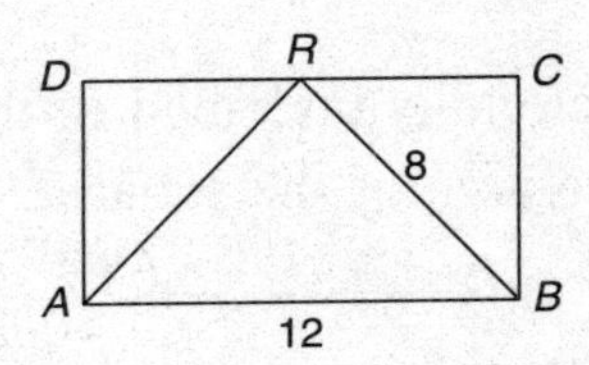

7. What is the value of x?

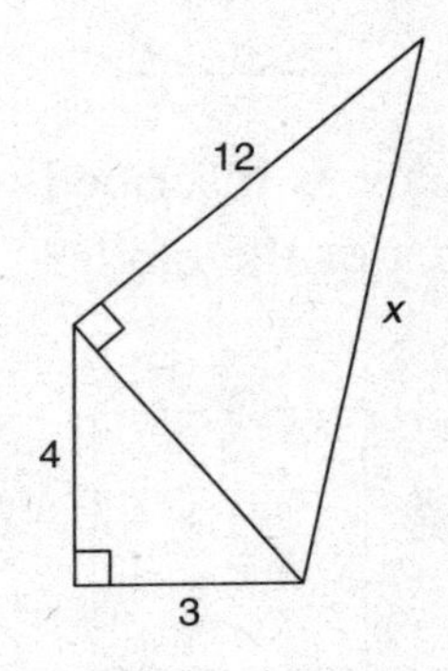

8. A rectangle and a circle have the same area. If the radius of the circle and the length of the rectangle are each 12, what is the width of the rectangle?

9. What is the area of the figure shown?

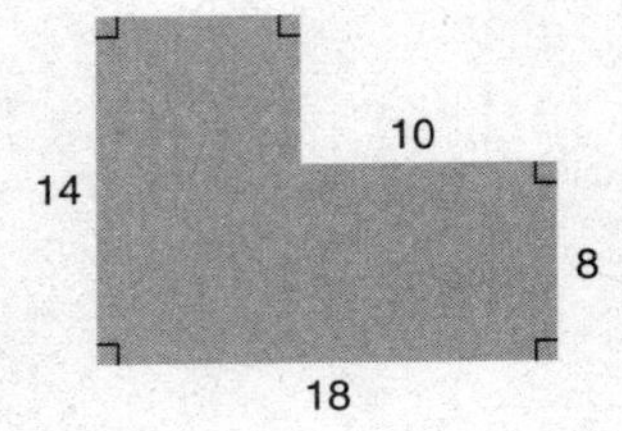

10. A certain manufacturing belt is shown in the following diagram. If the height of each gear is 18 and the length of the horizontal portion of the belt is 28, what is the total length of the belt?

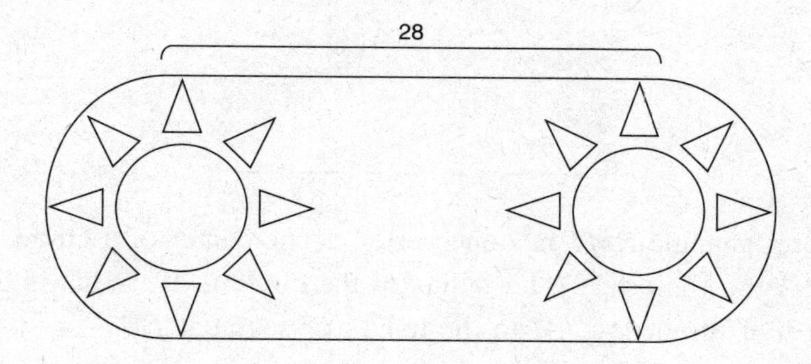

INTERMEDIATE

11. The figure shown is composed of 3 squares and 2 semicircles. Each square has a side of length 8. What is the perimeter of the entire figure?

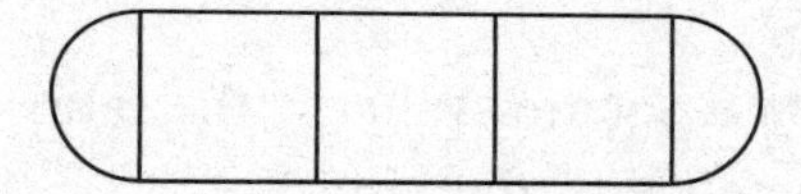

12. In the figure shown, what is the value of y?

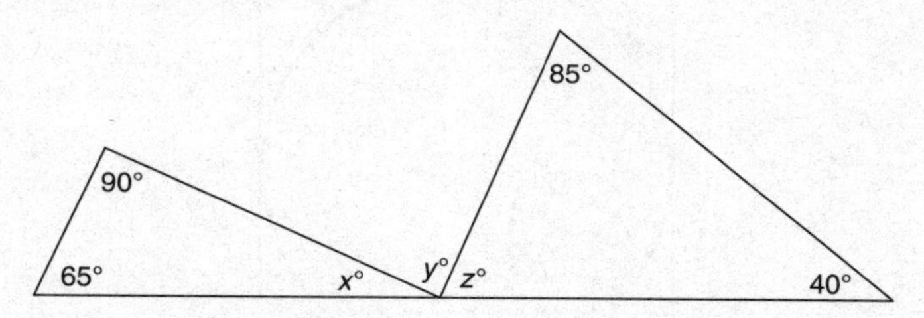

13. The arc in the figure shown is a quarter-circle with endpoints at the vertices of a square. If the area of the quarter circle is 4π, what is the area of the square?

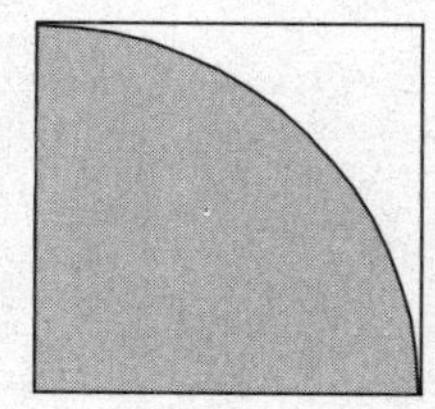

14. In the figure shown, the hypotenuse of the triangle coincides with the diameter of the semicircle. What is the circumference of the semicircle?

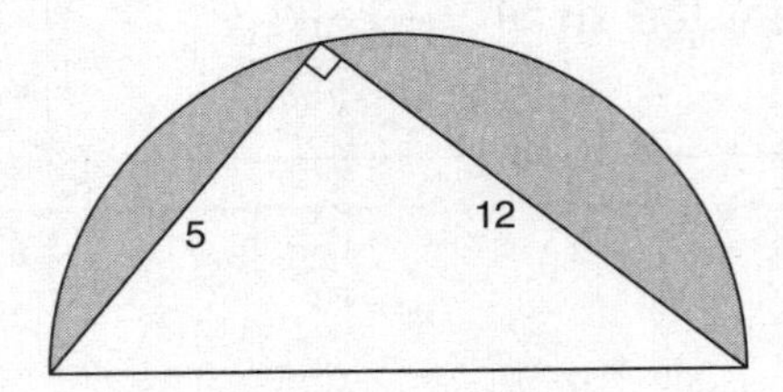

15. Equilateral triangle *RMT* has one vertex at the center of a circle. Its other vertices lie on the circle. The radius of the circle is 10. What is the ratio of the length of minor arc *TM* to the length of segment *TM*?

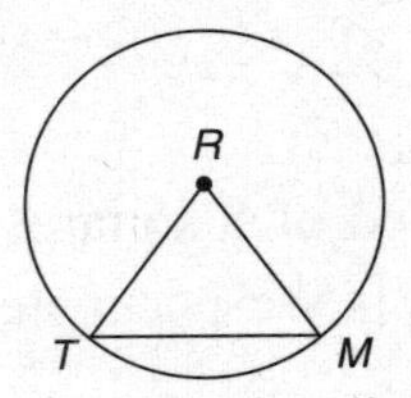

16. A circle is inscribed in a square. What is the ratio of the perimeter of the square to the circumference of the circle?

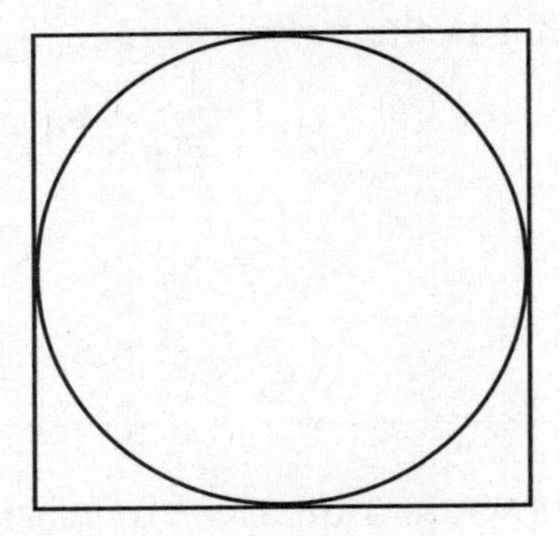

17. The following figure shows concentric circles. What is the area of the shaded region?

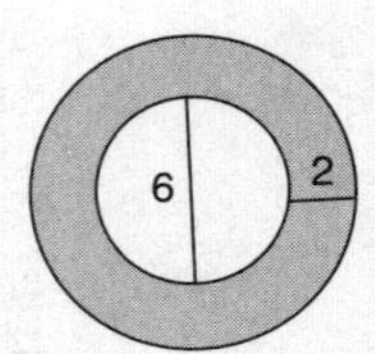

18. A company's logo consists of semicircles constructed on the sides of a right isosceles triangle. What is the total perimeter of the logo?

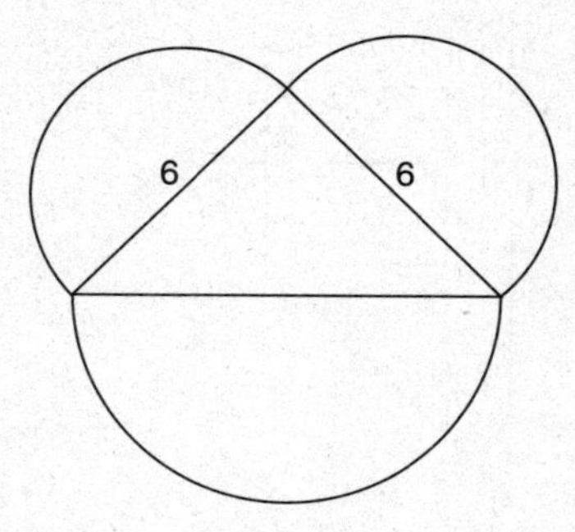

19. An equilateral triangle is constructed on each side of a square. What is the sum of the measures of the angles marked?

20. Each circle in the figure shown has a diameter of 10. What is the area of the shaded region?

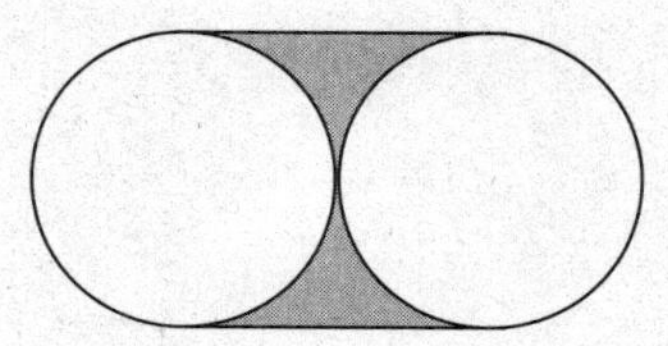

Advanced

21. The figure shows an equilateral triangle where each vertex is the center of a circle. Each circle has a radius of 20. What is the area of the shaded region?

22. What is the circumference of the semicircle in the figure shown?

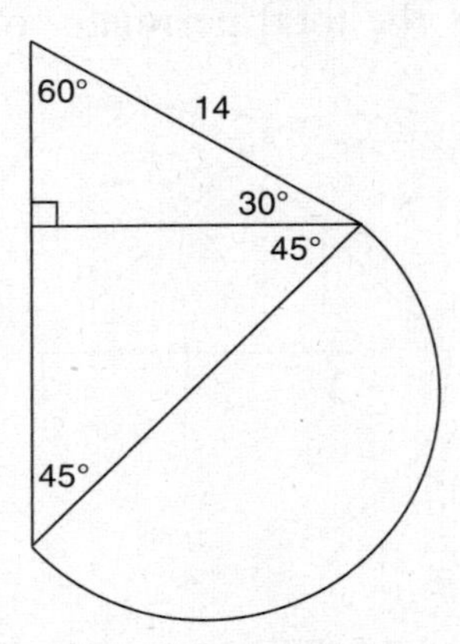

23. The hypotenuse of right triangle ABC is the diameter of the circle. If the diameter of the circle is 18, what is the area of the shaded region?

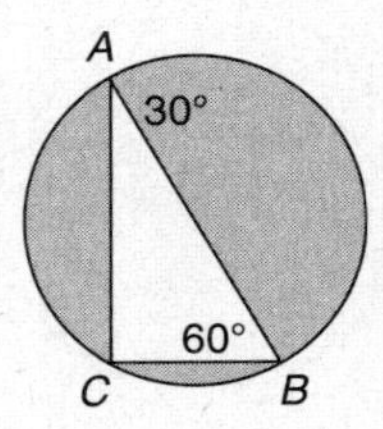

24. The shaded area in the figure shown consists of quarter circles. If each quarter circle has radius r, what is the area of the unshaded region in terms of r?

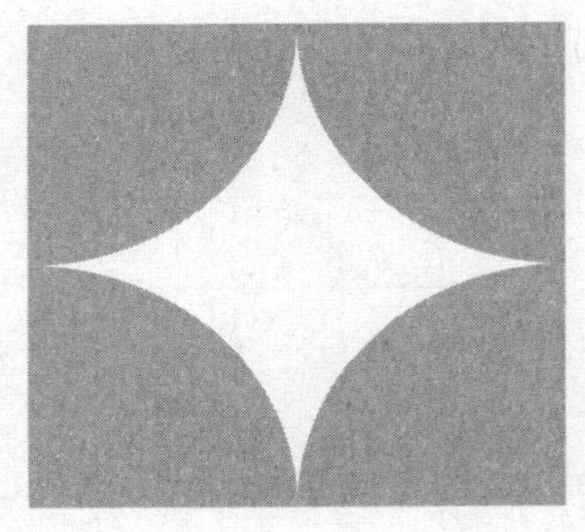

25. A circle is inscribed in an equilateral triangle. Radii to two of the points of tangency are shown. What is x?

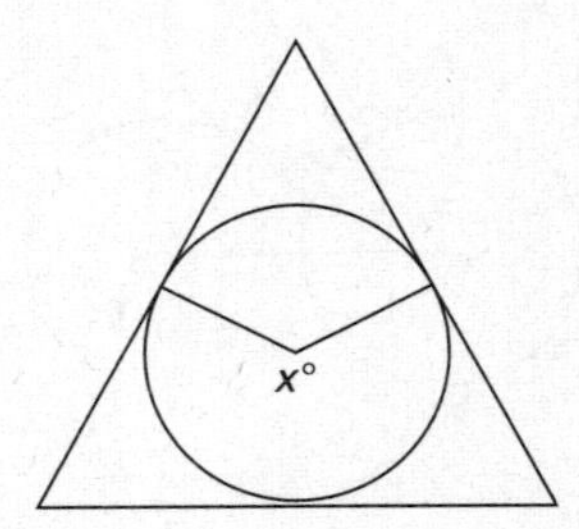

26. The length of the arc PQ in the quarter circle shown is 10π. If RS is 12, what is the area of the rectangle?

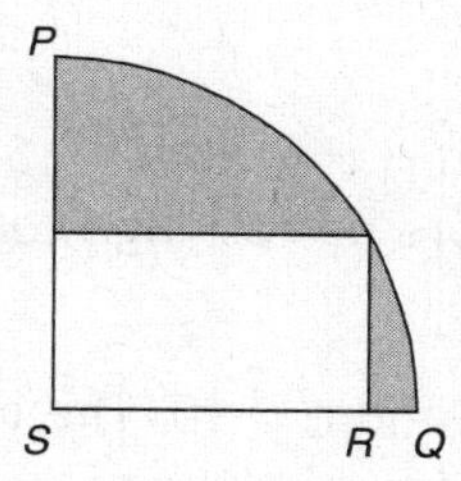

27. In the figure shown, H is the center of the semicircle. Triangles DHF and GHF each have two vertices on the circle. The measure of $\angle DHG$ is 120°. What is the measure of $\angle FHG$?

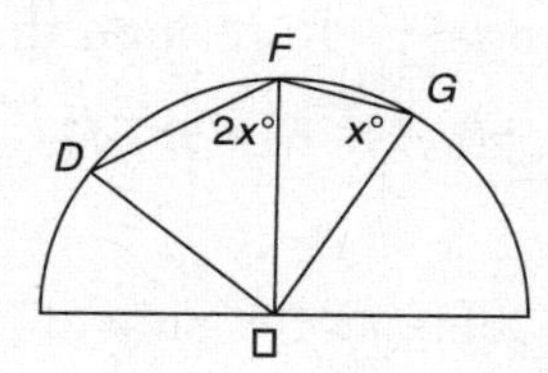

28. A cube with a side of length 4 has the same volume as a rectangular solid with a length of 6 and width of 2. What is the height of the rectangular solid?

29. In the figure shown, the arc of the quarter circle has length 6π. The rectangle has a perimeter of 60. What is the perimeter of the shaded region?

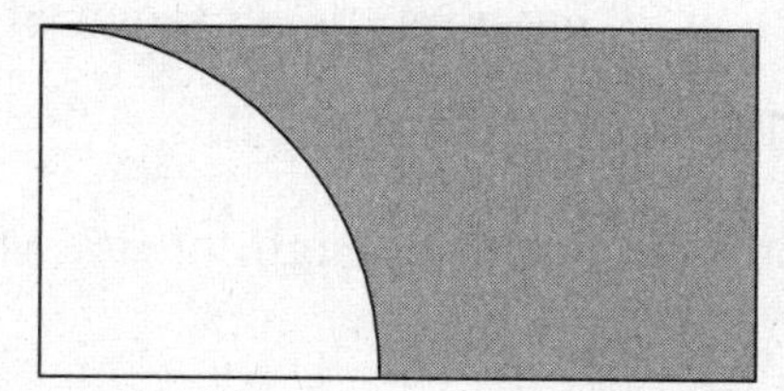

30. In the figure shown, a square is inscribed in a circle. Within the square are four circles, each with two points of tangency on the square and with two points of tangency with two other circles. The diameter of each smaller circle is $2x$. What is the circumference of the outer circle in terms of x?

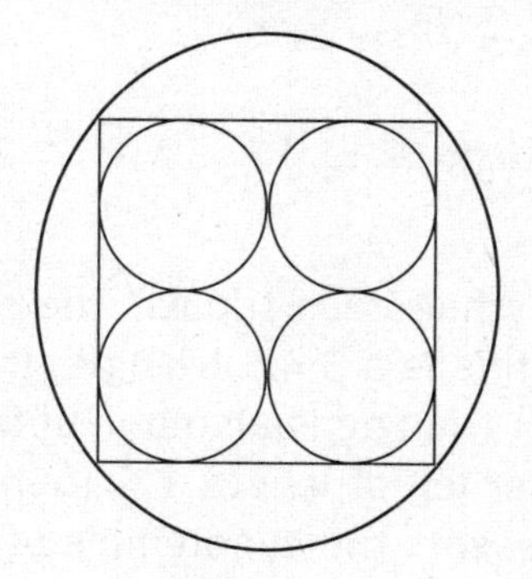

MULTIPLE FIGURES EXERCISES ANSWER KEY

BASIC

1. 6

If the area of the circle is 9π, then $r^2 = 9$ and $r = 3$. The diagonal of the square is a diameter of the circle and therefore has a length of 6.

2. 400

If $A = \pi r^2 = 100\pi$, then $r^2 = 100$ and $r = 10$. The length of each side of the square is the same as the diameter of the circle, so $s = 20$. The area of the square is $A = s^2 = 20^2 = 400$.

3. 50π

Use the 45°-45°-90° relationship to find the diagonal of the square (which is also the diameter of the circle). The diagonal is $10\sqrt{2}$, and therefore the radius of the circle is $5\sqrt{2}$. The area of the circle is $\pi(5\sqrt{2})^2 = \pi(5)^2(\sqrt{2})^2 = \pi(25)(2) = 50\pi$.

4. 4π

The diameter of the circle is the same as the side of the square. The diameter of the circle is 4. Therefore, its circumference is $C = \pi d = \pi(4) = 4\pi$.

5. 12π

The diagonal of the rectangle is the diameter of the circle. If the diameter is 12, then the circumference is $C = \pi d = \pi(12) = 12\pi$.

6. $24\sqrt{7}$

Since triangle ABR is isosceles, R must be the midpoint of DC.

$$RC = \frac{1}{2}DC = \frac{1}{2}AB = \frac{1}{2}(12) = 6.$$

Use the Pythagorean theorem to find the length of BC, which is also the height of the rectangle.

$$RC^2 + BC^2 = RB^2$$
$$6^2 + BC^2 = 8^2$$
$$36 + BC^2 = 64$$
$$BC^2 = 28$$
$$BC = \sqrt{28} = \sqrt{4}\sqrt{7} = 2\sqrt{7}$$

The area of the rectangle $= AB \times BC = 12 \times 2\sqrt{7} = 24\sqrt{7}$.

7. 13

Use your knowledge of the Pythagorean triples. The smaller triangle has two legs of lengths 3 and 4. Because this is a 3:4:5 triangle, the hypotenuse must equal 5. The hypotenuse of the smaller triangle is also one of the legs of the larger triangle. The larger triangle has another leg of length 12. Therefore, the larger triangle is a 5:12:13 triangle. Since x represents the hypotenuse of the larger triangle, $x = 13$.

8. 12π

The area of the circle is $A = \pi r^2 = \pi(12)^2 = 144\pi$. If the area of the rectangle is 144π, then

$$\begin{aligned} A &= lw \\ 144\pi &= 12w \\ \frac{144\pi}{12} &= \frac{12w}{12} \\ 12\pi &= w \end{aligned}$$

The question asks for the width, so the answer is 12π.

9. 192

First, find the area of the larger outlined rectangle. Then, subtract the unshaded area.

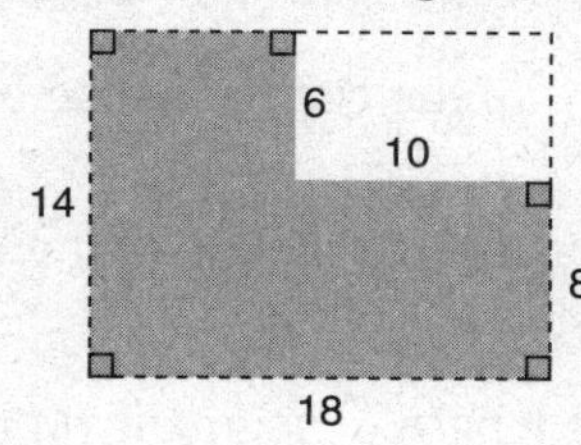

The area of the larger outlined rectangle is the product of its length and its width.

$14 \times 18 = 252$

The area of the unshaded region is $6 \times 10 = 60$.

The area of the figure is $252 - 60 = 192$.

10. $56 + 18\pi$

The belt can be thought of as consisting of two horizontal portions and two semicircular portions. The length of each horizontal portion is 28. Each gear has diameter 18; therefore, the length of each semicircular portion is $\frac{1}{2}(\pi d) = \frac{1}{2}(18\pi) = 9\pi$. The total length of the four portions is $28 + 28 + 9\pi + 9\pi = 56 + 18\pi$.

INTERMEDIATE

11. $48 + 8\pi$

The perimeter consists of 6 segments (each with length 8) and two semicircles (each with diameter 8). The perimeter is $6 \times 8 + 2 \times \frac{1}{2}(\pi d) = 48 + \pi(8) = 48 + 8\pi$.

12. 100°

The sum of the three angles in any triangle is 180°, and the sum of three angles forming a straight line must also be 180°. Find x and z and then use them to find y.

$$\begin{aligned} 65 + 90 + x &= 180 \\ 155 + x &= 180 \\ x &= 25 \end{aligned} \qquad \begin{aligned} 85 + 40 + z &= 180 \\ 125 + z &= 180 \\ z &= 55 \end{aligned} \qquad \begin{aligned} 25 + y + 55 &= 180 \\ y + 80 &= 180 \\ y &= 100 \end{aligned}$$

13. 16

The area of the quarter-circle is 4π. Therefore, the area of the entire circle would be $4 \times 4\pi = 16\pi$. If $16\pi = \pi r^2$, then $r^2 = 16$ and $r = 4$. The radius of the quarter-circle coincides with one side of the square, so each side of the square is 4. The area of the square is $A = 4 \times 4 = 16$.

14. $\mathbf{\frac{13\pi}{2}}$

The inscribed right triangle is a 5:12:13 triangle. Therefore, the diameter is 13. The circumference of the semicircle is: $\frac{1}{2}\pi d = \frac{1}{2}\pi(13) = \frac{13\pi}{2}$.

15. $\mathbf{\pi:3}$

Because the triangle is equilateral, it has three 60° angles, and all sides have length 10. The length of minor arc *TM* is $\frac{60}{360}$ of the circumference of the entire circle. So, minor arc *TM* has length $\frac{60}{360} \times 2\pi r = \frac{1}{6} \times 2\pi(10) = \frac{10\pi}{3}$. The length of segment *TM* is 10. The ratio of the length of arc *TM* to the length of segment *TM* is $\frac{\frac{10\pi}{3}}{10} = \frac{10\pi}{3} \times \frac{1}{10} = \frac{\pi}{3}$.

16. $\mathbf{4:\pi}$

Call the radius of the circle *r*. The perimeter of the square is $2r + 2r + 2r + 2r = 8r$. The circumference of the circle is $2\pi r$. The ratio is $\frac{8r}{2\pi r}$, or $\frac{4}{\pi}$.

17. $\mathbf{16\pi}$

The shaded region is the area of the outer circle (diameter 10, radius 5) minus the area of the inner circle (diameter 6, radius 3): $\pi(5)^2 - \pi(3)^2 = 25\pi - 9\pi = 16\pi$.

18. $\mathbf{6\pi + 3\pi\sqrt{2}}$

Use the 45°-45°-90° relationship to find that the hypotenuse of the triangle has length $6\sqrt{2}$. Each smaller semicircle has circumference $\frac{1}{2}\pi d = \frac{1}{2}\pi(6) = 3\pi$. The larger semicircle has circumference $\frac{1}{2}\pi d = \frac{1}{2}\pi(6\sqrt{2}) = 3\pi\sqrt{2}$. Therefore, the total perimeter is $3\pi + 3\pi + 3\pi\sqrt{2} = 6\pi + 3\pi\sqrt{2}$.

19. 600°

Label each interior angle of each square and triangle using polygon properties: 90° for interior angles of a square and 60° for interior angles of an equilateral triangle. At each marked angle, the sum of the four angles must be 360°. Therefore, each marked angle measures 360°– 60°– 60°– 90° = 150°. The sum of the measures of the four marked angles is $4 \times 150° = 600°$.

20. $\mathbf{100 - 25\pi}$

Imagine a rectangle whose top and bottom coincide with those of the figure and whose left and right sides are diameters of the circles. This rectangle is a square because its length and width are each 10. The shaded area is the area of the square minus the area of the two semicircles: $(10 \times 10) - 2 \times \frac{1}{2}\pi(5)^2 = 100 - 25\pi$.

ADVANCED

21. $400\sqrt{3} - 200\pi$

The shaded area is the area of the equilateral triangle minus the areas of three 60° sectors. The area of the triangle is $\frac{1}{2}bh = \frac{1}{2}(40)(20\sqrt{3}) = 400\sqrt{3}$ (use the 30°-60°-90° relationship to find the height of the triangle). The area of each sector is $\frac{60}{360} \times \pi r^2 = \frac{1}{6}\pi(20)^2 = \frac{400\pi}{6}$. The total area of the three sectors is $3 \times \frac{400\pi}{6} = 200\pi$. Therefore, the shaded area is $400\sqrt{3} - 200\pi$.

22. $\frac{7\pi\sqrt{6}}{2}$

Use 30°-60°-90° and 45°-45°-90° relationships to find the diameter. The diameter is $7\sqrt{6}$, so the circumference of the semicircle is $\frac{1}{2} \times \pi d = \pi\left(\frac{7\sqrt{6}}{2}\right) = \frac{7\pi\sqrt{6}}{2}$.

23. $81\pi - \frac{81\sqrt{3}}{2}$

The shaded region is the area of the circle minus the area of the triangle. The radius of the circle is 9. Therefore, the area of the circle is $\pi(9)^2 = 81\pi$. Use the 30°-60°-90° relationship to find all sides of the triangle so you can find its area. The area of the triangle is $\frac{1}{2} \times 9 \times 9\sqrt{3} = \frac{81\sqrt{3}}{2}$.

The shaded area is $81\pi - \frac{81\sqrt{3}}{2}$.

24. $4r^2 - \pi r^2$

The unshaded region is the area of the square minus the area of four quarter-circles. The area of the square is $2r \times 2r = 4r^2$. The area of each quarter circle is $\frac{1}{4}\pi r^2$; the sum of the areas of four quarter circles is πr^2. The difference is $4r^2 - \pi r^2$.

25. 240°

The radii are two sides of a quadrilateral. A line tangent to a circle meets a radius at the point of tangency at a 90° angle. Each angle of the equilateral triangle is 60°. Therefore, the remaining angle of the quadrilateral measures 120°. A circle has 360°. So, the measure of $\angle x$ can be found using subtraction: 360° – 120° = 240°.

26. 192

The quarter circle has arc length 10π, so the entire circle has circumference 40π, diameter 40, and radius 20. The unshaded rectangle has base (RS) 12 units. Its diagonal is a radius of the quarter circle and, therefore, has length 20. Find the

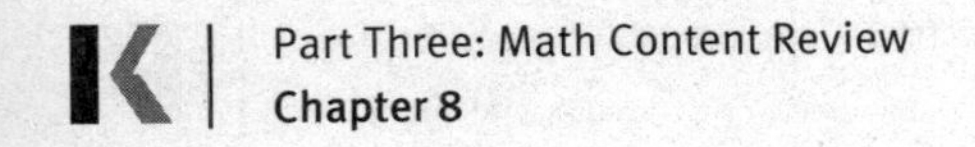

height of the rectangle using the Pythagorean triple 12:16:20 (a multiple of 3:4:5). The area of the rectangle is $12 \times 16 = 192$.

27. 100°

Radii *HD*, *HF*, and *HG* have the same length and form isosceles triangles *DHF* and *GHF*. Isosceles triangles have congruent base angles, so $\angle HFG = x°$ and $\angle HDF = 2x°$. The sum of the four angles of quadrilateral *HDFG* must be 360°. So:

$$\begin{aligned} 2x + 2x + x + x + 120 &= 360 \\ 6x + 120 &= 360 \\ 6x &= 240 \\ x &= 40 \end{aligned}$$

Triangle *FHG* has two 40° angles. Since a triangle has three angles that sum to 180°, the third angle must be 100°. So, the measure of $\angle FHG$ is 100°.

28. $\mathbf{\frac{16}{3}}$

The volume of a cube can be found by s^3, in which *s* represents the length of one side, since cubes have the same length for all edges. The cube has a length of 4, so the volume is $4^3 = 4 \times 4 \times 4 = 64$. This is also the volume of the rectangular solid with the length 6 and width 2. To find the height, plug in the numbers into the formula for the volume of a rectangular solid:

$$\begin{aligned} V &= l \times w \times h \\ 64 &= 6 \times 2 \times h \\ 64 &= 12h \\ \frac{64}{12} &= h \\ \frac{16}{3} &= h \end{aligned}$$

29. $\mathbf{36 + 6\pi}$

If the quarter circle arc has length 6π, then the entire circle would have a circumference of 24π. Since $C = 2\pi r = 24\pi$, $r = 12$. The radius of the quarter circle is the height of the rectangle. Since the perimeter of the rectangle is 60, the length of the rectangle must be 18. And the bottom portion of the shaded area will have a length of 18 minus the circle's radius, or 6. So, the perimeter of the shaded area is $6 + 12 + 18 + 6\pi$, or $36 + 6\pi$.

30. $\mathbf{4\pi x\sqrt{2}}$

If the diameter of each smaller circle is $2x$, then the side of the square is $2x + 2x = 4x$.

Use the 45°-45°-90° relationship to find the diagonal of the square, which is $4x\sqrt{2}$. The diagonal of the square is equal to the diameter of the outer circle.

The circumference of the outer circle is $4\pi x\sqrt{2}$.

THREE-DIMENSIONAL FIGURES (UNIFORM SOLIDS)

A **solid** is a three-dimensional figure (a figure having length, width, and height) and therefore may be rather difficult to represent accurately on a two-dimensional page. Figures are drawn "in perspective," giving them the appearance of depth. If a diagram represents a three-dimensional figure, it will be specified in the accompanying text.

Fortunately, only a few types of solids appear with any frequency on the GRE: rectangular solids (including cubes) and cylinders. You can help your understanding of such figures by spending some time with solids that you can find around your home, such as boxes for rectangular solids and soup cans for cylinders. This will help you visualize what the problems are describing and asking.

Other types, such as spheres, cones, and pyramids, may appear, but these questions typically will only involve understanding the solid's properties. Here are the terms used to describe the common solids:

Vertex: The vertices of a solid are the points at its corners. For example, a cube has eight vertices.

Edge: The edges of a solid are the line segments that connect the vertices and form the sides of each face of the solid. A cube has twelve edges.

Face: The faces of a solid are the polygons that form the outside of the solid. A rectangular prism has six faces, all rectangles. A cube (which is a rectangular prism) has six faces, all squares.

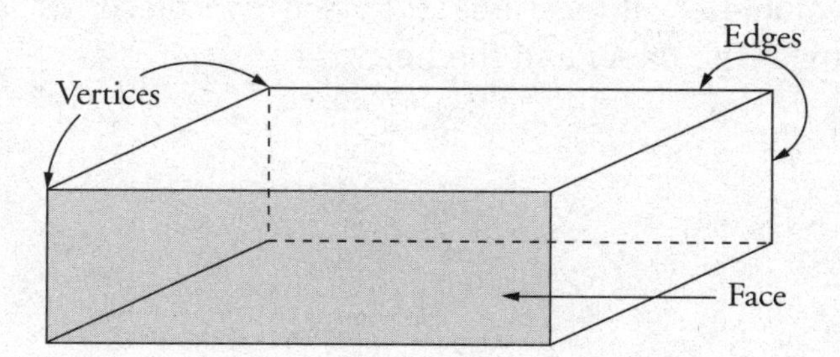

Volume: The volume of a solid is the amount of space enclosed by that solid. The volume of any uniform solid is equal to the area of its base times its height. Volume is expressed in cubic units.

Surface area: In general, the surface area of a solid is equal to the sum of the areas of the solid's faces.

Rectangular solid: A solid with six rectangular faces (all edges meet at right angles). Examples are cereal boxes, bricks, etc.

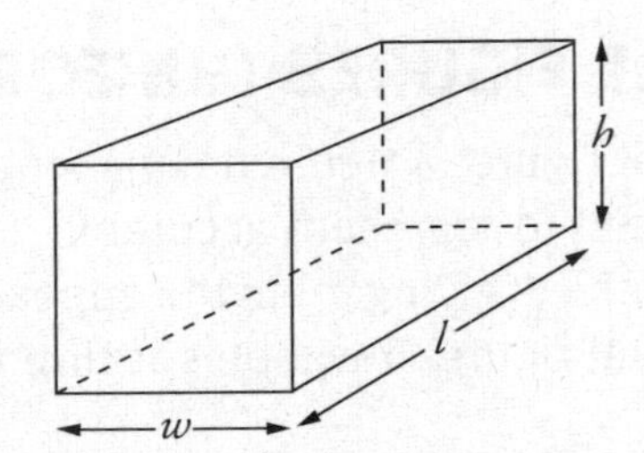

$$\text{Volume} = \text{area of base} \times \text{height} = \text{length} \times \text{width} \times \text{height} = l \times w \times h$$
$$\text{Surface area} = \text{sum of areas of faces} = 2lw + 2lh + 2wh$$

Cube: A special rectangular solid with all edges equal ($l = w = h$), such as a die. All faces of a cube are squares. Because each edge is of equal length, the edge is generally known as the variable e, as in the example shown.

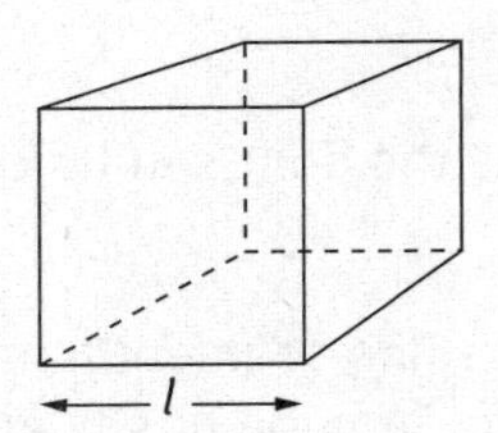

$$\text{Volume} = \text{area of base} \times \text{height} = l \times w \times h = e^3$$
$$\text{Surface area} = \text{sum of areas of faces} = 6e^2$$

Cylinder: A uniform solid whose base is a circle; for example, a classic soup can. To calculate the volume or surface area, we need two pieces of information for a cylinder: the radius of the base, and the height.

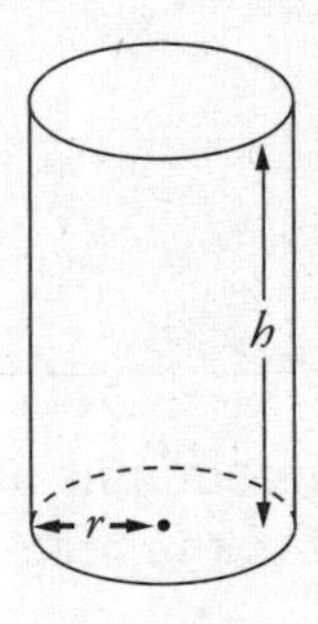

$$\text{Volume} = \text{area of base} \times \text{height} = \pi r^2 h$$
$$\text{Total surface area} = 2 \times \text{area of base} + \text{area of rest of shell (lateral surface area)}$$
$$\text{Total surface area} = 2(\pi r^2) + 2\pi rh$$

You can think of the surface area of a cylinder as having two parts: one part is the top and bottom (the circles), and the other part is the lateral surface. In a can, for example, the area of both the top and the bottom is just the area of the circle, or lid, which represents the top; hence, πr^2 for the top and πr^2 for the bottom, yielding a total of $2\pi r^2$. For the lateral surface, the area around the can, think of removing the can's label. When unrolled, it's actually in the shape of a rectangle. One side is the height of the can, and the other side is the distance around the circle, or circumference. Hence, its area is $h \times (2\pi r)$, or $2\pi rh$. And so, the total surface area is $2\pi r^2 + 2\pi rh$.

Sphere: Occasionally, a question might require you to understand what a sphere is. A sphere is made up of all the points in space a certain distance from a center point; it's like a three-dimensional circle. The distance from the center to a point on the sphere is the radius of the sphere. A basketball is a good example of a sphere. A sphere is not a uniform solid; the cross sections are all circles, but they are of different sizes. (In other words, a slice of a basketball from the middle is bigger than a slice from the top.)

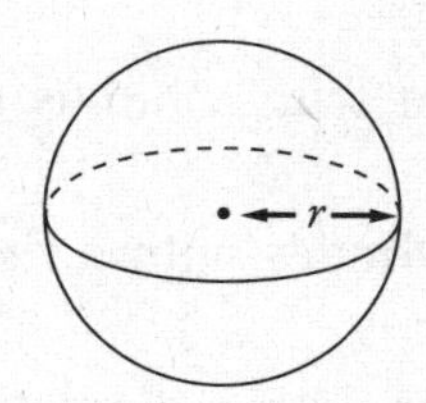

THREE-DIMENSIONAL FIGURES (UNIFORM SOLIDS) EXERCISES

BASIC

1. Find the volume of a rectangular solid with dimensions 10, 4, and 9.
2. Find the surface area of a rectangular solid with dimensions $\frac{1}{2}$, 6, and 12.
3. Find the volume of a cube with an edge of 8.
4. Find the surface area of a cube with an edge of 1.5.
5. Find the volume of a cylinder with a radius of 10 and a height of 4.
6. Find the surface area of a cylinder with a radius of 3 and a height of 12.
7. Find the volume of a cylinder with a diameter of 12 and a height of 3.
8. A rectangular solid has a volume of 72. Its width is 3 and its height is 4. What is the length?
9. A cube has a volume of 27. What is the length of one edge?
10. A cube with an edge of 4 has the same volume as a rectangular solid with a length and width of 2. What is the height of the rectangular solid?

INTERMEDIATE

11. What is the surface area of a cylinder with a radius of 5 and a height of 8?
12. The dimensions of a cube are each doubled, from 2 to 4. By what factor is the surface area of the cube increased?

13. The dimensions of the rectangular solid shown are each tripled. By what factor is the volume of the rectangular solid increased?

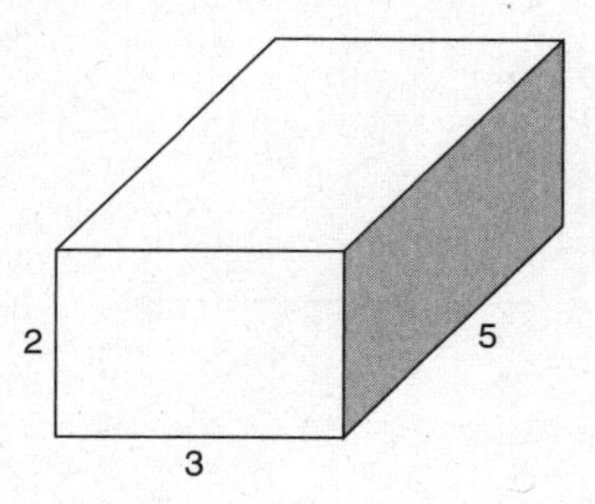

14. A cylinder has a surface area of 22π. If the cylinder has a height of 10, what is its radius?

15. What is the ratio of the surface area of a cube with an edge of 10 to the surface area of a rectangular solid with dimensions 2, 4, and 6?

16. What is the length of segment AR?

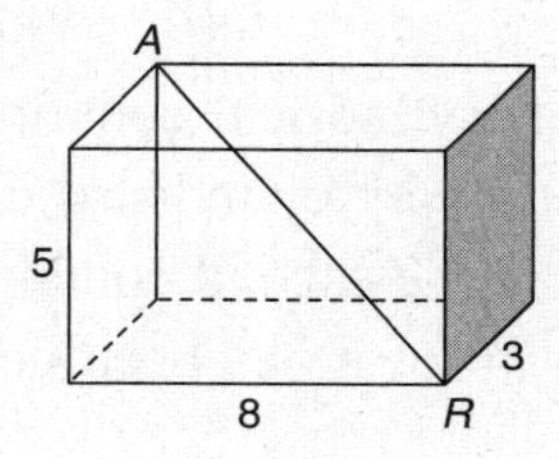

17. How many times can a cube-shaped container with an edge of 2 be emptied into a rectangular prism container with dimensions 4, 8, and 15?

18. The cylinders in the figure shown have the same volume. What is h?

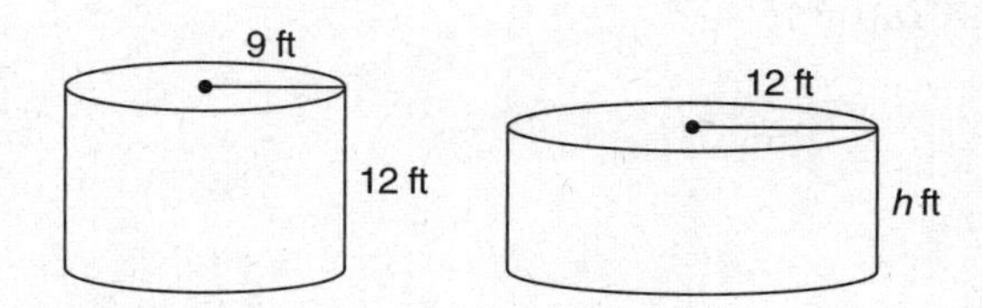

19. The length and width of a rectangular solid are reduced by 10%. The new volume is what percent of the original volume?

20. The solid shown is half a rectangular solid. What is the volume of the solid shown?

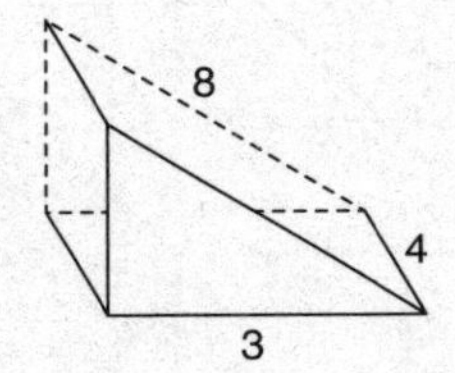

ADVANCED

21. A container in the shape of a rectangular solid with dimensions 5, 5, and 12 has four times the volume of a cylindrical container with a radius of 3. What is the height of the cylinder?

22. A brick with dimensions 10, 15, and 25 weighs 1.5 kg. A second brick (with the same density) has dimensions 12, 18, and 30. Given that weight is proportional to volume in objects of equivalent density, what is the weight of the second brick?

23. Water is poured from a full cylindrical container with a radius of 3 and a height of 5 into a different empty cylindrical container with a radius of 5. The water will not overflow the second container. How many inches high will the water reach when all of it has been poured?

24. Each dimension of a rectangular solid is an integer less than 11. The volume of the solid is 30. If the height of the solid is 10, what is the surface area of the solid?

25. The height of a cylinder is twice its radius. If the volume of the cylinder is 128π, what is the radius?

26. The diagonal AC of the rectangular solid forms a 60° angle with the diagonal of its base. Given the dimensions in the figure, what is the volume of the rectangular solid?

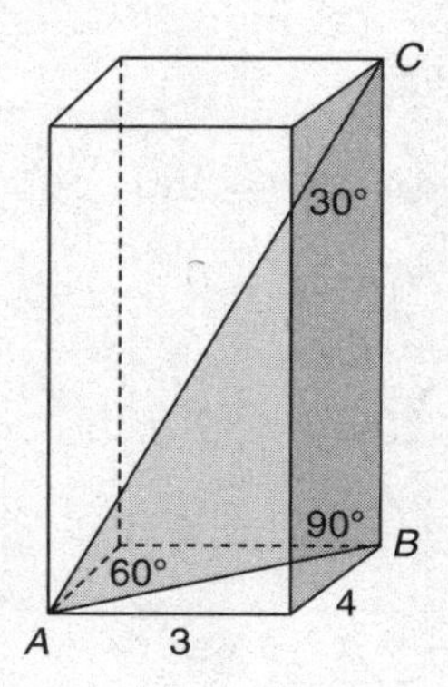

27. Cylinder A has twice the radius but half the height of Cylinder B. What is the ratio of the volume of Cylinder A to the volume of Cylinder B?

28. A cube of ice has edges of length 10. What is the volume of the largest cylinder that can be carved from the cube?

29. A brick with dimensions 3, 4, and 9 has three holes bored completely through it as shown. Each hole has a diameter of 2. What is the net volume of the brick?

30. A solid metal cylinder with a radius of 6 and a height of 3 is melted down, and all of the metal is used to recast a new solid cylinder with a radius of 3. What is the height of the new cylinder?

THREE-DIMENSIONAL FIGURES (UNIFORM SOLIDS) EXERCISES ANSWER KEY

BASIC

1. 360

$$V = lwh = 10 \times 4 \times 9 = 360$$

2. 162

$$\begin{aligned} SA &= 2(lw) + 2(wh) + 2(lh) \\ &= 2\left(\frac{1}{2} \times 6\right) + 2(6 \times 12) + 2\left(\frac{1}{2} \times 12\right) \\ &= 2(3) + 2(72) + 2(6) \\ &= 6 + 144 + 12 \\ &= 162 \end{aligned}$$

3. 512

$$\begin{aligned} V &= lwh \\ &= 8 \times 8 \times 8 \\ &= 512 \end{aligned}$$

4. 13.5

The surface area is the sum of the areas of all six faces. Each face is a square with area $A = lw = (1.5)(1.5) = 2.25$. Therefore the surface area of the cube is $6(2.25) = 13.5$.

5. 400π

$$\begin{aligned} V &= \pi r^2 h \\ &= \pi(10)^2(4) \\ &= \pi(100)(4) \\ &= 400\pi \end{aligned}$$

6. 90π

$$\begin{aligned} SA &= 2\pi r^2 + 2\pi rh \\ &= 2\pi(3)^2 + 2\pi(3)(12) \\ &= 2\pi(9) + 2\pi(36) \\ &= 18\pi + 72\pi \\ &= 90\pi \end{aligned}$$

7. 108π

If the diameter is 12, then the radius is 6.

$$\begin{aligned} V &= \pi r^2 h \\ &= \pi(6)^2(3) \\ &= \pi(36)(3) \\ &= 108\pi \end{aligned}$$

8. 6

$$\begin{aligned} V &= lwh \\ 72 &= l(3)(4) \\ 72 &= 12l \\ 6 &= l \end{aligned}$$

9. 3

$$\begin{aligned} V &= e^3 \\ 27 &= e^3 \\ 3 &= e \end{aligned}$$

10. 16

The cube has volume $V = e^3 = 4^3 = 64$. Use the volume 64 to solve for the height of the rectangular solid.

$$\begin{aligned} V &= lwh \\ 64 &= 2 \times 2 \times h \\ 64 &= 4h \\ 16 &= h \end{aligned}$$

INTERMEDIATE

11. 130π

$$\begin{aligned} SA &= 2\pi r^2 + 2\pi rh \\ &= 2\pi(5)^2 + 2\pi(5)(8) \\ &= 2\pi(25) + 2\pi(40) \\ &= 50\pi + 80\pi \\ &= 130\pi \end{aligned}$$

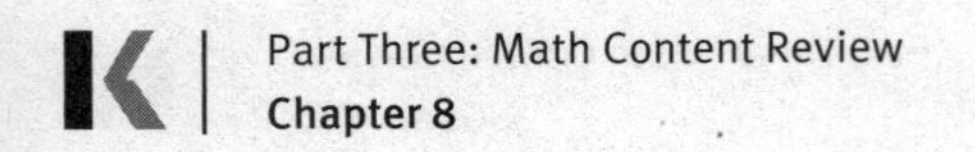

12. 4

The original surface area is $6e^2 = 6(2)^2 = 6 \times 4 = 24$. The new surface area is $6 \times e^2 = 6 \times 4^2 = 6 \times 16 = 96$. Since $96 \div 24 = 4$, the surface area is increased by a factor of 4.

13. 27

The original volume is $V = lwh = 3 \times 2 \times 5 = 30$. The new volume is $V = 9 \times 6 \times 15 = 810$. Since $810 \div 30 = 27$, the volume has increased by a factor of 27.

14. 1

Use the surface area formula:

$$\begin{aligned} SA &= 2\pi r^2 + 2\pi rh \\ 22\pi &= 2\pi r^2 + 2\pi r(10) \\ 22\pi &= 2\pi r^2 + 20\pi r \end{aligned}$$

Look closely at the last equation. The equation is true only when $r = 1$.

$$\begin{aligned} 22\pi &= 2\pi(1)^2 + 20\pi(1) \\ 22\pi &= 2\pi(1) + 20\pi(1) \\ 22\pi &= 2\pi + 20\pi \end{aligned}$$

15. 75:11

The cube has surface area $SA = 6e^2 = 6(10)^2 = 6 \times 100 = 600$.

The rectangular solid has surface area

$$\begin{aligned} SA &= 2(lw) + 2(wh) + 2(lh) \\ &= 2(2 \times 4) + 2(4 \times 6) + 2(2 \times 6) \\ &= 2(8) + 2(24) + 2(12) \\ &= 16 + 48 + 24 \\ &= 88 \end{aligned}$$

The ratio of the cube's surface area to the rectangular solid's surface area is $\frac{600}{88} = \frac{75}{11}$.

16. $7\sqrt{2}$

Segment AR is the hypotenuse of a right triangle with height 5. The base of the right triangle is also a diagonal of the base of the rectangular solid. The base has length 8 and width 3, so the diagonal has length $\sqrt{3^2 + 8^2} = \sqrt{9 + 64} = \sqrt{73}$.

Therefore the length of segment AR is $\sqrt{\left(\sqrt{73}\right)^2 + 5^2} = \sqrt{73 + 25} = \sqrt{98} = 7\sqrt{2}$.

17. 60

The rectangular container has volume $V = lwh = 4 \times 8 \times 15 = 480$. The cube has volume $V = e^3 = 2^3 = 8$. Since $\frac{480}{8} = 60$, the cubical container can be emptied 60 times into the rectangular prism container.

18. 6.75 or $6\frac{3}{4}$

The volume of the known cylinder is $V = \pi r^2 h = \pi(9)^2\,(12) = \pi(81)(12) = 972\pi$. Now use 972π to find h in the unknown cylinder.

$$\begin{aligned} V &= \pi r^2 h \\ 972\pi &= \pi(12)^2 h \\ 972\pi &= 144\pi h \\ 972 &= 144h \\ 6.75 &= h \end{aligned}$$

19. 81%

The volume of the original solid is $V = lwh$. The volume of the solid with reduced length and width is $V = (0.9l)(0.9w)(h) = (0.81)lwh$. Therefore the solid with reduced length and width has 81% of the volume of the original solid.

20. $6\sqrt{55}$

The rectangular solid has a front face with a width of 3 and a diagonal of 8. The height of the front face can be found using the Pythagorean theorem:

$$\begin{aligned} a^2 + b^2 &= c^2 \\ 3^2 + b^2 &= 8^2 \\ 9 + b^2 &= 64 \\ b^2 &= 55 \\ b &= \sqrt{55} \end{aligned}$$

The whole rectangular solid has dimensions 3, 4, and $\sqrt{55}$. Its volume is $V = lwh = 3 \times 4 \times \sqrt{55} = 12\sqrt{55}$. Half its volume is $6\sqrt{55}$.

ADVANCED

21. $\frac{25}{3\pi}$

The volume of the rectangular prism is $V = lwh = 5 \times 5 \times 12 = 300$. Therefore, the volume of the cylinder is $300 \div 4 = 75$. Use $V = 75$ to find the height.

$$V = \pi r^2 h$$
$$75 = \pi(3)^2 h$$
$$75 = 9\pi h$$
$$\frac{75}{9\pi} = h$$
$$\frac{25}{3\pi} = h$$

22. 2.592 kg

Use the fact that weight is proportional to volume. The volume of the first brick is $V = lwh = 10 \times 15 \times 25 = 3{,}750$. The volume of the second brick is $V = lwh = 12 \times 18 \times 30 = 6{,}480$.

Write and solve a proportion to find the weight of the second brick.

$$\frac{3{,}750}{1.5} = \frac{6{,}480}{x}$$
$$3{,}750x = (6{,}480)(1.5)$$
$$3{,}750x = 9{,}720$$
$$x = 2.592$$

23. $\frac{9}{5}$ or 1.8

The volume of water poured is $V = \pi r^2 h = \pi(3)^2\ (5) = \pi(9)(5) = 45\pi$.

Use 45π to find the height of a cylinder with that volume and a radius of 5.

$$V = \pi r^2 h$$
$$45\pi = \pi(5)^2 h$$
$$45\pi = 25\pi h$$
$$45 = 25h$$
$$\frac{45}{25} = h$$
$$\frac{9}{5} = h\text{, or } 1.8$$

24. 86

The volume is 30 and height is 10. Using the volume formula,

$$30 = lw(10)$$
$$3 = lw$$

If the product of length and width is 3 and each dimension is an integer, then these two dimensions must be 1 and 3. Therefore the dimensions of the solid are 1, 3, and 10.

$$\begin{aligned} SA &= 2(lw) + 2(wh) + 2(lh) \\ &= 2(1 \times 3) + 2(3 \times 10) + 2(1 \times 10) \\ &= 2(3) + 2(30) + 2(10) \\ &= 6 + 60 + 20 \\ &= 86 \end{aligned}$$

25. 4

Let $h = 2r$.

$$\begin{aligned} V &= \pi r^2 h \\ 128\pi &= \pi r^2(2r) \\ 128 &= 2r^3 \\ 64 &= r^3 \\ 4 &= r \end{aligned}$$

26. $60\sqrt{3}$

The diagonal of the base of the solid is the hypotenuse of a 3:4:5 triangle. Use the 30°-60°-90° relationship to find the height of the shaded triangle. Since the leg with a length of 5 is across from the angle that is 30 degrees, the leg across from the 60 degree angle is $5\sqrt{3}$, which is also the height of the solid. The solid has dimensions 3, 4, and $5\sqrt{3}$, so the volume is $V = lwh = 3 \times 4 \times 5\sqrt{3} = 60\sqrt{3}$.

27. $\frac{2}{1}$

If Cylinder B has radius r and height h, then Cylinder A has radius $2r$ and height $\frac{1}{2}h$.

The volume of Cylinder B is $V_B = \pi r^2 h$.

The volume of Cylinder A is $V_A = \pi(2r)^2\left(\frac{1}{2}h\right) = \pi(4r^2)\left(\frac{1}{2}h\right) = 2\pi r^2 h$.

The ratio is $\frac{V_A}{V_B} = \frac{2\pi r^2 h}{\pi r^2 h} = \frac{2}{1}$.

Picking Numbers would also be a great strategy here. Let the radius and height of Cylinder B be 2 and 4, respectively. Then the radius and height of Cylinder A are 4 and 2, respectively. $V_B = \pi r^2 h = 16\pi$. $V_A = \pi r^2 h = 32\pi$. The volume of Cylinder A is twice that of Cylinder B, and the ratio is 2:1.

28. 250π

The largest cylinder would have diameter 10 and height 10 (each equal to the edge of the cube). When diameter is 10, radius is 5. Therefore, the volume is $V = \pi r^2 h = \pi(5)^2\ (10) = \pi(25)(10) = 250\pi$.

29. $108 - 9\pi$

The net volume is the volume of the rectangular solid minus the volume of the three cylindrical holes. The volume of the rectangular solid is $V = lwh = 3 \times 4 \times 9 = 108$. The volume of each hole is $V = \pi r^2 h = \pi(1)^2(3) = 3\pi$.

The net volume is $108 - 3(3\pi) = 108 - 9\pi$.

30. 12

The volume of metal melted down is $V = \pi r^2 h = \pi(6)^2\ (3) = \pi(36)(3) = 108\pi$. Use 108π to find h in the recast cylinder.

$$
\begin{aligned}
V &= \pi r^2 h \\
108\pi &= \pi(3)^2 h \\
108\pi &= 9\pi h \\
108 &= 9h \\
12 &= h
\end{aligned}
$$

CHAPTER 9

Data Interpretation

UNDERSTANDING DATA ANALYSIS

The GRE will test your ability to draw conclusions about sets of objects or data. This chapter explores selecting and arranging members of a group, using statistics to describe the behavior of data, evaluating various presentations of data, and calculating the likelihood of events occurring based on specified manipulations of objects or data.

COUNTING METHODS

Whether counting the number of members in a set or determining the number of possible selections or arrangements that can be made from the set, understanding counting methods is a key element in data analysis. When a data set is too large for you to simply list and count its objects, the objects can often be arranged systematically. There are several counting methods you can then apply to determine the number of elements in the set or the number of possible selections from the set.

Before considering the details of counting data sets, it is helpful to understand the nature of the data. A data set is a group of objects, such as all prime numbers or players on a basketball team, that are presented together regardless of order. The objects in the set are called **members** or **elements**. A set can be either **finite**, with a number of elements that can be counted, or **infinite**. A **subset** is a grouping of the members within the set based on a shared characteristic. For example, the set of digits {0, 1, 2, 3, 4, 5, 6, 7, 8, and 9} is a finite subset of the set of real numbers. The **empty** set, a set with no members, is denoted by the symbol $\varnothing$.

Inclusion-Exclusion Principle

When you consider the relationship between two data sets, a **Venn diagram** can be useful. The following diagram represents three data sets, A, B, and C.

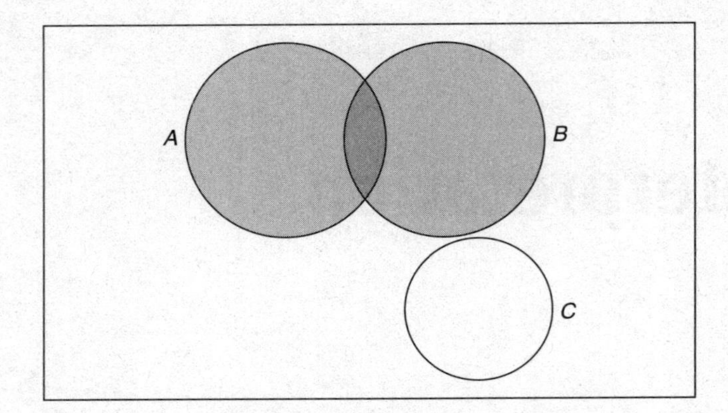

The shaded circles represent the set of elements that are found in *either A or B*. This is the **union** of A and B, or $A \cup B$. The dark shading of the overlapping section of the two circles A and B represents the set of elements that is found in *both A and B*. This is the **intersection** of A and B, or $A \cap B$. Since A and C do not overlap, $A \cap C$ is the empty set, and A and C are considered **mutually exclusive**.

The **inclusion-exclusion principle** for two finite sets states that the number of elements in $A \cup B$ equals the sum of the number of elements in A and the number of elements in B minus the number of elements in $A \cap B$. The principle is written as follows:

$$|A \cup B| = |A| + |B| - |A \cap B|$$

Subtracting the intersection accounts for the double-counting of the overlapping section of A and B when all of the elements within A are added to all of the elements within B.

Multiplication Principle

When two events occur sequentially, and the first does not influence the second, the **multiplication principle** states that the number of possible outcomes is $m \times n$, where m is the number of possible outcomes for the first event, and n is the number of possible outcomes for the second event. Similarly, the probability of two independent events occurring is $P(A \cap B) = P(A) \times P(B)$. The probability of the intersection of A and B—that is, of both A and B occurring—is equal to the product of the probability of A occurring by itself and the probability of B occurring by itself.

Example: How many possible outcomes are there if a fair six-sided die is rolled three times?

Each roll has 6 possible outcomes. So, there are $(6)(6)(6) = 216$ possible outcomes.

Example: Two cards are chosen sequentially from a standard deck of 52 distinct cards. The first card is not returned to the deck before drawing the second card. What is the number of possible outcomes?

The first selection has 52 possible outcomes. Since the first selection is not returned to the deck, the choices are not independent, but the multiplication principle can still be applied. In this case, the second selection only has 51 possible outcomes. There are $(52)(51) = 2{,}652$ possible outcomes.

Permutations

Suppose you have a set of n objects and you want to determine the number of possible orders, or **permutations**, of all of the objects. When the first object is assigned to its position, there are n possibilities. For the second object, there are $n - 1$ remaining positions. For the third, there are $n - 2$, and so on. The number of permutations can be found by applying the multiplication principle. The number of possible orders of n objects is as follows:

$$n\,(n - 1)(n - 2) \ldots (2)(1)$$

This product is written as $n!$ and pronounced ***n* factorial**.

Example: You are arranging 4 trophies in a row on a shelf. How many distinct ways are there to arrange the trophies?

In this case, $n = 4$. When you place the first trophy, there are 4 possible positions. For the second trophy, there are only 3 remaining positions. Once the second trophy is placed and you are placing the third trophy, there are only 2 positions left. For the last trophy, there is only one possible position. Therefore, there are $n! = 4! = (4)(3)(2)(1) = 24$ possible arrangements for 4 trophies.

Now suppose you have a set of n objects, but you only want to order some of them. Let the number of objects you are choosing for the subset be k. Again, there are n possibilities for the first selection, $n - 1$ for the second, $n - 2$ for the third, and so on. The difference here is that this only continues k times. The formula for the number of **permutations of *n* objects taken *k* at a time** is this:

$$P(n,k) = {}_nP_k = \frac{n!}{(n - k)!}$$

Example: Five runners are in a race. A gold, silver, and bronze medal will be awarded to the first, second, and third place winners. How many different ways could the medals be awarded to the runners?

For three different medals, there are 5 possible winners for the gold medal, 4 possible winners for the silver medal, and 3 possible winners for the bronze medal. There are $(5)(4)(3) = 60$ possible arrangements of the medals. Using the formula gives the same result:

$$P(5,3) = \frac{5!}{(5-3)!} = \frac{5!}{2!} = \frac{5 \times 4 \times 3 \times \cancel{2 \times 1}}{\cancel{2 \times 1}} = 5 \times 4 \times 3 = 60.$$

Note that when all objects are included in the arrangement, then $k = n$. The denominator of the formula becomes 0!, which is defined as 1. The formula then becomes the formula used for the previous example:

$$P(n,n) = \frac{n!}{0!} = n!$$

Combinations

Now suppose you have a set of n objects and you still want to select some number, k, of them, but their order does not matter. The number of possible groups chosen is called the number of **combinations of *n* objects taken *k* at a time** and is given by the formula:

$$C(n,k) = {}_nC_k = \frac{n!}{k!(n-k)!}$$

Combinations are often referenced as ***n* choose *k***.

Example: A choir director randomly selects 3 of his 6 members to form a group. How many possible groups of 3 members are there?

$${}_6C_3 = \frac{6!}{3!(6-3)!} = \frac{6!}{3!3!} = \frac{6 \times 5 \times 4 \times \cancel{3 \times 2 \times 1}}{3 \times 2 \times 1 \times \cancel{3 \times 2 \times 1}} = \frac{6 \times 5 \times 4}{3 \times 2 \times 1} = 20$$

COUNTING METHODS EXERCISES

BASIC

1. In how many different ways can the letters in the word DANCE be ordered?

2. Jake went to a football game with 4 of his friends. There are 120 different ways in which they can sit together in a row of 5 seats, with one person per seat. In how many of those ways is Jake sitting in the middle seat?

3. How many 3-digit positive integers are even and do not contain the digit 4?

4. From a box of 12 candles, you are to remove 5. How many different sets of 5 candles could you remove?

For Exercises 5 and 6, use the following information:

There are 10 finalists for the school spelling bee. A first, second, and third place trophy will be awarded with no ties or duplications.

5. In how many different ways can the judges award the three prizes?

6. How many different groups of 3 people can get prizes?

7. If you toss a fair coin 6 times, what is the number of total possible outcomes?

8. What is the value of 4! equal to?

9. In how many different ways can the letters of the word MEXICO be arranged?

10. Evaluate the following: $\frac{7!}{5!}$

INTERMEDIATE

11. There are 6 doors leading into a library. How many different ways could a student enter the room by one door and then leave it using a different door?

12. In how many distinct ways can the letters of the word METHODS be rearranged such that M and S occupy the first and last position respectively?

13. A quiz has 10 true/false questions. If each question is answered with "true" or "false" and none of them are left blank, in how many ways can the quiz be answered?

14. A pizza place offers a dinner combo consisting of pizza, salad, dessert, and a drink from the following menu.

    Pizza: cheese, pepperoni, vegetable
    Salad: Caesar, house
    Dessert: brownie, cookie
    Drink: tea, coffee, soda, lemonade

    How many possible dinner combos are there that include all four choices?

15. A company places a 6-symbol code on each product. The code consists of the letter T, followed by 3 numerical digits, and then 2 consonants. How many different codes are possible? (Consider Y a consonant.)

16. Gloria has 7 shirts to display on 7 possible mannequins in her boutique. If she has already placed the first shirt on mannequin 1, how many different ways can she display the rest of the shirts?

17. If $P(A) = 0.3$, $P(A \cap B) = 0.12$, and the events A and B are independent, then find $P(B)$.

18. There are 7 routes from Springfield to London and 12 routes from London to Fairview. The road a driver takes into London has no effect on the roads that driver can take out of London, and vice versa. Find the number of different ways to travel from Springfield to Fairview that pass through London.

19. If $P(A) = 0.8$, $P(A \cap B) = 0.16$, and the events A and B are independent, what is the value of $P(B)$?

20. Nine coins are tossed simultaneously. In how many of the outcomes will the fourth coin tossed show heads?

ADVANCED

21. In how many ways can 4 students be selected from a group of 12 students to represent a school in a swimming competition?

22. A student has 3 different science classes, 4 different math classes, 2 different history classes, 2 different English classes, and 5 different electives to choose from. How many different 5-course selections can she make if she needs to enroll in one of each type of class?

23. During practice, a basketball player can successfully complete a foul shot once in 3 throws on average. If the basketball player throws 3 balls in succession, what is the probability that the basketball player will be unsuccessful in all three attempts?

24. How many 4-digit codes can be made using the following digits: 1, 2, 3, 4, 5, 6, 7, 8, 9? (Repetition of a digit is allowed.)

25. If a coin is tossed twice, what is the probability that the coin lands tails up on both tosses?

26. A test has 2 sections, each containing 5 multiple-choice questions. Each question of the first section has 4 answer options, and each question of the second section has 3 answer options. In how many different ways can the test be answered without leaving any questions blank?

27. How many distinct combinations of 3 socks could be made by randomly selecting socks from a drawer containing 8 differently colored socks?

28. At a high school, 300 students are members of the band, chess club, or both. If 200 students are members of the band only and 50 students are members of the band and chess club, what is the probability that a student chosen at random is a member of only the chess club?

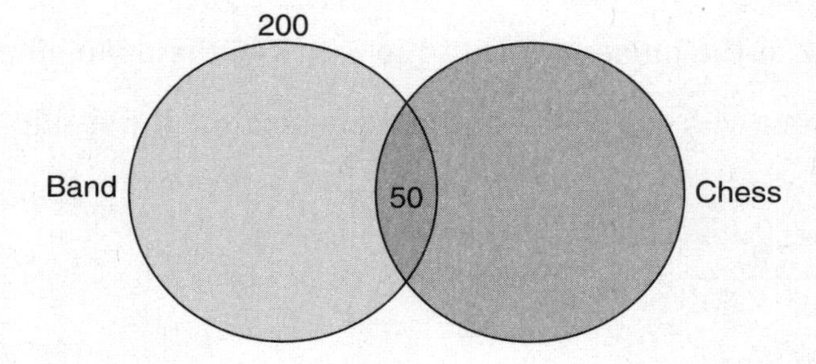

29. Two six-sided dice and one quarter are tossed. How many results are possible?

30. If 8 schools are all in the same conference, how many soccer games are played during the season if the teams all play each other exactly once?

COUNTING METHODS ANSWERS KEY

Basic

1. 120

Any 5-letter word with none of its letters repeating can be rearranged in 5! ways.

$$5 \times 4 \times 3 \times 2 \times 1 = 120$$

There are 120 ways.

2. 24

If Jake is in the middle, then that removes him from the permutation equation. You're really looking for the number of ways to arrange four people into four seats. So there are 4! ways, with Jake in the middle: $4 \times 3 \times 2 \times 1 = 24$.

3. 288

There are three events to consider: digit 1, digit 2, and digit 3.

There are eight possibilities for the first digit, since you can't use 0 or 4. There are nine possibilities for the second digit, since you can't use the digit 4. There are only four possibilities for the third digit, since you can only use even digits (the final 3-digit number must be even) and not the digit 4.

$$8 \times 9 \times 4 = 288$$

There are 288 possibilities.

4. 792

The order in the sets of 5 candles doesn't matter, so the number of possible 5-candle sets you can remove from a box of 12 equals $\frac{12!}{5!(12-5)!} = \frac{12!}{5!7!} = \frac{12 \times 11 \times 10 \times 9 \times 8}{5 \times 4 \times 3 \times 2 \times 1}$. Divide out the 12 in the numerator with the 4×3 in the denominator, and divide out the 10 in the numerator with the 5×2 in the denominator. The result is $\frac{11 \times 9 \times 8}{1} = 792$.

5. 720

$$\frac{10!}{(10-3)!} = \frac{10!}{7!} = 10 \times 9 \times 8 = 720$$

6. 120

$$\frac{10!}{3!(10-3)!} = \frac{10!}{3!7!} = \frac{10 \times 9 \times 8}{3 \times 2 \times 1} = 120$$

7. 64

There are 2 possible outcomes for each toss, so after 6 tosses there is a total of $2^6 = 64$ possible outcomes.

8. 24

$$4! = 4 \times 3 \times 2 \times 1 = 24$$

9. 720

Any 6-letter word with none of its letters repeating can be rearranged in 6! ways.

$$6 \times 5 \times 4 \times 3 \times 2 \times 1 = 720$$

10. 42

$$\frac{7 \times 6 \times \cancel{5 \times 4 \times 3 \times 2 \times 1}}{\cancel{5 \times 4 \times 3 \times 2 \times 1}} = \frac{7 \times 6}{1} = 42$$

INTERMEDIATE

11. 30

There are 6 doors to the library. The student needs to enter the library using one of those 6 doors. If the student has to leave the library using a different door, then the student has to take one of the remaining 5 doors. Therefore, the total number of ways a student could enter the library and leave using a different door is $6 \times 5 = 30$ ways.

12. 120

The word METHODS is a 7-letter word with none of the letters repeating. Any 7-letter word with none of its letters repeating can be rearranged in 7! ways. However, M and S must occupy the first and last position respectively, so the first and last positions can each be filled in only one way. The remaining 5 positions between M and S can be filled with the 5 other letters in $5! = 120$ ways.

13. 1,024

There are 10 events: question 1, question 2, question 3, question 4, question 5, question 6, question 7, question 8, question 9, and question 10.

There are 2 choices for each question.

$$2 \times 2 \times 2 \times 2 \times 2 \times 2 \times 2 \times 2 \times 2 \times 2 \text{ or } 2^{10} = 1{,}024$$

14. 48

There are 4 events: choosing a pizza, choosing a salad, choosing a dessert, and choosing a drink. There are 3 choices for the pizza, 2 choices for the salad, 2 choices for the dessert, and 4 choices for the drink.

$$3 \times 2 \times 2 \times 4 = 48$$

There are 48 dinner combos available.

15. 441,000

There are 6 events: letter 1, digit 1, digit 2, digit 3, letter 2, and letter 3.

We start with letter 1. Letter 1 is limited to the letter T, so there is only one possibility for letter 1. There are no restrictions on digits 1 through 3, so digit 1, digit 2, and digit 3 have

10 possibilities each. Letter 2 and letter 3 have to be consonants, so there are 21 possible letters for each.

$$1 \times 10 \times 10 \times 10 \times 21 \times 21 = 441{,}000$$

16. 720

There are 7 events: shirt 1, shirt 2, shirt 3, shirt 4, shirt 5, shirt 6, and shirt 7. There is only one possibility for the first shirt. That leaves 6 possibilities for shirt 2, 5 for shirt 3, 4 for shirt 4, 3 for shirt 5, 2 for shirt 6, and 1 for shirt 7.

$$1 \times 6 \times 5 \times 4 \times 3 \times 2 \times 1 = 720$$

17. 0.4

$$P(A \cap B) = P(A) \times P(B)$$
$$0.12 = 0.3 \times P(B)$$
$$0.4 = P(B)$$

18. 84

The number of different ways to reach Fairview from Springfield by passing through London $= 7 \times 12 = 84$.

19. 0.2

$$P(A \cap B) = P(A) \times P(B)$$
$$0.16 = 0.8 \times P(B)$$
$$0.2 = P(B)$$

20. 256

When a coin is tossed once, there are two outcomes, heads or tails. When 9 coins are tossed simultaneously, the total number of outcomes $= 2^9$. If the fourth coin has to show heads, then the number of possibilities for the fourth coin is only 1. So the number of outcomes where the fourth coin would show heads would be: $2 \times 2 \times 2 \times 1 \times 2 \times 2 \times 2 \times 2 \times 2$ or $2^8 = 256$.

ADVANCED

21. 495

The number of ways to select 4 students out of 12 students is an instance of selecting without replacement and without ordering. So, one can select 4 students out of 12 students in ${}_{12}C_4$ ways.

$$\frac{12!}{4!(12-4)!} = \frac{12!}{4! \times 8!}$$
$$= \frac{12 \times 11 \times 10 \times 9 \times 8!}{4 \times 3 \times 2 \times 1 \times 8!} = \frac{12 \times 11 \times 10 \times 9}{4 \times 3 \times 2 \times 1}$$

Cancel out common factors.

$$\frac{11 \times 5 \times 9}{1} = 495$$

22. 240

Since the student selects one of each type of class, you multiply the number of possibilities for each class. There are 3 different science classes, 4 different math classes, 2 different history classes, 2 different English classes, and 5 different electives to pick from: $3 \times 4 \times 2 \times 2 \times 5 = 240$.

23. $\mathbf{\frac{8}{27}}$

The basketball player throws the ball 3 times. It is possible that he could be successful one, two, or three times in those three throws. The probability that he will not make the foul shot in one throw $= 1 - \frac{1}{3} = \frac{2}{3}$. The probability that he will not make the foul shot in all three attempts $= \frac{2}{3} \times \frac{2}{3} \times \frac{2}{3} = \frac{8}{27}$.

24. 6,561

The number of permutations of 9 digits taken 4 at a time, if repetition of digits is allowed, is $9 \times 9 \times 9 \times 9$ or $9^4 = 6{,}561$.

25. $\mathbf{\frac{1}{4}}$

$$P(A \text{ and } B) = P(A) \times P(B)\text{: } \frac{1}{2} \times \frac{1}{2} = \frac{1}{4}.$$

26. 248,832

The 5 questions on the first section have 4 answer options each. These can be answered in 4^5 different ways. The 5 questions on the second section have 3 answer options each. These can be answered in 3^5 different ways. The test can be completed in $4^5 \times 3^5 = 1{,}024 \times 243 = 248{,}832$ different ways.

27. 56

n is the number of different socks to choose from, so $n = 8$. k is the number of socks in each possible combination, so $k = 3$.

$$\begin{aligned}
{}_nC_k &= \frac{n!}{k!(n-k)!} \\
{}_8C_3 &= \frac{8!}{3!(8-3)!} \\
&= \frac{8 \times 7 \times 6 \times \cancel{5 \times 4 \times 3 \times 2 \times 1}}{(3 \times 2 \times 1)(\cancel{5 \times 4 \times 3 \times 2 \times 1})} \\
&= \frac{8 \times 7 \times \cancel{6}}{\cancel{3 \times 2} \times 1} \\
&= 8 \times 7 \\
&= 56
\end{aligned}$$

There are 56 different possible combinations.

28. $\frac{1}{6}$

If 200 students are in the band only, then the total number of students in the chess club is $300 - 200 = 100$, since there are 300 students who are members of the band, chess club, or both. The number of students who are only in the chess club is $100 - 50 = 50$, since there are 50 students who are in both the band and chess club. Therefore, the probability that a student chosen at random from the 300 students is a member of only the chess club is as follows:

$$\frac{50}{300} = \frac{1}{6}$$

29. 72

There are 3 stages or events: two dice and one quarter. Each die has 6 possible outcomes. The quarter has 2 possible outcomes, either tails or heads.

$$6 \times 6 \times 2 = 72$$

30. 28

n is the number of teams, so $n = 8$.

k is the number of teams at a time, so $k = 2$.

$$\begin{aligned}
{}_nC_k &= \frac{n!}{(n-k)!k!} \\
{}_8C_2 &= \frac{8!}{(8-2)!2!} \\
&= \frac{8!}{6!2!} \\
&= \frac{8 \times 7 \times 6!}{6!2!} \\
&= \frac{8 \times 7}{2 \times 1} \\
&= \frac{56}{2} \\
&= 28
\end{aligned}$$

DESCRIPTIVE STATISTICS

Statistics is a way to describe and characterize a set of data. Most data sets can be described using measures of central tendency and measures of dispersion.

Measures of Central Tendency

The main measures of central tendency are the arithmetic mean, median, and mode. These values characterize where a data set is centered. Let's review the definitions of mean and median found in chapter 6 on Arithmetic.

The average (arithmetic mean) of a group of numbers is defined as the sum of the values divided by the number of values.

$$\textit{Average value} = \frac{\textit{Sum of values}}{\textit{Number of values}}$$

The median of a set is the middle term when all the terms in the set are listed in sequential order. When there is an even number of terms in a set, the median is the average of the two middle terms.

Since the mean incorporates every data value into a sum, it can be shifted up or down significantly when one or several data values, called **outliers**, are extremely high or extremely low compared to the rest of the values. In these cases, the median is often a more descriptive characteristic of what is typical in the data set.

The third measure of central tendency, **mode**, is the number that appears most frequently in a list of numbers.

Example: For the list of numbers—5, 2, 6, 2, 3, 5, 21, 4—the measures of central tendency are all different.

The mean is 6:

$$\frac{5+2+6+2+3+5+21+4}{8} = \frac{48}{8} = 6.$$

The median is 4.5: 2, 2, 3, 4, 5, 5, 6, 21 (The average of the middle two numbers, 4 and 5, is 4.5.)

The list has 2 modes: 2 and 5, because they both appear twice.

Measures of Dispersion

The main measures of dispersion are the range, interquartile range, and standard deviation. These values characterize the spread of a data set.

The **range** of a data set is the difference between the greatest number and the least number.

The **interquartile range** describes the spread of the middle half of the data. Finding the interquartile range involves dividing the data set into four sections, or quartiles. The **first quartile,** Q_1, is the median of all of the numbers below the median. The **second quartile,** Q_2, is the median of the entire data set. The **third quartile,** Q_3, is the median of all of the numbers above Q_2. The **interquartile range** is the difference between Q_3 and Q_1.

The range and interquartile range are often displayed on a **box-and-whisker plot,** also called a **box plot.**

Let's look at an example of the measures of dispersion considered so far.

Example: 2, 2, 3, 4, 5, 5, 6, 21

The median calculated previously, is 4.5. This is also Q_2.

The range is $21 - 2 = 19$.

Q_1 is the median of 2, 2, 3, and 4, which is 2.5.

Q_3 is the median of 5, 5, 6, and 21, which is 5.5.

The interquartile range is $5.5 - 2.5 = 3$. Notice how much larger the range is than the interquartile range because the range is affected by the outlier, 21.

A box-and-whisker-plot of the data is shown below.

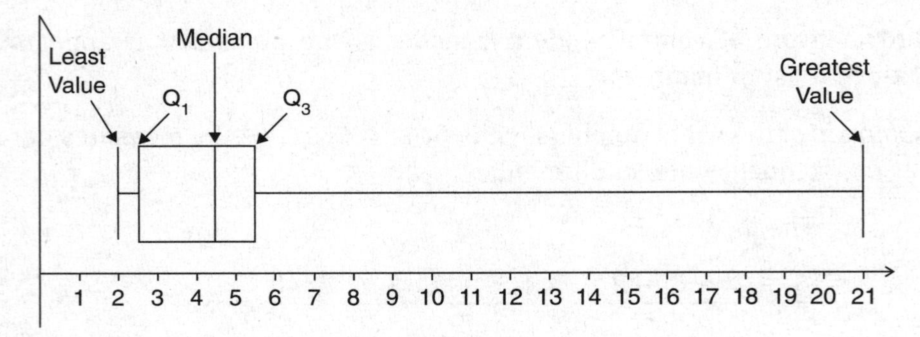

The third measure of dispersion on the GRE is the **standard deviation.** It is a measure of how spread out a set of numbers is (how much the numbers deviate from the mean). The greater the spread, the higher the standard deviation. You will rarely have to calculate the standard deviation on Test Day (although this skill may be necessary for some high-difficulty questions), but you will need a general understanding of what standard deviation is, so here's how it's calculated:

- Find the average (arithmetic mean) of the set.
- Find the differences between the mean and each value in the set.
- Square each of the differences.
- Find the average of the squared differences.
- Take the positive square root of the average.

Here's what the formula looks like if you follow the above steps to calculate the standard deviation of a set of n elements, in which x is the mean:

$$\text{Standard deviation} = \sqrt{\frac{(n_1 - x)^2 + (n_2 - x)^2 + \ldots + (n_n - x)^2}{n}}$$

Example: For the 5-day listing that follows, which city had the greater standard deviation in high temperatures?

High temperatures, in °F, in 2 cities over 5 days

September	1	2	3	4	5
City A	54	61	70	49	56
City B	62	56	60	67	65

Even without calculating, you can see that City A has the greater spread in temperatures and, therefore, the greater standard deviation in high temperatures. If you were to calculate the standard deviations for each city following the steps described above, you would find that the standard deviation in high temperatures for City A $= \sqrt{\frac{254}{5}} \approx 7.1$, while the same for City B $= \sqrt{\frac{74}{5}} \approx 3.8$.

Standard deviation is a critical measurement in normal distributions, sometimes referred to as "bell curves" because of their shape. In normal distributions, data points tend to cluster around the mean. At one standard deviation away from the mean, there are significantly fewer data points, and at two standard deviations away from the mean, there are fewer still. The GRE will likely refer to the standard deviation as d and the mean as m.

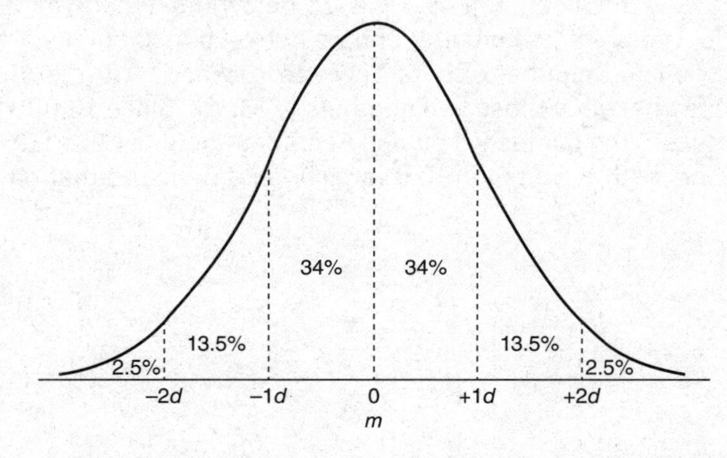

Any normal distribution is defined by its mean and standard deviation, so knowing good estimates for these values is critical to solving problems involving normal distribution. The mean is at the center and highest point on the curve, and because the distribution is symmetrical about the mean, the mean is also the median and mode. The standard deviation, since it is a measure of dispersion, determines the width of the normal distribution: the greater the standard deviation, the wider and flatter the curve. There are many real-world examples of variables that can be approximated by normal distribution: measurement errors, heights of homogeneous populations, random short term movements of stock prices, and standardized test scores.

Some problems that you encounter may deal with areas beneath a normal distribution curve, which are often expressed as percentages of the total area. The values shown in the diagram above for the various key areas of the curve are approximate, but they are sufficiently accurate for the questions you are likely to see on the GRE. To find the approximate number of data points that fall within one standard deviation above the mean, for instance, you would multiply the total number of data points by 34%. Moreover, note that the percentages of the area beneath the curve equate to probabilities. For example, the probability of a data point falling within one standard deviation above the mean is 34%. The actual formula for computing a normal distribution will not be tested, but you should familiarize yourself with the key values shown in the figure. Use those values to work through this example:

Example: A food manufacturer produces energy bars that have a mean weight of 50.0 grams. All bars are weighed before they are wrapped and shipped, and 68 percent of the bars weigh between 49.5 and 50.5 grams. If a given day's production is 10,000 energy bars, how many of those bars would be expected to weigh between 49.0 and 49.5 grams? (Assume that the weights are normally distributed.)

The problem provides the mean, 50 g, and states that 68% of the bars are between 49.5 and 50.5 g. Since 68% corresponds to the area represented by $\pm 1d$, the standard deviation is 0.5 g. So 49.0 g is $2d$ below the mean and 49.5 g is $1d$ below the mean. The question asks for the predicted number of bars between 49.0 and 49.5 g, that is, the predicted number of bars between $-2d$ and $-1d$ from the mean. The area between those two numbers is 13.5%. Since 10,000 bars are produced, the company could expect that about $0.135 \times 10{,}000 = 1{,}350$ bars with a weight in that range would be made that day.

DESCRIPTIVE STATISTICS EXERCISES

BASIC

For Exercises 1–4, use the following set of data:

8, 11, 12, 36, 45, 21, 9, 8, 7, 4

1. Find the average (arithmetic mean).
2. Find the median.
3. Find the mode.
4. Find the range.
5. Find the median for the following data set, which shows the scores of 11 students in Ms. Evans's class: 85, 78, 95, 92, 74, 98, 64, 99, 72, 88, 95.
6. The following table gives the ages of students in a class at a high school. Find the mean.

Age	14	15	16	17	18
Number of students	4	8	7	1	2

7. Of the following data set, which number, if any, is an outlier: 81, 73, 85, 79, 90, 155, 76, 89?
8. The sum of 30 measurements of data is 1,680. What is the average (arithmetic mean) of these measurements?
9. What is the third quartile of the following data set: 63, 68, 70, 63, 82, 58, 44?
10. Javier's test scores are 78, 89, 44, 98, 85, 72, 71, and 92. What is the range of his scores?

INTERMEDIATE

11. The average weight of 10 packages increases by 2 ounces when the weight of a book is added to one of them. If the average weight of the packages, after including the weight of the book, is 34 ounces, what is the weight of the book in ounces?

12. Bianca's average wage for 11 days was $110 per day. During the first 5 days, her average wage was $90 per day, and her average wage during the last 5 days was $120 per day. What was her wage on the sixth day?

13. The average of five numbers is 8. The average of three of the numbers is 7. What is the average of the remaining two numbers?

For Exercises 14–16, use the following information:

The numbers of books on each of 7 shelves were 13, 5, 11, 8, 19, 7, and 10.

14. Find the mean, median, mode, range, and interquartile range of the numbers.

15. If each bookshelf had 3 times as many books, what would be the mean, median, range, and interquartile range?

16. If each shelf had 3 fewer books than the initial numbers, what would be the interquartile range?

17. The daily attendance of an art exhibit for the past 8 days was 90, 95, 77, 78, 87, 81, 85, and 93. What are the mean, median, and mode of these attendance numbers?

For Exercises 18 and 19, use the following information:

The daily temperatures, in degrees Fahrenheit, for 9 days in June were 68, 73, 75, 76, 78, 81, 72, 83, and 81.

18. Find the mean, median, mode, and range of the temperatures.

19. If each day had been 6 degrees warmer, what would have been the mean, median, mode, and range of those 9 temperatures?

20. For one week, eight students recorded the numbers of hours they each studied. The mean of their data is 6 hours. If an outlier data value of 18 is removed from the set, what is the mean of the new data set, to the nearest hundredth?

ADVANCED

For Exercises 21 and 22 use the following information:

Test Scores							
Class A	89	92	78	96	86	80	90
Class B	98	86	70	82	92	89	88

21. Which is greater, the mean for Class A or the median of Class B?

22. If Class C has a mean of 86.5, order the classes' means from greatest to least.

23. For five consecutive days, the number of hours a student spends on schoolwork per day is 6, 8, 10, 6, and 7. Find the range, first quartile, and third quartile.

24. If a student's scores for the first four tests are 77, 98, 82, and 93, what is the minimum possible score on the fifth test that will allow his overall mean to be at least 90?

25. If the mean of numbers 18, x, 45, 79, and 98 is 56, then what is the mean of 148, 155, 232, 99, and x?

26. The sale prices of 8 houses are listed below. What is the range of the prices of the 8 houses?

$399,900	$289,500	$650,000	$399,900
$218,000	$250,000	$241,850	$189,000

For Exercises 27 and 28, use the following information:

The temperatures for 365 days were recorded and summarized in the following box plot.

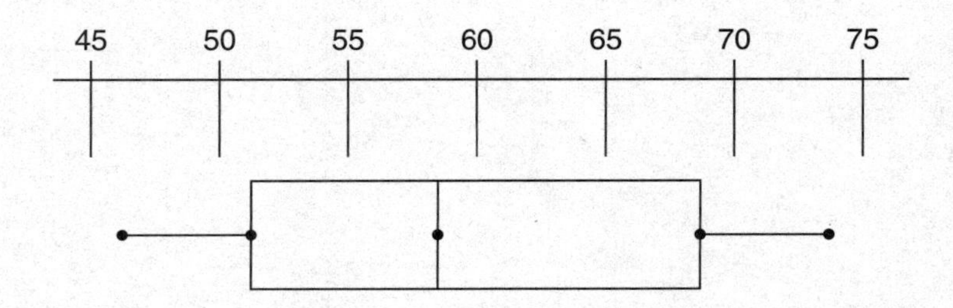

27. What are the range, the three quartiles, and the interquartile range of the measurements?

28. If the 80th percentile of the measurements is 72°, about how many measurements are between 69° and 72°? (Round your answer to the nearest whole number.)

29. A group of 40 data points has a mean of 34.7. Another group of 50 points has a mean of 42.1. What is the mean of the 90 points together, to the nearest tenth?

30. Fly With Us Airline keeps track of the deviations from scheduled departure times. If a flight leaves early, the deviation is negative. If a flight is late, the deviation is positive. Each employee gets a $2,000 bonus for every whole number below zero for the mean *and* the median of the deviation data. Given the following deviation values, would a bonus be given, and if so, what would it be per employee? {–3, 2, 0, 0, –4, 8, –6, –2, –5, 0}

DESCRIPTIVE STATISTICS ANSWERS KEY

Basic

1. 16.1

$$\frac{8 + 11 + 12 + 36 + 45 + 21 + 9 + 8 + 7 + 4}{10} = \frac{161}{10} = 16.1$$

2. 10

Arrange the numbers in ascending order: 4, 7, 8, 8, 9, 11, 12, 21, 36, 45. Find the average of the two middle numbers. $\frac{9 + 11}{2} = 10$.

3. 8

The mode is the number that occurs the most frequently. In this case, that is 8.

4. 41

The range is the largest number minus the smallest number: $45 - 4 = 41$.

5. 88

Place the data in ascending order: 64, 72, 74, 78, 85, **88**, 92, 95, 95, 98, 99. The median is the middle number, which here is 88.

6. 15.5

There are four 14s, eight 15s, seven 16s, one 17, and two 18s in the set. Instead of writing out all of the individual numbers, you can evaluate the following:

$4(14) + 8(15) + 7(16) + 1(17) + 2(18) = 341$

There are 22 numbers, so the mean is $\frac{341}{22}$, which equals 15.5.

The mean age is 15.5.

7. 155

The number 155 is much higher than the other numbers in the data set; it is an outlier.

8. 56

$$Mean = \frac{\textit{Sum of values}}{\textit{Number of values}} = \frac{1{,}680}{30} = 56$$

The mean is 56.

9. 70

Put the numbers in ascending order. Find the median, or second quartile: 44, 58, 63, **63**, 68, 70, 82. Q_2 is 63. The third quartile is the median of all the numbers above Q_2. The third quartile is 70.

10. 54

The range is the positive difference between the highest and the lowest numbers in a set. The highest score is 98, and the lowest score is 44, so the range is $98 - 44$, which equals 54.

INTERMEDIATE

11. 20

The new average weight of the packages after including the book is 34 ounces. The new average is 2 ounces more than the old average. The old average without including the book is 32 ounces. The total weight of the 10 packages, without including the weight of the book, is $10 \times 32 = 320$. After adding the book to one package, the average weight of the packages increases to 34 ounces. The total weight of the packages after including the weight of the book is $10 \times 34 = 340$ ounces. So, the book weighs $340 - 320 = 20$ ounces.

Another way to think of it is this: If the average is increased by 2 when the book is added and there are 10 packages total, the weight of the book is $2 \times 10 = 20$.

12. $160

Total wages $= 11 \times \$110 = \$1{,}210$.

The total wages for the first 5 days $= 5 \times 90 = \$450$.

The total wages for the last 5 days $= 5 \times 120 = \$600$.

Total wages earned during the 11 days = wages during first 5 days + wage on 6th day + wages during the last 5 days.

1,210 = 450 + wage on 6th day + 600.

Wage on 6th day $= 1210 - 450 - 600 = \$160$.

13. 9.5

The average of 5 quantities is 8. The sum of the 5 quantities is $5 \times 8 = 40$. The average of 3 of these 5 quantities is 7. The sum of these 3 quantities $= 3 \times 7 = 21$. The sum of the remaining two quantities $= 40 - 21 = 19$.

$$\text{Average of these two quantities} = \frac{19}{2} = 9.5.$$

14. Mean: 10.4286
Median: 10
Mode: none
Range: 14
Interquartile range: 6

Put the data values in order: 5, 7, 8, 10, 11, 13, 19.

$$\textit{Mean} = \frac{\textit{Sum of values}}{\textit{Number of values}} = \frac{5 + 7 + 8 + 10 + 11 + 13 + 19}{7} = \frac{73}{7} \approx 10.4286$$

There is an odd number of data values, so the median is the middle value when the values are arranged in numerical order: 5, 7, 8, **10**, 11, 13, 19. The median is 10. The range is the difference between the greatest number and the least number: $19 - 5 = 14$.

The interquartile range is the difference between Q_3 and Q_1: $13 - 7 = 6$.

15. Mean: 31.2857
Median: 30
Range: 42
Interquartile range: 18

New numbers with 3 times as many books: 39, 15, 33, 24, 57, 21, 30

Numbers in order: 15, 21, 24, 30, 33, 39, 57

$$Mean = \frac{Sum\ of\ values}{Number\ of\ values} = \frac{15 + 21 + 24 + 30 + 33 + 39 + 57}{7} = \frac{219}{7} \approx 31.2857$$

There is an odd number of data values, so the median is the middle value when the values are arranged in numerical order: 15, 21, 24, **30**, 33, 39, 57. The median is 30.

The range is the difference between the greatest number and the least number: $57 - 15 = 42$.

The interquartile range is the difference between Q_3 and Q_1: $39 - 21 = 18$.

16. 6
New numbers with 3 fewer books on each shelf: 10, 2, 8, 5, 16, 4, 7

Numbers in order: 2, 4, 5, 7, 8, 10, 16

Interquartile range: $Q_3 - Q_1 = 10 - 4 = 6$

Note that the interquartile range will stay the same if the same value is added to or subtracted from all of the data points in the set because the distance between the data points does not change.

17. Mean: 85.75
Median: 86
Mode: none

Numbers in order: 77, 78, 81, 85, 87, 90, 93, 95

$$Mean = \frac{Sum\ of\ values}{Number\ of\ values} = \frac{77 + 78 + 81 + 85 + 87 + 90 + 93 + 95}{8} = \frac{686}{8} = 85.75$$

There is an even number of data values, so the median is the average of the middle two values when they are arranged in numerical order: 77, 78, 81, **85**, **87**, 90, 93, 95. Then $(85 + 87) \div 2 = 86$. The median is 86.

All of the numbers appear with equal frequency, so there is no mode for this data set.

18. Mean: $76.3\overline{3}$
Median: 76
Mode: 81
Range: 15

Numbers in order: 68, 72, 73, 75, 76, 78, 81, 81, 83

$$Mean = \frac{Sum\ of\ values}{Number\ of\ values} = \frac{68 + 72 + 73 + 75 + 76 + 78 + 81 + 81 + 83}{9} = \frac{687}{9} = 76.3\overline{3}$$

There is an odd number of data values, so the median is the middle value when the values are arranged in numerical order: 68, 72, 73, 75, **76**, 78, 81, 81, 83. The median is 76. The number 81 occurs twice, which is more frequently than any other number in the data set, so the mode is 81.

The range is the difference between the greatest number and the least number: 83 − 68 = 15.

19. Mean: $82.3\overline{3}$
Median: 82
Mode: 87
Range: 15

Note that because the entire data set has essentially been shifted 6 places to the right, the mean, median, and mode will all increase by 6, and the range will not change.

New numbers at 6 degrees warmer in order: 74, 78, 79, 81, 82, 84, 87, 87, 89

$$Mean = \frac{Sum\ of\ values}{Number\ of\ values} = \frac{74 + 78 + 79 + 81 + 82 + 84 + 87 + 87 + 89}{9} = \frac{741}{9} = 82.3\overline{3}$$

There is an odd number of data values, so the median is the middle value when the values are arranged in numerical order: 74, 78, 79, 81, **82**, 84, 87, 87, 89. The median is 82.

The number 87 occurs twice, which is more frequently than any other number in the data set, so the mode is 87.

The range is the difference between the greatest number and the least number: 89 − 74 = 15.

20. 4.29

To find the new mean, you must first determine what the sum of the original data values was based on the information given.

$$Original\ mean = \frac{Sum\ of\ values}{Number\ of\ values} = \frac{?}{8} = 6$$

$$Sum\ of\ values = 6 \times 8 = 48$$

The original sum of values was 48. Now, you know that a data value has been removed (18), so reduce the sum of values by 18 and the number of values by 1.

$$New\ mean = \frac{Sum\ of\ values}{Number\ of\ values} = \frac{48 - 18}{8 - 1} = \frac{30}{7} \approx 4.29$$

ADVANCED

21. Median of Class B

The mean of Class A is

$$\frac{\textit{Sum of values}}{\textit{Number of values}} = \frac{89 + 92 + 78 + 96 + 86 + 80 + 90}{7} = \frac{611}{7} \approx 87.29$$

There is an odd number of data values, so the median of Class B is the middle value when the values are arranged in numerical order: 70, 82, 86, **88**, 89, 92, 98. The median of Class B is 88. So the median of Class B is greater.

22. Class A, Class C, Class B

The mean of Class C is 86.5. The mean of Class A to the nearest hundredth is 87.29. The mean of Class B is

$$\frac{98 + 86 + 70 + 82 + 92 + 89 + 88}{7} = \frac{605}{7} \approx 86.43.$$

So, the mean of Class A is greatest, and then Class C, and finally Class B.

23. Range: 4
Q_1: 6
Q_2: 9

Numbers in order: 6, 6, 7, 8, 10

Range = highest number − lowest number = 10 − 6 = 4

The median of the data is 7, so the first quartile is 6 and the third quartile is 9.

24. 100

$$\begin{aligned}
90 &= \frac{77 + 98 + 82 + 93 + x}{5} \\
90 &= \frac{350 + x}{5} \\
(5)90 &= \frac{350 + x}{5}(5) \\
450 &= 350 + x \\
450 - 350 &= 350 - 350 + x \\
100 &= x
\end{aligned}$$

The student needs a score of 100 on the fifth test for his overall mean to be at least a 90.

25. 134.8

The average (arithmetic mean) of the 5 numbers 18, x, 45, 79, and 98 is 56.

Therefore, the sum of these 5 numbers is $56 \times 5 = 280$.
$18 + 45 + 79 + 98 = 240$
So, $x = 280 - 240 = 40$.

The average of 148, 155, 232, 99, and x is $\frac{148 + 155 + 232 + 99 + 40}{5} = \frac{674}{5} = 134.8$.

26. 461,000

The range is the positive difference between the highest and the lowest number in a set. The highest value is 650,000, and the lowest value is 189,000. To find the range, subtract the lowest from the highest.

$650{,}000 - 189{,}000 = 461{,}000$

27. Range: 74 – 46 = 28
Q_1: 52
Q_2: 59
Q_3: 69
Interquartile range: 69 – 52 = 17

28. 18

Because 69° represents the 75th percentile, from 69° to 72° is 5% of the measurements. Then $5\% \times 365 = 18.25$, which rounds to 18.

29. 38.8

The mean of the 90 data points together is $\frac{40(34.7) + 50(42.1)}{90} = \frac{3493}{90} = 38.8\bar{1}$.

30. Yes; $4,000

Data values in order: −6, −5, −4, −3, −2, 0, 0, 0, 2, 8

The mean of the data is

$$\frac{(-6) + (-5) + (-4) + (-3) + (-2) + 0 + 0 + 0 + 2 + 8}{10} = \frac{-10}{10} = -1.$$

The median is also −1. This means the employees get $2,000 for being 1 below 0 for the mean and an additional $2,000 for the median being 1 below 0. So, each employee would receive a $4,000 bonus.

DATA GRAPHS AND TABLES

There are several different kinds of charts and graphs that can show up on the test. The most common kinds are tables, bar graphs, line graphs, and pie charts.

Tables

Tables share many of the characteristics of graphs, except for the visual advantages—you can estimate values from a graph, but not from a table. Tables can be enormous and complicated, but the basic structure is always the same: columns and rows. Here's an example of a simple one (much simpler than those that appear on the GRE):

JOHN'S INCOME, 2004–2008	
YEAR	INCOME
2004	$20,000
2005	$22,000
2006	$18,000
2007	$15,000
2008	$28,080

An easy question on this table might ask for the sum of the three lowest incomes. In this case, you would simply look up the amounts and then add. A harder question might ask for John's average income per year over the five given years; then you would have to add up the five incomes, and divide to find the average.

Bar Graphs

These can be used to display the information that would otherwise appear in a table. On a bar graph, the *height* of each column shows its value. Here's the information from the table above presented as a bar graph:

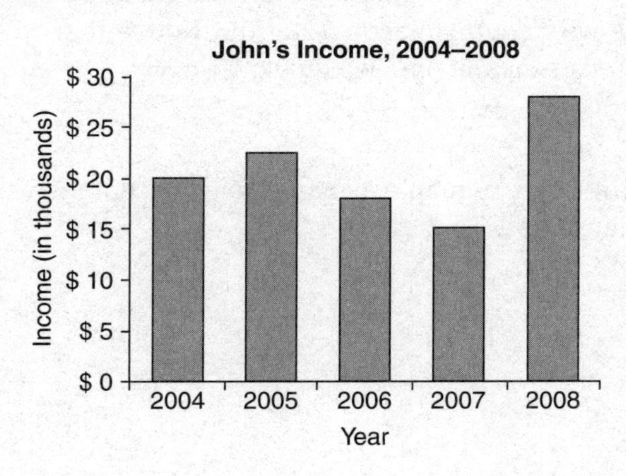

What's handy about a bar graph is that you can see the relative values by looking at their heights. By glancing at the graph shown, for example, it's easy to see that John's income in 2008 was almost double his income in 2007. But you can see this only because the scale starts at zero; the scale could just as easily start somewhere *other* than zero.

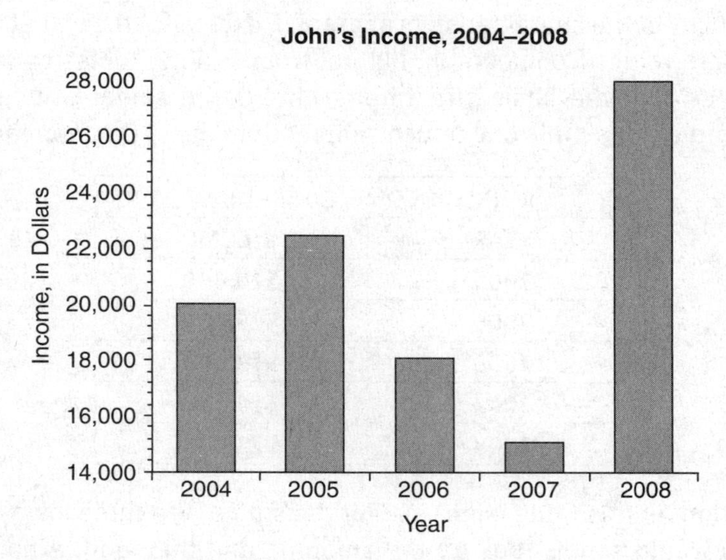

This graph presents the same information, but now you cannot estimate as simply as you did before. You can still tell at a glance that his 2008 income exceeded 2007, but you cannot quickly estimate the ratio.

In order to find some numerical value from a bar graph, find the correct bar (such as 2008 income) and move horizontally across from the top of the bar to the value on the scale on the left or, sometimes, on the right. Don't worry about getting too precise a value; usually, a close approximation will be good enough. For example, John's 2008 income was approximately $28,000. Notice that this is different from the table, where John's *exact* income was given.

Line Graphs

Line graphs follow the same principle as bar graphs, except the values are presented as points, rather than bars.

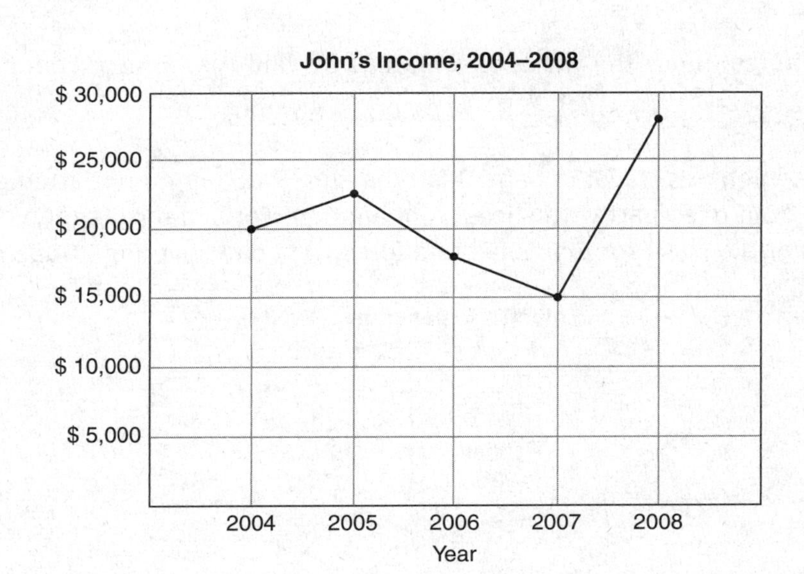

As with bar graphs, the value of a particular year is the vertical distance from the bottom of the graph to the line. And also as with bar graphs, you can see the relative value of the described amounts by looking at their heights—with the caution we mentioned before, that the base must be zero in order to estimate ratios.

Pie Charts

A pie chart shows how things are distributed; the fraction of a circle occupied by each piece of the "pie" indicates what fraction of the whole it represents. Usually, the pie chart will identify what percent of the whole each piece represents, with the whole being 100 percent.

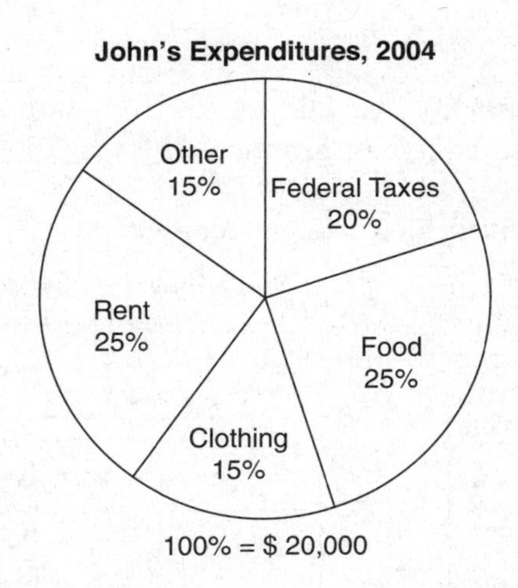

The total size of the whole pie is usually given as "TOTAL = $3,547" or "100% = $5.9 billion" or something of that nature. If we were asked to find the approximate value of a particular piece of the pie, we would multiply the appropriate percent by the whole. For instance, to find the amount John paid in federal taxes in 2004, we find the slice labeled "Federal Taxes," and we see that federal taxes represented 20% of his expenditures. Since the whole is \$20,000, we find that John's federal taxes for 2004 were 20% of \$20,000 or $\frac{1}{5} \times \$20{,}000 = \$4{,}000$.

Pie charts sometimes travel in pairs. If so, be sure that you do not attempt to compare slices from one chart with slices from another. For instance, suppose we were given another pie chart for John's expenditures, this one covering 2008.

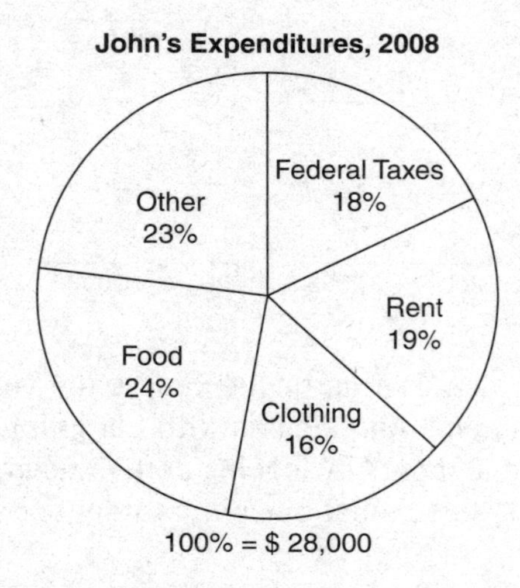

A careless glance might suggest that John paid less in federal taxes in 2008 than in 2004. Not true! His 2004 taxes were a greater percentage of his income than were his 2008 taxes, but his 2008 income was much greater than his 2004 income. In fact, he paid about \$1,000 *more* in taxes in 2008 than in 2004. Since the totals for the two charts are different, the pieces of the pie are not directly comparable.

Double Graphs

Very often the GRE will present two graphs for the same set of questions, or one graph and a table. The two charts or graphs will be related in some way.

Stuck Elevators in Country X

Year	Stuck Elevators
2009	432
2010	459
2011	621
2012	645

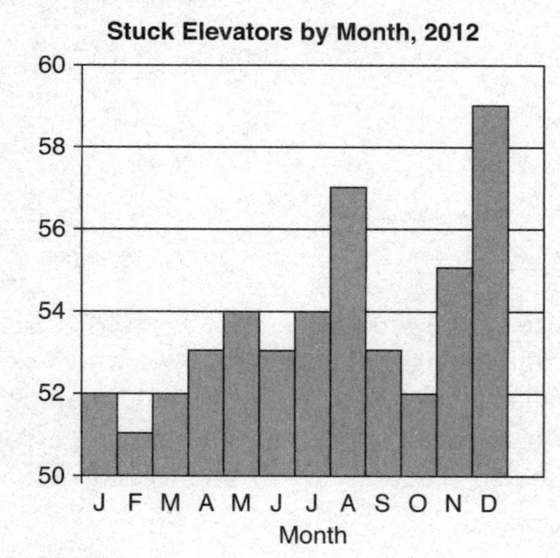

Here, the table covers stuck elevators for four years, while the accompanying bar graph breaks the information down by month for just one year, 2012.

What can be more complex is when the testmaker gives you two graphs (either two line graphs or two bar graphs) occupying the same space. Sometimes both graphs will refer to the same vertical scale; other times, one graph will refer to a scale on the left, the other graph to a scale on the right.

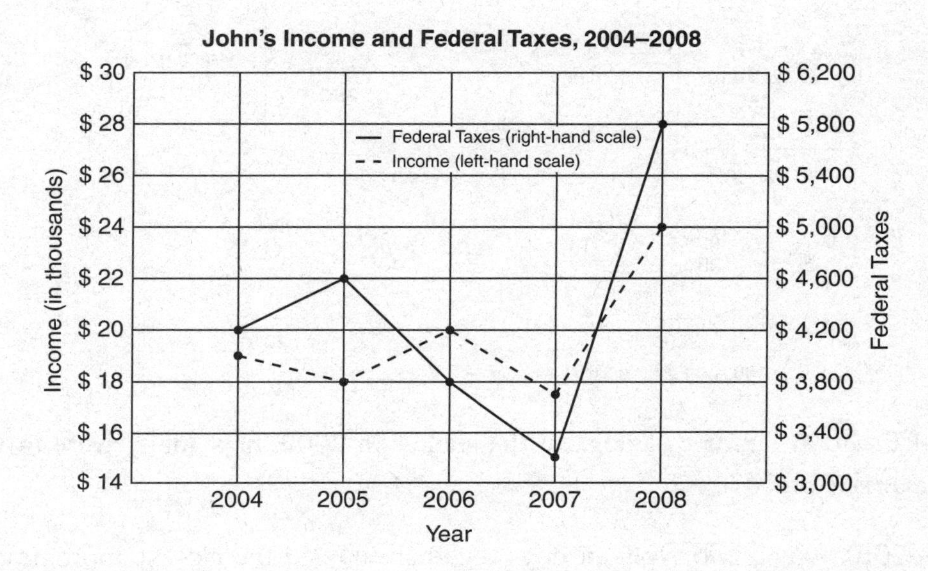

Here is the same graph of John's income, but with new information added. You now see at a glance not only John's income for a given year, but also the amount of federal taxes for that year. The income refers to the left-hand scale; the taxes to the right-hand scale. At this point, the number of potential questions that may be asked of the test taker has risen dramatically.

Double graphs are not really any more difficult than single graphs as long as you don't mix up the scales. Learn to double-check that you're using the correct scale when working with double graphs. If you find yourself getting confused, slow down and give yourself a chance to sort things out.

DATA GRAPHS AND TABLES EXERCISES

BASIC

Use the following graphs to answer Exercises 1–2.

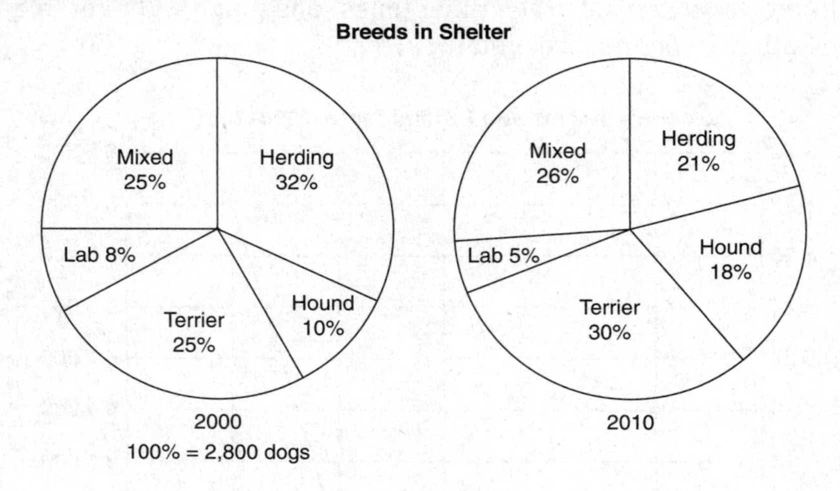

1. Of the total number of dogs in the shelter in 2000, how many were mixed or terrier breed?

2. In 2010, which two types of dogs together equaled the closest approximation of half of the total number of dogs that entered the shelter in 2010?

Use the following graph to answer Exercises 3–4.

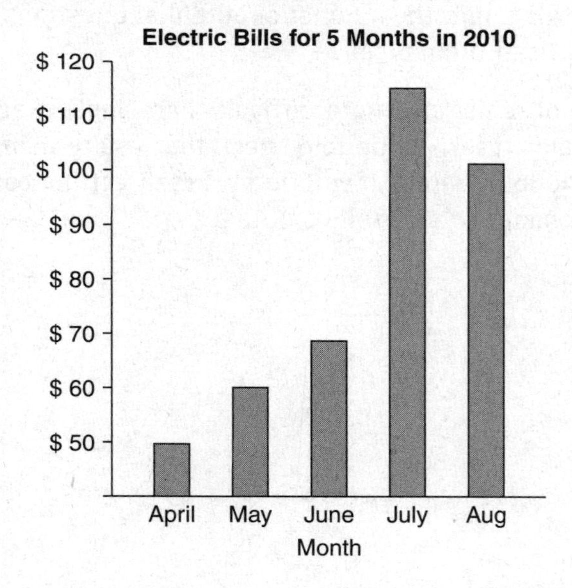

3. If the electric bill in September is half of the bill in August, will it be more or less than the bill in May?

4. How much greater was the July bill than the May bill?

Use the following graph to answer Exercises 5–6.

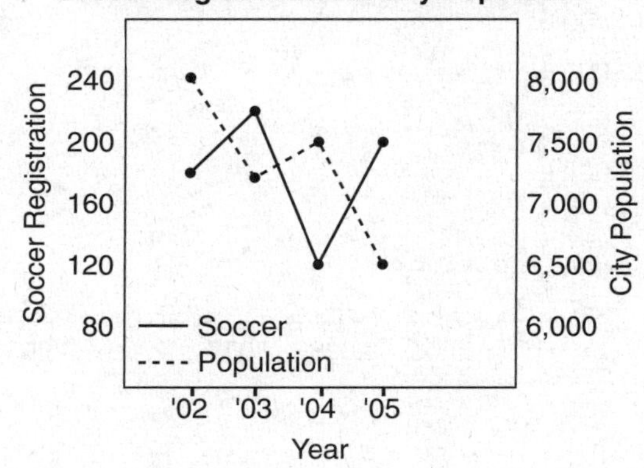

5. Did the number of soccer registrations increase or decrease from 2003 to 2004?

6. What was the approximate change in the city population from 2002 to 2004?

Use the following graph to answer Exercises 7–8.

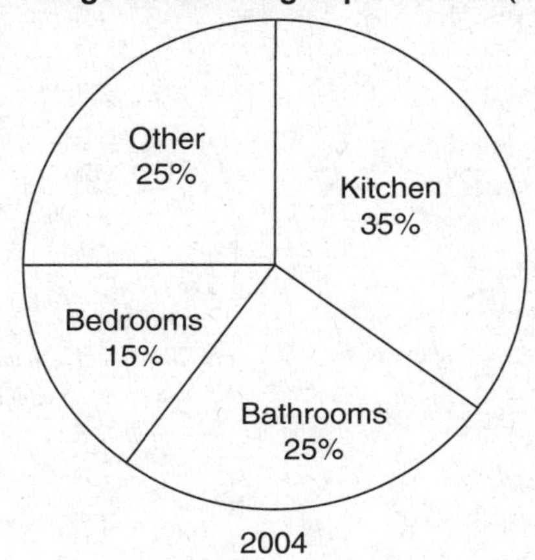

7. How much greater, as a percent of total expenditures, was the percent spent on remodeling kitchens than the percent spent on bedrooms?

8. If an average homeowner spends a total of $10,000 on remodeling and allocates the money according to the expenditure chart shown, how much was spent remodeling the bathrooms?

Use the following graph to answer Exercises 9–10.

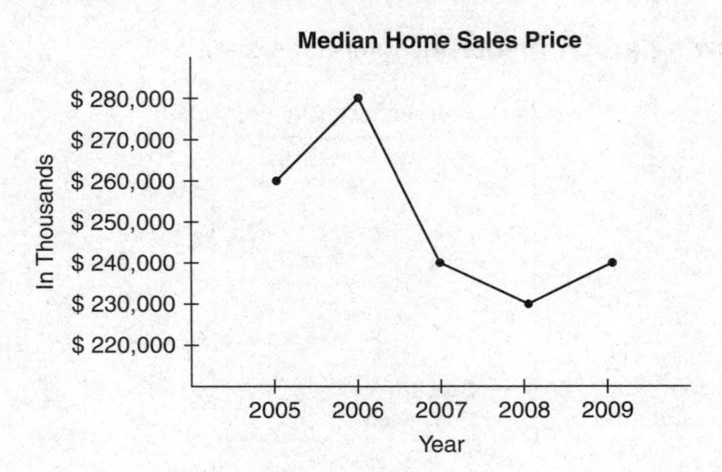

9. What was the difference in the median sales price between 2006 and 2007?

10. What was the percent increase in median sales prices from 2008 to 2009?

INTERMEDIATE

Use the following graphs to answer Exercises 11–12.

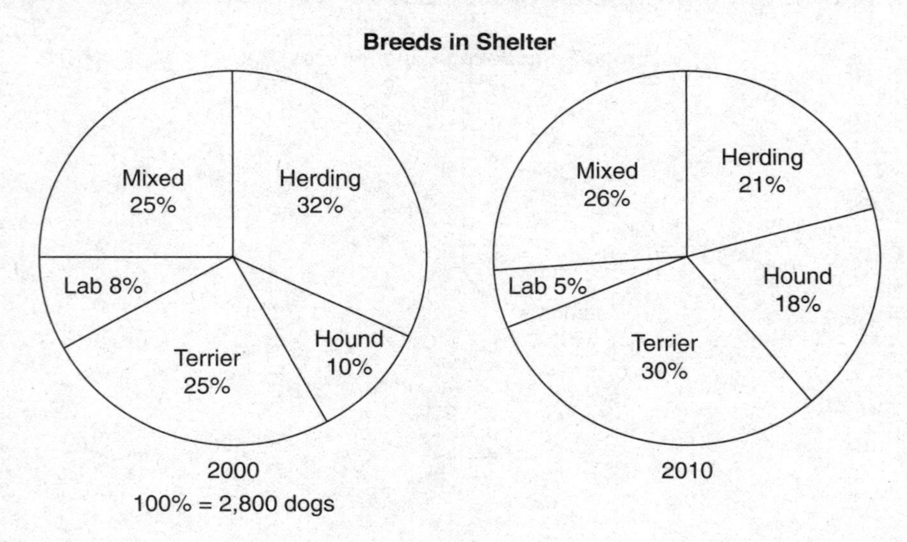

11. In 2000, how much greater of a percentage did labs and terriers represent than herding breeds? Express your answer as a percent of the total number of dogs in the shelter that year.

12. In 2010, if there were 850 herding dogs in the shelter, was this more or less than the number of herding dogs that entered the shelter in 2000?

Use the following graph to answer Exercises 13–14.

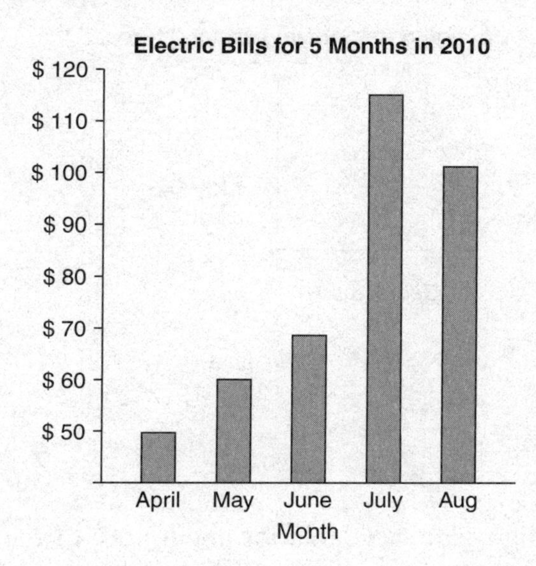

13. The bills shown represent 65% of the customer's total cost of electricity for the year. What was the total cost for the year, to the nearest dollar?

14. What is the percentage change of the cost of the electric bills from June to July?

Use the following graph to answer Exercises 15–16.

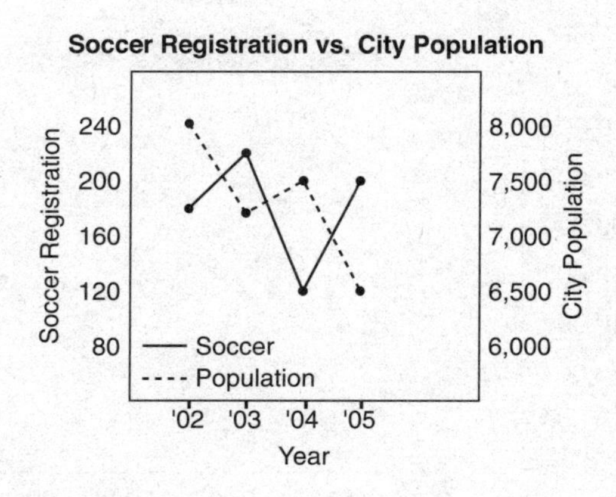

15. Between which two years did the greatest change in soccer registrations occur?

16. If the city population grew in 2006 and the yearly relationship between soccer registration and city population remained the same as it had been since 2002, would the number of soccer registrations in 2006 likely be greater than or less than it was in 2005?

Use the following graph to answer Exercises 17–18.

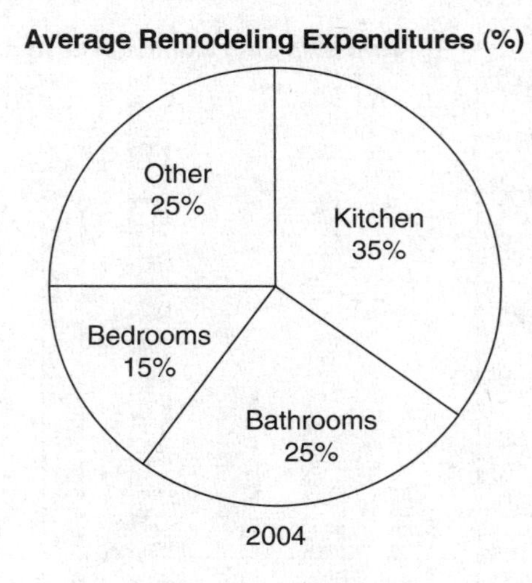

17. Remodeling media rooms accounts for about 40% of the "Other" section, accounting for $500,000 in expenditures. What is the total amount spent on remodeling "Other" rooms?

18. In a given year, if 12% of the total expenditures allocated to kitchen remodeling in 2004 were reallocated evenly across the remaining three categories, what would be the new percentage amounts for each category?

Use the following graph to answer Exercises 19–20.

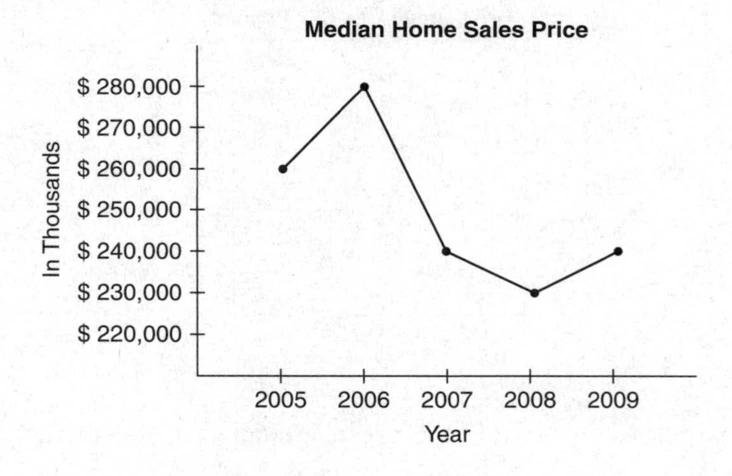

19. What is the percent decrease, to the nearest tenth, of the median home sales price from 2007 to 2008?

20. If the same price increase from 2008 to 2009 happened between 2009 and 2010, what was the median sales price in 2010?

ADVANCED

Use the following graph to answer Exercises 21–22.

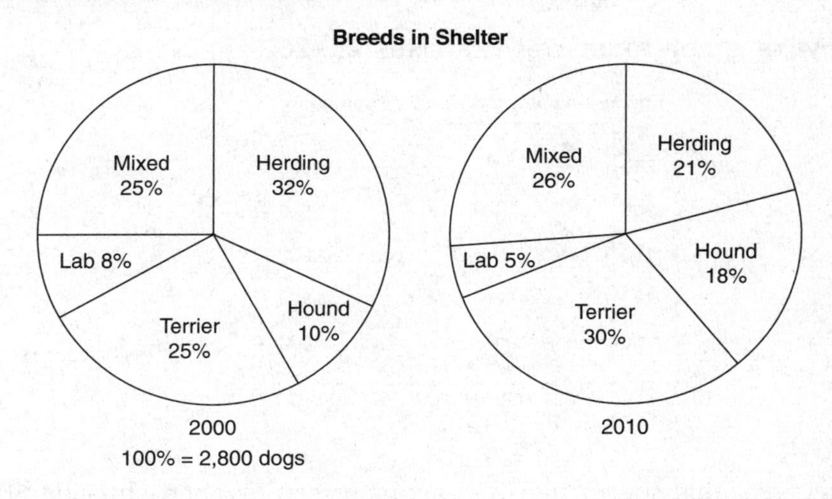

21. In 2010, if there were 850 herding dogs in the shelter, what was the total number of hound breed dogs that entered the shelter in 2000 and 2010 combined?

22. In 2010, there were 850 herding dogs in the shelter. If 40% of the mixed-breed dogs that entered the shelter in 2010 were smaller than 15 pounds, how many mixed-breed dogs were over 15 pounds?

Use the following graph to answer Exercises 23–24.

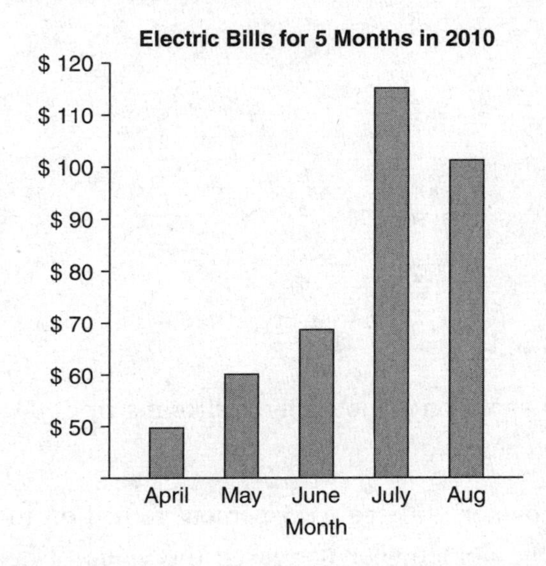

23. After the July electric bill, the homeowner increased the average temperature for the air conditioner by 2 degrees from 76 to 78 degrees. His electric company claims that every 2-degree increase will decrease the electric bill by 4%. Assuming the electric company's claim is correct, what would the electric bill in August have been without this change (to the nearest dollar)?

24. Some electric companies offer a billing option where you pay an average monthly bill. If this customer were paying an average of the five months listed, what would each month's bill have been?

Use the following graph to answer Exercises 25–26.

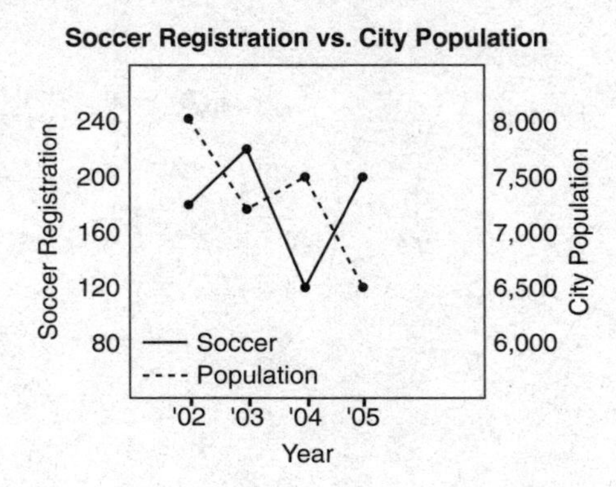

25. During the year that soccer registrations increased by approximately 80, what was the concomitant change in population?

26. During the only year in which the city population increased, what was the percent decrease in soccer registrations, rounded to the nearest whole number?

Use the following graph to answer Exercises 27–28.

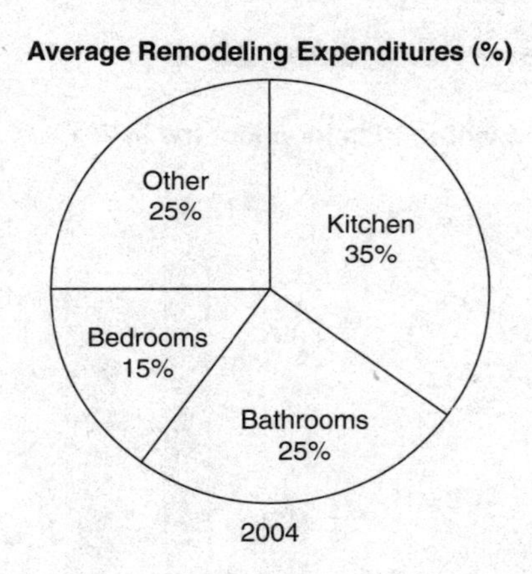

27. If $720 million was spent on remodeling bedrooms in 2004, how much was spent remodeling kitchens?

28. On average, a homeowner will see a 75 percent return on investment for kitchen remodels. One homeowner increased the value of her home by $6,000 by remodeling her kitchen. Assuming that she gets the same percent return for bathroom remodels and that her spending follows 2004 averages, how much more would her home increase in value (to the nearest dollar) if she also remodeled the bathrooms?

Use the following graph to answer Exercises 29–30.

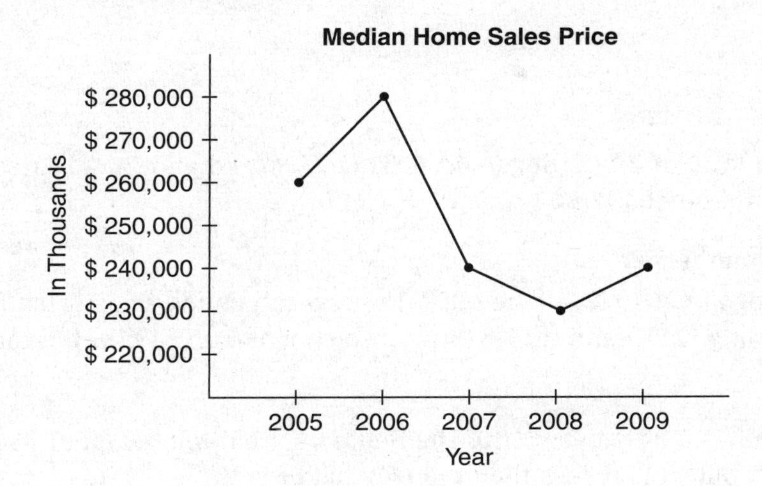

29. In 2009, a homeowner put his house on the market. If the average drop from listing price to sales price is 6%, what was the minimum price (in whole dollars) that the house could have been listed at so that it sold at the price greater than or equal to the median price?

30. A real estate agent says the percent change between 2007 and 2008 is the same as between 2008 and 2009. Is this true or false?

DATA GRAPHS AND TABLES ANSWERS KEY

Basic

1. 1,400

There were a total of 2,800 dogs and 50% (25% mixed and 25% terrier) belong to the two desired breeds. $2,800 \times 0.50 = 1,400$.

2. Herding and terrier

Half of the dogs in 2010 would be 50%. The two categories whose sum is closest to 50% are herding (21%) and terrier (30%), which make up 51% of the total.

3. Less

The August bill is a little over $100. The September bill will be about $50 if it is half of the August bill. This is less than the May bill of $60.

4. $55

To find how much greater the July bill was, subtract: $\$115 - \$60 = \$55$.

5. Decrease

The solid line represents soccer registrations, and it decreased from 2003 to 2004.

6. Decrease of 500

The population in 2002 was about 8,000, and the population in 2004 was about 7,500. Subtracting, we find the change in population was a decrease of 500.

7. 20%

Kitchen remodels accounted for 35% of the total money spent, while bedrooms accounted for 15%. Subtract to find how much more money was spent remodeling kitchens: $35\% - 15\% = 20\%$.

8. $2,500

Since the graph shows the average way money was spent, we can find the average amount spent on bathroom remodeling by multiplying the total spent by the percent typically spent on bathrooms: $\$10,000 \times 25\% = \$2,500$.

9. $40,000 price drop

The median sales price changed from $280,000 in 2006 to $240,000 in 2007, so there was a $40,000 drop.

10. 4.3%

The percent increase can be found using the formula $\frac{\textit{Amount of increase}}{\textit{Original whole}}$; substitute the known values and solve: $\frac{(240 - 230)}{230}$ or 4.3%.

Intermediate

11. 1%

The labs and terriers represented 33% (25% + 8%), and the herding breeds represented 32% of the total. Subtract: 33% − 32% = 1%. Of the total, 1 percent more labs and terriers entered the shelter in 2000 than herding breeds.

12. Less

In 2000, a total of 2,800 dogs were admitted to the shelter. Of that number, 32% were herding breed dogs: 2,800 × 0.32 = 896. The 850 herding breed dogs admitted in 2010 were less than the 896 admitted in 2000.

13. $603

You know that the total amount of the bills for April through August represents 65% of the total electricity costs for the year. Then 48 + 60 + 67 + 115 + 102 = 392, and $392 is equal to 65% of the total. Divide $392 by 65% to find the total cost for the year: $392 ÷ 0.65 ≈ $603.

14. 72%

To find percent change, use the formula $\frac{\textit{Amount of change}}{\textit{Original whole}}$.

Evaluating, this gives $\frac{(115-67)}{67} = \frac{48}{67} \approx 0.72$

15. 2003 and 2004

The greatest change in the solid line (soccer registrations) happened between 2003 and 2004.

16. Less

Given the relationship between population and soccer registration shown on the graph, if the population increases in 2006, the number of soccer registrations would likely have decreased.

17. $1,250,000

If $500,000 is 40% of the total "Other" remodeling, then dividing $500,000 by 40% will give the total for the "Other" section: $500,000 ÷ 0.40 = $1,250,000.

18. Kitchens: 23%; Bathrooms: 29%; Bedrooms: 19%; Other: 29%

Subtract 12 percent from Kitchens to get 23 percent. Then, divide the 12 percent evenly among the other three categories, which means adding 4 percent to each category's percentage.

19. 4.2%

In 2007, the median sales price was $240,000, and in 2008, the median sales price was $230,000. The difference in price is $10,000, and the original price is $240,000. To find the percent change, take the change in price, divide by the original, and multiply by 100%:

$$\frac{10{,}000}{240{,}000} \times 100\% = \frac{1}{24} \times 100\% = \frac{100}{24}\% \approx 4.2\%$$

20. \$250,000

From 2008 to 2009, the median sales price increased \$10,000. If this same increase occurs from 2009 to 2010, then the median sales price will be \$240,000 + \$10,000, or \$250,000.

ADVANCED

21. 1,009

First, you need to find the total number of dogs that entered the shelter in 2010: $850 \div 0.21 \approx 4{,}048$. The hound breed made up 18% of that number: $4{,}048 \times 0.18 \approx 729$. In 2000, you know that 10% of the 2,800 dogs that entered were hound breed dogs: $2{,}800 \times 0.1 = 280$. Now add the two totals together: $729 + 280 = 1{,}009$ total hound breed dogs.

22. 631

First, you need to find the total number of dogs that entered the shelter in 2010: $850 \div 0.21 \approx 4{,}048$. Of that total, 26% were mixed breed dogs: $4{,}048 \times 0.26 = 1{,}052$. If 40% of those dogs weighed less than 15 pounds, then 60% of them weighed more than 15 pounds: $1{,}052 \times 0.60 \approx 631$.

23. \$106

The current August bill, \$102, is 4% less than what the bill would have been without the adjustment. So if the bill would have been x, then $x - 0.04x = 102$. Solving this equation for x yields $x = 106.25$, which is \$106 rounded to the nearest dollar.

24. \$78.40

The sum of the bills is approximately \$392. Divide by 5 months to get about \$78.40.

25. 1,000

There was about an 80-person increase in soccer registrations from 2004-2005. During that same year, the population decreased by about 1,000.

26. 45%.

City population increased between 2003 and 2004. During that same period, soccer registrations dropped from about 220 to about 120, for an actual decrease of $220 - 120 = 100$. Because 100 is a bit less than half of 220, that's a percent decrease of just under 50%. Apply the percent change formula to find the actual value:

$$\textit{Percent change} = \frac{\textit{change}}{\textit{original}} \times 100\%$$

$$\textit{Percent change} = \frac{100}{220} \times 100\% \approx 45\%$$

27. \$1.68 Billion

\$720 million was 15% of the total. To find the total, use the formula *Percent* × *Whole* = *Part.* 15% × *Whole* = \$720 million.

\$720 million ÷ 0.15 = \$4.8 billion. Now you need to find 35% of this total: 35% × \$4.8 billion = \$1.68 billion.

28. \$4,286

Use the formula *Percent* × *Whole* = *Part* to find the total investment in the kitchen. If \$6,000 is 75% of the total, substitute the known values and find the whole: 75% × *Whole* = \$6,000, so the total investment was \$8,000. According to the chart, the ratio of money spent on bathrooms to money spent on kitchens is 25:35, or 5:7. So the homeowner spent $\$8{,}000 \times \frac{5}{7} = \$5{,}714$. Multiply by .75 to get \$4,286.

It's not really necessary to calculate the amount of money the homeowner spent on bathroom remodeling, though. If returns are identical for different types of remodeling, then the returns will also be in a ratio of 5:7. Thus, $\$6{,}000 \times \frac{5}{7} = \$4{,}286$.

29. \$255,320

To find the list price, we need to find the price that when 6% of it is subtracted will equal \$240,000, the median price in 2009. Call the listing price *L*. Then 0.94*L* = \$240,000. Divide both sides by 0.94 to get a listing price of approximately \$255,319.15. Since the question asks for the minimum whole dollar amount for the asking price that, when reduced by 6%, will yield a value greater than or equal to \$240,000, the asking price must be rounded up to the nearest whole dollar, \$255,320.

30. False

Use the formula $\frac{\text{Amount of change}}{\text{Original amount}}$ to compare the changes.

Change from $2007 \text{ to } 2008: \frac{(230 - 240)}{240} = -4.17\%$.

Change from $2008 \text{ to } 2009: \frac{(240 - 230)}{230} = 4.35\%$. Because $4.35 \neq 4.17$, the statement is false. The percent increase from 2008 to 2009 is actually more than the percent decrease in the previous year.

You don't need to do all the calculations for this problem to see that the real estate agent's statement is false. The numerators in the percent change equations for both sets of years have the same absolute value, but the denominators are different, so you know right away that the percent change for the two sets of years cannot be the same.

PROBABILITY

Probability involves situations that have a finite number of outcomes.

$$\textit{Probability} = \frac{\textit{Number of desired outcomes}}{\textit{Number of total possible outcomes}}$$

Example: If you have 12 shirts in a drawer and 9 of them are white, the probability of picking a white shirt at random is $\frac{9}{12} = \frac{3}{4}$. The probability can also be expressed as 0.75 or 75%.

Many probability questions involve finding the probability of a certain outcome after multiple repetitions of the same experiment or different experiments (a coin being tossed several times, etc.). These questions come in two forms: those in which each individual event must occur a certain way, and those in which individual events can have different outcomes.

To determine multiple-event probability where each individual event must occur a certain way:

- Figure out the probability for each individual event.
- Multiply the individual probabilities.

Example: If 2 students are chosen at random from a class with 5 girls and 5 boys, what's the probability that both students chosen will be girls? The probability that the first student chosen will be a girl is $\frac{5}{10} = \frac{1}{2}$, and since there would be 4 girls left out of 9 students, the probability that the second student chosen will be a girl is $\frac{4}{9}$. So the probability that both students chosen will be girls is $\frac{1}{2} \times \frac{4}{9} = \frac{1}{\not{2}_1} \times \frac{\not{4}^2}{9} = \frac{2}{9}$.

To determine multiple-event probability where individual events can have different types of outcomes, find the *total number of possible outcomes*. Do that by determining the number of possible outcomes for each individual event and multiplying these numbers together. Find the *number of desired outcomes* by listing out the possibilities.

Example: If a fair coin is tossed 4 times, what is the probability that at least 3 of the 4 tosses will come up heads?

There are 2 possible outcomes for each toss, so after 4 tosses there is a total of $2 \times 2 \times 2 \times 2 = 16$ possible outcomes. List all the possibilities where at least 3 of the 4 tosses come up heads:

H, H, H, T	H, T, H, H	H, H, H, H
H, H, T, H	T, H, H, H	

There's a total of 5 possible desired outcomes. So the probability that at least 3 of the 4 tosses will come up heads is $\frac{5}{16}$.

Probabilities range from 0, for an event that will certainly not occur, to 1, for an event that is certain to occur.

The sum of the probability that an event will occur and the probability that it will not occur is always equal to 1. Sometimes it is simpler to determine the probability of an event occurring by calculating the probability that an event will not occur and subtracting it from 1.

Example: What is the probability that at least 1 of the 4 coin tosses from the previous example will come up heads?

In this case, to list the desired outcomes, you would have to include all the ways to get 1 head, 2 heads, 3 heads, and 4 heads. That is a lot to consider, would become quite time-consuming, and could risk your overlooking one of the desired outcomes. Instead, by finding the probability that at least 1 of the coin tosses will *not* land on heads, you only need to list ways to get 0 heads. Only one outcome has 0 heads (T, T, T, T). Therefore, out of 16 total possible outcomes, as determined in the previous example, 1 outcome is not the desired outcome. So, the answer is:

$$1 - \frac{1}{16} = \frac{15}{16}$$

PROBABILITY EXERCISES

BASIC

Use this information to answer Exercises 1–2:

In a bag there are 6 red marbles, 7 blue marbles, and 3 black marbles.

1. What is the probability of drawing a blue marble?

2. What is the probability of drawing a red or black marble?

3. Find the probability of tossing at least 1 tail with 2 tosses of a fair coin.

4. Find the probability of randomly selecting 2 blue socks, one at a time without replacement, from a drawer with 14 white and 8 blue socks.

5. If there are 8 people (3 women and 5 men) under consideration for 2 positions on the city council, what is the probability of both positions being filled by women if the people to fill the positions are chosen randomly?

6. If an integer is randomly selected from all positive 2-digit integers, what is the probability that the integer chosen has a 3 in the tens place?

7. Charles says the probability of tossing 3 heads in a row is $\frac{3}{2}$, or $3\left(\frac{1}{2}\right)$. Is he correct?

8. Write the probability 0.80 as a reduced fraction.

9. The weather forecast says there's a 40% chance of rain tomorrow. What is the probability of it raining?

10. Colorado is said to have about 300 days of sunshine per year. If a day is randomly selected, what is the probability that it will be sunny? (Assume 365 days in a year.)

INTERMEDIATE

Use this information to answer Exercises 11–12:

In a bag there are 6 red marbles, 7 blue marbles, and 3 black marbles.

11. What is the probability of NOT drawing a black marble?

12. If two marbles are drawn without replacing the first, what is the probability of drawing a blue and then a red marble?

13. In a pet store, there are 8 puppies and 12 kittens. If a staff member randomly chooses 3 pets for this Saturday's adoption time, what is the probability that all 3 will be puppies?

14. A fair coin is tossed 3 times. What is the probability of NOT getting zero tails across all the three tosses?

15. What is the probability of rolling a 1, 2, or 3 when tossing one fair 6-sided die?

16. Find the probability of getting a sum of 6 when tossing 2 fair 6-sided dice.

17. Teri's work cafeteria offers a lunch with an entrée and a side. If there are 3 entrees (spaghetti, turkey, and pizza) and 4 sides (potatoes, French fries, broccoli, and salad), what is the probability of Teri randomly choosing pizza and salad?

18. A basketball player shoots two free throws, and each free throw is worth 1 point. If the team needs 2 points to win and the probability of the player making a basket is $\frac{1}{2}$, what is the probability that the team will win?

19. Each of 48 employees has entered for a chance to win 3 prizes. If the probability of drawing any given employee is the same, and if winning a prize makes the employee ineligible for the remaining prizes, what is the probability of any given employee winning a prize?

20. Twelve people—3 men and 9 women—want to attend a conference. If 4 people are chosen at random from this group, what is the probability that no men will attend the conference?

ADVANCED

Use this information to answer Exercises 21 and 22:

A bag has only 6 red marbles, 7 blue marbles, and 3 black marbles.

21. What is the probability of NOT drawing a red marble?

22. What is the probability of NOT drawing any red marble in three draws, when the marbles are replaced after each is drawn?

23. A spinner has the following six colored sections of equal size: 1 red, 1 green, 3 blue, and 1 white. With two spins, what is the probability of the spinner landing on white twice?

24. A fair 12-sided die is rolled once. What is the probability of it landing on 5, 7, or 9?

25. If a fair coin is tossed 3 times, what is the probability that the coin will land heads up exactly 2 times?

26. If there is a 30% chance of rain on Tuesday and a 40% chance of rain on Wednesday, what is the probability that there will be NO rain on both Tuesday and Wednesday?

27. In the game of bingo, each of the numbers from 1 to 60 is written on a ball and placed in a round cage. What is the probability that the first 2 balls that are drawn out of the cage will be 1 even and then 1 odd number, assuming that the balls are not replaced until the next game's drawing?

28. Julia is picking 4 books randomly from a bookshelf. If she has 18 science fiction books, 6 drama books, and 8 how-to books, what is the probability that she will pick 3 science fiction books and then 1 how-to book?

29. A mechanic has 4 metric, 6 standard, and 3 dual wrenches in his pail. If 3 wrenches are grabbed at random and not replaced, what is the probability that one metric, then one standard, and then one dual wrench will be picked?

30. Ten people, 3 women and 7 men, are running for 4 school board positions. If all candidates have an equal chance to win any of the 4 positions, what is the probability that NO women will get a position?

PROBABILITY ANSWERS KEY

Basic

1. $\frac{7}{16}$

$$Probability = \frac{\text{Number of desired outcomes}}{\text{Number of total possible outcomes}}$$

There are 7 blue marbles, or 7 desired outcomes, and 16 marbles in all, or 16 total possible outcomes. This means the probability of selecting a blue marble is $\frac{7}{16}$.

2. $\frac{9}{16}$

$$Probability = \frac{\text{Number of desired outcomes}}{\text{Number of total possible outcomes}}$$

There are 6 red marbles and 3 black marbles for a total of 9 desired outcomes, and 16 marbles in all, or 16 total possible outcomes. This means the probability of selecting a red or black marble is $\frac{9}{16}$.

3. $\frac{3}{4}$

There are 2 possible outcomes for each toss, so after 2 tosses there is a total of $2 \times 2 = 4$ total possible outcomes. List the possibilities where at least one of the coins comes up tails:

TT

TH

HT

There are 3 possible outcomes where there is at least 1 tail. So, the probability that at least one of the tosses will come up tails is $\frac{3}{4}$.

4. $\frac{4}{33}$

$$Probability = \frac{\text{Number of desired outcomes}}{\text{Number of total possible outcomes}}$$

There is a total of 22 socks, so the total number of possible outcomes is 22 for the first sock picked. The first draw has a probability of $\frac{8}{22}$. Since there would then be 7 blue socks remaining out of 21 left in the drawer, the probability that the second sock selected would be blue is $\frac{7}{21}$. Because the selections of socks are independent events, to find the combined probability, multiply the two together: $\frac{8}{22} \times \frac{7}{21} = \frac{56}{462} = \frac{4}{33}$.

5. $\mathbf{\frac{3}{28}}$

The probability of a woman randomly getting the first position is $\frac{3}{8}$. The probability of a woman randomly getting the second position, given that the first position has also been filled by a woman, is $\frac{2}{7}$. To find the probability of both events happening, multiply the two probabilities: $\frac{3}{\cancel{8}_4} \times \frac{\cancel{2}^1}{7} = \frac{3}{28}$.

6. $\mathbf{\frac{1}{9}}$

There can be 9 different digits in the tens place. So, the probability of the integer chosen being a 3 is $\frac{1}{9}$.

7. No

Remember that probability is never less than zero or greater than 1. The probability would actually be $\frac{1}{2} \times \frac{1}{2} \times \frac{1}{2} = \frac{1}{8}$.

8. $\mathbf{\frac{4}{5}}$

To write the probability as a fraction, convert the decimal 0.80 to a fraction, $\frac{8}{10}$, and reduce.

9. $\mathbf{\frac{2}{5}}$ **or 0.4**

If there is a 40% chance of rain, this is the same as a probability of $\frac{40}{100}$ or $\frac{2}{5}$ or 0.4.

10. $\mathbf{\frac{60}{73}}$

The number of desired outcomes is 300. The total number of possible outcomes is 365. So, the probability of the desired outcome is $\frac{300}{365}$ or $\frac{60}{73}$.

Intermediate

11. $\mathbf{\frac{13}{16}}$

$$\textit{Probability} = \frac{\textit{Number of desired outcomes}}{\textit{Number of total possible outcomes}}$$

Because $6 + 7 = 13$ of the marbles in the bag are not black, there are 13 desired outcomes. There are $6 + 7 + 3 = 16$ marbles in all, or 16 total possible outcomes. This means the probability of NOT selecting a black marble is $\frac{13}{16}$.

12. $\mathbf{\frac{7}{40}}$

The probability of drawing a blue marble in the first draw is $\frac{7}{16}$. Since that blue marble is not returned to the bag, there are only 15 marbles remaining, but all 6 red marbles are still there as possible desired outcomes for the second draw. That means the probably of selecting a red marble on the second draw is $\frac{6}{15}$. To find the probability of both happening, multiply the probabilities:

$$\frac{7}{\cancel{16}_8} \times \frac{\cancel{6}^3}{15} = \frac{21}{120} = \frac{7}{40}$$

13. $\mathbf{\frac{14}{285}}$

The probability of getting a puppy on the first selection is $\frac{8}{20}$ or $\frac{2}{5}$. Since there are 7 puppies out of 19 pets left, the probability of getting a puppy on the second selection is $\frac{7}{19}$. Now for the third selection there are 6 puppies out of 18 pets left, so the probability of getting a puppy on this selection is $\frac{6}{18}$ or $\frac{1}{3}$. To find the total probability, multiply the 3 fractions together: $\frac{2}{5} \times \frac{7}{19} \times \frac{1}{3} = \frac{14}{285}$.

14. $\mathbf{\frac{7}{8}}$

The question essentially asks for the probability of not getting all heads. The probability of getting a head on the first toss is $\frac{1}{2}$. In fact, the probability of getting a head on any of the 3 tosses is $\frac{1}{2}$, so the probability of getting 3 heads (or no tails) is $\frac{1}{2} \times \frac{1}{2} \times \frac{1}{2} = \frac{1}{8}$. Since the question asks for the probability that the coin does NOT land on heads for all three tosses, subtract the probability of the non-desired outcome (all heads, no tails) from 1: $1 - \frac{1}{8} = \frac{7}{8}$.

15. $\mathbf{\frac{1}{2}}$

$$\textit{Probability} = \frac{\textit{Number of desired outcomes}}{\textit{Number of total possible outcomes}}$$

There are six total possible outcomes when rolling a 6-sided die. The number of desired outcomes in this case is 3: rolling a 1, a 2, or a 3. So, the probability of rolling a 1, 2, or 3 is $\frac{3}{6}$ or $\frac{1}{2}$.

16. $\frac{5}{36}$

There are six possible outcomes for each die that is rolled. Since there are 2 dice, there are $6 \times 6 = 36$ total possible outcomes. List the possibilities where the sum of the 2 dice would add up to 6:

$1 + 5$

$2 + 4$

$3 + 3$

$4 + 2$

$5 + 1$

There are 5 possible outcomes where the sum of the 2 dice is 6, so the probability of rolling 2 dice whose sum is 6 is $\frac{5}{36}$.

17. $\frac{1}{12}$

The probability of randomly selecting pizza (or any given entrée) is $\frac{1}{3}$. The probability of randomly selecting salad (or any given side) is $\frac{1}{4}$. To find the total probability, multiply the two fractions together: $\frac{1}{3} \times \frac{1}{4} = \frac{1}{12}$.

18. $\frac{1}{4}$

There are 2 possible outcomes for each shot, successful or unsuccessful. The probability of the player making the first shot is $\frac{1}{2}$, and the probability of making the second shot is also $\frac{1}{2}$. To find the probability of both events occurring, multiply the two fractions together: $\frac{1}{2} \times \frac{1}{2} = \frac{1}{4}$.

19. $\frac{1}{16}$

The best way to find the probability that any given employee will win one of the prizes is to find the probability that he or she wins *none* of the prizes, then subtract that from 1. The probability of not winning the first prize is $\frac{47}{48}$. Then, since the winner of each prize is made ineligible, the chances of not winning the second and third prizes are $\frac{46}{47}$ and $\frac{45}{46}$, respectively. To combine these probabilities, multiply them:

$$\frac{47}{48} \times \frac{46}{47} \times \frac{45}{46} = \frac{\cancel{47}}{48} \times \frac{\cancel{46}}{\cancel{47}} \times \frac{45}{\cancel{46}} = \frac{45}{48} = \frac{15}{16}$$

Because the probability of losing all three prizes is $\frac{15}{16}$, the probability of winning one of them is $1 - \frac{15}{16} = \frac{1}{16}$.

20. $\frac{14}{55}$

The probability of no men attending is the same as that of 4 women attending. The probability of a woman being randomly selected for the first spot is $\frac{9}{12}$. Then the probabilities for a woman to be chosen for the second, third, and fourth spots are $\frac{8}{11}$, $\frac{7}{10}$, and $\frac{6}{9}$, respectively. Multiply the fractions together to find the total probability: $\frac{9}{12} \times \frac{8}{11} \times \frac{7}{10} \times \frac{6}{9} = \frac{14}{55}$.

Advanced

21. $\frac{5}{8}$

There are 16 total marbles in the bag. Of these 16, 7 are blue and 3 are black; the rest are red. The probability of drawing a marble that is NOT red is $\frac{7 + 3}{16} = \frac{10}{16} = \frac{5}{8}$.

22. $\frac{125}{512}$

The probability of NOT drawing a red on the first draw is $\frac{10}{16}$, which can be simplified to $\frac{5}{8}$. Then, since the marble is placed back in the bag after each draw, the probability of not drawing a red on the second and third draws remain $\frac{5}{8}$ for each. Multiply the fractions together: $\frac{5}{8} \times \frac{5}{8} \times \frac{5}{8} = \frac{125}{512}$.

23. $\frac{1}{36}$

The probability of landing on white with the first spin is $\frac{1}{6}$. Then, the probability of landing on white again is $\frac{1}{6}$. Multiply the fractions to find the total probability: $\frac{1}{36}$.

24. $\frac{1}{4}$

There are 3 desired outcomes: rolling a 5, a 7, or a 9. There are 12 total possible outcomes. The probability is then $\frac{3}{12}$, or $\frac{1}{4}$.

25. $\mathbf{\frac{3}{8}}$

There are 2 possible outcomes for each toss, so after 3 tosses there are $2 \times 2 \times 2 = 8$ total possible outcomes. To determine how many combinations exist where there are exactly 2 heads, use the combination formula: $\frac{3!}{2!(3-2)!} = \frac{3 \times 2 \times 1}{2 \times 1 \times 1} = \frac{6}{2} = 3$. So there are 3 ways of obtaining exactly 2 heads (shown below) out of 8 total possible arrangements. The probability is $\frac{3}{8}$.

HTH

HHT

THH

26. 42%

Since there is a 30% chance of rain on Tuesday, there is a 70% chance of NO rain on Tuesday. Similarly, if there is a 40% chance of rain on Wednesday, there is a 60% chance of NO rain on Wednesday. To find the probability that there will be no rain for *both* days, multiply the two probabilities together: $70\% \times 60\% = 0.7 \times 0.6 = 0.42 = 42\%$.

27. $\mathbf{\frac{15}{59}}$

The probability of the first ball being even is $\frac{30}{60}$ or $\frac{1}{2}$. Then, the probability of the second ball being odd is $\frac{30}{59}$. So, the total probability is $\frac{1}{2} \times \frac{30}{59}$, or $\frac{15}{59}$.

28. $\mathbf{\frac{204}{4495}}$

The probability of the first book being a science fiction book is $\frac{18}{32}$. Then, the subsequent probabilities are $\frac{17}{31}$ and $\frac{16}{30}$ for selecting science fiction books. Then, the probability of selecting a how-to book is $\frac{8}{29}$. Multiplying them together and reducing gives $\frac{204}{4495}$.

29. $\mathbf{\frac{6}{143}}$

The probability of the first wrench being metric is $\frac{4}{13}$. Then, the probability of the next being standard is $\frac{6}{12}$ or $\frac{1}{2}$, and then the probability of the next being dual is $\frac{3}{11}$. Multiplying them together and reducing gives $\frac{6}{143}$.

30. $\mathbf{\frac{1}{6}}$

No woman getting a position is the same as all the positions being filled by men. The probability of the first position being filled by a man is $\frac{7}{10}$, and the probabilities that the subsequent 3 positions would be filled by men are $\frac{6}{9}$ or $\frac{2}{3}$, $\frac{5}{8}$, and $\frac{4}{7}$, respectively. To find the total probability of 4 men being randomly selected to fill the positions, multiply the fractions: $\frac{7}{10} \times \frac{2}{3} \times \frac{5}{8} \times \frac{4}{7} = \frac{1}{6}$.

PART FOUR

Advanced Math Practice

CHAPTER 10

High-Difficulty Question Sets

INTRODUCTION

In this chapter, you'll find four 20-question practice sets consisting entirely of questions most test takers find challenging. Following each practice set, you'll find the answer key, as well as complete explanations for every question.

Remember that because the GRE is a Multi-Stage Test (MST), a strong performance on the first scored Quantitative section will cause you to see a high-difficulty second section. This high-difficulty section will likely include a mix of medium- and high-difficulty questions, with a few on the very high end of the difficulty scale. If you are aiming for an exceptionally high score on the Quantitative part of the exam, then you will want to be sure you can handle even these very toughest questions.

That's exactly what the four practice sets in this chapter are for: to help you prepare for the most advanced math questions the GRE may present to you on Test Day. Many of the questions in these practice sets ask you to combine multiple concepts in order to arrive at a solution, and many may take longer than two minutes to answer. Again, on Test Day, even a high-difficulty Quantitative section will not consist entirely of questions like the ones here; many will be less challenging. Therefore, we suggest that you work through these practice sets without timing yourself. Instead, focus on mastering the various ways the GRE can present math content to make questions more difficult.

If you are not yet comfortable with at least some of the exercises marked "advanced" in the preceding chapters of this book, then continue to review the arithmetic, algebra, geometry, and other topics presented there until you are finding the correct answers to at least half of those advanced exercises. Then you'll be ready to tackle the tougher questions here.

As always, be sure to ***review the explanation*** to every problem you do in this chapter. There is often more than one path to the correct answer; with that in mind, we've included multiple solutions to a number of these questions. Noting a different way to work through a problem, even if you got that problem right, is a powerful way to enhance your critical thinking skills—and it's ultimately those skills that you will be relying on in order to achieve a very high score on Test Day.

Ready to take your Quantitative score to the next level? Then turn the page and begin work!

ADVANCED MATH QUESTIONS PRACTICE SET 1

1.

$$bc < 0$$
$$ab > 0$$
$$cd < 0$$

Quantity A	Quantity B
ac	bd

(A) Quantity A is greater.
(B) Quantity B is greater.
(C) The two quantities are equal.
(D) The relationship cannot be determined from the information given.

2.

$$x > 0$$

The probability that event A will occur is x, and the probability that event A will not occur is y, where $y > 3x$.

Quantity A	Quantity B
x	$\frac{1}{5}$

(A) Quantity A is greater.
(B) Quantity B is greater.
(C) The two quantities are equal.
(D) The relationship cannot be determined from the information given.

3. List A contains 7 consecutive multiples of 4 and nothing else. The average (arithmetic mean) of the 3 greatest integers in list A is 80.

Quantity A	Quantity B
The average (arithmetic mean) of the 5 smallest integers in list A	68

(A) Quantity A is greater.
(B) Quantity B is greater.
(C) The two quantities are equal.
(D) The relationship cannot be determined from the information given.

4.

$$b > 1$$

$$\frac{\left(b^4\right)^x\left(b^{30}\right)}{\left(b^2\right)^7} < \left(b^x\right)\left(b^{37}\right)$$

Quantity A	Quantity B
x	8

(A) Quantity A is greater.
(B) Quantity B is greater.
(C) The two quantities are equal.
(D) The relationship cannot be determined from the information given.

5. The random variable X has a normal distribution with a mean of 70.

Quantity A	Quantity B
The probability that $-30 \leq X \leq 150$	The probability that $-60 \leq X \leq 120$

(A) Quantity A is greater.
(B) Quantity B is greater.
(C) The two quantities are equal.
(D) The relationship cannot be determined from the information given.

6. z is a positive integer.

$$x = 12y + 25$$
$$y = 7z + 5$$

Quantity A	Quantity B
The remainder when x is divided by 42	The remainder when y is divided by 2

(A) Quantity A is greater.
(B) Quantity B is greater.
(C) The two quantities are equal.
(D) The relationship cannot be determined from the information given.

7.

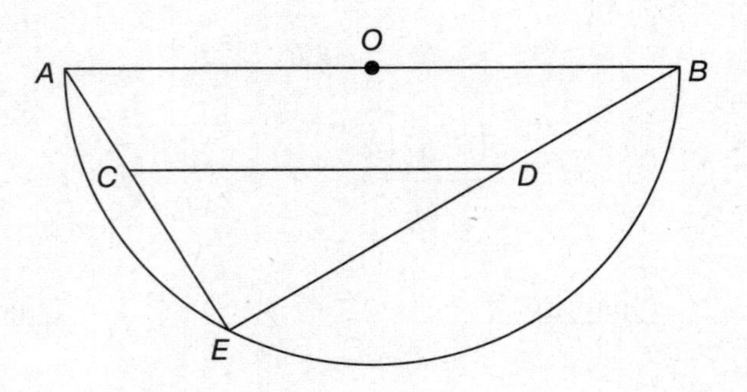

The area of semicircle O is 16π.

$$\angle CDE = \angle ABE = 30^\circ$$
$$AC = \sqrt{2}$$

Quantity A	Quantity B
DE	$3\sqrt{5}$

Ⓐ Quantity A is greater.
Ⓑ Quantity B is greater.
Ⓒ The two quantities are equal.
Ⓓ The relationship cannot be determined from the information given.

8. $a{:}b = 3{:}1$ $\quad a{:}d = 15{:}2$ $\quad c{:}e = 7{:}9$ $\quad c{:}d = 10{:}3$

If $e = 60$, what is the value of b?

Ⓐ 20
Ⓑ 35
Ⓒ 70
Ⓓ 180
Ⓔ The value cannot be determined from the information given.

9. What is the area of the figure bounded by lines described by the following equations?

$$y + 2 = 2x$$

$$\frac{x^3 - 5}{11} = 2$$

$$\frac{y - 5}{2} = x$$

$$(x + 2)^5 = 0$$

(A) 6
(B) 12
(C) 17
(D) 35
(E) 56

10. Both x and y are integers. Which of the following expressions must be the square of an integer?

Indicate <u>all</u> such expressions.

[A] $(x + y)(x - y) + 8xy + 17y^2$

[B] $9x^4 - 12x^2y^2 + 4y^4$

[C] $x^6 + 2x^3y^3 + y^6$

11. Sequence S is the sequence of numbers $a_1, a_2, a_3, \ldots, a_n$. For each positive integer n, the nth number a_n is defined by $a_n = \frac{n+1}{3n}$. What is the product of the first 53 numbers in sequence S?

(A) $\frac{2}{3^{53}}$

(B) $\frac{2}{3^{50}}$

(C) $\frac{2}{3^{49}}$

(D) $\frac{3}{2^{50}}$

(E) $\frac{2}{3^{25}}$

12. A group of 8 machines that work at the same constant rate can complete 14 jobs in 7 hours. How many hours would it take 17 of these machines to complete 34 of these jobs?

(A) 4
(B) 6
(C) 8
(D) 12
(E) 16

13. Let $M = 10!$ and let $N = M^3$. If a is the greatest integer value of x such that 2^x is a factor of N, and b is the greatest integer value of y such that 3^y is a factor of N, then what is the value of $a + b$?

(A) 18
(B) 24
(C) 33
(D) 36
(E) 54

14.

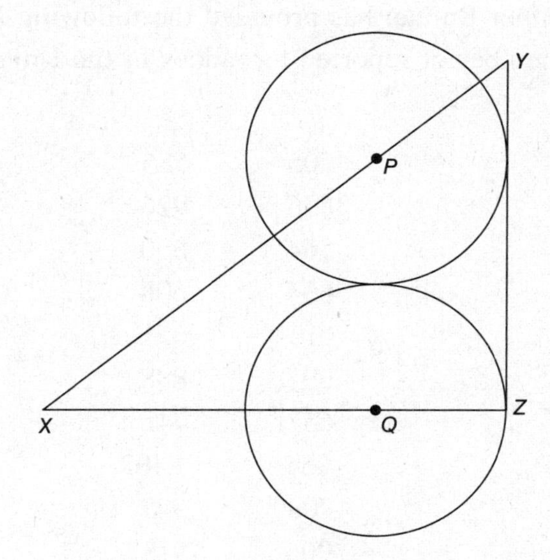

Circles P and Q each have a circumference of 10π and are each tangent to YZ. If $XZ = 17$, then $YZ =$

Ⓐ 10

Ⓑ $\frac{85}{6}$

Ⓒ $\frac{43}{3}$

Ⓓ 15

Ⓔ 17

15. A furniture company has warehouses in two cities: Madison and York. The Madison warehouse stocks desks, tables, and chairs in a ratio of 5:9:15. If all the desks and tables are transferred from the Madison warehouse to the York warehouse, increasing that warehouse's inventory of items by 20 percent, which of the following could be the number of items in the York warehouse after the transfer?

Indicate all such numbers.

[A] 168

[B] 280

[C] 290

[D] 336

[E] 504

[F] 600

16. The U.S. Weather Bureau has provided the following information about the total annual number of reported tornadoes in the United States for the years 1956 to 1975:

505	585
856	926
564	660
604	608
616	653
697	888
657	741
464	1,102
704	947
906	918

The average number of tornadoes per reported year is 730, and the standard deviation of the set is 168.

If one of the numbers in the data set is chosen at random, what is the probability that the number is less than the median and is not within one standard deviation of the mean?

Ⓐ 10%
Ⓑ 15%
Ⓒ 25%
Ⓓ 30%
Ⓔ 40%

17. Liquid *A* is 32 percent iodine and the rest water. Liquid *B* is 17 percent iodine and the rest water.

Quantity A	Quantity B
The percentage of a 23 percent iodine mixture of liquid *A* and liquid *B* that is liquid *A*	38%

Ⓐ Quantity A is greater.
Ⓑ Quantity B is greater.
Ⓒ The two quantities are equal.
Ⓓ The relationship cannot be determined from the information given.

Questions 18–20 refer to the following graphs.

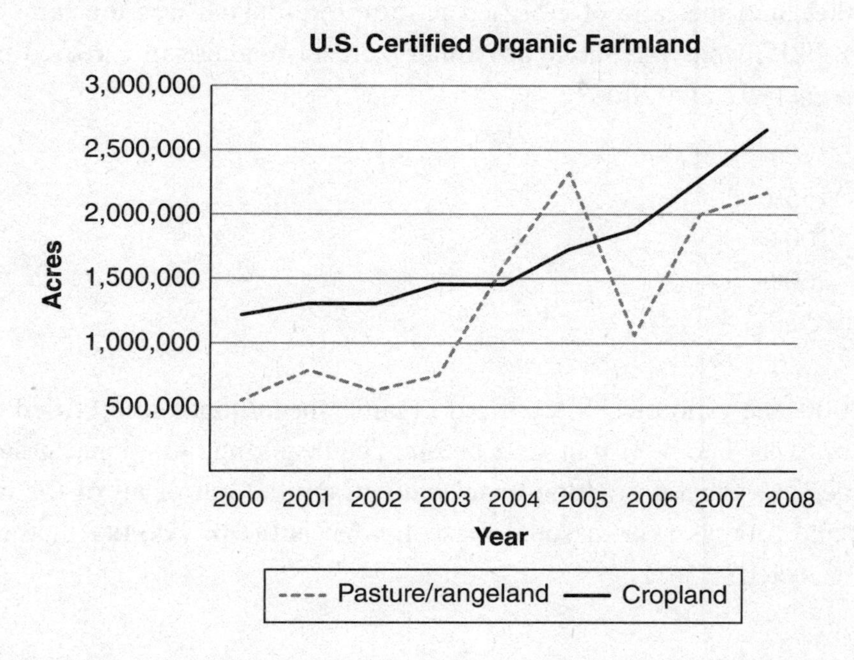

Source: U.S. Department of Agriculture, Economic Research Service, Table 2—U.S. certified organic farmland acreage, livestock numbers, and farm operations, 1992–2011 (based on information from USDA-accredited state and private organic certifiers), http://www.ers.usda.gov/data-products/organic-production.aspx

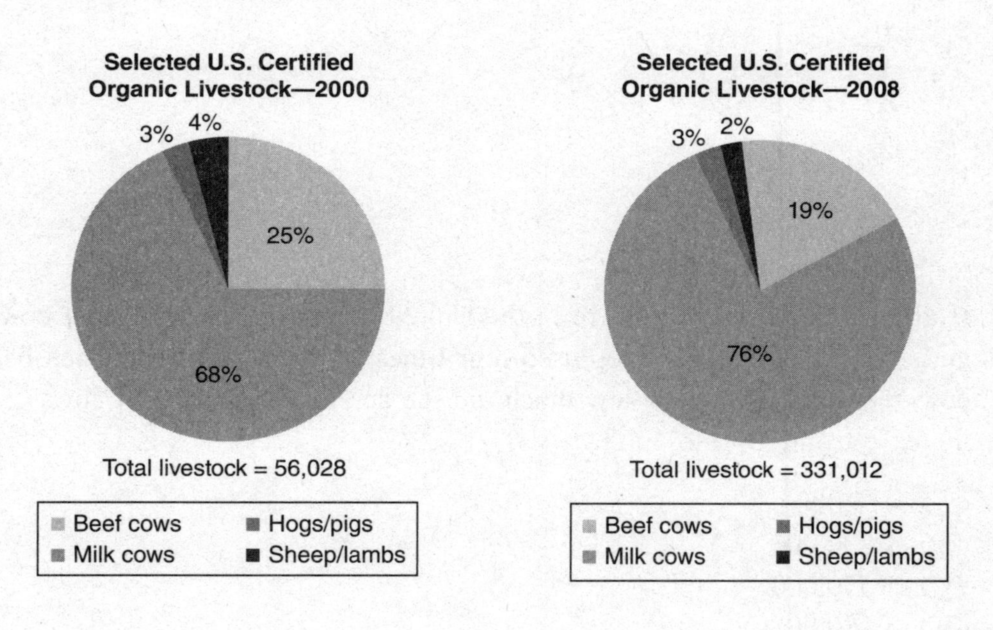

18. If U.S. certified organic farmland acreage increased 42 percent from 1997 to 2002, and the ratio of certified pasture to cropland was the same in 1997 as in 2006, what was the approximate percent increase in certified pasture between 1997 and 2008?

(A) 75%
(B) 155%
(C) 200%
(D) 260%
(E) 355%

19. In 2000, the ratio of U.S. certified organic sheep/lambs to certified organic turkeys was 1 to 4. If that year before Thanksgiving, a national organic retailers' association randomly selected an animal out of all of the hogs/pigs and turkeys as a seasonal mascot, what is the probability that a turkey was selected?

(A) $\frac{16}{3}$
(B) $\frac{19}{16}$
(C) $\frac{16}{19}$
(D) $\frac{3}{4}$
(E) $\frac{8}{11}$

20. If all certified organic pasture in the United States was used by beef cows and sheep, and if each sheep used four times as much pasture as each beef cow, then approximately how much did the acreage used by beef cows increase from 2000 to 2008?

(A) 475,000
(B) 1,200,000
(C) 1,550,000
(D) 2,100,000
(D) 2,500,000

ADVANCED MATH QUESTIONS PRACTICE SET 1 ANSWER KEY

1. B
2. D
3. C
4. B
5. A
6. D
7. A
8. B
9. D
10. A, B, C
11. B
12. C
13. D
14. B
15. A, D, E
16. A
17. A
18. E
19. C
20. B

ADVANCED MATH QUESTIONS PRACTICE SET 1 ANSWERS AND EXPLANATIONS

1. B

Start by analyzing the centered information. None of the individual variables can be isolated in a way that establishes its sign definitively. However, as ab is positive, there are only two possibilities: a and b are either both positive or both negative.

If both a and b are positive, then c must be negative in order to make $bc < 0$, in which case d must be positive so that $cd < 0$. Given those signs, Quantity A, ac, must be negative, and Quantity B, bd, must be positive.

If both a and b are negative, then c must be positive so that $bc < 0$, in which case d must be negative so that $cd < 0$. In this case, Quantity A, ac, is still negative, and Quantity B, bd, is still positive. Thus, in both cases, Quantity B is greater, and the correct answer is **(B)**.

2. D

The sum of the probability that an event will occur and the probability that the event will not occur is 1. So here, $x + y = 1$. Since $y > 3x$, you know that $x + y > x + 3x$, or, simplified, $x + y > 4x$. Since $x + y = 1$, you also know that $1 > 4x$, or $4x < 1$. Dividing both sides of this inequality by 4, you get $x < \frac{1}{4}$. Since $x > 0$, you know that $0 < x < \frac{1}{4}$. This is the range of values that Quantity A must fall into. Quantity B is $\frac{1}{5}$, which is within the range of values that Quantity A can be equal to; there are values in that range that are both smaller and larger than $\frac{1}{5}$. Since more than one relationship between the quantities is possible, the relationship between them cannot be determined.

Choice **(D)** is correct.

3. C

Start with the centered information: a list of 7 consecutive multiples of 4. Draw seven dashes to represent the numbers in the list, with a view to filling in what you know:

— — — — — — —

The last three numbers in the list are the greatest, and their average is 80. Because they are evenly spaced, 80 is not just the average of the three, but also their median. Because the three numbers are multiples of 4, the other two are 76 and 84:

— — — — 76 80 84

Now fill in the rest of List *A*, going backward from 76, with consecutive multiples of 4. The result looks like this:

60 64 68 72 76 80 84

Again, the average of an evenly spaced list is also its median. So to find Quantity A, just take the middle number of the first five elements in the list, which are

60 64 68 72 76

The median, which is also the average, is 68.

Quantity A is 68 and Quantity B is 68. The quantities are equal, and choice **(C)** is correct.

4. B

Use these exponent rules to solve this problem:

(i) $(a^c)^d = a^{cd}$

(ii) $a^c a^d = a^{c+d}$

(iii) $\frac{a^c}{a^d} = a^{c-d}$

Start by simplifying the inequality:

$$\frac{(b^4)^x(b^{30})}{(b^2)^7} < (b^x)(b^{37})$$

$$\frac{(b^{4x})(b^{30})}{b^{14}} < b^{x+37}$$

$$\frac{b^{4x+30}}{b^{14}} < b^{x+37}$$

$$b^{4x+30-14} < b^{x+37}$$

$$b^{4x+16} < b^{x+37}$$

Because $b > 1$, the exponent on the left side of the inequality must be less than the one on the right, so you can conclude that $4x + 16 < x + 37$. Subtracting 16 from both sides of the inequality produces $4x < x + 21$. Subtracting x from both sides of this inequality, you have $3x < 21$. Dividing both sides by 3 yields $x < 7$. Quantity A, which is x, is less than 7. Quantity B is 8. So Quantity B is greater. Choice **(B)** is correct.

5. A

The range of the variables in Quantity A is $150 - (-30) = 180$. This is the same as the range of the variables in Quantity B, $120 - (-60) = 180$. The mean of the distribution, 70, falls within both of those ranges. However, the range for Quantity B lies farther to the left side of the bell curve, therefore encompassing more of the tail of the normal distribution. Since values in the tail occur much less frequently than those closer to the mean, the probability that a random value of X falls within the range associated with Quantity B is less than the probability that X will fall within the range associated with Quantity A.

Choice **(A)** is correct.

6. D

To evaluate Quantity A, start by substituting $7z + 5$ for y in the equation $x = 12y + 25$ to produce $x = 12(7z + 5) + 25 = 84z + 60 + 25 = 84z + 85$. Because 84 is a multiple of 42, and because z is a positive integer, $84z$ is also a multiple of 42. So when $84z$ is divided by 42, there is no remainder. When 85 is divided by 42, the quotient is 2 and the remainder is 1. So when x, which equals $84z + 85$, is divided by 42, the remainder is $0 + 1 = 1$.

Evaluating Quantity B is a bit simpler. If an even integer is divided by 2, the remainder is zero. If an odd integer is divided by 2, the remainder is 1. So is y even or odd? It depends on z. If z is even, then $7z$ is also even and $7z + 5$ is odd, since an even plus an odd is an odd. But if z is odd, then $7z$ is also odd and $7z + 5$ is even, since an odd plus an odd is an even. Because you don't know the value of z, there's no way to tell whether the remainder when y is divided by 2 will be 0 or 1.

Quantity A is 1 and Quantity B is either 0 or 1. So the two quantities could be equal, or alternatively, Quantity A could be greater. Because more than one relationship is possible, choice **(D)** is correct.

7. A

If you're not sure how to solve a complex geometry problem, a good first step is to fill in as many deductions as you can and then reassess the situation. First, note that any triangle formed by the diameter of a circle and a point on the circle will always be a right triangle. Because triangles AEB and CED each have a right angle and a 30 degree angle, it follows that they are both 30-60-90 right triangles. Next, since you're given the area of the semicircle, you can find the diameter of semicircle O—which is also the hypotenuse of ΔABE.

If the area of half a circle is 16π, then the area of the whole circle is 32π:

$$\begin{aligned} A &= \pi r^2 \\ 32\pi &= \pi r^2 \\ 32 &= r^2 \\ r &= \sqrt{32} = 4\sqrt{2} \end{aligned}$$

This is the radius. The diameter is double the radius, or $8\sqrt{2}$, which is also the hypotenuse of ΔABE. Recall that the sides of a 30-60-90 triangle are in the proportion $x : x\sqrt{3} : 2x$. Since the hypotenuse is $8\sqrt{2}$, it follows that the short leg (AE) is $4\sqrt{2}$ and the long leg (BE) is $4\sqrt{6}$.

The last piece of information you haven't used yet is the fact that $AC = \sqrt{2}$. Subtracting this from AE gives CE:

$$CE = AE - AC = 4\sqrt{2} - \sqrt{2} = 3\sqrt{2}$$

CE is the shorter leg of a 30-60-90 triangle, and DE (the value in Quantity A) is the longer leg. Thus,

$$DE = CE\sqrt{3} = 3\sqrt{2}\sqrt{3} = 3\sqrt{6}$$

This is greater than Quantity B, so choice **(A)** is the answer.

8. B

The problem provides a series of 4 ratios pertaining to 5 different variables and also gives the value of one of the variables (e), so there will be enough information to solve for the values of every variable, and (E) will not be the correct answer. However, the question only requires that you find the value of one variable, b. You could try calculating variables one by one until you arrive at a value for b, but that could become cumbersome very quickly; you would start with the proportion $\frac{c}{60} = \frac{7}{9}$, which would leave you with a fraction if you solved for c. Backsolving would be similarly difficult. Often the most efficient way to solve problems such as this is to chain the given ratios algebraically, multiplying the appropriate ones until you are left with a ratio that is either e:b or b:e.

The variables b and e are each contained in only one of the given ratios, but they are both in the denominator, so one of the ratios needs to be inverted to allow the formation of a chain of ratios in which everything except b and e will cancel out. Start with this: $\left(\frac{a}{b}\right)\left(\frac{e}{c}\right) = \frac{a \times e}{b \times c}$. The a in the numerator can be eliminated by multiplying this expression by $\frac{d}{a}$, and the c in the denominator will be eliminated if the expression is multiplied by $\frac{c}{d}$, like this: $\frac{a \times e \times d \times c}{b \times c \times a \times d}$. Everything but $\frac{e}{b}$ cancels out, so $\left(\frac{a}{b}\right)\left(\frac{e}{c}\right)\left(\frac{d}{a}\right)\left(\frac{c}{d}\right) = \frac{e}{b}$. Now substitute the known values:

$$\left(\frac{3}{1}\right) \times \left(\frac{9}{7}\right) \times \left(\frac{2}{15}\right) \times \left(\frac{10}{3}\right) = \frac{12}{7} = \frac{e}{b}$$

Plug in 60 for *e:*

$$\frac{12}{7} = \frac{60}{b}$$

Cross multiply to get $12b = 420$ and divide both sides by 12 to get $b = 35$, or think like this: $5(12) = 60$, so $b = 5(7) = 35$.

The correct answer is **(B).**

9. D

A couple of the equations have exponents, but it turns out that all four equations describe straight lines, so this problem isn't as scary as it looks at first glance. Start by simplifying, with the $y = mx + b$ form as your goal. The first and third equations become $y = 2x - 2$ and $y = 2x + 5$, respectively. The second and fourth equations present a bit more of a challenge. Here are the algebraic steps to simplify the second equation:

$$\begin{aligned} \frac{x^3 - 5}{11} &= 2 \\ x^3 - 5 &= 22 \\ x^3 &= 27 \\ x &= 3 \end{aligned}$$

Here are the steps to simplify the fourth equation:

$$\begin{aligned} (x + 2)^5 &= 0 \\ (x + 2)(x + 2)(x + 2)(x + 2)(x + 2) &= 0 \end{aligned}$$

The only way for all five identical factors to multiply to zero is for each factor to equal zero, so:

$$\begin{aligned} x + 2 &= 0 \\ x &= -2 \end{aligned}$$

With the four equations in these forms, you can now see that $y = 2x - 2$ and $y = 2x + 5$ have the same slope and are therefore parallel. Moreover, $x = 3$ and $x = -2$ are both parallel to the y-axis and thus parallel to each other. Since the opposite sides of this four-sided polygon are parallel, it is a parallelogram. The area of a parallelogram is base times height.

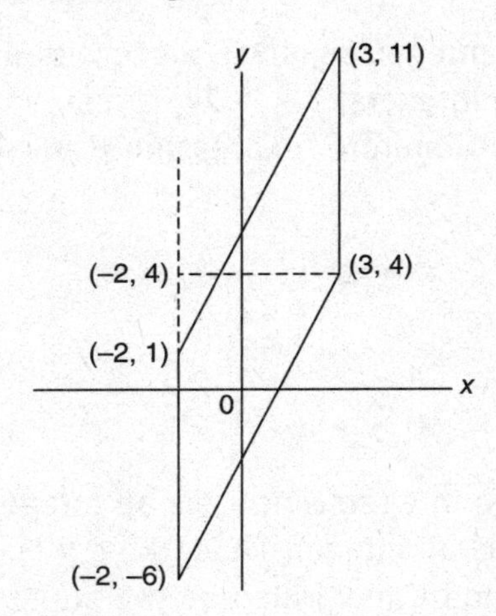

You would have to use the Pythagorean theorem to calculate the lengths of the non-vertical sides and the vertical height between those sides. However, if you consider the vertical sides as the bases, then you can calculate the height easily by constructing a perpendicular to the extension of one of those bases, as shown above. The length of each vertical base is $1 + |-6| = 7$. The horizontal height is the distance between the two vertical bases: $3 + |-2| = 5$.

Plug into the area formula for a parallelogram: base x height $= 7 \times 5 = 35$. Choice **(D)** is correct.

10. A, B, C

First, evaluate expression A: $(x + y)(x - y) + 8xy + 17y^2$.

Multiply the first two terms together:

$$(x + y)(x - y) = x^2 - y^2$$

Plug this into the initial equation and simplify:

$$x^2 - y^2 + 8xy + 17y^2 = x^2 + 8xy + 16y^2$$

Reverse FOIL the equation:

$$x^2 + 8xy + 16y^2 = (x + 4y)(x + 4y) = (x + 4y)^2$$

Since y is an integer, $4y$ is an integer. Since x and $4y$ are both integers, $x + 4y$ is an integer. So $(x + 4y)^2$ is the square of an integer, and it follows that expression A must be the square of an integer.

Next, consider expression B: $9x^4 - 12x^2y^2 + 4y^4$.

Reverse FOIL the equation:

$$9x^4 - 12x^2y^2 + 4y^4 = (3x^2 - 2y^2)(3x^2 - 2y^2) = (3x^2 - 2y^2)^2$$

Since x is an integer, x^2 is an integer, as is $3x^2$. Since y is an integer, y^2 is an integer, as is $2y^2$. Since $3x^2$ and $2y^2$ are integers, $3x^2 - 2y^2$ is also an integer. Thus, $(3x^2 - 2y^2)^2$ is the square of an integer. Therefore, expression B must also be the square of an integer.

Lastly, evaluate expression C: $x^6 + 2x^3y^3 + y^6$.

Reverse FOIL the equation:

$$(x^3)^2 + 2(x^3)(y^3) + (y^3)^2 = (x^3 + y^3)(x^3 + y^3) = (x^3 + y^3)^2$$

Since x is an integer, x^3 is an integer. Since y is an integer, y^3 is an integer. Since x^3 and y^3 are integers, $x^3 + y^3$ is an integer. Thus, $(x^3 + y^3)^2$ is the square of an integer. So it turns out that expression C must also be the square of an integer.

Choices **(A)**, **(B)**, and **(C)** are all correct.

11. B

Write out a description of the product of the first 53 numbers of sequence S. Call the product P. There's no need to write out every term; just write out enough to recognize the pattern:

$$P = \left[\frac{1+1}{3(1)}\right] \times \left[\frac{2+1}{3(2)}\right] \times \left[\frac{3+1}{3(3)}\right] \times \left[\frac{4+1}{3(4)}\right] \times \ldots \times \left[\frac{51+1}{3(51)}\right] \times \left[\frac{52+1}{3(52)}\right] \times \left[\frac{53+1}{3(53)}\right]$$

$$= \left[\frac{2}{3(1)}\right] \times \left[\frac{3}{3(2)}\right] \times \left[\frac{4}{3(3)}\right] \times \left[\frac{5}{3(4)}\right] \times \ldots \times \left[\frac{52}{3(51)}\right] \times \left[\frac{53}{3(52)}\right] \times \left[\frac{54}{3(53)}\right]$$

$$= \left(\frac{1}{3^{53}}\right) \times \left(\frac{2}{1}\right) \times \left(\frac{3}{2}\right) \times \left(\frac{4}{3}\right) \times \left(\frac{5}{4}\right) \times \ldots \times \left(\frac{52}{51}\right) \times \left(\frac{53}{52}\right) \times \left(\frac{54}{53}\right)$$

$$= \left(\frac{1}{3^{53}}\right) \times 54$$

$$= \frac{54}{3^{53}}$$

None of the answer choices is written as $\frac{54}{3^{53}}$. Note that instead, every answer choice has a 2 in the numerator. So rewrite $\frac{54}{3^{53}}$ using prime factorization:

$$54 = 2 \times 27 = 2 \times 3 \times 9 = 2 \times 3 \times 3 \times 3$$

The prime factorization of 54 is $2 \times 3 \times 3 \times 3$, or 2×3^3.

Rewrite the answer as $\frac{54}{3^{53}} = \frac{2 \times 3^3}{3^{53}}$. Cancel the three 3s in the numerator with three 3s in the denominator to get $\frac{2}{3^{50}}$. Or, think like this: if $b \neq 0$, then $\frac{b^a}{b^c} = b^{a-c}$. So:

$$\frac{2 \times 3^3}{3^{53}} = 2 \times \frac{3^3}{3^{53}} = 2 \times 3^{3-53} = 2 \times 3^{-50} = 2 \times \frac{1}{3^{50}} = \frac{2}{3^{50}}.$$

Choice **(B)** is correct.

12. C

Work = Rate × Time.

If each machine works at the rate of r jobs per hour, then together, 8 machines work at the rate of $8r$ jobs per hour.

Using the information given in the question, set up an equation and solve for r.

$$\begin{aligned} \frac{8r \text{ jobs}}{\text{hour}} \times 7 \text{ hours} &= 14 \text{ jobs} \\ 56r &= 14 \\ r &= \frac{1}{4} \end{aligned}$$

Thus, each machine works at the rate of $\frac{1}{4}$ jobs per hour. The question asks how many hours it takes 17 machines to do something. Since one machine works at the rate of $\frac{1}{4}$ jobs per hour, 17 machines work at the rate of $17 \times \frac{1}{4} = \frac{17}{4}$ jobs per hour.

Use the work formula again, with t representing the time in hours, to solve for t.

$$\begin{aligned} \frac{17 \text{ jobs}}{4 \text{ hours}} \times t \text{ hours} &= 34 \text{ jobs} \\ t \text{ hours} &= 34 \text{ jobs} \times \frac{4 \text{ hours}}{17 \text{ jobs}} \\ t &= 34 \div 17 \times 4 = 2 \times 4 = 8 \end{aligned}$$

The correct answer is **(C)**.

13. D

When you see a problem that looks as if it is going to involve manipulation of very large numbers, using prime factors is often an efficient strategy.

First write out the prime factorization of 10!:

$$
\begin{aligned}
&(1)(2)(3) \quad (4) \quad (5) \quad (6) \quad (7) \quad (8) \quad (9) \quad (10) \\
= \; &(1)(2)(3) \quad [(2)(2)] \quad (5) \quad [(2)(3)] \quad (7) \quad [(2)(2)(2)] \quad [(3)(3)] \quad [(2)(5)]
\end{aligned}
$$

Count the number of times each number (greater than 1) appears in the prime factorization.

$2 = 8$ times

$3 = 4$ times

$5 = 2$ times

$7 = 1$ time

So you can write 10! like this: $10! = (2^8)(3^4)(5^2)(7^1)$

And you can write N like this:

$$
\begin{aligned}
N &= M^3 = (10!)^3 \\
&= \left[\left(2^8\right)\left(3^4\right)\left(5^2\right)\left(7^1\right)\right]^3 \\
&= \left[\left(2^8\right)^3\right]\left[\left(3^4\right)^3\right]\left[\left(5^2\right)^3\right]\left[\left(7^1\right)^3\right] \\
&= \left(2^{8(3)}\right)\left(3^{4(3)}\right)\left(5^{2(3)}\right)\left(7^{1(3)}\right) \\
&= \left(2^{24}\right)\left(3^{12}\right)\left(5^6\right)\left(7^3\right)
\end{aligned}
$$

Since the factor with the base of 2 is 2^{24}, the greatest possible value of x such that 2^x is a factor of $(10!)^3$ is 24. (Remember that the greatest factor of any number is that number itself.) So $a = 24$.

By the same logic, since the factor with the base of 3 is 3^{12}, the greatest possible value of y such that 3^y is a factor is $(10!)^3$ is 12. So $b = 12$.

Finally, add a and b to get to the answer: $a + b = 24 + 12 = 36$.

Choice **(D)** is correct.

14. B

Take a moment to analyze the figure and write in whatever information you have. You know the circumference of both circles, so you can easily find their radii:

$$\begin{aligned} C &= 2\pi r = 10\pi \\ 2r &= 10 \\ r &= 5 \end{aligned}$$

Next, you know that YZ is tangent to both circles. This means that angle Z is a right angle, so triangle XYZ is a right triangle. This leaves a lot of unknowns in the diagram, however. Whenever a complex figure seems to be missing a lot of information, a good strategy is to look for hidden right triangles. Here, it's natural to add line segment PQ to create right triangle PQX:

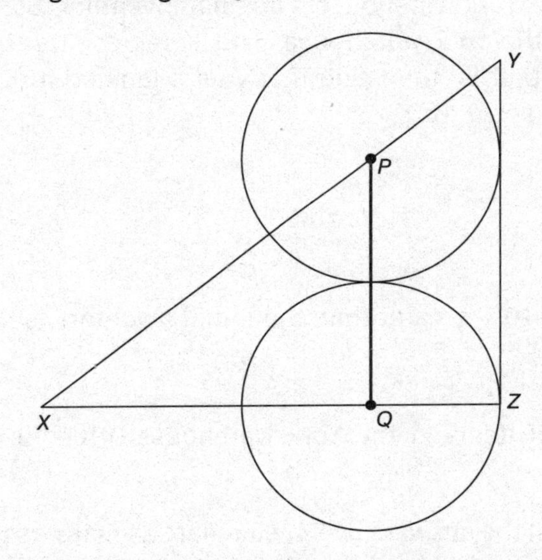

Consider the dimensions of this new right triangle. Leg PQ is equal to two radii of the circles, or $2 \times 5 = 10$. Leg XQ is equal to $XZ - QZ$. The question states that $XZ = 17$, and you know that QZ is a radius, so it's 5. Thus, $XQ = 17 - 5 = 12$. Now, since triangles XPQ and XYZ are similar (they're both right triangles and they share angle X), their sides must be proportional. Set up a proportion and solve:

$$\begin{aligned} \frac{YZ}{PQ} &= \frac{XZ}{XQ} \\ \frac{YZ}{10} &= \frac{17}{12} \end{aligned}$$

$$YZ = \frac{17 \times 10}{12} = \frac{170}{12} = \frac{85}{6}.$$

Choice **(B)** is correct.

15. A, D, E

While it would be possible to Backsolve, the fact that this is an "all that apply" question makes that approach considerably less efficient—you'd have to check every answer choice (unless your Backsolving work helped you to recognize what pattern is at play in the problem). So solve this one using straightforward math and reasoning.

First, note that the only types of furniture being moved are desks and tables, so these are the ratio elements of interest. Since the ratio of the number of desks to the number of tables is 5 to 9, you can say that the number of desks is $5x$ and the number of tables is $9x$, where x is a positive integer. So the total number of desks and tables transferred to the York warehouse is $5x + 9x = 14x$, where x is a positive integer. In other words, the total number of desks and tables transferred to York is a multiple of 14.

Using the second piece of information in the question stem, you know that $14x$ represents 20% of the York warehouse's original inventory. So, if Y_o represents the original inventory in the York warehouse, and Y_f represents the final inventory in the York warehouse (that is, the value that you're looking for), you can set up the following equation to solve for Y_o:

$$\frac{1}{5}Y_o = 14x$$

$$Y_o = 70x$$

The final inventory is 20% greater than the initial inventory, so:

$$Y_f = 1.2Y_o = 84x$$

Hence, the number of items in the York warehouse after the transfer must be a multiple of 84.

Rather than dividing all the answers by 84, eliminate as many as possible by checking for divisibility by 3, since 84 is evenly divisible by 3. That quick process eliminates choices (B) and (C) merely by summing their digits and noting that the sums are not divisible by 3. Choice (A) is readily identifiable as 2×84, (D) is 2×168, and (E) is the sum of (A) and (D), so those numbers are all divisible by 84. Since 504 is divisible by 84 and $504 + 84 = 588$, it's clear that 600 is not divisible by 84. Only choices **(A), (D),** and **(E)** are correct. Going through a quick analysis such as this eliminates the need to open up the calculator and enter several numbers, but that alternative is always available as well.

16. A

Because the question mentions the median, write the numbers in ascending order to make the median easy to spot:

464	505	564	585	604	608	616	653	657	660
697	704	741	856	888	906	918	926	947	1,102

The list contains an even number of numbers, so the median is the average of the middle two: $(660 + 697) \div 2 = 1{,}357 \div 2 = 678.5$.

The standard deviation is a measure of the absolute value of the distance from the mean. Therefore, the range of all terms within one standard deviation of the mean goes from $730 - 168 = 562$ to $730 + 168 = 898$. The numbers less than the median that are *not* within this range are the following:

464 505

That's 2 numbers out of 20, so the probability is $2 \div 20$, or 10%, which is choice **(A)**.

17. A

The trickiest part of this question is translating Quantity A. There is a mixture of liquids *A* and *B* that is 23% iodine. Quantity A essentially asks, "What percent of this mixture is liquid *A*?" To figure this out, imagine pouring equal parts (ounces, liters, tablespoons—it doesn't matter) of liquid *A* and liquid *B* into an empty flask until you reach the correct percentage of iodine. Each part of liquid *A* that you pour adds $32\% - 23\% = 9\%$ *more* iodine than you're looking for. Each part of liquid *B* that you pour into the flask adds $23\% - 17\% = 6\%$ *less* iodine than you're looking for. The key is to get the +9s and the −6s to cancel out to 0; it's at that point that you'll know you've got the percent of iodine just right. To do this, keep adding parts of liquid *A* (+9s) until you hit a multiple of 6, then add as many parts of liquid *B* (−6s) as needed to reach 0.

$$+9 + 9 - 6 - 6 - 6 = +18 - 18 = 0$$

Thus, there are three parts of liquid *B* for every two parts of Liquid *A* in a 23% iodine mixture. This means that the ratio of liquid *A* to liquid *B* is 2:3, so $\frac{2}{2+3} = \frac{2}{5} = 40\%$ of the mixture is liquid *A*. This is greater than Quantity B, 38%, so the answer is **(A)**.

18. E

This question asks about acres of certified organic farmland, so consult the line graph.

You're told that certified farmland increased 42 percent from 1997 to 2002. The graph shows that in 2002, there were about 600,000 acres of pasture and 1,300,000 acres of cropland, for a total of about 1,900,000 acres. That's 42 percent more than in 1997: $1{,}900{,}000 = 1.42x$; $x = 1{,}338{,}028$ acres in 1997.

The ratio of pasture to cropland was the same in 1997 as in 2006. In 2006, there were just a little over 1,000,000 acres of pasture—call it about 1,050,000 acres—and 1,900,000 acres of cropland, for a ratio of 10.5 to 19 or, eliminating the decimal, 21 to 38. Therefore, the ratio of pasture to all acreage is $21{:}(21 + 38)$ or 21:59. That can also be expressed as $\frac{21}{59}$ or about 0.36. Total acreage in 1997 was 1,338,028. Multiply: $1{,}338{,}028 \times 0.36 = 481{,}690$.

For ease of calculation, round this to 500,000. The answer choices are far enough apart that rounding will not lead you to an incorrect answer. According to the graph, there were about 2,200,000 acres of pasture in 2008. Your final step is to calculate the percent increase from 1997 to 2008:

$$\text{Percent change} = 100 \times \frac{\text{Difference in values}}{\text{Original value}}$$

$$100 \times \frac{2{,}200{,}000 - 500{,}000}{500{,}000} = 340\%$$

The denominator has been rounded up, so the actual percent change will be slightly greater than 340. The correct answer is **(E).**

19. C

This question concerns livestock in 2000, so look at the pie chart for that year.

According to the question, the ratio of sheep to turkeys is 1:4. In the pie chart, 4% of all the livestock represented is sheep, and 3% of the same total is hogs. Therefore, the ratio of hogs to sheep is 3:4. The two ratios both contain sheep, so set the number of units of sheep equal to compare hogs to turkeys:

$$\frac{\text{Hogs}}{\text{Sheep}} = \frac{3}{4}; \frac{\text{Sheep}}{\text{Turkeys}} = \frac{1}{4} \rightarrow \frac{4}{16}$$

$$\frac{\text{Hogs}}{\text{Turkeys}} = \frac{3}{16}$$

The association is selecting an animal from among hogs and turkeys, and you are asked for the probability that the selected animal is a turkey. Probability is expressed as a fraction, number of desired outcomes over the total number of outcomes, in its

most reduced form—in other words, a ratio. The ratio of turkeys to hogs plus turkeys is 16:(3 + 16), or 16:19. Thus, the correct answer is **(C).**

20. B

This question asks about pasture acreage, which is in the line graph. It also involves cows and sheep in 2000 and 2008, so you'll also use data about livestock in the two pie charts.

In 2000, beef cows represented 25% of all livestock in the pie chart. For purposes of setting up a ratio, think of this as 25 parts of the whole. Sheep were 4% of the livestock, or 4 parts of the whole. However, sheep use four times as much pasture as beef cows. To weight the sheep's pasture acreage accordingly, multiply their 4 parts by 4 to get 16 parts. You now have a total of $25 + 16 = 41$ parts, of which the beef cows used 25 parts. Put another way, the beef cows used $\frac{25}{41}$ of the pasture, or about 61%.

In 2000, there were about 550,000 acres of pasture, so beef cows used about $550{,}000 \times 0.61 = 335{,}500$ acres.

Now do the same calculations for 2008. Beef cows were 19% of the livestock (19 parts of the whole), and sheep were 2%. Again, sheep use four times as much pasture, so multiply their 2 parts of the whole by 4 to get 8 parts. There was a total of $19 + 8 = 27$ parts, and the beef cows used 19 parts. So the beef cows used $\frac{19}{27}$ of the pasture, or about 70%.

In 2008, there were about 2,200,000 acres of pasture, so beef cows used about $2{,}200{,}000 \times 0.7 = 1{,}540{,}000$ acres.

Finally, the question asks for the approximate increase from 2000 to 2008: $1{,}540{,}000 - 335{,}500 = 1{,}204{,}500$. The correct answer is **(B).**

ADVANCED MATH QUESTIONS PRACTICE SET 2

1.

$$\frac{x^2 + 4}{5} = \frac{x + 8}{3}$$

Quantity A	Quantity B
x	x^2

(A) Quantity A is greater.
(B) Quantity B is greater.
(C) The two quantities are equal.
(D) The relationship cannot be determined from the information given.

2. Five sixth-grade classes are competing against one another in an effort to raise money for a local charity. No class raised less than $60, and one class raised at least five times as much money as all the other classes combined.

Quantity A	Quantity B
The minimum average (arithmetic mean) amount raised by the five classes	$240

(A) Quantity A is greater.
(B) Quantity B is greater.
(C) The two quantities are equal.
(D) The relationship cannot be determined from the information given.

3.

Weight (kg)	Percent of sample
7	5%
8	10%
9	20%
10	10%
11	10%
12	5%
13	25%
14	10%
15	5%

A veterinarian is working with a dog breeders' association to analyze the weights of a sample of 240 healthy two-year-old purebred dachshunds. The veterinarian collects the above data.

Quantity A	Quantity B
median weight of the dogs in this sample	mean weight of the dogs in this sample

(A) Quantity A is greater.
(B) Quantity B is greater.
(C) The two quantities are equal.
(D) The relationship cannot be determined from the information given.

4.

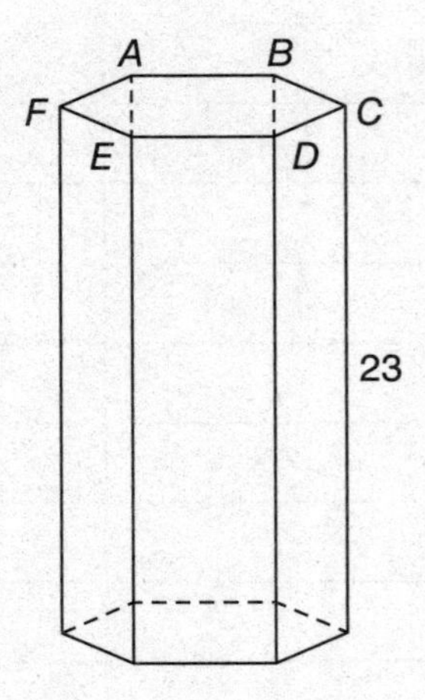

Hexagon *ABCDEF* is a regular hexagon with a perimeter of 48. What is the volume of the uniform prism shown?

(A) 96
(B) $96\sqrt{3}$
(C) 1,104
(D) $1{,}472\sqrt{3}$
(E) $2{,}208\sqrt{3}$

5.

$$3x^{\frac{2}{3}} + 2\sqrt[3]{x} - 8 = 0$$

Quantity A	Quantity B
x	$\frac{64}{25}$

(A) Quantity A is greater.
(B) Quantity B is greater.
(C) The two quantities are equal.
(D) The relationship cannot be determined from the information given.

6. Top-Notch Landscaping must mow sixteen 0.75-acre lots and twelve 1.5-acre lots to complete a certain job. Each of the company's landscapers can mow at a rate of 20 minutes per 0.5 acre.

Quantity A	Quantity B
The minimum number of landscapers needed to complete the job in 6 hours	4

(A) Quantity A is greater.
(B) Quantity B is greater.
(C) The two quantities are equal.
(D) The relationship cannot be determined from the information given.

7.

$$abcd \neq 0$$
$$ad > bc$$
$$bd < 0$$

Quantity A	Quantity B
$\frac{b}{a}$	$\frac{d}{c}$

(A) Quantity A is greater.
(B) Quantity B is greater.
(C) The two quantities are equal.
(D) The relationship cannot be determined from the information given.

8. The perimeters of rectangle Y and rectangle Z are equal. The lengths of the sides of rectangle Y are $x^2 + 21$ and $7x - 5$. The lengths of the sides of rectangle Z are 41 and 5. What is the area of rectangle Y ?

(A) 245
(B) 480
(C) 540
(D) 578
(E) 720

9. The average (arithmetic mean) of a, b, c, and d is 12. The average of b, c, d, and e is 17. What is the value of $3(e - a)$?

   Ⓐ $\frac{240}{7}$
   Ⓑ 48
   Ⓒ 54
   Ⓓ 58
   Ⓔ 60

10. The student population of a certain school increased $a\%$ from 1995 to 2005 and $b\%$ from 2005 to 2015. If the student population increased by 80% from 1995 to 2015 and the increase in the number of students in the second decade was three times the increase in the number of students in the first decade, what is the value of b?

    Ⓐ 20
    Ⓑ $26\frac{2}{3}$
    Ⓒ 50
    Ⓓ 60
    Ⓔ 75

11. The only items on a shelf are 8 green bins and 4 orange bins. If 3 bins are to be selected from the shelf, one after the other, at random and without replacement, what is the probability that at least one green bin is selected?

    Give your answer as a fraction.

12. A store calls the original price of an item the X-level price. The store reduces the X-level price of an item by 30%, resulting in what the store calls the Y-level price of the item. The store then reduces the Y-level price of the item by 20%, resulting in what the store calls the Z-level price of the item. The Z-level price of the item is less than $49. Which of the following statements must be true?

    Indicate all such statements.

    [A] The X-level price of the item is less than $86.00.
    [B] When the Z-level price of the item is subtracted from the Y-level price of the item, the result is less than $16.00.
    [C] The Y-level price of the item is a percent greater than the Z-level price of the item, where $0 < a < 30$.

13. The ratio of w to x is 5:4. The ratio of x to y is 8:1. If the ratio of y:z is 1:3, what is the ratio of $w + x$ to z ?

    (A) 13:3
    (B) 6:1
    (C) 13:2
    (D) 12:1
    (E) 18:1

14. ♥ x ♥ y ♥ $= (2x - y)^2$ and ☻a☻b☻ $= (a + 2b)^2$. If ☻ 5 ☻ $(n-1)$ ☻ $= 5n^2 + 4n$ and $n < 6$, then ♥ $2n$ ♥ 2 ♥ =

    (A) 4
    (B) 9
    (C) 16
    (D) 36
    (E) 49

15. 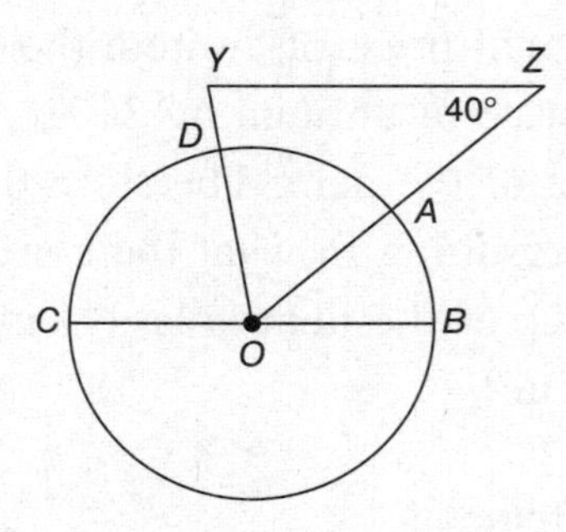

$\overline{BC}$ is a diameter of circle O and is parallel to $\overline{YZ}$. If the length of arc ABC is 22π and the length of minor arc CD is 11π, then what is the positive difference between the degree measures of $\angle OYZ$ and $\angle YOZ$?

16. In the sequence $a_1, a_2, a_3, \ldots, a_n, \ldots$, each term after the first is equal to r times the previous term, where $a_1 > 0$ and $r > 1$. If $a_1 + a_2 + a_3 = 21$ and $a_2 + a_3 = 18$, what is the value of $a_3 + a_4 + a_5$?

17. Blanca took a non-stop car trip that encompassed three different sections of roadways. Each of the three sections covered the same distance. Blanca averaged 45 miles per hour over the first section, 60 miles per hour over the second section, and 54 miles per hour for the entire trip. What was her average speed for the third section to the nearest mile per hour?

Questions 18–20 refer to the following data.

Percent of Housing Units by Year Built

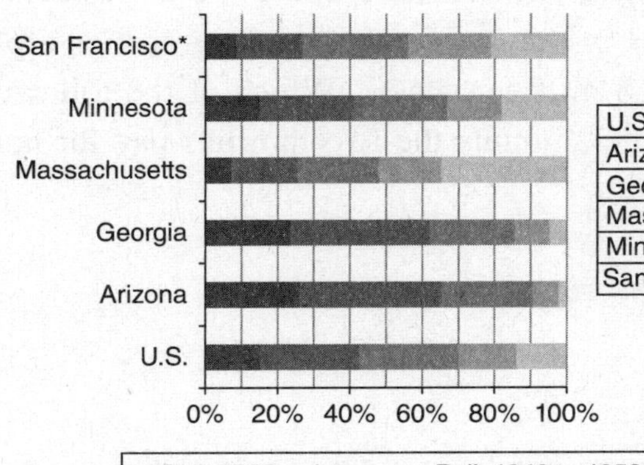

Total Housing Units

U.S.	132,057,804
Arizona	2,859,768
Georgia	4,094,812
Massachusetts	2,808,549
Minnesota	2,353,932
San Francisco area	1,747,506

■ Built 2000 or later ■ Built 1940 to 1959
■ Built 1980 to 1999 ■ Built 1939 or earlier
■ Built 1960 to 1979

Source: 2009–2013 American Community Survey 5-Year Estimates, DP04: Selected Housing Characteristics, census.gov

U.S. Housing Cost as a Percent of Income: Owners with a Mortgage

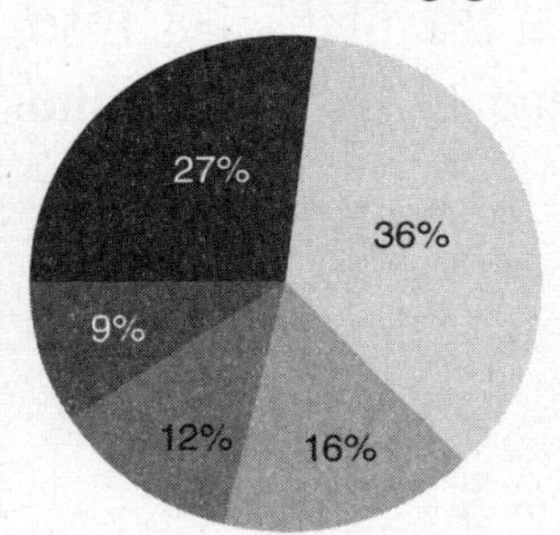

Total: 49.5 million

■ < 20.0% ■ 30.0% to 34.9%
■ 20.0% to 24.9% ■ ≥ 35.0%
■ 25.0% to 29.9%

U.S. Housing Cost as a Percent of Income: Renters

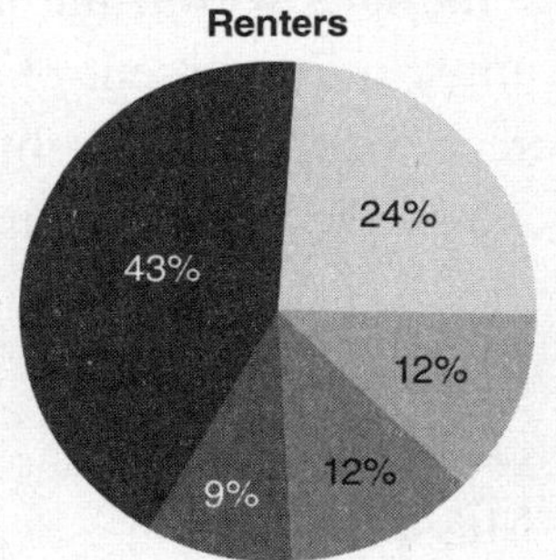

Total: 37.5 million

■ < 20.0% ■ 30.0% to 34.9%
■ 20.0% to 24.9% ■ ≥ 35.0%
■ 25.0% to 29.9%

18. Family A earns $60,000 and rents its residence, and this family is in the category of housing cost as a percent of income that contains about 3.5 million renters. Family B makes $85,000 and owns its residence with a mortgage, and this family is in the category of housing cost as a percent of income that contains about 8 million owners. Which of the following could be the difference between the amounts the two families pay for housing?

    Indicate all such values.

    [A] $14,500
    [B] $8,200
    [C] $4,000
    [D] $3,700
    [E] $3,165
    [F] $0

19. A real estate investment firm is evaluating housing units built after 1939 and before 1980 for their suitability for profitable renovation and resale. Of the markets within the United States shown, where are there the least such units, if 20 percent of all housing built in the 1940s and 1950s and 60 percent of all housing built in the 1960s and 1970s is not within the price range the firm is considering?

    (A) Arizona
    (B) Georgia
    (C) Massachusetts
    (D) Minnesota
    (E) San Francisco

20. A national survey conducted in 2014 randomly selected participants who either rent or have a mortgage and asked them: “Do you pay 30 percent or more of your income on housing?” Participants who responded yes completed survey A. Those who responded no completed survey B. Then a prize winner was drawn randomly from all survey participants. What is the approximate ratio of the probability that the prize winner completed survey A and lived in a residence more than 75 years old at the time of the survey to the probability that the prize winner completed survey B and lived in a residence built in the previous 15 years? Assume that percent of income spent on housing is proportionally distributed across residents of housing of varying ages.

(A) $\frac{9}{20}$

(B) $\frac{2}{3}$

(C) $\frac{9}{10}$

(D) $\frac{13}{10}$

(E) $\frac{3}{2}$

ADVANCED MATH QUESTIONS PRACTICE SET 2 ANSWER KEY

1. B
2. A
3. B
4. E
5. B
6. C
7. D
8. B
9. E
10. C
11. $\frac{54}{55}$
12. B, C
13. B
14. D
15. 80
16. 84
17. 60
18. D, E, F
19. A
20. B

ADVANCED MATH QUESTIONS PRACTICE SET 2 ANSWERS AND EXPLANATIONS

1. B

Begin by solving the equation $\frac{x^2 + 4}{5} = \frac{x + 8}{3}$ for the possible values of x.

Cross-multiply to get $3(x^2 + 4) = 5(x + 8)$. Multiplying out each side produces $3x^2 + 12 = 5x + 40$.

Subtracting $5x$ from both sides, you have $3x^2 - 5x + 12 = 40$. Subtracting 40 from both sides gives $3x^2 - 5x - 28 = 0$.

Now factor $3x^2 - 5x - 28$ into a product of binomials of the form $(x + a)(3x + b)$, where a and b are constants. With some testing, it turns out that $3x^2 - 5x - 28 = (x - 4)(3x + 7)$.

So $(x - 4)(3x + 7) = 0$. When the product of a group of factors is 0, at least one of the factors must be 0, so either $x - 4 = 0$ or $3x + 7 = 0$. If $x - 4 = 0$, then $x = 4$. If $3x + 7 = 0$, then $3x = -7$, and $x = -\frac{7}{3}$.

Quantity A could be 4 or $-\frac{7}{3}$. Quantity B could be 16 or $\frac{49}{9} = 5.\bar{4}$. Because the smallest possible value for Quantity B is greater than the largest possible value for Quantity A, Choice **(B)** is correct.

2. A

From the centered information, you know that four of the five classes—excluding the high-performing class—raised at least \$240 altogether. The high-performing class raised at least five times as much as that. So to find the minimum amount raised by the high-performing class, multiply \$240 by 5.

To calculate the value of Quantity A, you'll simply find the minimum total amount raised by all five classes and divide by 5. That's

$$\frac{\$240 + (5)(\$240)}{5} = \$288$$

Note that a bit of critical thinking obviates this final calculation, though. Once you recognize that you'll be multiplying \$240 by 6 to get the total amount raised, then dividing by 5 to get the average, you know that the average must be greater than \$240. Compare, don't calculate!

The correct answer is **(A)**.

3. B

The question presents data about the relative frequency of 240 dogs' weights.

Quantity A, the median weight of this sample, is the average weight of the 120th and 121st dogs when their weights are in order from least to greatest. The table is already in order from least to greatest, so add the percentages until you find where the midpoint of the sample, 50%, falls. That's $5 + 10 + 20 + 10 = 45\%$—not there yet. So keep going: $45 + 10 = 55\%$. That's over the midpoint, so the dogs whose weights you'd average to get the median weight are in the last category added, which is the 11 kg category. The median weight of dogs in this sample is 11 kg.

Calculating Quantity B, the mean weight, requires calculating a weighted average. Instead of working with the percentages of 240 dogs given, which would take a lot of time, remember that percentages represent ratios out of a whole of 100. Thus, $5\% = \frac{5}{100} = \frac{1}{20}$. Every 5% is 1 part out of 20, so represent 5% as 1, 10% as 2, 15% as 3, 20% as 4, and 25% as 5 parts. This approach maintains the proportionality of the categories while making the numbers easier to work with. In your weighted average formula, divide by the total of 20 parts.

$$\frac{1 \times 7 + 2 \times 8 + 4 \times 9 + 2 \times 10 + 2 \times 11 + 1 \times 12 + 5 \times 13 + 2 \times 14 + 1 \times 15}{20} =$$

$$\frac{7 + 16 + 36 + 20 + 22 + 12 + 65 + 28 + 15}{20} =$$

$$\frac{221}{20}$$

You can easily estimate that this fraction is slightly over 11 (since $\frac{220}{20} = 11$). Quantity A is exactly 11, so Quantity B is greater.

The correct answer is **(B)**.

4. E

The solid shown is a uniform hexagonal prism. While this is certainly an unusual shape, the volume formula is the same for all uniform solids: *Volume* = *Area of Base* × *height*. (Note: a uniform solid is one whose dimensions do not change over the height of the solid. An example of a *non*-uniform solid would be a pyramid or a cone, which is wide at the base and narrow at the top.) For this question, then, if you knew the area of hexagon *ABCDEF*, you could multiply it by 23 to find the volume of the solid.

The question states that *ABCDEF* is regular, which means that all its angles are equal and all its sides are the same length. Since the perimeter is 48, each side must have a length $48 \div 6 = 8$. To find the area of *ABCDEF*, note that a regular hexagon divides into 6 congruent equilateral triangles, like this:

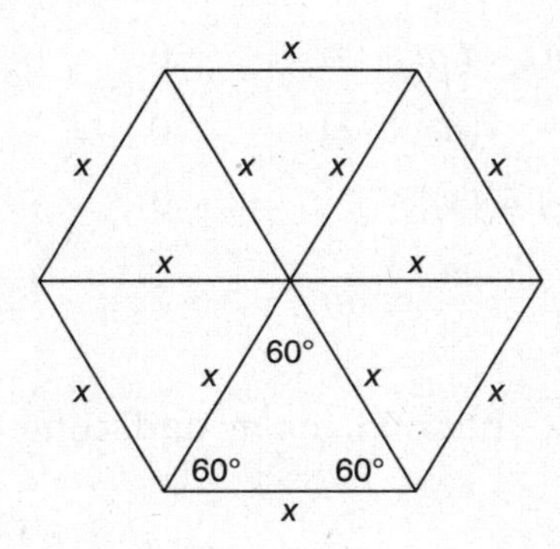

To find the area of the whole hexagon, find the area of one of the triangles and multiply it by 6 (since there are six triangles and they're all the same). In this question, each triangle looks like this:

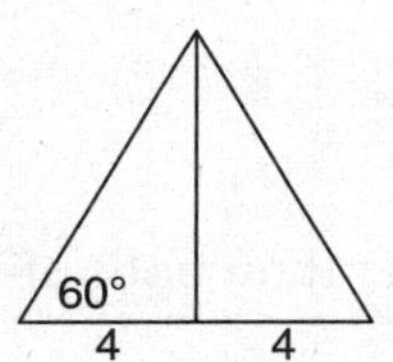

By the properties of 30-60-90 triangles, the height must be $4\sqrt{3}$, so the area of each triangle is $\frac{1}{2}(8)(4\sqrt{3}) = 16\sqrt{3}$. Multiply by 6 to get the total area of the hexagon: $6 \times 16\sqrt{3} = 96\sqrt{3}$.

To find the volume of the solid, multiply by 23:

$$\text{Volume} = 23 \times 96\sqrt{3} = 2{,}208\sqrt{3}$$

Choice **(E)** is correct.

5. B

Strange-looking equations on the GRE are often quadratics in disguise. Note that $x^{\frac{2}{3}}$ is actually just $\sqrt[3]{x}$ squared. To make the equation look more manageable, try letting $\sqrt[3]{x} = m$. Then the equation becomes $3m^2 + 2m - 8 = 0$. Factoring this equation is made more challenging by the coefficient in front of the m^2 term. To factor an expression of the form $ax^2 + bx + c$, look for two numbers that multiply to ac and sum to b. In this case, the two numbers that multiply to $(3)(-8) = -24$ and sum to 2 are 6 and -4. Rewrite the equation, substituting $6m - 4m$ for $2m$. Then factor and solve for m:

$$\begin{aligned} 3m^2 + 2m - 8 &= 0 \\ 3m^2 + 6m - 4m - 8 &= 0 \\ 3m(m+2) - 4(m+2) &= 0 \\ (3m-4)(m+2) &= 0 \\ m = \frac{4}{3} \text{ or } m &= -2 \end{aligned}$$

At this point, substitute $\sqrt[3]{x}$ back in for m and cube each equation to find the possible values of x:

$$\begin{aligned} \sqrt[3]{x} &= \frac{4}{3} \text{ or } \sqrt[3]{x} = -2 \\ x &= \left(\frac{4}{3}\right)^3 \text{ or } x = (-2)^3 \\ x &= \frac{64}{27} \text{ or } x = -8 \end{aligned}$$

Because Quantity B, $\frac{64}{25}$, is greater than both of the possible values of x, **(B)** is the correct answer.

6. C

This is a combined work problem—the fact that several people are working together to complete a task tells you that. So you need to know how big the task is, and at what rate the workers can tackle it. The first sentence tells you that the landscapers have to mow a total of $16(0.75) + 12(1.5) = 12 + 18 = 30$ acres, and the second sentence provides each landscaper's mowing rate.

One way to figure out how many landscapers are needed is to find out how long it would take a single landscaper to complete the task. There are various ways to do that calculation. You might think like this: 20 minutes per 0.5 acre = 40 minutes per 1 acre, so the entire task takes 40 minutes times 30 acres, or 1,200 minutes, which is 20 hours.

Alternatively, you could convert the given rate to its inverse. The inverse of 20 minutes per half-acre is three half-acres per hour. If you consider one "job" to be a half-acre of mowing, then there are 60 "jobs" that must be completed:

$$\frac{60 \text{ jobs}}{\frac{3 \text{ jobs}}{1 \text{ hour}}} = 60 \text{ jobs}\left(\frac{1 \text{ hour}}{3 \text{ jobs}}\right) = 20 \text{ hours}$$

From here, you can simply test Quantity B to see whether it's too small, too big, or just right. If one landscaper could finish the job in 20 hours, then 4 landscapers working together at the same rate could finish the job in 5 hours; that's a bit faster than needed, so it might look as though the answer is (B). But what happens if you reduce the number of landscapers to 3?

$$\frac{20\text{ hours}}{3} = 6\frac{2}{3}\text{ hours}$$

...so 3 landscapers is too few. At least 4 landscapers are needed to finish the job in 6 hours or less, so the correct answer is **(C).**

7. D

The first line of the centered information indicates that none of the variables is equal to 0. The second line shows a relationship between ad and bc, but neither of those is one of the expressions being compared. The third line provides the additional information that b and d must have different signs since their product is negative. Because the compared quantities are given as fractions, you will need to use division in your calculations. And since you're dealing with inequalities, when you divide, you must take into account whether you are dividing by a positive or a negative number because the latter will "flip" the inequality.

Since you know that $bd < 0$, you can divide $ad > bc$ by that term, flipping the inequality accordingly, so that $\frac{ad}{bd} < \frac{bc}{bd}$ and $\frac{a}{b} < \frac{c}{d}$. That's starting to look a lot like the values in the two columns.

But there's one more complication here: the question asks for the relationship between $\frac{b}{a}$ and $\frac{d}{c}$. As long as both fractions have the same sign, taking the reciprocal of both sides of an inequality changes the relationship. For instance, $\frac{1}{2} < \frac{3}{2}$, but of course $\frac{2}{1} > \frac{2}{3}$. If both fractions are negative, the same relationship still holds: $-\frac{2}{3} < -\frac{2}{5}$, but $-\frac{3}{2} > -\frac{5}{2}$. Before you choose (A), though, consider that one fraction might be negative and the other positive: say $a = 1$, $b = -2$, $c = 1$ and $d = 1$. Now, $\frac{b}{a} = -2$ and $\frac{d}{c} = 1$, making Quantity B greater. In fact, as long as the fraction on the left is negative and the fraction on the right is positive, taking the reciprocal will not alter the relationship. So more than one relationship is possible, and the answer is **(D).**

Note: Another way to see that multiple relationships are possible is to divide both sides of the centered inequality by a and c:

$$ad > bc$$
$$\frac{d}{c} \; ? \; \frac{b}{a}$$

The centered information tells you that b and d have opposite signs, but it's silent about the signs of a and c. Thus, after dividing both sides of the inequality by ac, you don't know whether or not you should flip the inequality sign. This is another way to determine that it's impossible to tell which of Quantity A or B is greater, and the correct answer must be **(D).**

8. B

The question states that the perimeters of rectangles Y and Z are equal. You can calculate the perimeter of rectangle Z given that the sides are 41 and 5:

$$\begin{aligned} \text{Perimeter } Z &= 2(41) + 2(5) \\ &= 82 + 10 \\ &= 92 \end{aligned}$$

Thus, the perimeter of rectangle Y must also equal 92. Set up the equation for the perimeter and simplify.

$$\begin{aligned} 2\left[\left(x^2 + 21\right) + (7x - 5)\right] &= 92 \\ 2\left(x^2 + 21 + 7x - 5\right) &= 92 \\ 2\left(x^2 + 7x + 16\right) &= 92 \\ x^2 + 7x + 16 &= 46 \\ x^2 + 7x - 30 &= 0 \end{aligned}$$

Factoring the equation yields $(x - 3)(x + 10)$. So x comes out to -10 or 3. The lengths of the sides of rectangle Y are $x^2 + 21$ and $7x - 5$.

If $x = 3$, then the lengths of the sides of rectangle Y are $x^2 + 21 = 3^2 + 21 = 9 + 21 = 30$ and $7x - 5 = 7(3) - 5 = 21 - 5 = 16$

If $x = -10$, then the lengths of the sides of rectangle Y are $x^2 + 21 = (-10)^2 + 21 = 100 + 21 = 121$ and $7x - 5 = 7(-10) - 5 = -70 - 5 = -75$

So if $x = -10$, the length of one of the sides would be negative, which is not possible. It follows that x must equal 3 and the lengths of the sides of rectangle Y must be 30 and 16.

The area of any rectangle is length times width. The area of rectangle Y is $30 \times 16 = 480$. The correct answer is **(B).**

9. E

The average formula is: Average $= \frac{\text{Sum of terms}}{\text{Number of terms}}$.

Since the average of a, b, c, and d is 12, you can write the following equation: $\frac{a + b + c + d}{4} = 12$. Eliminate the fraction by multiplying both sides of this equation by 4 to yield $a + b + c + d = 48$.

Use the fact that the average of b, c, d, and e is 17 to write another equation: $\frac{b + c + d + e}{4} = 17$. Again, eliminate the fraction by multiplying both sides of this equation by the 4 in the denominator: $b + c + d + e = 68$.

You now have the two equations $a + b + c + d = 48$ and $b + c + d + e = 68$. The fastest way to the correct answer at this point is to use combination. Subtract the first equation from the second equation: $(b + c + d + e) - (a + b + c + d) = 68 - 48$. That yields $e - a = 20$. Since the question asks for the value of $3(e - a)$, multiply 20 by 3. That's 60. Choice **(E)** is correct.

10. C

Since you're starting with an unknown value and taking percent increases, this problem represents an excellent opportunity to Pick Numbers. If the starting population (1995) were 100, then the final population (2015) would be 180, since there's an overall percent increase of 80%. Now, the second population increase was three times the first population increase, so the total increase of 80 students would equal the population increase from 1995 to 2005 plus three times that increase:

$x + 3x = 80$, where $x =$ the increase from 1995 to 2005.

So if x, the first population increase, is 20, then the increase from 2005 to 2015 is 3 times that, or 60. Apply the percent change formula (be careful to use the appropriate original value—the population in 2005—in the denominator):

$$\frac{180 - 120}{120} \times 100\% = \frac{60}{120} \times 100\% = 50\%$$

Double check that this represents the value that you're looking for—the second percent increase—before moving on.

You could certainly also solve this using algebra, though it would very likely take longer:

Let p represent the initial population. There are two population increases such that the second is three times the first, and the sum of the two population increases is equal to 80% of p. Use $\frac{a}{100}$ to represent the first percent increase. That means that the first population increase is $\frac{pa}{100}$. The second population increase is 3 times the first, or $\frac{3pa}{100}$. The first and second population increases sum to 80% of p, so you can set up this equation:

$$\frac{pa}{100} + \frac{3pa}{100} = \frac{80p}{100}$$

This equation simplifies to $a + 3a = 80$ (note the similarity to the first equation in the solution above), allowing you to solve for a, the first percent increase. Now that you know that $a = 20$, you can write an equation that represents the change in school population from 2005 to 2015:

$$1.2p + 1.2p\left(\frac{b}{100}\right) = 1.8p$$

Divide all terms by p and then subtract 1.2 from both sides of the equation to simplify: $\frac{1.2b}{100} = 0.6$, so $1.2b = 60$, and $b = 50$.

Choice **(C)** is correct.

11. $\mathbf{\frac{54}{55}}$

The sum of the probability that an event occurs and the probability that it doesn't occur is equal to 1.

In this case, it is easier to find the probability of the opposite of what the question is asking for, and then subtract that probability from 1. The opposite of at least one green bin being selected is that no green bins are selected. That is, all selected bins are orange.

There are 4 orange bins out of a total of $8 + 4 = 12$ bins. The probability that the first bin selected is orange is $\frac{4}{12} = \frac{1}{3}$.

If the first bin selected is orange, there are 8 green bins and 3 orange bins left on the shelf. The probability that the second bin selected is also orange is $\frac{3}{11}$.

If the first two bins selected are orange, the probability that the third bin is also orange is $\frac{2}{10} = \frac{1}{5}$.

To find the probability that all 3 bins selected are orange, multiply these three probabilities:

$$\frac{1}{\cancel{3}} \times \frac{\cancel{3}}{11} \times \frac{1}{5} = \frac{1}{1} \times \frac{1}{11} \times \frac{1}{5} = \frac{1}{55}.$$

Since that is the opposite of the event the question asks for, subtract this probability from 1 to get the answer:

$$1 - \frac{1}{55} = \frac{55}{55} - \frac{1}{55} = \frac{55 - 1}{55} = \frac{54}{55}$$

The correct answer is $\mathbf{\frac{54}{55}}$.

12. B, C

Set a variable for the original price, the X-level price, such as T.

After the reduction of the X-level price by 30% to the Y-level price, the Y-level price of the item, in dollars, is $T - 0.3T = 0.7T$.

After the reduction of the Y-level price by 20% to the Z-level price, the Z-level price of the item, in dollars, is $0.7T - 0.2(0.7T) = 0.7T - 0.14T = 0.56T$.

Thus, the prices can be written as follows

X-level = T dollars
Y-level = $0.7\ T$ dollars
Z-level = $0.56\ T$ dollars

Consider statement A.

The question stem says that the Z-level price of the item is less than 49 dollars. Using $0.56T$ for the Z-level price, $0.56T < 49$.

$T < \frac{49}{0.56}$; since $\frac{49}{0.56} = 87.50$, it follows that $T < 87.50$.

So the original price was less than $87.50. However, the original price was not necessarily less than $86.00. Statement A is not necessarily true.

Next, consider statement B.

Find the result of subtracting the Z-level price from the Y-level price, in terms of T:

$0.7T - 0.56T = 0.14T$

Since the Z-level price of the item is less than $49, you know that $0.56T < 49$. Notice that 0.14 is exactly $\frac{1}{4}$ of 0.56.

Divide both sides of the inequality $0.56\,T < 49$ by 4:

$$\frac{0.56\,T}{4} < \frac{49}{4}$$
$$0.14\,T < 12.25$$

So the the result of subtracting the Z-level price from the Y-level price is less than \$12.25. That means it's certainly less than \$16. Statement B must be true.

Finally, consider statement C.

Use the percent change formula to determine what percent greater than the Z-level price the Y-level price is (the calculation is identical to finding the percent increase from the Z-level price to the Y-level price):

$$\frac{0.7\,T - 0.56\,T}{0.56\,T} \times 100\% = \frac{0.14\,T}{0.56\,T} \times 100\% = \frac{14}{56} \times 100\% = \frac{1}{4} \times 100\% = 25\%.$$

So $a = 25$, and it is true that $0 < a < 30$. Statement C must be true.

(B) and **(C)** are the correct answers.

13. B

You are given 3 ratios of various combinations of *w*, *x*, *y*, and *z* and asked to manipulate them in some manner to determine the value of $\frac{w+x}{z}$. One way to approach this task would be to find the ratios of *w*:*z* and *x*:*z* individually, then combine them. Note that $\left(\frac{w}{x}\right)\left(\frac{x}{y}\right)\left(\frac{y}{z}\right) = \frac{w}{z}$ because the *x* and *y* terms cancel out. Substituting the known values, $\left(\frac{5}{4}\right)\left(\frac{8}{1}\right)\left(\frac{1}{3}\right) = \frac{10}{3}$. Similarly, $\left(\frac{x}{y}\right)\left(\frac{y}{z}\right) = \frac{x}{z} = \left(\frac{8}{1}\right)\left(\frac{1}{3}\right) = \frac{8}{3}$.

The fraction $\frac{w+x}{z}$ can be expressed as $\frac{w}{z} + \frac{x}{z} = \frac{10}{3} + \frac{8}{3} = \frac{18}{3} = \frac{6}{1}$.

Another approach to solving this problem would be to look for commonalities among the given ratios. Since *x*:*y* is 8:1 and *z*:*y* is 3:1, it follows that *x*:*y*:*z* = 8:1:3. The ratio of *w* to *x* is stated to be 5:4. That's the same as 10:8, which results in a common value for *x*. Therefore, *w*:*x*:*y*:*z* = 10:8:1:3, and $\frac{w+x}{z} = \frac{10+8}{3} = \frac{18}{3} = \frac{6}{1}$.

Finally, you might consider Picking Numbers. Say $y = 1$. Then $x = 8$, $z = 3$, and $w = 10$. Then the answer is, again, $\frac{w+x}{z} = \frac{10+8}{3} = \frac{18}{3} = \frac{6}{1}$.

Choice **(B)** is correct.

14. D

Symbolism questions frighten a lot of students, but symbols are just odd-looking functions, and questions featuring them are typically best solved by substitution. Start by replacing the wacky shapes with simple function notation. Use an "h" for the hearts and an "s" for the smileys:

$$\begin{aligned} h(x, y) &= (2x - y)^2 \\ s(a, b) &= (a + 2b)^2 \end{aligned}$$

Now the question becomes: given that $s(5, n - 1) = 5n^2 + 4n$ and $n < 6$, what is $h(2n, 2)$? Begin by plugging 5 and $n - 1$, respectively, into function s:

$$s(5, n - 1) = [5 + 2(n - 1)]^2 = (5 + 2n - 2)^2 = (2n + 3)^2 = 4n^2 + 12n + 9$$

The question says that this equals $5n^2 + 4n$, so set the quantities equal and solve the quadratic:

$$\begin{aligned} 4n^2 + 12n + 9 &= 5n^2 + 4n \\ n^2 - 8n - 9 &= 0 \\ (n - 9)(n + 1) &= 0 \\ n &= 9 \text{ or } n = -1 \end{aligned}$$

The question states that $n < 6$, so n must equal -1. Now that you know that $n = -1$, you can find $h(2n, 2)$:

$$\begin{aligned} h(2n, 2) &= h[(2)(-1), 2] = h(-2, 2) \\ h(-2, 2) &= [2(-2) - 2]^2 = (-4 - 2)^2 = (-6)^2 = 36 \end{aligned}$$

The correct answer is **(D)**.

15. 80

The key to beating a complex figures problem like this one is to take a deep breath and start filling in what you know. It follows from the properties of parallel lines crossed by a transversal that $\angle AOB$ must be 40°. This is a handy fact because it means that arc ABC spans $180° + 40° = 220°$ of the circle. You now know arc ABC and its corresponding central angle. The question provides the length of arc CD (11π), which means that you can set up a proportion to solve for arc CD's central angle (COD):

$$\begin{aligned} \frac{\angle COD}{11\pi} &= \frac{220}{22\pi} \\ \angle COD &= \frac{220 \times 11\pi}{22\pi} = 220 \div 2 = 110 \end{aligned}$$

Since $\angle COD = 110°$, it follows from the properties of parallel lines crossed by a transversal that $\angle OYZ = 110°$ as well. The final angle of the triangle, $\angle YOZ$, is therefore $180° - 40° - 110° = 30°$. After doing all this work, take care to answer the right question! You need the positive difference between $\angle OYZ$ and $\angle YOZ$, which is $110 - 30 = 80$. The correct answer is **80.**

Note: Remember that GRE figures are not necessarily drawn to scale. The one in this question isn't. If you tried to estimate based on the appearance of the figure, you most probably did not get the correct answer. Remember to use geometric reasoning to arrive at your answers on Test Day rather than estimation.

16. 84

Each term after the first term is r times the previous term, so:

$$\begin{aligned} a_2 &= ra_1 \\ a_3 &= r(ra_1) = r^2a_1 \\ a_4 &= r\left(r^2a_1\right) = r^3a_1 \\ a_5 &= r\left(r^3a_1\right) = r^4a_1 \end{aligned}$$

Subtract the second equation given in the problem from the first equation to solve for a_1:

$$\begin{aligned} a_1 + a_2 + a_3 &= 21 \\ -(a_2 + a_3 &= 18) \\ \hline a &= 3 \end{aligned}$$

The next four terms are as follows:

$$\begin{aligned} a_2 &= 3r \\ a_3 &= 3r^2 \\ a_4 &= 3r^3 \\ a_5 &= 3r^4 \end{aligned}$$

Next, solve for r using the equation $a_1 + a_2 + a_3 = 21$.

$$\begin{aligned} 3 + 3r + 3r^2 &= 21 \\ 3(1 + r + r^2) &= 21 \\ 1 + r + r^2 &= 7 \\ -6 + r + r^2 &= 0 \\ (r + 3)(r - 2) &= 0 \\ r &= -3 \text{ or } 2 \end{aligned}$$

The question states that $r > 1$, so $r = 2$.

Thus,

$$\begin{aligned} a_2 &= 6 \\ a_3 &= 12 \\ a_4 &= 24 \\ a_5 &= 48 \end{aligned}$$

The value of $a_3 + a_4 + a_5 = 12 + 24 + 48 =$ **84.**

Alternatively, once you've figured out that $a_1 = 3$, you can find r very quickly by trial and error. The first three terms sum to 21, so r can't be all that big, and the fact that both 18 and 21 are integers suggests that r is probably also an integer. It has to be greater than 1, so try $r = 2$. That produces $a_2 = 6$ and $a_3 = 12$. Score! The fact that $3 + 6 + 12 = 21$ confirms that $r = 2$, and now it's just a matter of calculating two more terms and adding $a_3 + a_4 + a_5$, as above.

17. 60

When solving multi-part journey problems, creating a DiRT box table (D = *distance*, R = *rate of speed*, T = *time spent traveling*) with the known variables filled in can greatly simplify the problem. This particular problem provides very little directly quantifiable information, but fill in what's given, as shown in this table:

	Distance	Rate	Time
Section 1		45 mph	
Section 2		60 mph	
Section 3		?	
Total		54 mph	

Since the distance is the same for each section, simply pick a number for a distance that will work efficiently with the numbers provided in the problem. The most convenient number for the distance will be the least common multiple (LCM) of the given rates on the legs of the journey. The LCM of 45 and 60 is 180. Plugging 180 in as the distance for each of the sections yields:

	Distance	Rate	Time
Section 1	180 m	45 mph	
Section 2	180 m	60 mph	
Section 3	180 m	?	
Total	540 m	54 mph	

Using the Distance formula, D = RT, allows easy calculation of the time for each leg of the journey as well as the cumulative time for the total journey:

	Distance	Rate	Time
Section 1	180 m	45 mph	4 h
Section 2	180 m	60 mph	3 h
Section 3	180 m	?	?
Total	540 m	54 mph	10 h

Since the total time is 10 hours, solving for the time it takes to complete the Section 3 leg simply means subtracting the time of the first two legs from the total time: $10 - 4 - 3 = 3$ hours.

Now use D = RT to find that the 180 miles traveled in the third section's 3 hours yields an average rate for that section of 180 miles ÷ 3 hours = **60 mph.**

Algebraic Explanation

The problem can also be solved algebraically, by use of the same helpful table for establishing a firm basis to find the unknown x:

	Distance	Rate	Time
Section 1		45 mph	
Section 2		60 mph	
Section 3		x	
Total		54 mph	

Since the distance is the same for each section, but is not identified, use a variable to stand in for that distance. Using y for each section's distance yields:

	Distance	Rate	Time
Section 1	y	45 mph	
Section 2	y	60 mph	
Section 3	y	x	
Total	$3y$	54 mph	

The time column is still blank, but the times for each section can be calculated using the formula $\text{Time} = \frac{\text{Distance}}{\text{Rate}}$.

	Distance	Rate	Time
Section 1	y	45 mph	$y/45$
Section 2	y	60 mph	$y/60$
Section 3	y	x	y/x
Total	$3y$	54 mph	

Since the Time column of the DiRT box is cumulative, the total time for the trip is the sum of the 3 individual sections' times, and that total will satisfy the **Average Rate = Total Distance ÷ Total Time** equation, so you can write the following equation:

$$\frac{y}{45} + \frac{y}{60} + \frac{y}{x} = \frac{3y}{54}$$

Divide both sides by y and reduce:

$$\frac{1}{45} + \frac{1}{60} + \frac{1}{x} = \frac{3}{54} = \frac{1}{18}$$

Multiply both sides by the LCM, $180x$:

$$\frac{180x}{45} + \frac{180x}{60} = \frac{180x}{x} = \frac{180x}{18}$$

Simplify:

$$\begin{aligned} 4x + 3x + 180 &= 10x \\ 7x + 180 &= 10x \\ 180 &= 3x \end{aligned}$$

So x, which represents the average rate of speed for the third section in miles per hour, equals **60.**

18. D, E, F

This question concerns housing cost as a percent of income, so look at the pie charts.

Family A rents—look at the pie chart that represents renters. This family is in the category that contains about 3.5 million people. Total renters are 37.5 million, and 3.5 million is a little less than 10% of that total. Only one category fits, the category with 9% of renters. Therefore, family A pays 30.0% to 34.9% of its income on housing. Family A earns \$60,000, so it pays between $0.30 \times \$60{,}000 = \$18{,}000$ and (round 34.9% to 35%) $0.35 \times \$60{,}000 = \$21{,}000$ on housing.

Family B owns a home with a mortgage—look at the pie chart that represents owners. This family is in the category that contains about 8 million people. The total of owners with a mortgage is 49.5 million. To make the calculation easy, round that to 50 million and set up a proportion to find what percent 8 million is of the total:

$$\frac{8}{50} = \frac{x}{100}$$

You double 50 to turn it into 100, so double 8 to find that $x = 16$.

Family B is in the category that contains about 16% of owners with a mortgage, so it pays 20.0% to 24.9% of income on housing. This family earns $85,000, so it pays between $0.20 \times \$85{,}000 = \$17{,}000$ and (round 24.9% to 25%) $0.25 \times \$85{,}000 = \$21{,}250$ on housing.

The ranges of the two families' housing costs overlap, so it's possible that there is no difference between what the families pay for housing. The maximum difference occurs when family A pays the greatest amount ($21,000) and family B pays the least amount ($17,000): $\$21{,}000 - \$17{,}000 = \$4{,}000$. Remember that you rounded 34.9% up a little to calculate the $21,000, so the actual possible difference is a little less than $4,000 (if you didn't round, you know it's $3,940). The correct answers are from $0 to a little less than $4,000. That's **(D), (E),** and **(F).**

19. A

This question concerns the year housing units were built, so look at the stacked bar graph for information. The question asks about units built between 1940 and 1979, inclusive, so look at the third and fourth segments of the bars. You need to find the area with the least units in this date range that also meet the price range criterion, meaning you will consider only $100 - 20 = 80\%$ of the 1940–1959 units and $100 - 60 = 40\%$ of the 1960–1979 units.

Note that Arizona and Massachusetts have very similar total units. However, Massachusetts has a much greater percentage of its units in the relevant decades (about 40% to Arizona's 30%), and of those a much greater proportion are in the more heavily weighted 1940–1959 category. Therefore, without doing any calculations, you can eliminate **(C)** Massachusetts; Arizona must have fewer units.

Also consider Georgia and Arizona, which have very similar percentages of their housing units built in the decades of interest. Georgia, however, has many more total housing units, so **(B)** Georgia can be eliminated.

Do the calculations for Arizona, Minnesota, and San Francisco:

Arizona:

1940–1959: $0.80 \times 8\% = 6.4\%$

1960–1979: $0.40 \times 25\% = 10.0\%$

Total % of units: $6.4\% + 10.0\% = 16.4\%$

0.164×2.86 million units $= 469{,}040$ units

Minnesota:

1940–1959: $0.80 \times 15\% = 12.0\%$

1960–1979: $0.40 \times 26\% = 10.4\ \%$

Total % of units: $12.0\% + 10.4\% = 22.4\%$

$0.224 \times 2{,}353{,}932$ units $= 527{,}281$ units

So far, Arizona is the clear winner with fewer units.

San Francisco:

1940–1959: $0.80 \times 23\% = 18.4\%$

1960–1979: $0.40 \times 29\% = 11.6\%$

Total % of units: $18.4\% + 11.6\% = 30\%$

$0.30 \times 1{,}747{,}506$ units $= 524{,}252$ units

Arizona still has the fewest number of units of interest to the real estate firm. The correct answer is **(A).**

20. B

This question concerns the percent of income spent on housing as well as the age of that housing. Use information from the stacked bar graph and both pie charts to calculate your answer. Probability is the ratio of the number of desired outcomes over the total possible outcomes. The question asks for the approximate ratio and the answer choices are relatively far apart, so rounding is a good strategy to make calculations easier.

There are a total of $49.5 + 37.5 = 87$ million people who either rent or have a mortgage. Of the 49.5 million mortgage holders, $9\% + 27\% = 36\%$ spend 30% or more of their income on housing; that's $0.36 \times 49.5 = 17.82$ million (round to 18 million). Of the 37.5 million renters, $9\% + 43\% = 52\%$ pay 30% of more of their income on housing; that's $0.52 \times 37.5 = 19.5$ million. Therefore, the probability that a survey participant completed survey A is $\frac{18 + 19.5}{87} = \frac{37.5}{87} \approx 43\%$.

This was a national survey, so look at the bottommost bar on the bar chart to find percent ownership by age of housing in the United States. "More than 75 years old" in 2014 means the housing was built before 1940. About 13% of U.S. housing was built before 1940. So multiply: $0.13 \times 0.43 \approx 0.056$. So 5.6% of the survey participants completed survey A and lived in a unit more than 75 years old. Stated differently, there is a 5.6% probability that a survey participant completed survey A and lived in a unit more than 75 years old.

The probability that a survey participant completed survey B is about $100\% - 43\% = 57\%$. In 2014, "built in the previous 15 years" means built in 2000 or later. About 15% of U.S. housing was built in or after 2000, so multiply: $0.15 \times 0.57 \approx 0.086$. So 8.6% of the survey participants completed survey B and lived in a unit no more than 15 years old. Stated differently, there is an 8.6% probability that a survey participant completed survey B and lived in a unit no more than 15 years old.

The ratio of the probabilities is $\frac{5.6\%}{8.6\%}$ or approximately $\frac{2}{3}$. The correct answer is **(B).**

Note that because the total of 87 million remains the same throughout the problem, you are really being asked for the ratio of $\frac{\frac{\text{Group A}}{\text{Total}}}{\frac{\text{Group B}}{\text{Total}}}$. That means that you could have also solved for the ratio of the *number* of those who took survey A and lived in a unit more than 75 years old to the *number* of those who took survey B and lived in a unit no more than 15 years old. That ratio is about 4.9 million to 7.4 million, or about $\frac{2}{3}$.

ADVANCED MATH QUESTIONS PRACTICE SET 3

1. a is a prime number
b is a positive factor of $a - 1$

Quantity A	Quantity B
a^2	The product of b and the least non-prime integer greater than a

Ⓐ Quantity A is greater.
Ⓑ Quantity B is greater.
Ⓒ The two quantities are equal.
Ⓓ The relationship cannot be determined from the information given.

2. In rectangle *A*, the ratio of the length to the width is 3:1. In rectangle *B*, the ratio of the length to the width is 4:1. The area of rectangle *A* is 27 percent of the area of rectangle *B*. The perimeter of rectangle *A* is v percent of the perimeter of rectangle *B*.

Quantity A	Quantity B
v	54

Ⓐ Quantity A is greater.
Ⓑ Quantity B is greater.
Ⓒ The two quantities are equal.
Ⓓ The relationship cannot be determined from the information given.

3. The probability that event *A* occurs and event *B* does not occur is greater than $\frac{1}{4}$. The probability that event *B* occurs and event *A* does not occur is greater than $\frac{1}{4}$. The probability that neither of the events *A* and *B* occurs is greater than $\frac{1}{8}$.

Quantity A	Quantity B
The probability that both of the events A and B occur	$\frac{1}{4}$

Ⓐ Quantity A is greater.
Ⓑ Quantity B is greater.
Ⓒ The two quantities are equal.
Ⓓ The relationship cannot be determined from the information given.

4.

$$\frac{x - 1}{(x - 2)^2 + 25} = \frac{2}{17}$$

Quantity A	Quantity B
x	8

Ⓐ Quantity A is greater.
Ⓑ Quantity B is greater.
Ⓒ The two quantities are equal.
Ⓓ The relationship cannot be determined from the information given.

5. The arithmetic mean of $\{6, 1, 11, x, 2\}$ is x.
The median of $\{2, y, 11, 1, 5\}$ is y.

Quantity A	Quantity B
x	y

Ⓐ Quantity A is greater.
Ⓑ Quantity B is greater.
Ⓒ The two quantities are equal.
Ⓓ The relationship cannot be determined from the information given.

6. A survey measures the heights of 900 people, which are found to be normally distributed. The mean height is 5' 5", and 150 people in the survey have a height between 5' 1" and 5' 3".

Quantity A	Quantity B
The number of people in the survey who are taller than 5' 9"	The number of people in the survey who are more than 2 standard deviations above the mean

Ⓐ Quantity A is greater.
Ⓑ Quantity B is greater.
Ⓒ The two quantities are equal.
Ⓓ The relationship cannot be determined from the information given.

7.

$$0 < p < 1$$

When an experiment is conducted, events A, B, and C are all independent of one another. Each of events A, B, and C has a probability p of occurring.

Quantity A	Quantity B
The probability that events A and B occur and event C does not occur	$p(1 - p)$

8. Both of the points $P(17, -20)$ and $Q(25, t)$ are in the xy-plane. Which of the following statements alone give(s) sufficient additional information to determine whether $t > -18$?

Indicate all such statements.

[A] The slope of the line that goes through the points $P(17, -20)$ and $Q(25, t)$ is $\frac{3}{4}$.
[B] The distance between the points $P(17, -20)$ and $Q(25, t)$ is 10.
[C] The point $Q(25, t)$ is the midpoint of the line segment whose endpoints are $P(17, -20)$ and $R(33, 3t + 34)$.

9. If $x = (3^8)(81^5)$ and $y = (9^7)(27^{12})$, then $xy =$

Ⓐ 3^{48}
Ⓑ 3^{54}
Ⓒ 3^{61}
Ⓓ 3^{68}
Ⓔ 3^{78}

10. What is the area of a triangle that has two sides that each have a length of 10, and whose perimeter is equal to that of a square whose area is 81?

(A) 30
(B) 36
(C) 42
(D) 48
(E) 60

11. How many different committees of 2 men and 2 women can be formed from a group of 12 people, half of whom are men?

(A) 225
(B) 450
(C) 495
(D) 900
(E) 2,970

12. Yaire had 50 pieces of candy, 64% of which were gummy worms. She has eaten several gummy worms, and no other candies, so that now her gummy worms make up 60% of her candies. How many pieces of candy does she have left?

(A) 48
(B) 46
(C) 45
(D) 36
(E) 32

13. Water can be pumped from a 100,000 gallon tank at a uniform rate by opening valve V, valve W, or both. The rate at which water is pumped out by opening valve V is 10,000 gallons per hour. If the tank is completely full and both valves are opened, the tank can be drained in 6 hours. How long would it take to empty the tank if it were 80% full and only valve W were opened?

(A) 6 hours
(B) 8 hours
(C) 12 hours
(D) 15 hours
(E) 18 hours

14. For which of the following sets of integers is the difference between the range and the interquartile range greater than or equal to 10?

    Select all such sets of integers.

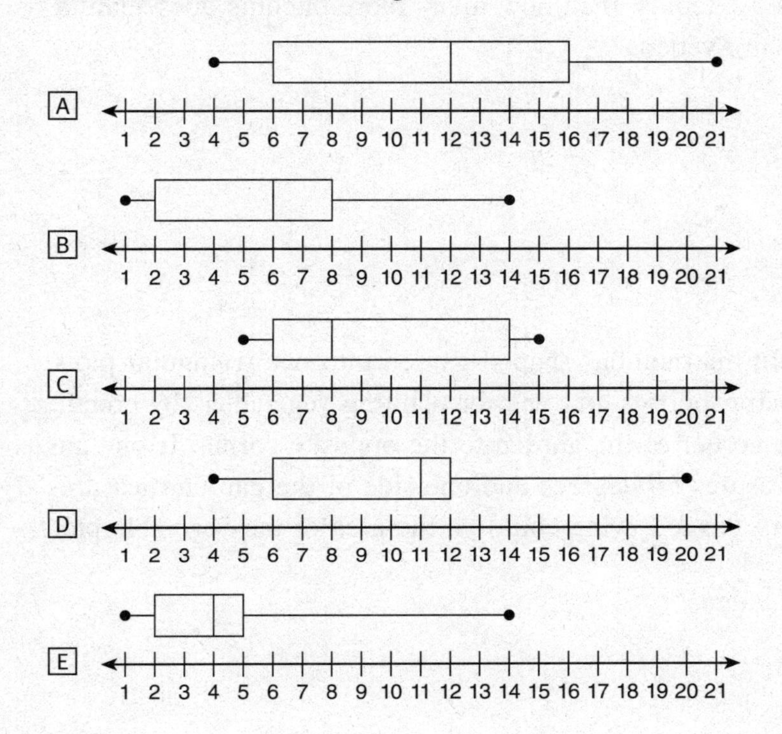

15. The ratio of five-dollar bills to one-dollar bills in Armando's wallet is 2:3. When he made a $7 purchase, he paid for that purchase with only five-dollar bills, and with the smallest possible number of five-dollar bills. He received his change in one-dollar bills. After completing that transaction, the ratio of five-dollar bills to one-dollar bills was 1:3. How many one-dollar bills did Armando have before his purchase?

16. If Juanita gives half of her bitcoins in a 1:2 ratio to Pat and Svetlana, respectively, then Pat will have one fourth as many bitcoins as will Svetlana, who will in turn have twice as many bitcoins as will Juanita. If Pat currently has 2 bitcoins, then how many more bitcoins does Juanita currently have than Svetlana?

(A) 8
(B) 12
(C) 20
(D) 22
(E) 24

17. Uwe wants to split his rhombus-shaped garden into two triangular plots, one for planting strawberries and one for planting vegetables, by erecting a fence from one corner of the garden to the opposite corner. If one angle of the garden measures 60 degrees and one side of the garden measures x meters, which of the following could be the area of the vegetable plot?

Indicate all such values.

[A] $\frac{x^2\sqrt{3}}{2}$

[B] $\frac{x^2\sqrt{3}}{4}$

[C] $\frac{x^2\sqrt{3}}{8}$

[D] $\frac{x^2}{4}$

[E] $\frac{x^2}{2}$

Questions 18–20 refer to the following graphs.

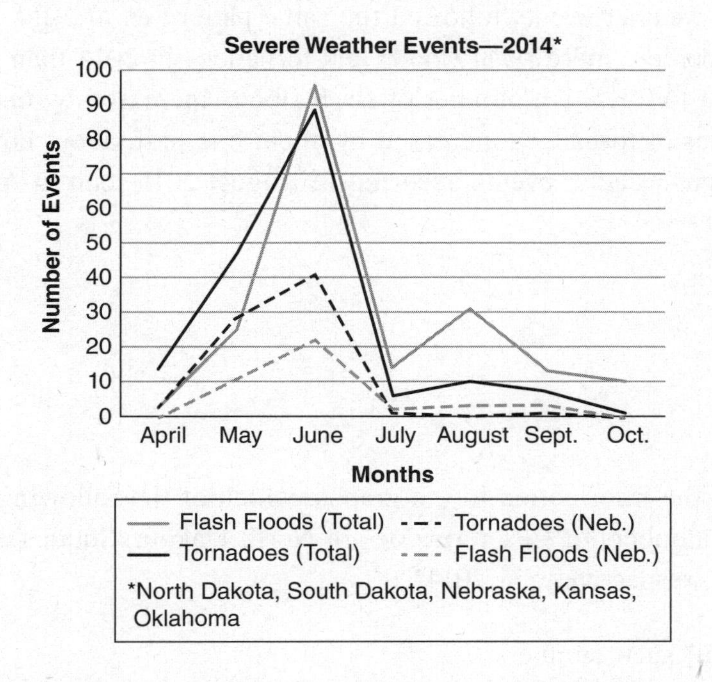

Source: National Atmospheric and Oceanic Administration (NOAA) National Climactic Data Center, Storm Events Database, http://www.ncdc.noaa.gov/stormevents/

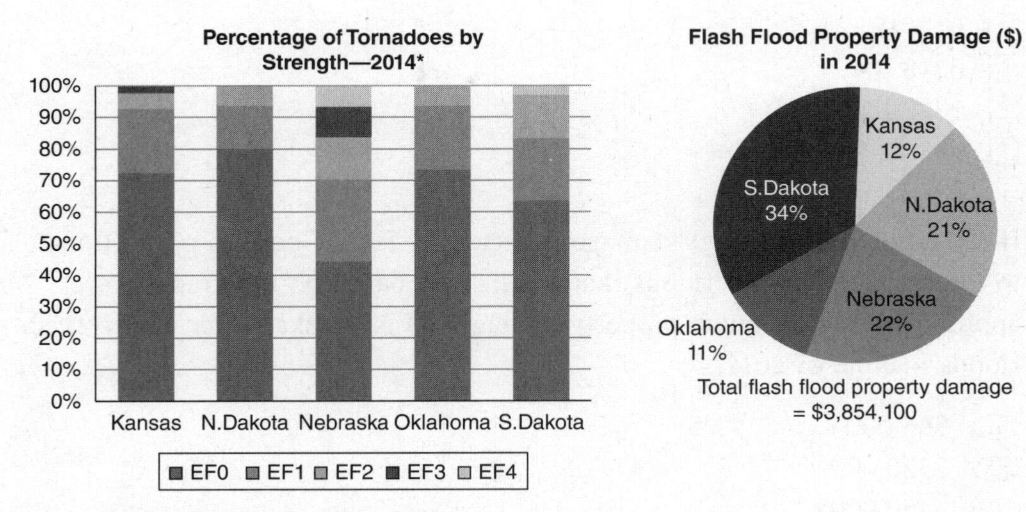

*As measured by the Enhanced Fujita (EF) Scale.

18. If in the years 2013, 2014, and 2015, in the five states shown, the number of severe weather events followed the same pattern as in 2014, but there were 25 percent more flash floods and tornadoes in 2014 than in 2013 and, from 2014 to 2015, the number of flash floods increased by one-third, and the number of tornadoes increased by about one-half, about how many more severe weather events occurred in August 2015 than in August 2013?

Ⓐ 5
Ⓑ 23
Ⓒ 27
Ⓓ 33
Ⓔ 56

19. Based on the information in the graphs, which of the following could have been the numbers of EF1 tornadoes in North Dakota, South Dakota, and Nebraska, respectively, in 2014?

Indicate <u>all</u> such numbers.

[A] 2, 6, 19
[B] 4, 3, 26
[C] 6, 3, 19
[D] 6, 18, 19
[E] 13, 20, 26
[F] 39, 60, 78

20. If the amount of property damage inflicted by flash floods is proportional to the number of flash floods that occur, then based on the graphs, approximately how much property damage did Nebraska suffer due to flash floods in June of 2014?

Ⓐ $65,000
Ⓑ $320,000
Ⓒ $450,000
Ⓓ $735,000
Ⓔ $2,158,296

ADVANCED MATH QUESTIONS PRACTICE SET 3 ANSWER KEY

1. D	6. A	11. A	16. A
2. B	7. B	12. C	17. B
3. D	8. A, C	13. C	18. B
4. B	9. E	14. D, E	19. A, C
5. D	10. D	15. 9	20. C

ADVANCED MATH QUESTIONS PRACTICE SET 3 ANSWERS AND EXPLANATIONS

1. D

From the centered information, you can determine that the greatest number b can be is $a - 1$, as the greatest factor of any number is the number itself. So to make Quantity B as large as possible, substitute $a - 1$ for b, and multiply by the next non-prime greater than a. As the only consecutive prime numbers are 2 and 3, in nearly all cases, the least non-prime integer greater than a will be $a + 1$. Quantity B would thus be $(a - 1)(a + 1) = a^2 - 1$, making Quantity A, which is a^2, greater. However, consider the special case mentioned of the consecutive prime numbers 2 and 3. If a is 2, $a - 1 = 1$, and the next non-prime greater than a would be 4. In this specific case, Quantity A, a^2, would be 4 and Quantity B would also be 4. Since two relationships are possible between the quantities, the answer is **(D).**

Note that you could also solve this problem by Picking Numbers for a and b. The smallest prime number is 2, so try that first. In that case, b has to equal $2 - 1 = 1$, and the least non-prime integer greater than a is 4. So Quantity A equals $a^2 = 2^2 = 4$, and Quantity B equals $1 \times 4 = 4$. With these numbers, the two quantities are equal. However, they cannot always be equal, because for greater values of a, there will be many possibilities for b. For example, if $a = 5$ and $b = 4$, then Quantity A equals $5 \times 5 = 25$ and Quantity B equals $4 \times 6 = 24$. More than one relationship is possible, so again, the answer is **(D).**

2. B

Since the problem is stated in terms of ratios rather than actual dimensions, use variables and algebra to find the solution for the value of v. Set the width of rectangle A equal to x and the length to $3x$. By the same logic, the width and length of rectangle B can be represented by y and $4y$. The area of a rectangle is $l \times w$, so the areas of rectangles A and B, respectively, are $3x \times x = 3x^2$ and $4y \times y = 4y^2$. From the information given in the problem, $3x^2 = 0.27 \times 4y^2$, which means that $\frac{3x^2}{4y^2} = \frac{27}{100}$.

Divide both sides of the equation by 3 and multiply by 4 to obtain $\frac{x^2}{y^2} = \frac{9}{25}$. Take the square root of both sides, and voila, $\frac{x}{y} = \frac{3}{5}$. (Because the problem deals with geometric figures, you do not need to worry about the negative roots.)

The perimeter of a rectangle with length l and width w is $2(l + w)$. The perimeter of rectangle A is $2(3x + x) = 2(4x) = 8x$. The perimeter of rectangle B is $2(4y + y) = 2(5y) = 10y$. So the ratio of the perimeter of rectangle A to the perimeter of rectangle B is $\frac{8x}{10y} = \frac{4x}{5y}$.

Since there are now two independent variables, x and y, and two distinct linear equations, solve for the value of v. Use the fact that

$$\textit{Perimeter } A = \left(\frac{v}{100}\right) \textit{Perimeter } B$$

to set up the proportion $\frac{\textit{Perimeter } A}{\textit{Perimeter } B} = \frac{v}{100}$. From earlier calculations:

$$\frac{\textit{Perimeter } A}{\textit{Perimeter } B} = \frac{4x}{5y} = \left(\frac{4}{5}\right)\left(\frac{x}{y}\right).$$

Furthermore, $\frac{x}{y} = \frac{3}{5}$, so $\left(\frac{4}{5}\right)\left(\frac{3}{5}\right) = \frac{v}{100}$. Multiply both sides of the equation by 25 to get $\frac{4 \times 3}{1} = \frac{v}{4}$. Cross multiply to find that $v = 48$. So Quantity A is 48. Quantity B is given as 54. Quantity B is greater and choice **(B)** is correct.

3. D

There are four possibilities, of which exactly one can occur:

(i) Event A occurs and event B does not occur.
(ii) Event B occurs and event A does not occur.
(iii) Neither event A nor event B occurs.
(iv) Both events A and B occur.

The centered information provides a lower limit for the probabilities for each of the first three of the above possibilities. Sum these three known lower limits:

$$\frac{1}{4} + \frac{1}{4} + \frac{1}{8} = \frac{2}{8} + \frac{2}{8} + \frac{1}{8} = \frac{5}{8}.$$

Because this is a sum of lower limits, the probability that one of the first three possibilities occurs must be $> \frac{5}{8}$ and the unknown probability that both events A and B occur must be $< \frac{3}{8}$, since the total of the four probabilities must equal 1. Therefore, the range of the probability that both events A and B occur is $0 \leq P_{A \textit{ and } B} < \frac{3}{8}$.

Since the probability that both occur could be either greater or less than $\frac{1}{4}$, answer choice **(D)** is correct.

4. B

To solve the equation $\frac{x-1}{(x-2)^2+25}=\frac{2}{17}$ for the possible values of x, start by applying FOIL to the $(x-2)^2$ term in the left-hand denominator and combining like terms:

$$\frac{x-1}{x^2-4x+4+25}=\frac{2}{17}$$
$$\frac{x-1}{x^2-4x+29}=\frac{2}{17}$$

Now cross multiply:

$$(17)(x-1) = 2\left(x^2-4x+29\right)$$
$$17x-17 = 2x^2-8x+58$$

Subtract $17x - 17$ from both sides of the equation to produce a quadratic set equal to zero:

$$17x-17-(17x-17) = 2x^2-8x+58-(17x-17)$$
$$0 = 2x^2-25x+75$$

In order to reverse FOIL this equation, look for combinations of factors of 2 and 75 that will produce a sum of –25 when both factors of 75 have the same sign.

$$2(-5) = -10 \text{ and } 1(-15) = -15$$
$$-10+(-15) = -25$$

So the factored equation is $(2x-15)(x-5)=0$, and x must be either $7\frac{1}{2}$ or 5. Since Quantity B is 8, that quantity is greater than either possible value of x, and the correct answer is **(B)**.

5. D

In order to determine the value of x, use the average formula:

$$\text{Average} = \frac{\text{Sum of Values}}{\text{Number of Values}}.$$

Using the values given in the centered information, you can see that

$$x = \frac{(6 + 1 + 11 + x + 2)}{5} = \frac{20 + x}{5}.$$

Multiply both sides of the equation by 5 to get $5x = 20 + x$, then subtract x from both sides to get $4x = 20$. Divide both sides by 4 to obtain $x = 5$.

Since y is the median of the set, put the elements of the set in order, with y as the middle term: $\{1, 2, y, 5, 11\}$. Because of its position in the set, $2 \leq y \leq 5$.

Comparing x to y, given the allowable range, y could be either less than 5 or equal to 5 (the known value for x). Therefore, more than one relationship between the quantities is possible. The correct answer must be **(D).**

6. A

As with all Quantitative Comparison questions, you need the two values to be expressed in comparable terms, so think of Quantity A in terms of standard deviations above the mean. Because this is a standard deviation question, you might not be able to calculate the exact number of people who are taller than a certain height, but you should be able to approximate it as a percentage of the total.

The question stem states that the data are normally distributed, which means that about 68% of the heights of those involved in the study fall within 1 standard deviation of the mean. It also means that about 95% of those heights fall within 2 standard deviations of the mean. The question stem also says that 150 of the 900 people surveyed were between 5' 1" and 5' 3".

You don't know how great one standard deviation is, but you do know that a certain percentage of the total data lies between two points. Assume for a moment that one standard deviation is 2" in height. In that case, the people whose heights are between 5' 1" and 5' 3" would fall between 1 and 2 standard deviations below the mean of 5' 5". Half of 68% is 34%, which is the amount of data that falls between the mean and 1 standard deviation below the mean. Half of 95% is 47.5%, which is the amount of data that falls between the mean and 2 standard deviations below the mean. Use this information to calculate the percentage of data that would fall between 1 and 2 standard deviations below the mean: 47.5% – 34% = 13.5%.

However, the actual number of people whose heights fall between 5' 1" and 5' 3" is 150/900, or about 16.7%, which is more than 13.5%. Normal distribution means that data points are grouped more densely near the mean than farther away from it. Because more than 13.5% of the data falls between 5' 1" and 5' 3", you know this data selection is closer to the mean than the data selection between 1 and 2 standard deviations below the mean. In other words, you don't know exactly what the standard deviation is for this data set, but you do know it is greater than 2". That means that two standard deviations above the mean will be greater than 5' 9"; therefore, there are more people who are taller than 5' 9" than people who are 2 standard deviations above the mean.

Quantity A is greater, and the correct answer is **(A).**

Note: If you memorize 68%, 95%, 99.7%, and other common numbers for data sets with normal distribution, standard deviation problems will be much easier. For example, if you have memorized that about 13.5% of the data in a normal distribution falls between the 1 and 2 standard deviations below the mean, you don't need to calculate that number. Also note that you don't really need to calculate 150/900. 13.5% of 1,000 would be 135, and you can see at a glance that 150/900 is going to be larger than 135/1000.

7. B

The probability that an event does not occur is equal to 1 minus the probability that the event does occur. Since the probability that event C occurs is p, the probability that event C does not occur is $1 - p$. Since the events A, B, and C are independent of one another, the probability that event A occurs, event B occurs, and event C does not occur is equal to $p \times p \times (1 - p) = p^2(1 - p)$.

So Quantity A is $p^2(1 - p)$ and Quantity B is defined as $p(1 - p)$. The centered information that $0 < p < 1$ defines p as a positive fraction. It follows that $(1 - p)$ must also be a positive fraction. In Quantity A, three positive factors are multiplied; in Quantity B, two positive factors are multiplied. So both Quantities represent positive values. More specifically, Quantity A is Quantity B multiplied by p. In other words, Quantity A is Quantity B multiplied by a positive fraction less than 1.

It follows that Quantity B is greater than Quantity A, and the correct answer is **(B).**

Here's another way to see the answer: Because p and $(1 - p)$ are both positive, you can divide both quantities by p and $(1 - p)$, leaving p in Quantity A and 1 in Quantity B. Because $p < 1$, Quantity B is greater.

8. A, C

Consider statement A:

Plug into the slope formula for a line with a slope of $\frac{3}{4}$ that goes through the points $P(17, -20)$ and $Q(25, t)$:

$$\frac{t - (-20)}{25 - 17} = \frac{3}{4}$$

Based on this equation, there is just one value for t. So statement A gives sufficient additional information to determine whether $t > -18$.

For the record:

$$\begin{aligned} \frac{t - (-20)}{25 - 17} &= \frac{3}{4} \\ \frac{t + 20}{8} &= \frac{3}{4} \\ t + 20 &= 6 \\ t &= -14 \end{aligned}$$

Keep in mind that this additional calculation is not necessary. To use your time efficiently on Test Day, you would not actually need to solve for t.

Next, consider statement B:

Use the distance formula to try to solve for t, given the fact that the distance between the points $P(17, -20)$ and $Q(25, t)$ is equal to 10:

$$\begin{aligned} \sqrt{(25 - 17)^2 + (t - (-20))^2} &= 10 \\ \sqrt{8^2 + (t + 20)^2} &= 10 \\ \sqrt{64 + (t + 20)^2} &= 10 \\ \left(\sqrt{64 + (t + 20)^2}\right)^2 &= 10^2 \\ 64 + (t + 20)^2 &= 100 \\ (t + 20)^2 &= 36 \end{aligned}$$

Since $(t + 20)^2 = 36$, $t + 20 = 6$ or $t + 20 = -6$.

If $t + 20 = 6$, then $t = -14$ and $t > -18$.

If $t + 20 = -6$, then $t = -26$ and $t < -18$.

Statement B does not give sufficient additional information to determine whether or not $t > -18$.

Finally, consider statement C:

The midpoint of the line segment whose endpoints are $P(17, -20)$ and $R(33, 3t + 34)$ is the point

$$\left(\frac{17 + 33}{2}, \frac{-20 + (3t + 34)}{2}\right).$$

This simplifies to

$$\left(\frac{50}{2}, \frac{-20 + 3t + 34}{2}\right).$$

Simplify further and recognize that the point $Q\left(25, \frac{3t + 14}{2}\right)$ is the point $Q(25, t)$.

Set the y-coordinates equal to find that $\frac{3t + 14}{2} = t$.

Solve for t:

$$\begin{aligned} 3t + 14 &= 2t \\ t + 14 &= 0 \\ t &= -14. \end{aligned}$$

Statement C gives sufficient additional information to determine that $t > -18$. As was the case for statement A, you could have stopped after the first equation in the explanation because that equation was sufficient to solve for one value of t.

The correct answers are **(A)** and **(C)**.

9. E

Problems involving multiple bases are typically best solved by first converting all the terms to the same base. Here, 3 can be conveniently used as the base for each term:

$9 = 3 \times 3 = 3^2$
$27 = 3 \times 3 \times 3 = 3^3$
$81 = 3 \times 3 \times 3 \times 3 = 3^4$

Substitute 3^2 for 9, 3^3 for 27, and 3^4 for 81 to yield the following:

$x = (3^8)[(3^4)^5]$
$y = [(3^2)^7][(3^3)^{12}]$

Use the laws of exponents $(b^a)^c = b^{ac}$ and $b^a b^c = b^{a+c}$ to simplify:

$x = (3^8)(3^{4 \times 5}) = (3^8)(3^{20}) = 3^{8+20} = 3^{28}$

$y = (3^{2 \times 7})(3^{3 \times 12}) = (3^{14})(3^{36}) = 3^{14+36} = 3^{50}$

Finally, multiply x times y:

$xy = (3^{28})(3^{50}) = 3^{28+50} = 3^{78}$

Choice **(E)** is correct.

10. D

The area of a square is its side length squared. Since the area of this square is 81, each side is equal to $\sqrt{81}$ or 9. The perimeter of a square is 4 times the length of its side—in this case, 36. Since the perimeter of the triangle is equal to that of the square, subtract the two existing sides, 10 and 10, from 36, to get 16 for the third side. Here's a labeled diagram; remember to draw one if a problem doesn't provide it:

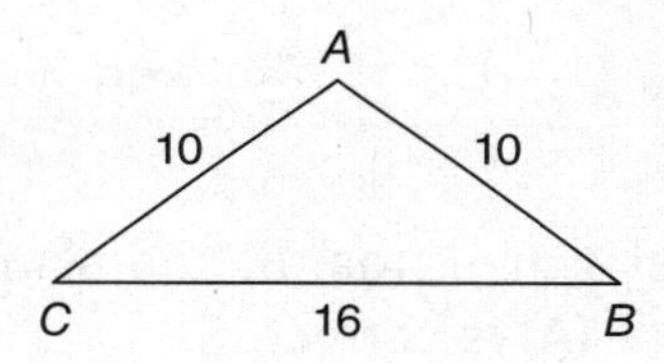

This triangle has two equal sides of length 10. Any triangle that has two equal sides is an isosceles triangle. Drop a perpendicular to divide it into 2 identical right triangles:

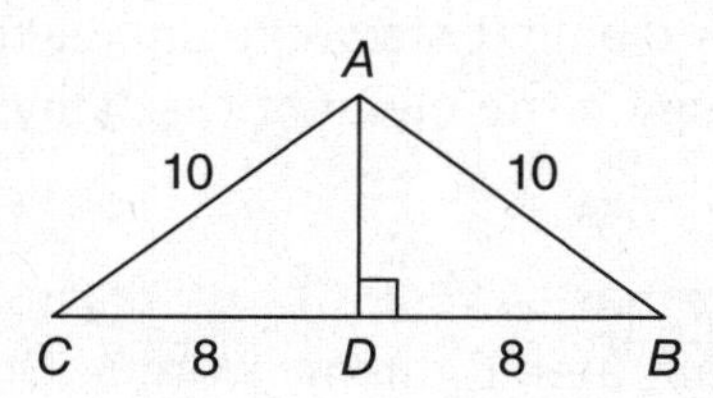

Here, point D divides side BC of the triangle into the two equal segments BD and DC, each equal to 8. To find the area of this triangle, you need the length of AD. Consider triangle ABD. Since the length of hypotenuse AB is 10 and the length of leg BD is 8, the side lengths of right triangle ABD are multiples of the 3-4-5 Pythagorean triple, with each part of the 3 to 4 to 5 ratio multiplied by 2, so $AD = 6$.

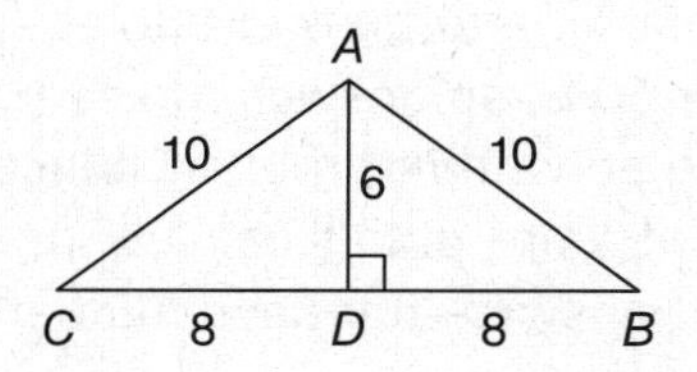

BC is the base of triangle ABC and AD is the height of the triangle drawn to base BC. The area of any triangle is one-half times base times height. Therefore, the area of triangle ABC is

$$\frac{1}{2} \times BC \times AD = \frac{1}{2} \times 16 \times 6 = 48.$$

Choice **(D)** is correct.

11. A

Split the group of 12 people into two groups of 6 men and 6 women each. Find the number of ways to choose 2 men from 6, and the number of ways to choose 2 women from 6, and multiply those two results together to find the answer.

Fortunately, the numbers are the same both times, so you'll need to apply the combination formula, ${}_nC_k = \frac{n!}{k!(n-k)!}$, just once:

$${}_6C_2 = \frac{6 \times 5 \times 4 \times 3 \times 2 \times 1}{2 \times 1 \times 4 \times 3 \times 2 \times 1} = \frac{30}{2} = 15.$$

So there are 15 ways to select 2 men from 6, and also 15 ways to select 2 women from 6. $15 \times 15 = 225$. Choice **(A)** is correct.

12. C

This is a variation on the classic mixture problem in which one element of the mixture changes, while the other element stays the same. You can simplify the solution of these problems by focusing on the element that stays the same; here, that's the other kind(s) of candy.

To start, you know that Yaire has $(64\%)(50) = (50\%)(64) = 32$ gummy worms. So she must have 18 candies that aren't gummy worms. Once she's eaten some of her worms, those 18 other candies now make up $100\% - 60\% = 40\%$ of the new total number of candies. So:

$18 = (0.4)(T_n)$, where T_n is the new total. Solving for T_n gives you the answer: 45 pieces of candy.

Alternatively, you could Backsolve this one. Answer choice (D) would give you a gummy worm percentage of 50%, and answer choice (B) would give you a gummy worm percentage of just over 60%, so answer choice **(C)** must be correct. However, there's an even faster way to solve this problem. Since the answer choices are all integers, you can deduce that Yaire didn't eat any partial gummy worms. So you know that 60% of the correct answer—her new number of gummy worms—must be an integer. And since 60% is equivalent to $\frac{3}{5}$, the only potentially correct answers would have to be divisible by 5.

Choice **(C)** is correct.

13. C

A quick way to answer this question is to use the simplified formula for calculating combined work, $T = \frac{AB}{A + B}$, where T represents the total time to do a defined task, and A and B represent the amount of time each person or machine or, in this case, valve takes to do the task. Since the task is to empty a 100,000 gallon tank and opening valve V results in the tank draining at 10,000 gallons per hour, opening V only would empty the tank in 10 hours. Filling in the known values, you get $6 = \frac{10W}{10 + W}$.

$$\begin{aligned} 6(10 + W) &= \frac{10W(10 + W)}{10 + W} \\ 60 + 6W &= 10W \\ 60 &= 4W \\ 15 &= W \end{aligned}$$

Opening only valve W empties the tank in 15 hours when the tank is full. However, the question asks how long it would take to drain the tank from 80% full. That's 0.8×15 hrs $= 12$ hrs. Choice **(C)** is correct.

14. D, E

The range is the difference between the highest point and the lowest point, which are represented by the furthest points on the plots.

The interquartile range is the difference between the third quartile and the first quartile, which are represented by the two vertical edges of the rectangle on the plot.

First, calculate the range and the interquartile range, then take difference between the two. The question is asking for all choices that have a result greater than or equal to 10.

A. Range $= 21 - 4 = 17$
   Interquartile Range $= 16 - 6 = 10$
   Difference $= 17 - 10 = 7$

B. Range $= 14 - 1 = 13$
   Interquartile Range $= 8 - 2 = 6$
   Difference $= 13 - 6 = 7$

C. Range $= 15 - 5 = 10$
Interquartile Range $= 14 - 6 = 8$
Difference $= 10 - 8 = 2$

(Note that the moment you notice the range is 10, there's no need to check the interquartile range, since 10 minus anything will be less than 10.)

D. Range $= 20 - 4 = 16$
Interquartile Range $= 12 - 6 = 6$
Difference $= 16 - 6 =$ **10**

E. Range $= 14 - 1 = 13$
Interquartile Range $= 5 - 2 = 3$
Difference $= 13 - 3 =$ **10**

Only answer choices **(D)** and **(E)** have a difference greater than or equal to 10; therefore, they are the only two correct answers.

15. 9

The 2:3 ratio of bills does not necessarily mean that Armando starts with 2 five-dollar bills and 3 one-dollar bills; he could have any multiple of 2 and 3—such as 4 and 6, 20 and 30, etc. Therefore, the ratio can be expressed, using a common multiplier, as $\frac{2x}{3x}$. After using 2 five-dollar bills for the transaction, Armando will have $2x - 2$ five dollar bills and, with the 3 one-dollar bills he receives in change, $3x + 3$ one dollar bills. Since the final ratio is 1:3, you can set up a proportion to find $\frac{1}{3} = \frac{2x - 2}{3x + 3}$.

Multiply both sides by the denominators to get $1(3x + 3) = 3(2x - 2)$. Simplify to $3x + 3 = 6x - 6$ and solve: $9 = 3x$, so $x = 3$. However, x is not the answer to the question asked, but merely the common multiplier used in the initial ratio. Since Armando started with $3x$ dollar bills, and since $x = 3$, it follows that he had $3(3) = 9$ one-dollar bills initially.

16. A

If the question asked for one value (say, Juanita's number of bitcoins), then Backsolving would be a terrific way to bypass most of the work on this complex translation question. Unfortunately, the question asks for the difference between two values, so backsolving won't work. Algebra it is!

Let J, P, and S represent the number of bitcoins currently in the hands of Juanita, Pat, and Svetlana, respectively. (For simplicity's sake, we'll refer to bitcoins as "coins" from here on out.) If Juanita gives half of her coins away, she'll have $\frac{J}{2}$ coins left. Next, consider the ratio. If Pat and Svetlana receive Juanita's coins in a 1:2 ratio, that means Pat will receive one third of those coins and Svetlana will receive the remaining two thirds. Thus, Pat will receive one third of one half, or one sixth, of Juanita's coins, or $\frac{J}{6}$. Svetlana will receive two thirds of one half, or $\frac{2}{3} \times \frac{1}{2} = \frac{1}{3}$ of Juanita's coins, or $\frac{J}{3}$. After Juanita makes the gift, she will have $\frac{J}{2}$ coins, Pat will have $P + \frac{J}{6}$, and Svetlana will have $S + \frac{J}{3}$. You're told that Pat's total will be one fourth of Svetlana's. Thus:

$$P + \frac{J}{6} = \frac{1}{4}\left(S + \frac{J}{3}\right)$$

Multiply this equation by 12 to eliminate the fractions, then simplify as much as you can:

$$\begin{aligned} 12P + 2J &= 3\left(S + \frac{J}{3}\right) \\ 12P + 2J &= 3S + J \\ 12P &= 3S - J \end{aligned}$$

You're also told that, after the gift, Svetlana's total will be double Juanita's. Thus:

$$\begin{aligned} S + \frac{J}{3} = 2\left(\frac{J}{2}\right) &= J \\ 3S + J &= 3J \\ 3S &= 2J \end{aligned}$$

Substitute $2J$ in place of $3S$ in the earlier equation to get:

$$\begin{aligned} &12P = 3S - J = 2J - J = J \\ &12P = J \end{aligned}$$

Finally, you can make use of the fact that Pat has 2 coins. This means that Juanita has $12(2) = 24$ coins. Because $3S = 2J$, it follows that $S = \frac{2}{3}J$, and Svetlana has $\frac{2(24)}{3} = 2(8) = 16$ coins. The question asks how many more coins Juanita has than Svetlana, so the answer is $24 - 16 = 8$, choice **(A)**.

17. B

Whenever a geometry problem comes without a figure, start by drawing one yourself:

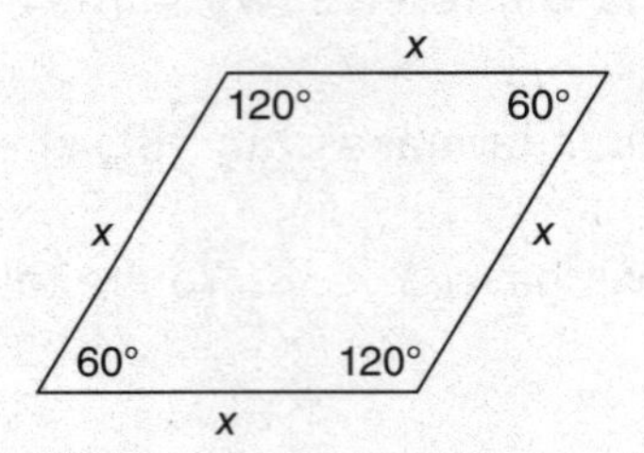

Uwe wants to split this garden into two triangles. What makes this question tricky is that there are two ways to split a rhombus into two triangles. Try it this way first:

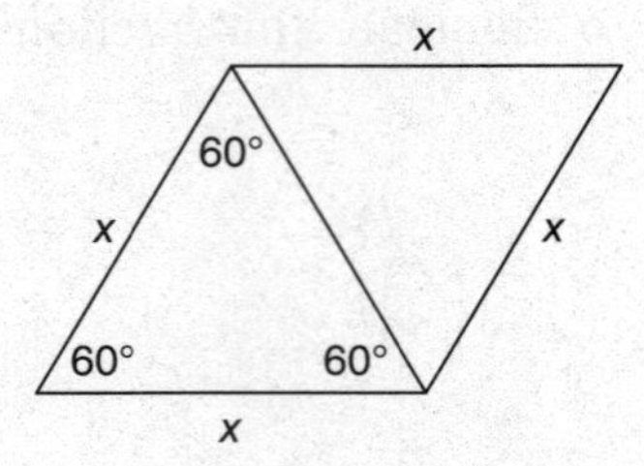

The area of one plot would then be the area of an equilateral triangle with side x. Use the properties of 30-60-90 triangles to deduce that the height of such a triangle must be $\frac{x\sqrt{3}}{2}$, which allows you to calculate the area as follows:

$$\text{Area} = \frac{1}{2}bh = \left(\frac{1}{2}\right)(x)\left(\frac{x\sqrt{3}}{2}\right) = \frac{x^2\sqrt{3}}{4}$$

This is one of the choices, but the garden might be split the other way:

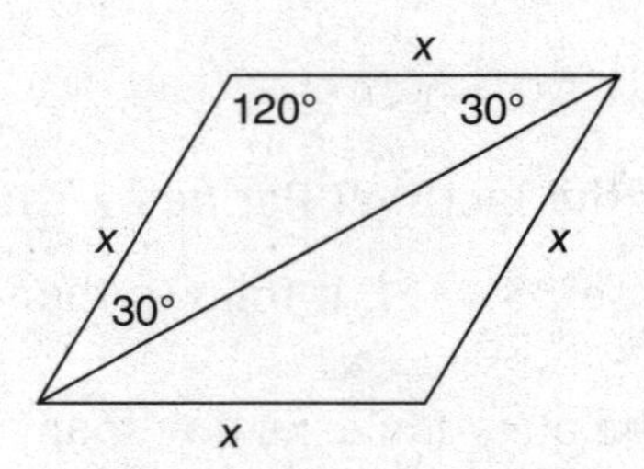

To find the area of one of these triangles, further subdivide it into two right triangles and use again the properties of 30-60-90 right triangles:

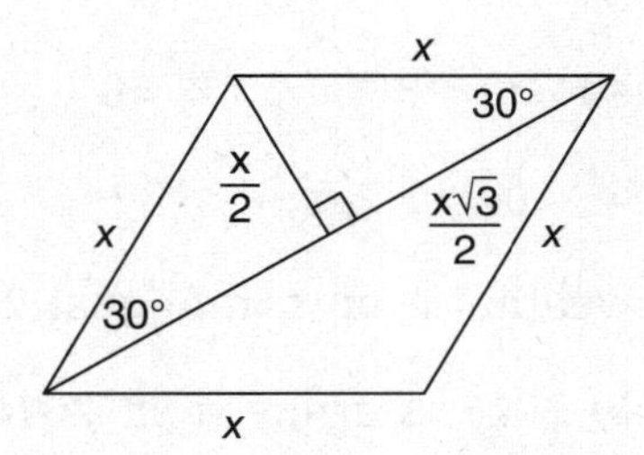

The base of the big triangle is therefore twice $\frac{x\sqrt{3}}{2}$, or $x\sqrt{3}$. The height is $\frac{x}{2}$, and the area is found as follows:

$$Area = \frac{1}{2}bh = \frac{1}{2}\left(x\sqrt{3}\right)\left(\frac{x}{2}\right) = \frac{x^2\sqrt{3}}{4}$$

This is the same as the previous area, so only one answer choice is correct: **(B)**.

Note that you can save some time by using critical thinking to deduce up front that there must only be one correct answer. Imagine breaking up a rhombus into four triangles rather than two:

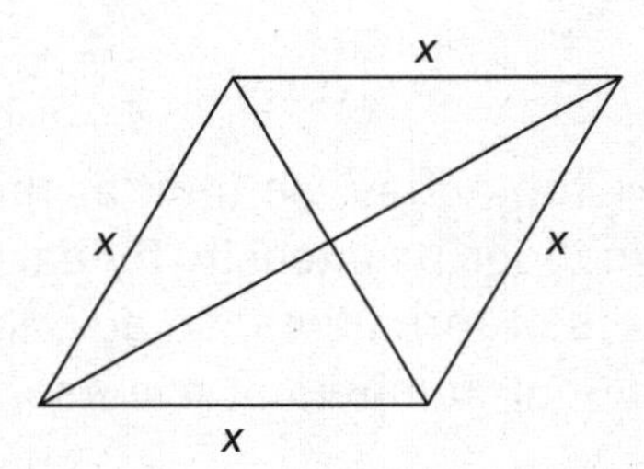

Because the starting shape is a rhombus, all four triangles are congruent. No matter which two triangles you combine to make a big triangle—be it the left and bottom triangle as we tried first, or the left and top triangle as we tried second—the total area will be the same.

18. B

This question refers to the number of flash floods and tornadoes for all five states, so read the line graph, specifically looking at August. The line graph tells you only about events in 2014, so you will need to calculate the number of events in 2013 and 2015. Since the question tells you that events followed the same pattern each year, you can infer that the percent increases in the question affect each month the same way.

In August 2014, there were about 30 flash floods and 10 tornadoes. Both numbers are a 25 percent increase over the previous year, so calculate the number of events in August 2013:

$$\text{Floods} + \text{Tornadoes} = 30 + 10 = 40$$

$$40 = e + 0.25e;\ 40 = 1.25e;\ e = 32$$

There were about 32 severe weather events in August 2013.

Now calculate the numbers in 2015, starting from 2014 as a base and increasing the floods by one-third and the tornadoes by one-half:

Floods:

$$f = 30 + \frac{1}{3}(30);\ f = 40$$

Tornadoes:

$$t = 10 + \frac{1}{2}(10);\ t = 15$$

Add to find that there were about $40 + 15 = 55$ severe weather events in August 2015. The question asks how many more events occurred in August 2015 than in August 2013, so subtract: $55 - 32 = 23$. The correct answer is **(B).**

19. A, C

This question asks about EF1 tornadoes, so look at the stacked bar graph, which tells you the proportion of tornadoes by intensity for each state. The question is asking for possible actual numbers of tornadoes, so also look at the line graph, which gives you some information about numbers of events.

The line graph gives you the number of tornadoes in Nebraska in 2014, so start there.

April	3
May	28
June	41
July	1
August	0
Sept	1
Oct	0
	74

Your reading of the graph may vary slightly due to estimating the numbers, but you should have a number reasonably close to 74. Then in the stacked bar graph, the

EF1 tornadoes are represented by the second bar from the bottom. For Nebraska, about 25% of the tornadoes were EF1, so $0.25 \times 74 \approx 19$ EF1 tornadoes. The correct answer must have a last number that is close to 19. Again, depending on your estimates from the graphs, you might have arrived at a slightly different number, but choices (B), (E), and (F) clearly do not have the right number for Nebraska and can be eliminated.

Now you can use Backsolving to see which answer choices fit the data. If you add up the tornadoes for all five states, there are 174 (again, your estimate may vary slightly but should be close enough to solve this problem). The stacked bar graph tells you that about 13% of North Dakota's tornadoes and 20% of South Dakota's tornadoes were EF1. According to choice (D), there were 6 EF1 tornadoes in North Dakota, representing about 13% of all North Dakota tornadoes: $6 = 0.13x$; $x \approx 46$ North Dakota tornadoes. Choice (D) also says there were 18 EF1 tornadoes in South Dakota, which would represent 20% of all the tornadoes in South Dakota: $18 = 0.2x$; $x = 90$. Add these to the 74 Nebraska tornadoes: $46 + 90 + 74 = 210$. That's too many total tornadoes, so eliminate choice (D).

Now keep in mind that this all-that-apply question asks what *could have been* the numbers of EF1 tornadoes in 2014. The stacked bar graph tells you what percent of North and South Dakota's tornadoes were EF1—13% of North Dakota's and 20% of South Dakota's tornadoes. However, nowhere are you told how many total tornadoes those states had. (You can deduce only that together they had fewer than 100 tornadoes, to allow for Nebraska's 74 and those occurring in Kansas and Oklahoma). North Dakota could have had more, or South Dakota could have had more. Therefore, a wide range of numbers are possible values. Choice **(A)** works if North Dakota had 15 tornadoes and South Dakota had 30, and choice **(C)** works if North Dakota had 45 tornadoes and South Dakota had 15. Choices **(A)** and **(C)** are correct.

20. C

This question concerns the amount of property damage due to flash floods (look at the pie chart) and the number of flash floods that occur (look at the line graph). The answer choices are far apart, so approximation is an efficient strategy for this question.

Of all the damage accounted for in the pie chart (\$3,854,100), 22% was in Nebraska, so calculate Nebraska's share of the damage: $0.22 \times \$3,854,100 = \$847,902$. For ease of calculation, round this to \$850,000.

Next, figure out how much of that damage occurred in June. The total number of flash floods in Nebraska was as follows:

Month	Floods
April	0
May	11
June	22
July	2
August	3
Sept	3
Oct	0
	41

Your reading of the graph may vary slightly due to estimating the numbers, but you should have a number close to 40. The number of floods in June was 22, or a little over half the total. You are told that the amount of damage is proportional to the number of events, so a bit more than half the damage in Nebraska happened in June. Half the damage is $850,000 ÷ 2 = $425,000, so the correct answer is the one that is a little more than that. (If you did the calculations precisely, you should have arrived at $454,972.) The correct answer is **(C)**.

ADVANCED MATH QUESTIONS PRACTICE SET 4

1. Set S is the set of all positive integers that are less than 100 that are the square of an integer. Set T is the set of all positive integers that are less than 100 that are equal to the cube of an integer.

Quantity A	Quantity B
The number of integers that are in at least one of the sets S and T	12

Ⓐ Quantity A is greater.
Ⓑ Quantity B is greater.
Ⓒ The two quantities are equal.
Ⓓ The relationship cannot be determined from the information given.

2. $$2x - 3y = 7$$

Quantity A	Quantity B
$4x^2 - 12xy + 9y^2 + 18$	70

Ⓐ Quantity A is greater.
Ⓑ Quantity B is greater.
Ⓒ The two quantities are equal.
Ⓓ The relationship cannot be determined from the information given.

3. The integer n is greater than or equal to 40. When the integer n is divided by 28, the remainder is 12.

Quantity A	Quantity B
The remainder when $2n$ is divided by 14	The remainder when $3n$ is divided by 14

Ⓐ Quantity A is greater.
Ⓑ Quantity B is greater.
Ⓒ The two quantities are equal.
Ⓓ The relationship cannot be determined from the information given.

4.

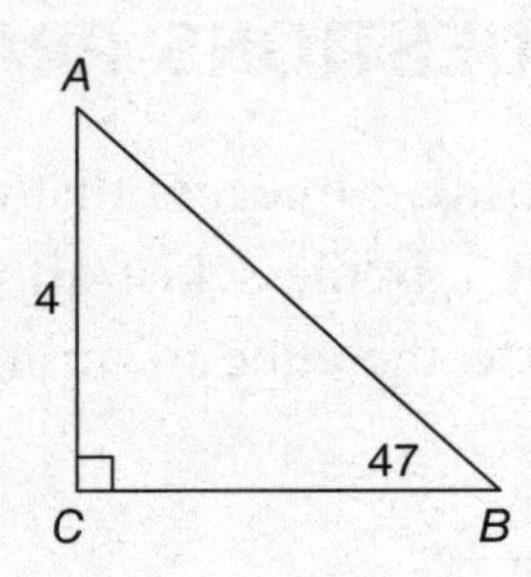

Quantity A	Quantity B
The area of triangle *ABC*	8

Ⓐ Quantity A is greater.
Ⓑ Quantity B is greater.
Ⓒ The two quantities are equal.
Ⓓ The relationship cannot be determined from the information given.

5. A certain pregnancy test comes back positive for 95 percent of pregnant women who take it. However, it also comes back positive for 5 percent of non-pregnant women who take it. Two percent of the female population of city A is pregnant. Marisa, a resident of city A, takes the pregnancy test and gets a positive result.

Quantity A	Quantity B
The probability that Marisa is pregnant	75%

Ⓐ Quantity A is greater.
Ⓑ Quantity B is greater.
Ⓒ The two quantities are equal.
Ⓓ The relationship cannot be determined from the information given.

6. Clarice, Yolanta, and Albert have among them some chocolates. One of them has triple the chocolates of another. Yolanta has 24 percent of the chocolates.

Quantity A	Quantity B
The number of chocolates Albert has if Clarice has more chocolates than Yolanta	The number of chocolates Clarice has if Yolanta has more chocolates than Albert

(A) Quantity A is greater.
(B) Quantity B is greater.
(C) The two quantities are equal.
(D) The relationship cannot be determined from the information given.

7. Researchers measured the incubation times of 100 emperor penguin eggs. The researchers' analysis showed that the data followed normal distribution, the mean was 64 days, and the standard deviation was 1 day. After recounting, the researchers found more eggs and added them to the original data set: 4 eggs took 64 days to hatch, 2 took 61 days, and 2 took 67 days.

Quantity A	Quantity B
1 day	1 standard deviation of the data set that includes all 108 eggs

(A) Quantity A is greater.
(B) Quantity B is greater.
(C) The two quantities are equal.
(D) The relationship cannot be determined from the information given.

8. The median of the 35 numbers in list L is 18. The mode of list L is not 18 and not 19.

Quantity A	Quantity B
The arithmetic mean of the 17 greatest numbers in list L	19

Ⓐ Quantity A is greater.
Ⓑ Quantity B is greater.
Ⓒ The two quantities are equal.
Ⓓ The relationship cannot be determined from the information given.

9. Lorenzo traveled 120 miles by automobile to visit his brother. The first segment of the trip required Lorenzo to drive on 10 miles of city roads before he reached the interstate highway and accelerated to a constant speed of 60 mph. The last segment of the trip was a 10 mile journey over country roads on which Lorenzo averaged 50 miles per hour. If the entire trip took Lorenzo 2 hours and 22 minutes, what was his average speed in miles per hour for the portion of the route between his home and the interstate highway?

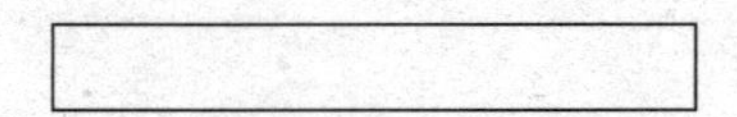

10. If $\frac{x}{3}$ is a positive odd integer and $x \leq 15$, which values could be the median of the list of numbers 1, 3, 5, 7, 9, 11, 13, 15, 17, x, $\frac{x}{3}$?

Indicate all such values.

[A] 5
[B] 6
[C] 7
[D] 9
[E] 11

11. Olga's swimming pool has three pipes connected to it. If the pool is empty, pipe A can fill it in 3 hours and pipe B can fill it in 4 hours. If the pool is at capacity, pipe C can empty it in 2 hours. The capacity of Olga's pool is 2,400 cubic meters. If all three pipes are activated when the pool is empty, how many hours will it take for the pool to be filled to 60 percent of capacity?

(A) 7.0
(B) 7.2
(C) 8.0
(D) 8.6
(E) 12.0

12. If $|8x - 7| > 3x + 8$, then each of the following could be true EXCEPT:

(A) $1 < x < 3$
(B) $-1 < x < 0$
(C) $2 < x < 4$
(D) $x = 3.5$
(E) $x = -1$

13. A fair coin is tossed six times. What is the probability that the result will be exactly three heads and three tails?

(A) $\frac{1}{6}$
(B) $\frac{1}{4}$
(C) $\frac{5}{16}$
(D) $\frac{1}{2}$
(E) $\frac{13}{24}$

14. The ratio of cranberry to apple to orange juice in a certain cocktail containing no other liquids is x:5:3. After 6 gallons of water are added to the cocktail, the ratio of orange juice to total liquid becomes 1:4. After another 24 gallons of water are added to the cocktail, the ratio of apple juice to total liquid becomes 1:4. What fraction of the original cocktail is cranberry juice?

(A) $\frac{1}{5}$

(B) $\frac{1}{4}$

(C) $\frac{1}{3}$

(D) $\frac{1}{2}$

(E) $\frac{2}{3}$

15. What is the probability of rolling a number less than 3 at least 3 times in 5 rolls of a six-sided die?

(A) $\frac{2}{15}$

(B) $\frac{17}{81}$

(C) $\frac{1}{3}$

(D) $\frac{2}{5}$

(E) $\frac{15}{32}$

16 Juliana has three options to invest her money for one year. Option A is discounted notes that she can buy for $98 and cash in for a face value of $100 a year later. Option B is a savings account that pays a 2% annual interest rate compounded semi-annually. Option C is an account that pays 2.03% simple interest for one year. Which of the answer choices represents the percent returns from the three choices in order from lowest to highest?

Ⓐ A, B, C
Ⓑ A, C, B
Ⓒ B, C, A
Ⓓ C, A, B
Ⓔ C, B, A

17. Rick's Sandwich Shop lets customers create their own sandwiches by choosing from among a certain number of ingredients. When ordering Rick's Rockin' Sandwich, the customer must choose 1 of 3 breads, 1 of 6 meats, and 1 of 4 cheeses. The customer can also add up to 3 of 8 different veggies. How many unique Rick's Rockin' Sandwiches could a customer order?

Ⓐ 4,032
Ⓑ 6,624
Ⓒ 6,696
Ⓓ 24,192
Ⓔ 28,932

Questions 18–20 refer to the following charts.

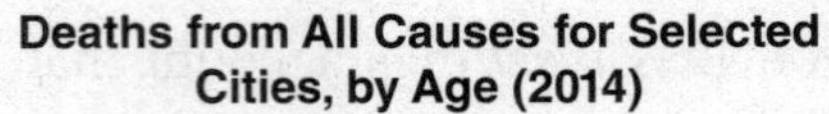

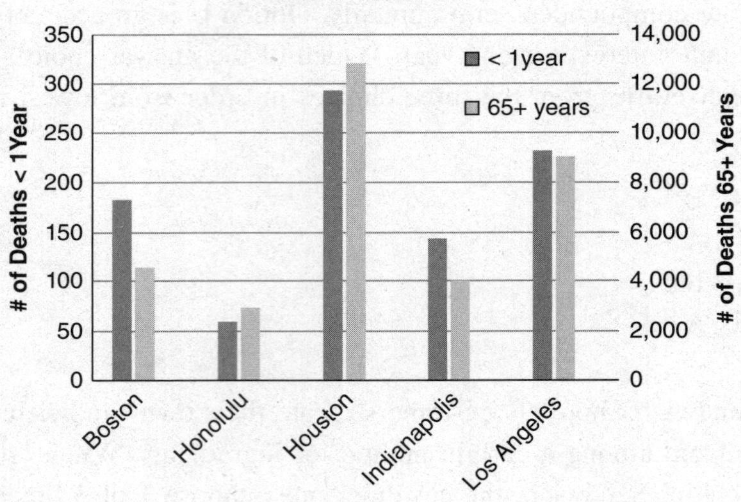

Source: Centers for Disease Control and Prevention, *Morbidity and Mortality Weekly Report* (*MMWR*), Table III: Mortality in 122 U.S. cities, data provided by National Notifiable Diseases Surveillance System (NNDSS), http://wonder.cdc.gov/mmwr/mmwrmort.asp.

Deaths from Pneumonia and Influenza for Selected U.S. Cities (2014)

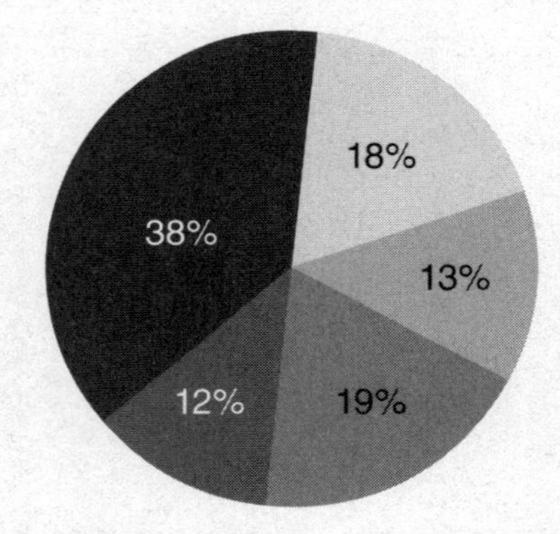

Total deaths from pneumonia/influenza for selected cities: 3,532

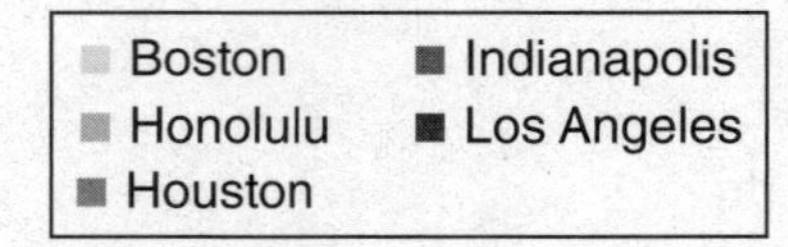

18. Imagine that at the beginning of 2014, Boston and Los Angeles implemented a public health program that reduced deaths of infants less than 1 year old by 20 percent, while the cities of Honolulu and Indianapolis terminated an identical program. What would have been the approximate total impact of these program changes on the number of infant deaths in these cities?

    (A) There would have been 600 more deaths.
    (B) There would have been 70 more deaths.
    (C) There would have been 30 fewer deaths.
    (D) There would have been 65 fewer deaths.
    (E) There would have been 150 fewer deaths.

19. If in 2014, the population of Los Angeles was 75 percent greater than the population of Houston, what is the ratio of the incidence of pneumonia- and influenza-related deaths (expressed as a percent of the city's population) in Houston to the incidence in Los Angeles in that year?

    (A) $\frac{1}{2}$
    (B) $\frac{7}{8}$
    (C) $\frac{15}{16}$
    (D) $\frac{8}{7}$
    (E) $\frac{4}{3}$

20. Assuming that 80 percent of all deaths due to pneumonia or influenza occur among the elderly, defined as those age 65 and over, in which city was the least proportion of all deaths among the elderly attributed to pneumonia or influenza?

    (A) Houston
    (B) Indianapolis
    (C) Honolulu
    (D) Boston
    (E) Los Angeles

ADVANCED MATH QUESTIONS PRACTICE SET 4 ANSWER KEY

1. B
2. B
3. A
4. B
5. B
6. B
7. B
8. D
9. 20
10. C, D
11. B
12. A
13. C
14. A
15. B
16. C
17. C
18. C
19. B
20. A

ADVANCED MATH QUESTIONS PRACTICE SET 4 ANSWERS AND EXPLANATIONS

1. B

Start by determining all the integers in each of sets *S* and *T*. These are the integers in set *S*:

$1^2 = 1 \times 1 = 1$

$2^2 = 2 \times 2 = 4$

$3^2 = 3 \times 3 = 9$

$4^2 = 4 \times 4 = 16$

$5^2 = 5 \times 5 = 25$

$6^2 = 6 \times 6 = 36$

$7^2 = 7 \times 7 = 49$

$8^2 = 8 \times 8 = 64$

$9^2 = 9 \times 9 = 81$

$10^2 = 10 \times 10 = 100$, but this is not included since the centered information specifies less than 100.

These are the integers in set *T*:

$1^3 = 1 \times 1 \times 1 = 1$

$2^3 = 2 \times 2 \times 2 = 4 \times 2 = 8$

$3^3 = 3 \times 3 \times 3 = 9 \times 3 = 27$

$4^3 = 4 \times 4 \times 4 = 16 \times 4 = 64$

Thus, the numbers in set *S* are the 9 integers 1, 4, 9, 16, 25, 36, 49, 64, and 81.

The numbers in set *T* are 1, 8, 27, and 64.

Two of the numbers in set *T*, 1 and 64, are also in set *S*, so don't count those.

It turns out that there are $9 + 2 = 11$ integers in at least one of sets *S* and *T*. Quantity A is 11 and Quantity B is 12. Choice **(B)** is correct.

2. B

You need to compare a quadratic to 70, and the only centered information is a linear equation. Try squaring both sides of the centered equation to see if you get anything similar to Quantity A. Squaring both sides of $2x - 3y = 7$ gives you $(2x - 3y)^2 = 49$. Using the identity $(a - b)^2 = a^2 - 2ab + b^2$, the equation $(2x - 3y)^2 = 49$ becomes $4x^2 - 12xy + 9y^2 = 49$. That looks awfully similar to Quantity A. In fact, Quantity

A is just the result of adding 18 to $4x^2 - 12xy + 9y^2$, which is one side of the equation $4x^2 - 12xy + 9y^2 = 49$. Adding 18 to both sides of this equation produces $4x^2 - 12xy + 9y^2 + 18 = 67$. The left side of this equation is Quantity A, so Quantity A equals 67. Quantity B is 70. Quantity B is greater, and **(B)** is correct.

3. A

The centered information says that when the integer n is divided by 28, the remainder is 12. This means that $n = 28m + 12$, where m is an integer such that $m \geq 1$.

Consider Quantity A. Since $n = 28m + 12$, $2n = 56m + 24$. Since 56 is a multiple of 14 ($56 = 4 \times 14$), and m is a positive integer, the remainder when $56m$ is divided by 14 is 0. When 24 is divided by 14, the quotient is 1 and the remainder is 10. Thus, the remainder when $2n$, which equals $56m + 24$, is divided by 14 is $0 + 10 = 10$. Quantity A is 10.

Now consider Quantity B. Since $n = 28m + 12$, $3n = 84m + 36$. Since 84 is a multiple of 14 ($84 = 6 \times 14$), and m is a positive integer, the remainder when $84m$ is divided by 14 is 0. When 36 is divided by 14, the quotient is 2 and the remainder is 8. You can conclude that the remainder when $3n$, or $84m + 36$, is divided by 14 is $0 + 8 = 8$. Quantity B is 8.

Quantity A is 10 and Quantity B is 8. Quantity A is greater and the correct answer is **(A)**.

4. B

The area of any triangle is $\frac{1}{2} \times \text{base} \times \text{height}$. The area of triangle ABC is $\frac{1}{2} \times AC \times BC$. Since $AC = 4$, the area of right triangle ABC is $\frac{1}{2} \times 4 \times BC = 2 \times BC$. To determine the value of Quantity A, you would need to know the length of BC, but there's no way to calculate that without using trigonometry. Fortunately, you can often answer Quantitative Comparison questions speedily without knowing precise values.

Consider what is known about triangle ABC. In any triangle, the sum of the measures of the 3 interior angles is 180°. You know that the measure of angle ACB is 90° and the measure of angle ABC is 47°. Subtract these two values from 180° to find that angle CAB thus equals 43°. Now in any triangle, a side that is opposite a greater angle is longer than a side that is opposite a smaller angle. So side AC, which is across from a 47° angle, is longer than side BC, which is across from a 43° angle. Since $AC = 4$, it must be that $BC < 4$.

The area of triangle ABC is $2(BC)$. Since $BC < 4$, the area of triangle ABC must be less than $2(4) = 8$. Quantity A is less than 8, and Quantity B is equal to 8. Quantity B is greater and choice **(B)** is correct.

5. B

This question succumbs quickly to Kaplan's Picking Numbers strategy. Although this is one of the hardest GRE questions you're ever likely to see, it's surprisingly straightforward if you apply the Picking Numbers strategy to it.

Suppose 1,000 women take the test. The question states that 2 percent, or $1{,}000 \times 0.02 = 20$, of those women are pregnant. It follows that $1000 - 20 = 980$ women are not pregnant. Of the 20 pregnant women who take the test, 95%, or $20 \times 0.95 = 19$, will see a positive result. Of the 980 non-pregnant women who take the test, 5%, or $980 \times 0.05 = 49$ will see a positive result.

Take a moment to paraphrase Quantity A. It asks, in essence, "If a woman takes the test and gets a positive result, what is the probability that she is actually pregnant?" You know that a total of $19 + 49 = 68$ women will have a positive test, only 19 of whom are in fact pregnant. You can stop here: there is no need to calculate the precise probability, because 19 out of 68 is clearly less than 75%. The correct answer is **(B).** (For the record, the probability is $\frac{19}{68} \approx 28\%$.)

6. B

Before diving into the quantities, consider the possible scenarios. There are three:

Scenario 1: Yolanta has triple the chocolates of someone else. In this case, Yolanta has 24% of the chocolates, someone else has $24\% \div 3 = 8\%$, and the third person has $100\% - 24\% - 8\% = 68\%$.

Scenario 2: Someone has triple the chocolates of Yolanta. In this case, Yolanta has 24% of the chocolates, someone else has $24\% \times 3 = 72\%$, and the third person has $100\% - 24\% - 72\% = 4\%$.

Scenario 3: Yolanta is not involved in the triple scenario. In that case, Clarice and Albert share the remaining $100\% - 24\% = 76\%$ of the chocolates in a ratio of 3:1.

$$\begin{aligned} 3x + x &= 76\% \\ 4x &= 76\% \\ x = 76\% \div 4 &= 19\% \\ 3x = 19\% \times 3 &= 57\% \end{aligned}$$

Thus, in Scenario 3, Yolanta has 24% of the chocolates, someone else has 19%, and the third person has 57%.

The key to this question is that in all three cases, Yolanta has the median number of chocolates. This insight allows you to conclude that in Quantity A, where Clarice has more chocolates than Yolanta, Albert has the fewest. Thus, Quantity A refers to the person with the fewest chocolates (i.e., less than the median of 24%). In Quantity B, where Yolanta has more chocolates than Albert, Clarice must have the most. Thus,

Quantity B refers to the person with the most chocolates (i.e., more than the median of 24%). Paraphrasing, the quantities say:

Quantity A	Quantity B
The number of chocolates the person with the fewest chocolates (less than 24%) has.	The number of chocolates the person with the most chocolates (greater than 24%) has.

The answer is **(B)**.

7. B

First, recognize that the mean does not change. The data added is symmetrical around the original mean of 64 days. The 4 new eggs with a 64-day incubation period will not affect the average. There are 2 new eggs with a 61-day incubation period and 2 new eggs with a 67-day incubation period. Both 61 and 67 are the same distance from the mean (3 days) and the number added is the same at both data points (2 eggs), so they will not change the average either.

Another way to confirm this is to find the average of the new eggs:

$$\begin{aligned}\text{Average} &= \frac{((64 \times 4) + (61 \times 2) + (67 \times 2))}{8} \\ &= \frac{256 + 122 + 134}{8} \\ &= \frac{512}{8} = 64\end{aligned}$$

Next, consider the original data set and how it changes with the addition of the 8 new eggs. If the standard deviation were to remain the same, the distribution of the new data added would have to mirror the normal distribution of the original data set. In other words, about 68% would be within 1 day, 95% within 2 days, and 99.7% within 3 days. The new data added, however, is more spread out. The new data set's standard deviation will be greater than the original data set's of 1 day; therefore, the correct answer is **(B)**.

Note: if you'd like to use this question to learn more about how normal distribution works, you can estimate how many eggs had a 64-day incubation period in the original data set. In the normal distribution, about 68% of the data is within one standard deviation of the mean. That means that of the 100 original eggs, about 68 had an incubation period greater than 63 days and less than 65 days. The new total of 64-day eggs is $68 + 4 = 72$. The new total number of eggs is $100 + 8 = 108$. The new percent of eggs within 1 day of the mean is $72/108 = 66.7\%$. This is less than 68%, which means that compared to the original data set, slightly more data is farther from the mean. The standard deviation would need to become greater in order to include 68% of the data; therefore, the new standard deviation will be more than 1 day.

8. D

The numbers in list *L* could be anything, as long as 18 is the median but not the mode and 19 is also not the mode. For example, imagine that each of the 17 greatest numbers in list *L* is 100, the median is 18, and each of the other 17 numbers is 0. This would be a permissible list because the median (but not the mode) would indeed be 18. Quantity A, the average of the 17 greatest numbers in list *L*, would then be 100, making it greater than Quantity B.

But it is also entirely possible that many of the 35 numbers in list *L* are very close to, but not quite equal to, 18. For example, imagine that each of the 17 greatest numbers is 18.01, the median is 18, and each of the other 17 numbers is 0. (Note that the question never stated that the numbers in List L had to be integers.) In this case, Quantity A would be 18.01, and Quantity B would be greater. Since more than one relationship between the quantities is possible, choice **(D)** is correct.

9. 20

The question describes a three-segment trip and provides the distances for the first and third segments, the average speeds for the second and third segments, the total distance traveled, and the total duration for the trip. The question asks you to calculate the average speed for the first segment. The formula to use is $\text{Speed} = \frac{\text{Distance}}{\text{Time}}$. The distance of the first segment is given in the problem, but you need to calculate time. You can do this by calculating the time required for the second and third segments of the trip and subtracting them from the total time.

For the interstate portion of the trip you know that Lorenzo's speed was 60 mph, but the problem does not give you the distance covered. Since the total distance was 120 miles and the first and third legs were 10 miles each, the distance covered driving on the interstate must have been $120 - 10 - 10 = 100$ miles. Rearrange the speed formula to obtain

$$\text{Time} = \frac{\text{Distance}}{\text{Speed}} = \frac{100 \text{ miles}}{\left(60 \frac{\text{miles}}{\text{hour}}\right)} = 1\frac{2}{3} \text{ hours,}$$

which converts to 100 minutes. Use this same formula to calculate the time it took Lorenzo to drive the third segment:

$$\text{Time} = \frac{10 \text{ miles}}{\left(50 \frac{\text{miles}}{\text{hour}}\right)} = \frac{1}{5} \text{ hour} = 12 \text{ minutes.}$$

Since the entire trip took 2 hours and 22 minutes, or 142 minutes, Lorenzo drove the first segment in 142 – 100 – 12 = 30 minutes. Now you have known values for both distance and time (convert 30 minutes to 0.5 hour) for the first segment:

$$\text{Speed} = \frac{10 \text{ miles}}{0.5 \text{ hours}} = 20 \text{ miles per hour.}$$

10. C, D

You are given a list with 9 integers and two terms containing a variable, for a total of 11 terms. The median will be the middle, or sixth, term in the list when the terms are ordered from least to greatest.

Because $\frac{x}{3}$ is known to be a positive integer, x must be a positive multiple of 3.

Because $\frac{x}{3}$ is odd, x must be odd. Moreover, because $0 < x \leq 15$, there are limited possible values for x. It could be that $x = 3$ and $\frac{x}{3} = 1$, or that $x = 9$ and $\frac{x}{3} = 3$, or that $x = 15$ and $\frac{x}{3} = 5$.

If $x = 3$ and $\frac{x}{3} = 1$, then the list is 1, 1, 3, 3, 5, 7, 9, 11, 13, 15, 17, and the median is 7.

If $x = 9$ and $\frac{x}{3} = 3$, then the list is 1, 3, 3, 5, 7, 9, 9, 11, 13, 15, 17, and the median is 9.

If $x = 15$ and $\frac{x}{3} = 5$, then the list is 1, 3, 5, 5, 7, 9, 11, 13, 15, 15, 17, and the median is 9.

The median is 7 or 9, so only **(C)** and **(D)** are correct.

11. B

Solving this question takes a lot of steps, but none of the steps are all that difficult. The key to beating questions like this is to stay calm and be methodical.

Pipe A can fill a 2,400 cubic meter pool in 3 hours. This means that pipe A's rate is $\frac{2,400}{3} = 800\,m^3$ per hour. Similarly, pipe B can fill the pool in 4 hours, so pipe B's rate is $\frac{2,400}{4} = 600\,m^3$ per hour. Since pipe C empties rather than fills the pool, you need to make sure its rate is negative, but you can find this rate just as easily as the other two: $-\frac{2,400}{2} = -1,200\,m^3$ per hour.

The easiest way to solve a question that has multiple rates is simply to add them up. In this case, the three pipes working together fill the pool at a rate of 800 + 600

$- 1{,}200 = 200\ m^3$ per hour. The question asks how long it will take to fill 60 percent of the pool. Sixty percent of 2,400 is $2{,}400 \times 0.6 = 1{,}440\ m^3$. So $1{,}440\ m^3$ need to be filled at a total rate of $200\ m^3$ per hour. Set up a proportion: $\frac{200}{1\text{ hr}} = \frac{1{,}440}{x\text{ hrs}}$. So $1{,}440 = 200x$. Do the final division to get the answer: $\frac{1{,}440}{200} = 7.2$ hours, choice **(B)**.

12. A

To solve an absolute value inequality (or equation), write out two possibilities: one in which the absolute value expression is positive, and one in which it's negative. Case 1 (positive): $(8x - 7) > 3x + 8$, so $5x > 15$, and $x > 3$.

Case 2 (negative): $-(8x - 7) > 3x + 8$, so $-8x + 7 > 3x + 8$, which means that $-1 > 11x$ and $x < -\frac{1}{11}$.

This means that the inequality holds true as long as either $x > 3$ or $x < -\frac{1}{11}$. Now consider the choices. The question asks for the choice that CAN'T be true, and that's choice **(A)**. If x is between 1 and 3, then it is neither greater than 3 nor less than $-\frac{1}{11}$.

13. C

Since probability is defined as $\frac{\text{Number of desired outcomes}}{\text{Number of possible outcomes}}$, a good place to start is to determine the total number of outcomes (that is, with the denominator). Each coin toss has two possible outcomes (heads or tails). Since the coin is tossed six times in this problem, the total number of possible outcomes is $2 \times 2 \times 2 \times 2 \times 2 \times 2 = 2^6 = 64$.

In order to determine how many of those outcomes have exactly 3 heads and 3 tails, use the combinations formula. Think of it as choosing those spaces, between 1 and 6, where the three heads will show up. Order doesn't matter because heads in 1, 2, and 5 is the same as heads in 5, 1, and 2. Calculating ${}_6C_3$ produces $\frac{6 \times 5 \times 4 \times 3 \times 2 \times 1}{3 \times 2 \times 1 \times 3 \times 2 \times 1}$.

The $3 \times 2 \times 1$ in the numerator and denominator of the fraction cancel. The additional $3 \times 2 \times 1$ in the denominator cancels with the 6 in the numerator, leaving $\frac{5 \times 4}{1} = 20$.

Returning to the probability formula, $\frac{\text{20 desired outcomes}}{\text{64 total outcomes}} = \frac{5}{16}$. Choice **(C)** is correct.

14. A

This problem tests your fundamental understanding of ratios. If the ratio of cranberry to apple to orange juice is x:5:3, what that means is that the volume of cranberry juice is a multiple of x, the volume of apple juice is that same multiple of 5, and the volume of orange juice is that same multiple of 3. You're not told what this multiple is, so call it n. Thus, there is a total of $xn + 5n + 3n = xn + 8n$ gallons of juice in the cocktail.

After 6 gallons of water are added, the amount of orange juice ($3n$) doesn't change. Since this amount, as a fraction of the total, is given as 1:4, you can set up a proportion:

$$\frac{3n}{xn + 8n + 6} = \frac{1}{4}$$

Cross multiply and simplify:

$$\begin{aligned} 12n &= xn + 8n + 6 \\ xn - 4n &= -6 \end{aligned}$$

Next, you learn that adding another 24 gallons to the cocktail (so, 30 gallons total) makes the ratio of apple juice to total liquid 1:4. The amount of apple juice ($5n$), like the amount of orange juice, hasn't changed, so you can again set up a proportion and solve:

$$\begin{aligned} \frac{5n}{xn + 8n + 30} &= \frac{1}{4} \\ 20n &= xn + 8n + 30 \\ xn - 12n &= -30 \end{aligned}$$

You now have a system of equations with two variables (x and n), so use combination to solve:

$$\begin{aligned} (xn - 4n &= -6) \\ -(xn - 12n &= -30) \\ \hline 0 + 8n &= 24 \\ n &= 3 \end{aligned}$$

Plug this value for n into one of the original equations to find x:

$$\begin{aligned} xn - 4n &= -6 \\ 3x - 4(3) &= -6 \\ 3x = -6 + 12 &= 6 \\ x &= 2 \end{aligned}$$

In this problem, x represents the portion of the ratio that corresponds to cranberry juice. Thus, the ratio of cranberry to apple to orange juice is 2:5:3, and the ratio of cranberry juice to total liquid is $\frac{2}{2+5+3} = \frac{2}{10} = \frac{1}{5}$, choice **(A)**.

15. B

Before performing any calculations, think through your strategic approach. Since 3 or 4 or 5 rolls coming up less than 3 is a favorable event, you will need to calculate all three of those probabilities and add them to get the final answer.

The probability of rolling a 1 or 2 on a single roll is $\frac{2}{6} = \frac{1}{3}$, so the probability of not rolling a number less than 3 is $1 - \frac{1}{3} = \frac{2}{3}$.

Start with the most straightforward event, 5 rolls all coming up less than 3. The probability of that occurring is $\left(\frac{1}{3}\right)^5 = \frac{1}{243}$, since the first roll and all subsequent rolls must be less than 3.

Any one particular set of 4 favorable and 1 unfavorable outcomes has a probability of $\left(\frac{1}{3}\right)^4\left(\frac{2}{3}\right)^1 = \frac{2}{243}$. However, there are 5 ways to obtain this result (the unfavorable outcome could be any one of the five rolls), so the probability of 4 favorable rolls is $5 \times \left(\frac{2}{243}\right) = \frac{10}{243}$.

Calculating the probability of 3 favorable rolls gets a bit trickier. The probability of any one set of outcomes producing that result is $\left(\frac{1}{3}\right)^3\left(\frac{2}{3}\right)^2 = \frac{4}{243}$. In order to determine how many different ways 5 rolls would result in 3 favorable outcomes, you can use the combinations formula:

$$\frac{n!}{k!(n-k)!} = \frac{5!}{3!(5-3)!} = \frac{5!}{3!2!} = \frac{5 \times 4 \times 3 \times 2 \times 1}{3 \times 2 \times 1 \times 2 \times 1} = 10.$$

Therefore, the probability of 3 favorable outcomes is

$$10 \times \left(\frac{4}{243}\right) = \frac{40}{243}.$$

Add these three probabilities to obtain $\frac{1 + 10 + 40}{243} = \frac{51}{243} = \frac{17}{81}$.

Choice **(B)** is correct.

16. C

Option A results in an increase in value of $2, but Juliana only has to invest $98, so the percent return is

$$\frac{\$2}{\$98} \approx 0.0204 \times 100\% = 2.04\%$$

per year. (At this point, if you're short on time, you can eliminate choices (A) and (B) because you know option A gives a better return than option C (a flat 2.03%)).

In option B, you can calculate the compound interest using the formula

$$\text{Final balance} = \text{Principal} \times \left(1 + \frac{\text{Interest rate}}{n}\right)^{(t)(n)}.$$

In this case $t = 1$ (because Juliana invests for 1 year) and $n = 2$, the number of times compounding occurs per year. Pick $100 for the principal to make things as simple as possible:

$$\text{Final balance} = \$100 \times \left(1 + \frac{.02}{2}\right)^2 = \$100 \times 1.01^2$$

$$\text{Final balance} = \$100 \times 1.01 \times 1.01 = \$102.01$$

Since the initial investment of $100 earned $2.01 in interest, the rate of return is 2.01% for one year.

Option C is given as 2.03% annually.

Therefore, the correct order from lowest to highest percent returns is B, C, A. The correct answer is **(C)**.

17. C

This question asks about the number of ways you can combine certain ingredients, which means it is a permutation/combination problem. The customer must choose 1 of 3 breads, 1 of 6 meats, and 1 of 4 cheeses. Multiply these together to find the number of possibilities for those three ingredients: $3 \times 6 \times 4 = 72$. The veggies create several more possibilities because the customer can choose up to 3 of 8 veggies. Start by considering the case when a customer chooses the maximum of 3 veggies. Since there is no difference between choosing lettuce, tomato, onion and choosing lettuce, onion, tomato, order does not matter; therefore, use the combination formula:

$$_nC_k = \frac{n!}{k!(n-k)!}$$

In this case, we are selecting 3 out of 8, so $k = 3$ and $n = 8$:

$$\begin{aligned} _8C_3 &= \frac{8!}{3!(8-3)!} \\ &= \frac{8!}{3! \times 5!} \\ &= \frac{8 \times 7 \times 6}{3 \times 2 \times 1} \\ &= 8 \times 7 \\ &= 56 \end{aligned}$$

Now, to find the number of total sandwich possibilities, multiply the number of combinations of 3 veggies by the number of combinations of the other ingredients: $56 \times 72 = 4{,}032$.

There are, however, three more possibilities. The customer might select only 2 of 8 veggies or 1 of 8 veggies, or the customer may select none. Add these possibilities to the total. If the customer chooses 2 of the 8 veggies, $k = 2$ and $n = 8$:

$$\frac{8!}{2!(8-2)!} = \frac{8 \times 7}{2!} = \frac{8 \times 7}{2} = 28$$

Again, multiply the number of combinations of veggies by the number of combinations of the other ingredients: $28 \times 72 = 2{,}016$.

If the customer chooses only 1 of the 8 veggies, there will be 8 possibilities. You can plug the numbers $k = 1$ and $n = 8$ into the combination formula, but it is faster to recognize that there are only 8 ways to select 1 out of 8 veggies. Yet again, multiply the number of combinations of veggies by the number of combinations of the other ingredients: $8 \times 72 = 576$.

And finally, you need to consider what happens if the customer does not choose any veggies. You can plug $k = 0$ and $n = 8$ into the combination formula (remembering that $0! = 1$), but this is unnecessary if you recognize that there is only 1 possibility: no veggies! One last time, multiply the number of combinations of veggies (or lack thereof) by the number of combinations of the other ingredients: $1 \times 72 = 72$.

The last step is to sum all of these possibilities:
$4{,}032 + 2{,}016 + 576 + 72 = 6{,}696$. The correct answer is **(C).**

18. C

This question asks about deaths of infants under age 1 year in certain cities. That information is found in the bar graph, specifically in the left bar of each cluster. Note that this graph has two *y*-axes, one on the left and one on the right. Deaths that occur before age 1 year are plotted against the *y*-axis on the left; be sure to read the correct bars against the correct axis.

Boston is represented by the bars on the far left and Los Angeles by the bars on the far right. Honolulu and Indianapolis are represented by the second and fourth pairs of bars, respectively.

Before doing any calculations, you might note that the numbers of infant deaths in both Boston and Los Angeles are significantly higher than in either Honolulu or Indianapolis. Therefore, the effect of introducing public health programs in Boston and Los Angeles will be greater than the effect of terminating programs in the other two cities. In other words, more deaths will be prevented than not prevented, and you can eliminate choices (A) and (B). On Test Day, even if you had no more time to invest in this problem, by using your critical thinking skills to eliminate answer choices, you would improve your odds of guessing the correct answer.

In 2014, Boston had just under 200 infant deaths, and Los Angeles had just over 200. That's about 400 infant deaths total for these cities. The question wants you to imagine that a public health program was introduced in these cities, dropping infant deaths by 20%. In that case, there would be $(0.2)(400) = 80$ fewer infant deaths in those two cities, or 320 infant deaths instead of the actual 400.

Honolulu had about 60 infant deaths, and Indianapolis had not quite 150—call it 140. That's $60 + 140 = 200$ deaths, and you're told to imagine what would have happened had a program reducing infant deaths by 20% been terminated. Without such a program acting to reduce infant deaths, the number would go up, but by how much? Call the number of infant deaths in Honolulu and Indianapolis d. The current 200 infant deaths represent 80% of what the number would be without the program, so you can write the following equation and solve: $0.8d = 200$; $d = 250$. There would have been 50 more deaths.

Therefore, had the stated program changes occurred, there would have been about $320 + 250 = 570$ deaths. There were actually about $400 + 200 = 600$ deaths. That means about 30 fewer deaths would have occurred. Alternatively, you could add the differences in the two cities to get the net difference: $-80 + 50 = 30$. Either way, the correct answer is **(C).**

19. B

This question concerns deaths from pneumonia/influenza, which you find in the pie chart.

The pie chart represents the 3,532 pneumonia/influenza deaths that occurred in five cities. Of these, 19% occurred in Houston, and 38% occurred in Los Angeles. That's a ratio of 19:38 or 1:2. However, the question is asking for the ratio of the *rate of incidence* of these deaths.

If Los Angeles's population were twice the size of Houston's, then the rate of incidence in the two cities would be the same, giving a ratio of 1:1. Because Los Angeles's population is less than twice Houston's, you know that the rate of incidence is higher in Los Angeles than in Houston. The ratio of Houston to Los Angeles must be less than 1. On the basis of this critical thinking, you can eliminate (D) and (E).

The question does not give you the actual populations of the cities, but it does tell you their relative populations. If you call the population of Houston p, then the population of Los Angeles is $1.75p$ or $\frac{7}{4}p$. As shown in the pie chart, for every 1 death from pneumonia/influenza in Houston, there were 2 in Los Angeles. In Houston, therefore, the rate of incidence of deaths can be represented by the ratio of 1 to p, or $\frac{1}{p}$. In Los Angeles, the rate of deaths can be represented as 2 to $\frac{7}{4}p$, or

$$\frac{2}{\frac{7p}{4}} = 2 \times \frac{4}{7p} = \frac{8}{7p}.$$

Now find the ratio of the rates of incidence:

$$\frac{\frac{1}{p}}{\frac{8}{7p}} = \frac{1}{p} \times \frac{7p}{8} = \frac{7}{8}.$$

The correct answer is **(B)**, $\frac{7}{8}$.

20. A

This question asks about deaths due to pneumonia/influenza, so look at the pie chart for those data. It also concerns people who died at age 65 or over, and that information is in the bar graph, specifically in the right-hand bar of each pair of bars. Remember that the deaths occurring at age 65 and over are plotted against the *y*-axis on the right side of the bar graph.

Only 80% of the pneumonia/influenza deaths occurred among the elderly. However, because this 80% applies equally to each city, you do not have to take it into account as you compare the percentages of deaths in each city attributed to pneumonia/influenza. If a city has the most such deaths when 100% of pneumonia/influenza deaths are counted, then it will also have the most such deaths when the number of pneumonia/influenza deaths is reduced by 20% for all cities.

If you compare cities using estimation, your thinking might go like this: Boston and Houston have about the same number of P/I deaths, making them easy to compare. Houston has far more deaths from all causes, so Houston's proportion of P/I deaths is lower than Boston's. Houston has about 50% more P/I deaths than Honolulu or Indianapolis, and it has more than four times (Honolulu) and three times (Indianapolis) the deaths from all causes. Therefore, Houston's proportion of P/I deaths is lower than Honolulu's or Indianapolis's. Finally, compare Houston to Los Angeles. Houston has more total deaths and far fewer P/I deaths than Los Angeles has, so its proportion of P/I deaths is definitely lower than Los Angeles's. Houston is the winner; choice **(A)** is correct.

You can also solve using calculation. The pie chart represents a total of 3,532 deaths, and it gives the percentage distribution among the five cities. Determine the number of pneumonia/influenza deaths for each city:

	Deaths from P/I
Boston	$3{,}532 \times 0.18 = 636$
Honolulu	$3{,}532 \times 0.13 = 459$
Houston	$3{,}532 \times 0.19 = 671$
Indianapolis	$3{,}532 \times 0.12 = 424$
Los Angeles	$3{,}532 \times 0.38 = 1{,}342$

Now find the number of deaths among the elderly from the bar graph. Boston is between 4,000 and 5,000, so say 4,500. Honolulu is at about 3,000, Houston is the tallest bar at about 13,000, Indianapolis is at 4,000, and the bar for Los Angeles is at about 9,000.

	Deaths from P/I	Deaths 65+
Boston	$3{,}532 \times 0.18 = 636$	4,500
Honolulu	$3{,}532 \times 0.13 = 459$	3,000
Houston	$3{,}532 \times 0.19 = 671$	13,000
Indianapolis	$3{,}532 \times 0.12 = 424$	4,000
Los Angeles	$3{,}532 \times 0.38 = 1{,}342$	9,000

You can perform calculations to determine the city with the lowest proportion of deaths from P/I, or you can compare cities using estimation. If you calculate the proportions, the result looks like this:

	Deaths from P/I	Deaths 65+	Proportion
Boston	$3{,}532 \times 0.18 = 636$	4,500	14%
Honolulu	$3{,}532 \times 0.13 = 459$	3,000	15%
Houston	$3{,}532 \times 0.19 = 671$	13,000	5%
Indianapolis	$3{,}532 \times 0.12 = 424$	4,000	11%
Los Angeles	$3{,}532 \times 0.38 = 1{,}342$	9,000	15%

The winner is Houston, and answer choice **(A)** is correct.

PART FIVE

GRE Resources

APPENDIX

Math Reference

The math on the GRE covers a lot of ground—from number properties and arithmetic to basic algebra and symbol problems to geometry and statistics. Don't let yourself be intimidated.

We've highlighted the 100 most important concepts that you need to know and divided them into three levels. The GRE Quantitative sections test your understanding of a relatively limited number of mathematical concepts, all of which you will be able to master.

Level 1 consists of foundational math topics. Though these topics may seem basic, review this list so that you are aware that these skills may play a part in the questions you will answer on the GRE. Look over the Level 1 list to make sure you're comfortable with the basics.

Level 2 is where most people start their review of math. Level 2 skills and formulas come into play quite frequently on the GRE. If the skills needed to handle Level 1 or 2 topics are keeping you from feeling up to the tasks expected on the GRE Quantitative section, you might consider taking the Kaplan GRE Math Refresher course.

Level 3 represents the most challenging math concepts you'll find on the GRE. Don't spend a lot of time on Level 3 if you still have gaps in Level 2, but once you've mastered Level 2, tackling Level 3 can put you over the top.

LEVEL 1

1. How to add, subtract, multiply, and divide WHOLE NUMBERS

You can check addition with subtraction.

$17 + 5 = 22 \qquad 22 - 5 = 17$

You can check multiplication with division.

$5 \times 28 = 140 \qquad 140 \div 5 = 28$

2. How to add, subtract, multiply, and divide FRACTIONS

Find a common denominator before adding or subtracting fractions.

$$\frac{4}{5} + \frac{3}{10} = \frac{8}{10} + \frac{3}{10} = \frac{11}{10} \text{ or } 1\frac{1}{10}$$
$$2 - \frac{3}{8} = \frac{16}{8} - \frac{3}{8} = \frac{13}{8} \text{ or } 1\frac{5}{8}$$

To multiply fractions, multiply the numerators first and then multiply the denominators. Simplify if necessary.

$$\frac{3}{4} \times \frac{1}{6} = \frac{3}{24} = \frac{1}{8}$$

You can also reduce before multiplying numerators and denominators. This keeps the products small.

$$\frac{5}{8} \times \frac{2}{15} = \frac{{}^{1}\cancel{5}}{{}_{4}\cancel{8}} \times \frac{\cancel{2}^{1}}{\cancel{15}_{3}} = \frac{1}{12}$$

To divide by a fraction, multiply by its reciprocal. To write the reciprocal of a fraction, flip the numerator and the denominator.

$$5 \div \frac{1}{3} = \frac{5}{1} \times \frac{3}{1} = 15 \qquad \frac{1}{3} \div \frac{4}{5} = \frac{1}{3} \times \frac{5}{4} = \frac{5}{12}$$

3. How to add, subtract, multiply, and divide DECIMALS

To add or subtract, align the decimal points and then add or subtract normally. Place the decimal point in the answer directly below existing decimal points.

$$\begin{array}{r} 3.25 \\ +4.4 \\ \hline 7.65 \end{array} \qquad \begin{array}{r} 7.65 \\ -4.4 \\ \hline 3.25 \end{array}$$

To multiply with decimals, multiply the digits normally and count off decimal places (equal to the total number of places in the factors) from the right.

$$2.5 \times 2.5 = 6.25$$
$$0.06 \times 2{,}000 = 120.00 = 120$$

To divide by a decimal, move the decimal point in the divisor to the right to form a whole number; move the decimal point in the dividend the same number of places. Divide as though there were no decimals, then place the decimal point in the quotient.

$$6.25 \div 2.5$$
$$= 62.5 \div 25 = 2.5$$

4. How to convert FRACTIONS TO DECIMALS and DECIMALS TO FRACTIONS

To convert a fraction to a decimal, divide the numerator by the denominator.

$$\frac{4}{5} = 0.8 \qquad \frac{4}{50} = 0.08 \qquad \frac{4}{500} = 0.008$$

To convert a decimal to a fraction, write the digits in the numerator and use the decimal name in the denominator.

$$0.003 = \frac{3}{1{,}000} \qquad 0.03 = \frac{3}{100} \qquad 0.3 = \frac{3}{10}$$

5. How to add, subtract, multiply, and divide POSITIVE AND NEGATIVE NUMBERS

When addends (the numbers being added) have the same sign, add their absolute values; the sum has the same sign as the addends. But when addends have different signs, subtract the absolute values; the sum has the sign of the greater absolute value.

$3 + 9 = 12$, but $-3 + (-9) = -12$
$3 + (-9) = -6$, but $-3 + 9 = 6$

In multiplication and division, when the signs are the same, the product/quotient is positive. When the signs are different, the product/quotient is negative.

$6 \times 7 = 42$ and $-6 \times (-7) = 42$
$-6 \times 7 = -42$ and $6 \times (-7) = -42$
$96 \div 8 = 12$ and $-96 \div (-8) = 12$
$-96 \div 8 = -12$ and $96 \div (-8) = -12$

6. How to plot points on the NUMBER LINE

To plot the point 4.5 on the number line, start at 0, go right to 4.5, halfway between 4 and 5.

To plot the point −2.5 on the number line, start at 0, go left to −2.5, halfway between −2 and −3.

7. How to plug a number into an ALGEBRAIC EXPRESSION

To evaluate an algebraic expression, choose numbers for the variables or use the numbers assigned to the variables.

Evaluate $4np + 1$ when $n = -4$ and $p = 3$.

$4np + 1 = 4(-4)(3) + 1 = -48 + 1 = -47$

8. How to SOLVE a simple LINEAR EQUATION

Use algebra to isolate the variable. Do the same steps to both sides of the equation.

$$28 = -3x - 5$$
$$28 + 5 = -3x - 5 + 5 \quad \text{Add 5.}$$
$$33 = -3x$$
$$\frac{33}{-3} = \frac{-3x}{-3} \quad \text{Divide by } -3.$$
$$-11 = x$$

9. How to add and subtract LINE SEGMENTS

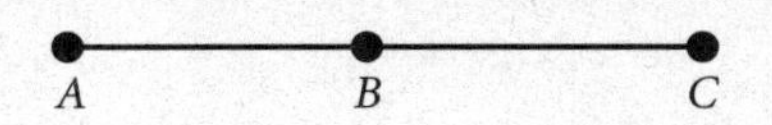

If $AB = 6$ and $BC = 8$, then $AC = 6 + 8 = 14$.
If $AC = 14$ and $BC = 8$, then $AB = 14 - 8 = 6$.

10. How to find the THIRD ANGLE of a TRIANGLE, given the other two angles

Use the fact that the sum of the measures of the interior angles of a triangle always equals 180°.

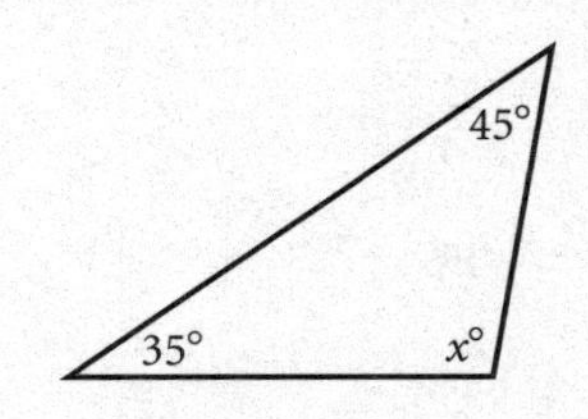

$$35 + 45 + x = 180$$
$$80 + x = 180$$
$$x = 100$$

LEVEL 2

11. How to use PEMDAS

When you're given a complex arithmetic expression, it's important to know the order of operations. Just remember PEMDAS (as in "Please Excuse My Dear Aunt Sally"). What PEMDAS means is this: Clean up **Parentheses** first (nested sets of parentheses are worked from the innermost set to the outermost set); then deal with **Exponents** (or **Radicals**); then do the **Multiplication** and **Division** together, going from left to right; and finally do the **Addition** and **Subtraction** together, again going from left to right.

Example:

$$9 - 2 \times (5 - 3)^2 + 6 \div 3 =$$

Begin with the parentheses:

$$9 - 2 \times (2)^2 + 6 \div 3 =$$

Then do the exponent:

$$9 - 2 \times 4 + 6 \div 3 =$$

Now do multiplication and division from left to right:

$$9 - 8 + 2 =$$

Finally, do addition and subtraction from left to right:

$$1 + 2 = 3$$

12. How to use the PERCENT FORMULA

Identify the part, the percent, and the whole.

$$Part = Percent \times Whole$$

Find the part.

Example:

What is 12 percent of 25?

Setup:

$$Part = \frac{12}{100} \times 25 = \frac{300}{100} = 3$$

Find the percent.

Example:

45 is what percent of 9?

Setup:

$$\begin{aligned} 45 &= \frac{\text{Percent}}{100} \times 9 \\ 4{,}500 &= \text{Percent} \times 9 \\ 500 &= \text{Percent} \end{aligned}$$

Find the whole.

Example:

15 is $\frac{3}{5}$ percent of what number?

Setup:

$$15 = \frac{3}{5}\left(\frac{1}{100}\right) \times Whole$$

$$15 = \frac{3}{500} \times Whole$$

$$Whole = 15\left(\frac{500}{3}\right) = \frac{7{,}500}{3} = 2{,}500$$

13. How to use the PERCENT INCREASE/DECREASE FORMULAS

Identify the original whole and the amount of increase/decrease.

$$Percent\ increase = \frac{Amount\ of\ increase}{Original\ whole} \times 100\%$$

$$Percent\ decrease = \frac{Amount\ of\ decrease}{Original\ whole} \times 100\%$$

Example:

The price goes up from \$80 to \$100. What is the percent increase?

Setup:

$$\begin{aligned} Percent\ increase &= \frac{20}{80} \times 100\% \\ &= 0.25 \times 100\% = 25\% \end{aligned}$$

14. How to predict whether a sum, difference, or product will be ODD or EVEN

Don't bother memorizing the rules. Just take simple numbers such as 2 for even numbers and 3 for odd numbers and see what happens.

Example:

If m is even and n is odd, is the product mn odd or even?

Setup:

Say $m = 2$ and $n = 3$.
$2 \times 3 = 6$, which is even, so mn is even.

15. How to recognize MULTIPLES OF 2, 3, 4, 5, 6, 9, 10, and 12

2: Last digit is even.
3: Sum of digits is a multiple of 3.
4: Last two digits are a multiple of 4.
5: Last digit is 5 or 0.
6: Sum of digits is a multiple of 3, and last digit is even.
9: Sum of digits is a multiple of 9.
10: Last digit is 0.
12: Sum of digits is a multiple of 3, and last two digits are a multiple of 4.

16. How to find a COMMON FACTOR of two numbers

Break both numbers down to their prime factors to see which they have in common. Then multiply the shared prime factors to find all common factors.

Example:

What factors greater than 1 do 135 and 225 have in common?

Setup:

First find the prime factors of 135 and 225; $135 = 3 \times 3 \times 3 \times 5$, and $225 = 3 \times 3 \times 5 \times 5$. The numbers share $3 \times 3 \times 5$ in common. Thus, aside from 3 and 5, the remaining common factors can be found by multiplying 3, 3, and 5 in every possible combination: $3 \times 3 = 9$, $3 \times 5 = 15$, and $3 \times 3 \times 5 = 45$. Therefore, the common factors of 135 and 225 are 3, 5, 9, 15, and 45.

17. How to find a COMMON MULTIPLE of two numbers

The product of two numbers is the easiest common multiple to find, but it is not always the least common multiple (LCM).

Example:

What is the least common multiple of 28 and 42?

Setup:

$$28 = 2 \times 2 \times 7$$
$$42 = 2 \times 3 \times 7$$

The LCM can be found by finding the prime factorization of each number, then seeing the greatest number of times each factor is used. Multiply each prime factor the greatest number of times it appears.

In 28, 2 is used twice. In 42, 2 is used once. In 28, 7 is used once. In 42, 7 is used once, and 3 is used once.

So you multiply each factor the greatest number of times it appears in a prime factorization:

$$\text{LCM} = 2 \times 2 \times 3 \times 7 = 84$$

18. How to find the AVERAGE or ARITHMETIC MEAN

$$\textit{Average} = \frac{\textit{Sum of terms}}{\textit{Number of terms}}$$

Example:

What is the average of 3, 4, and 8?

Setup:

$$\textit{Average} = \frac{3 + 4 + 8}{3} = \frac{15}{3} = 5$$

19. How to use the AVERAGE to find the SUM

$$\textit{Sum} = (\textit{Average}) \times (\textit{Number of terms})$$

Example:

17.5 is the average (arithmetic mean) of 24 numbers.

What is the sum of the 24 numbers?

Setup:

$$\textit{Sum} = 17.5 \times 24 = 420$$

20. How to find the AVERAGE of CONSECUTIVE NUMBERS

The average of evenly spaced numbers is simply the average of the smallest number and the largest number. The average of all the integers from 13 to 77, for example, is the same as the average of 13 and 77:

$$\frac{13 + 77}{2} = \frac{90}{2} = 45$$

21. How to COUNT CONSECUTIVE NUMBERS

The number of integers from A to B inclusive is $B - A + 1$.

Example:

How many integers are there from 73 through 419, inclusive?

Setup:

$$419 - 73 + 1 = 347$$

22. How to find the SUM OF CONSECUTIVE NUMBERS

Sum = (*Average*) × (*Number of terms*)

Example:

What is the sum of the integers from 10 through 50, inclusive?

Setup:

Average: $\frac{10 + 50}{2} = 30$

Number of terms: $50 - 10 + 1 = 41$

Sum: $30 \times 41 = 1{,}230$

23. How to find the MEDIAN

Put the numbers in numerical order and take the middle number.

Example:

What is the median of 88, 86, 57, 94, and 73?

Setup:

First, put the numbers in numerical order, then take the middle number:

57, 73, 86, 88, 94

The median is 86.

In a set with an even number of numbers, take the average of the two in the middle.

Example:

What is the median of 88, 86, 57, 73, 94, and 100?

Setup:

First, put the numbers in numerical order.

57, 73, 86, 88, 94, 100

Because 86 and 88 are the two numbers in the middle:

$$\frac{86 + 88}{2} = \frac{174}{2} = 87$$

The median is 87.

24. How to find the MODE

Take the number that appears most often. For example, if your test scores were 88, 57, 68, 85, 98, 93, 93, 84, and 81, the mode of the scores would be 93 because it appears more often than any other score. (If there's a tie for most often, then there's more than one mode. If each number in a set is used equally often, there is no mode.)

25. How to find the RANGE

Take the positive difference between the greatest and least values. Using the example under "How to find the MODE" above, if your test scores were 88, 57, 68, 85, 98, 93, 93, 84, and 81, the range of the scores would be 41, the greatest value minus the least value ($98 - 57 = 41$).

26. How to use actual numbers to determine a RATIO

To find a ratio, put the number associated with *of* on the top and the number associated with *to* on the bottom.

$$\text{Ratio} = \frac{\text{of}}{\text{to}}$$

The ratio of 20 oranges to 12 apples is $\frac{20}{12}$, or $\frac{5}{3}$.

Ratios should always be reduced to lowest terms. Ratios can also be expressed in linear form, such as 5:3.

27. How to use a ratio to determine an ACTUAL NUMBER

Set up a proportion using the given ratio.

Example:

The ratio of boys to girls is 3 to 4. If there are 135 boys, how many girls are there?

Setup:

$$\frac{3}{4} = \frac{135}{g}$$

$$3 \times g = 4 \times 135$$

$$3g = 540$$

$$g = 180$$

28. How to use actual numbers to determine a RATE

Identify the quantities and the units to be compared. Keep the units straight.

Example:

Anders typed 9,450 words in $3\frac{1}{2}$ hours. What was his rate in words per minute?

Setup:

First convert $3\frac{1}{2}$ hours to 210 minutes. Then set up the rate with words on top and minutes on bottom (because "per" means "divided by"):

$$\frac{9{,}450 \text{ words}}{210 \text{ minutes}} = 45 \text{ words per minute}$$

29. How to deal with TABLES, GRAPHS, AND CHARTS

Read the question and all labels carefully. Ignore extraneous information and zero in on what the question asks for. Take advantage of the spread in the answer choices by approximating the answer whenever possible and choosing the answer choice closest to your approximation.

30. How to count the NUMBER OF POSSIBILITIES

You can use multiplication to find the number of possibilities when items can be arranged in various ways.

Example:

How many three-digit numbers can be formed with the digits 1, 3, and 5 each used only once?

Setup:

Look at each digit individually. The first digit (or, the hundreds digit) has three possible numbers to plug in: 1, 3, or 5. The second digit (or, the tens digit) has two possible numbers, since one has already been plugged in. The last digit (or, the ones digit) has only one remaining possible number. Multiply the possibilities together: $3 \times 2 \times 1 = 6$.

31. How to calculate a simple PROBABILITY

$$\textit{Probability} = \frac{\textit{Number of desired outcomes}}{\textit{Number of total possible outcomes}}$$

Example:

What is the probability of throwing a 5 on a fair six-sided die?

Setup:

There is one desired outcome—throwing a 5. There are 6 possible outcomes—one for each side of the die.

$$\textit{Probability} = \frac{1}{6}$$

32. How to work with new SYMBOLS

If you see a symbol you've never seen before, don't be alarmed. It's just a made-up symbol whose operation is uniquely defined by the problem. Everything you need to know is in the question stem. Just follow the instructions.

33. How to SIMPLIFY BINOMIALS

A binomial is a sum or difference of two terms. To simplify two binomials that are multiplied together, use the **FOIL** method. Multiply the **F**irst terms, then the **O**uter terms, followed by the **I**nner terms and the **L**ast terms. Lastly, combine like terms.

Example:

$$(3x + 5)(x - 1) =$$
$$3x^2 - 3x + 5x - 5 =$$
$$3x^2 + 2x - 5$$

34. How to FACTOR certain POLYNOMIALS

A polynomial is an expression consisting of the sum of two or more terms, where at least one of the terms is a variable.

Learn to spot these classic polynomial equations.

$$ab + ac = a(b + c)$$
$$a^2 + 2ab + b^2 = (a + b)^2$$
$$a^2 - 2ab + b^2 = (a - b)^2$$
$$a^2 - b^2 = (a - b)(a + b)$$

35. How to solve for one variable IN TERMS OF ANOTHER

To find x "in terms of" y, isolate x on one side, leaving y as the only variable on the other.

36. How to solve an INEQUALITY

Treat it much like an equation—adding, subtracting, multiplying, and dividing both sides by the same thing. Just remember to reverse the inequality sign if you multiply or divide by a negative quantity.

Example:

Rewrite $7 - 3x > 2$ in its simplest form.

Setup:

$$7 - 3x > 2$$

First, subtract 7 from both sides:

$$7 - 3x - 7 > 2 - 7$$
$$-3x > -5$$

Now divide both sides by −3, remembering to reverse the inequality sign:

$$x < \frac{5}{3}$$

37. How to handle ABSOLUTE VALUES

The *absolute value* of a number n, denoted by $|n|$, is defined as n if $n \geq 0$ and $-n$ if $n < 0$. The absolute value of a number is the distance from zero to the number on the number line. The absolute value of a number or expression is always positive.

$$|-5| = 5$$

If $|x| = 3$, then x could be 3 or −3.

Example:

If $|x - 3| < 2$, what is the range of possible values for x?

Setup:

Represent the possible range for $x - 3$ on a number line.

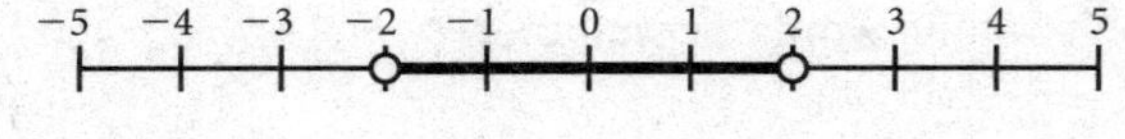

$|x - 3| < 2$, so $(x - 3) < 2$ and $(x - 3) > -2$
$x - 3 < 2$ and $x - 3 > -2$
$x < 2 + 3$ and $x > -2 + 3$
$x < 5$ and $x > 1$
So, $1 < x < 5$.

38. How to TRANSLATE ENGLISH INTO ALGEBRA

Look for the key words and systematically turn phrases into algebraic expressions and sentences into equations.

Here's a table of key words that you may have to translate into mathematical terms:

Operation	Key Words
Addition	sum, plus, and, added to, more than, increased by, combined with, exceeds, total, greater than
Subtraction	difference between, minus, subtracted from, decreased by, diminished by, less than, reduced by
Multiplication	of, product, times, multiplied by, twice, double, triple, half
Division	quotient, divided by, per, out of, ratio of _ to _
Equals	equals, is, was, will be, the result is, adds up to, costs, is the same as

39. How to find an ANGLE formed by INTERSECTING LINES

Vertical angles are equal. Angles along a line add up to 180°.

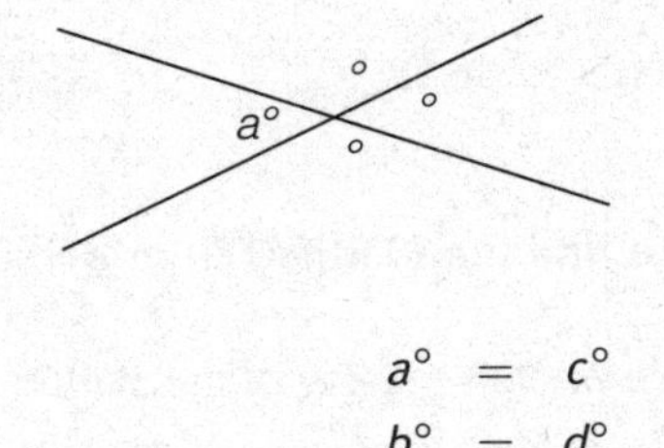

$$a^\circ = c^\circ$$
$$b^\circ = d^\circ$$
$$a^\circ + b^\circ = 180^\circ$$
$$a^\circ + b^\circ + c^\circ + d^\circ = 360^\circ$$

40. How to find an angle formed by a TRANSVERSAL across PARALLEL LINES

When a transversal crosses parallel lines, all the acute angles formed are equal, and all the obtuse angles formed are equal. Any acute angle plus any obtuse angle equals 180°.

Example:

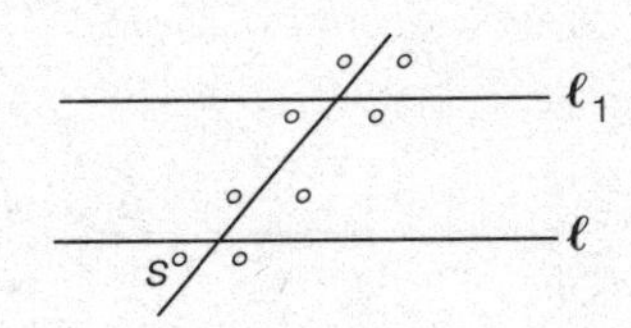

$$e^\circ = g^\circ = p^\circ = r^\circ$$
$$f^\circ = h^\circ = q^\circ = s^\circ$$
$$e^\circ + q^\circ = g^\circ + s^\circ = 180^\circ$$

41. How to find the AREA of a TRIANGLE

$$Area = \frac{1}{2}(Base)(Height)$$

Base and height must be perpendicular to each other. Height is measured by drawing a perpendicular line segment from the base—which can be any side of the triangle—to the angle opposite the base.

Example:

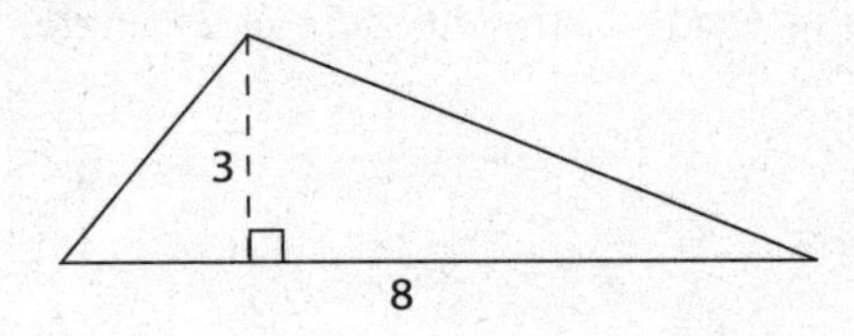

Setup:

$$Area = \frac{1}{2}(8)(3) = 12$$

42. How to work with ISOSCELES TRIANGLES

Isosceles triangles have at least two equal sides and two equal angles. If a GRE question tells you that a triangle is isosceles, you can bet that you'll need to use that information to find the length of a side or a measure of an angle.

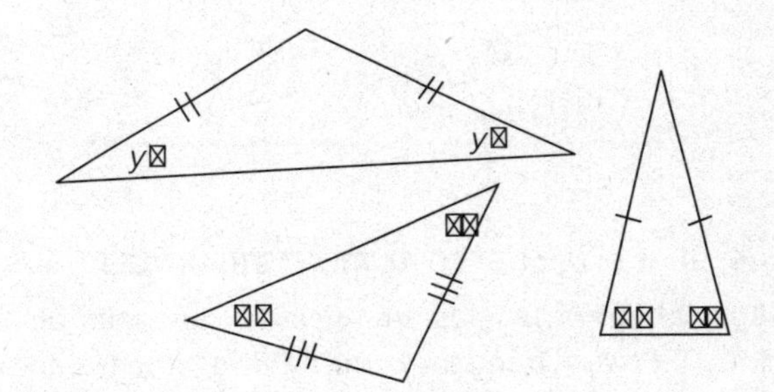

43. How to work with EQUILATERAL TRIANGLES

Equilateral triangles have three equal sides and three 60° angles. If a GRE question tells you that a triangle is equilateral, you can bet that you'll need to use that information to find the length of a side or the measure of an angle.

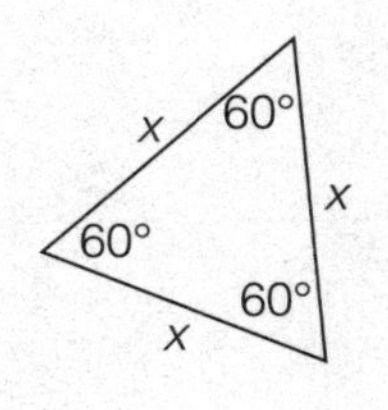

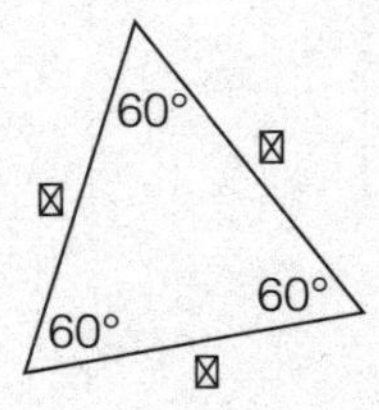

44. How to work with SIMILAR TRIANGLES

In similar triangles, corresponding angles are equal, and corresponding sides are proportional. If a GRE question tells you that triangles are similar, use the properties of similar triangles to find the length of a side or the measure of an angle.

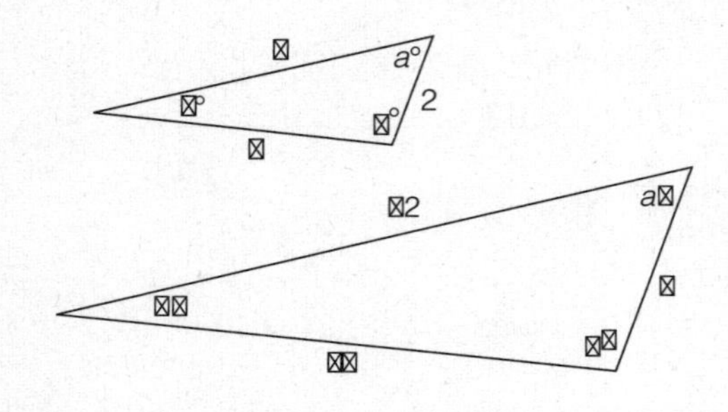

45. How to find the HYPOTENUSE or a LEG of a RIGHT TRIANGLE

For all right triangles, the Pythagorean theorem is $a^2 + b^2 = c^2$, where a and b are the legs and c is the hypotenuse.

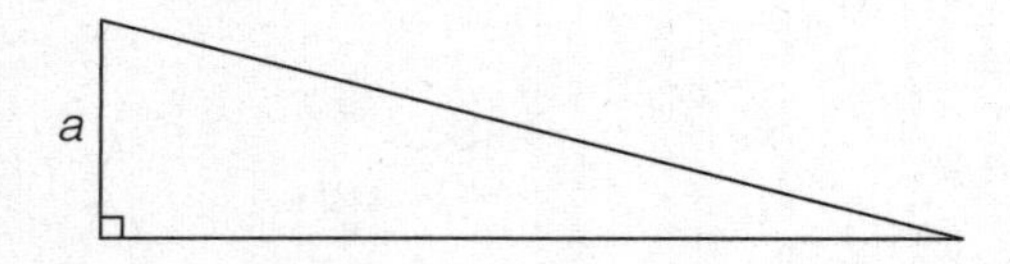

46. How to spot SPECIAL RIGHT TRIANGLES

Special right triangles are ones that are seen on the GRE with frequency. Recognizing them can streamline your problem solving.

3:4:5
5:12:13

These numbers (3, 4, 5 and 5, 12, 13) represent the ratio of the side lengths of these triangles.

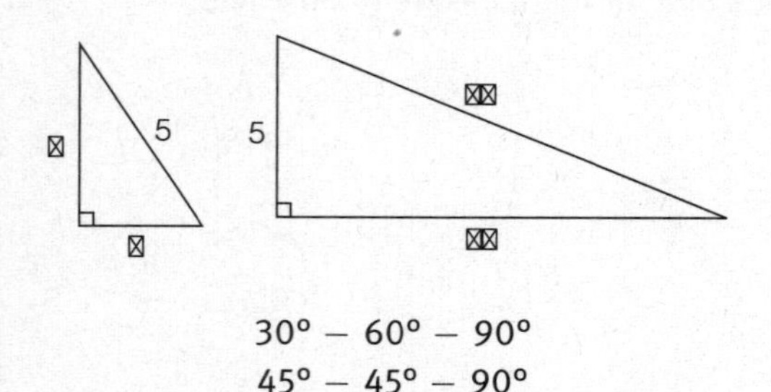

$30° - 60° - 90°$
$45° - 45° - 90°$

In a $30 - 60 - 90$ triangle, the side lengths are multiples of 1, $\sqrt{3}$, and 2, respectively. In a $45 - 45 - 90$ triangle, the side lengths are multiples of 1, 1, and $\sqrt{2}$, respectfully.

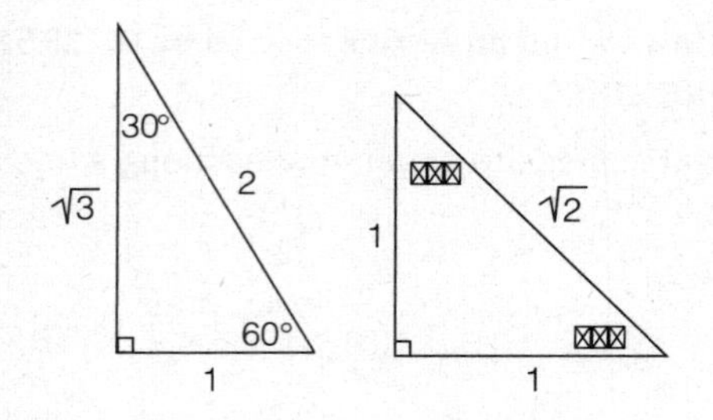

47. How to find the PERIMETER of a RECTANGLE

$$Perimeter = 2(Length + Width)$$

Example:

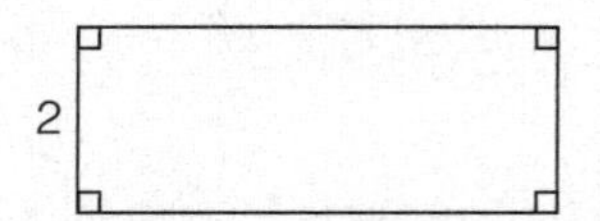

Setup:

$$Perimeter = 2(2 + 5) = 14$$

48. How to find the AREA of a RECTANGLE

$$Area = (Length)(Width)$$

Example:

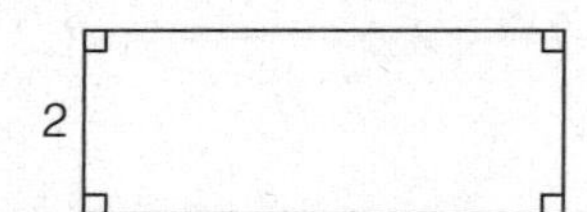

Setup:

$$Area = 2 \times 5 = 10$$

49. How to find the AREA of a SQUARE

$$Area = (Side)^2$$

Example:

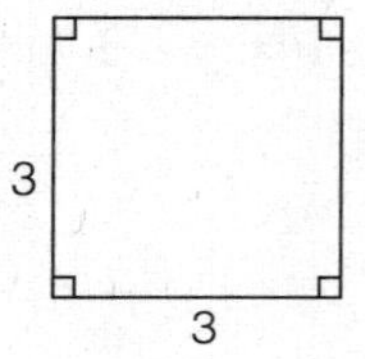

Setup:

$$Area = 3^2 = 9$$

50. How to find the AREA of a PARALLELOGRAM

$$Area = (Base)(Height)$$

Example:

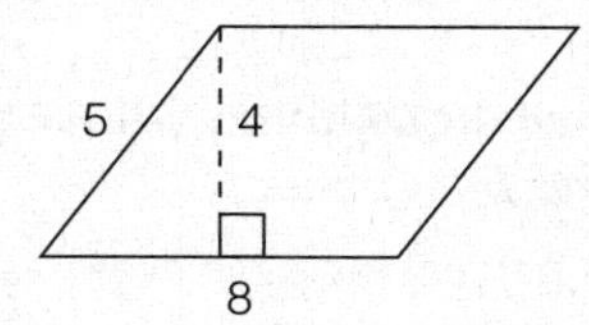

Setup:

$$Area = 8 \times 4 = 32$$

51. How to find the AREA of a TRAPEZOID

A trapezoid is a quadrilateral having only two parallel sides. You can always drop a perpendicular line or two to break the figure into a rectangle and a triangle or two triangles. Use the area formulas for those familiar shapes. Alternatively, you could apply the general formula for the area of a trapezoid:

$$Area = (Average\ of\ parallel\ sides) \times (Height)$$

Example:

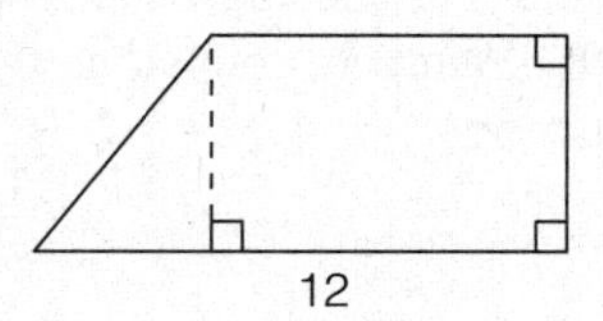

Setup:

$$Area\ of\ rectangle = 8 \times 5 = 40$$

$$Area\ of\ triangle = \frac{1}{2}(4 \times 5) = 10$$

$$Area\ of\ trapezoid = 40 + 10 = 50$$

$$Area\ of\ trapezoid = \left(\frac{8+12}{2}\right) \times 5 = 50$$

52. How to find the CIRCUMFERENCE of a CIRCLE

Circumference = 2□r, *where* r *is the radius*
Circumference = □d, *where* d *is the diameter*

Example:

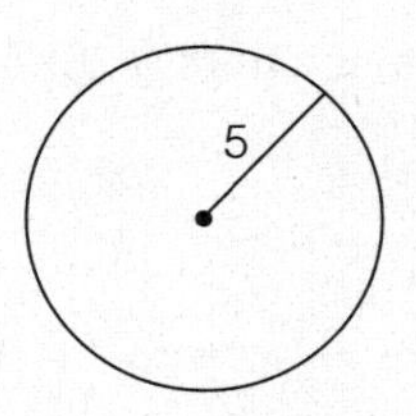

Setup:

$$Circumference = 2\pi(5) = 10\pi$$

53. How to find the AREA of a CIRCLE

$Area = \pi r^2$ *where* r *is the radius*

Example:

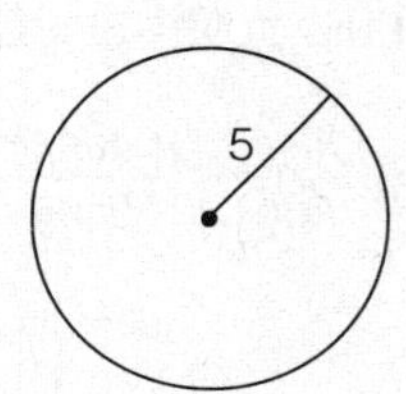

Setup:

$$Area = \pi \times 5^2 = 25\pi$$

54. How to find the DISTANCE BETWEEN POINTS on the coordinate plane

If two points have the same *x*-coordinates or the same *y*-coordinates—that is, they make a line segment that is parallel to an axis—all you have to do is subtract the numbers that are different. Just remember that distance is always positive.

Example:

What is the distance from (2, 3) to (–7, 3)?

Setup:

The *y*'s are the same, so just subtract the *x*'s: 2 – (–7) = 9.

If the points have different x-coordinates and different y-coordinates, make a right triangle and use the Pythagorean theorem or apply the special right triangle attributes if applicable.

Example:

What is the distance from (2,3) to (–1,–1)?

Setup:

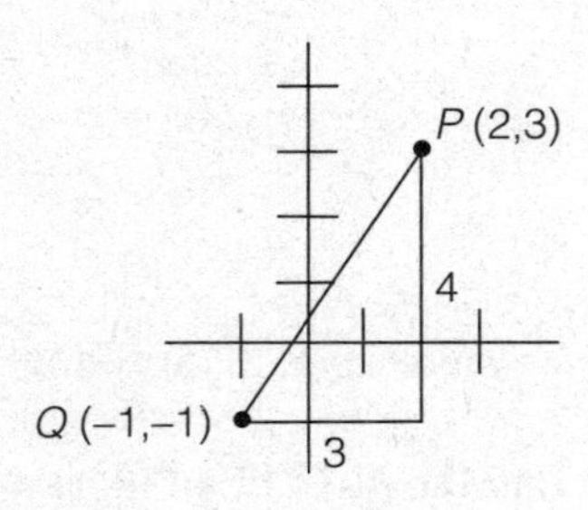

It's a 3:4:5 triangle!

$PQ = 5$

55. How to find the SLOPE of a LINE

$$Slope = \frac{Rise}{Run} = \frac{Change\ in\ y}{Change\ in\ x}$$

Example:

What is the slope of the line that contains the points (1,2) and (4,–5)?

Setup:

$$Slope = \frac{-5-2}{4-1} = \frac{-7}{3} = -\frac{7}{3}$$

LEVEL 3

56 How to determine COMBINED PERCENT INCREASE/DECREASE when no original value is specified

Start with 100 as a starting value.

Example:

A price rises by 10 percent one year and by 20 percent the next. What's the combined percent increase?

Setup:

Say the original price is \$100.

Year one:
$100 + (10% of 100) = 100 + 10 = 110

Year two:
110 + (20% of 110) = 110 + 22 = 132

From 100 to 132 is a 32 percent increase.

57. How to find the ORIGINAL WHOLE before percent increase/decrease

Think of a 15 percent increase over x as $1.15x$ and set up an equation.

Example:

After decreasing by 5 percent, the population is now 57,000. What was the original population?

Setup:

$0.95 \times (Original\ population) = 57{,}000$

Divide both sides by 0.95.

$Original\ population = 57{,}000 \div 0.95 = 60{,}000$

58. How to solve a SIMPLE INTEREST problem

With simple interest, the interest is computed on the principal only and is given by

$Interest = Principle \times rt$

In this formula, r is defined as the interest rate per payment period, and t is defined as the number of payment periods.

Example:

If \$12,000 is invested at 6 percent simple annual interest, how much interest is earned after 9 months?

Setup:

Since the interest rate is annual and we are calculating how much interest accrues after 9 months, we will express the payment period as $\frac{9}{12}$.

$$(12{,}000) \times (0.06) \times \frac{9}{12} = \$540$$

59. How to solve a COMPOUND INTEREST problem

If interest is compounded, the interest is computed on the principal as well as on any interest earned. To compute compound interest:

$$(\textit{Final balance}) = (\textit{Principal}) \times \left(1 + \frac{\textit{interest rate}}{c}\right)^{(\textit{time})(c)}$$

where c = the number of times the interest is compounded annually.

Example:

If $10,000 is invested at 8 percent annual interest, compounded semiannually, what is the balance after 1 year?

Setup:

Final balance

$$= (10{,}000) \times \left(1 + \frac{0.08}{2}\right)^{(1)(2)}$$
$$= (10{,}000) \times (1.04)^2$$
$$= \$10{,}816$$

Semiannual interest is interest that is distributed twice a year. When an interest rate is given as an annual rate, divide by 2 to find the semiannual interest rate.

60. How to solve a REMAINDERS problem

Pick a number that fits the given conditions and see what happens.

Example:

When n is divided by 7, the remainder is 5. What is the remainder when $2n$ is divided by 7?

Setup:

Find a number that leaves a remainder of 5 when divided by 7. You can find such a number by taking any multiple of 7 and adding 5 to it. A good choice would be 12. If $n = 12$, then $2n = 24$, which when divided by 7 leaves a remainder of 3.

61. How to solve a DIGITS problem

Use a little logic—and some trial and error.

Example:

If A, B, C, and D represent distinct digits in the addition problem below, what is the value of D?

$$\begin{array}{r} AB \\ +\,BA \\ \hline CDC \end{array}$$

Setup:

Two 2-digit numbers will add up to at most something in the 100s, so $C = 1$. B plus A in the units column gives a 1, and since A and B in the tens column don't add up to C, it can't simply be that $B + A = 1$. It must be that $B + A = 11$, and a 1 gets carried. In fact, A and B can be any pair of digits that add up to 11 (3 and 8, 4 and 7, etc.), but it doesn't matter what they are: they always give you the same value for D, which is 2:

$$\begin{array}{r} 47 \\ +\;74 \\ \hline 121 \end{array} \qquad \begin{array}{r} 83 \\ +\;38 \\ \hline 121 \end{array}$$

62. How to find a WEIGHTED AVERAGE

Give each term the appropriate "weight."

Example:

The girls' average score is 30. The boys' average score is 24. If there are twice as many boys as girls, what is the overall average?

Setup:

$$\textit{Weighted avg.} = \frac{(1 \times 30) + (2 \times 24)}{3} = \frac{78}{3} = 26$$

HINT: Don't just average the averages.

63. How to find the NEW AVERAGE when a number is added or deleted

Use the sum of the terms of the old average to help you find the new average.

Example:

Michael's average score after four tests is 80. If he scores 100 on the fifth test, what's his new average?

Setup:

Find the original sum from the original average:

$$\textit{Original sum} = 4 \times 80 = 320$$

Add the fifth score to make the new sum:

$$\textit{New sum} = 320 + 100 = 420$$

Find the new average from the new sum:

$$\textit{New average} = \frac{420}{5} = 84$$

64. How to use the ORIGINAL AVERAGE and NEW AVERAGE to figure out WHAT WAS ADDED OR DELETED

Use the sums.

Number added = (*New sum*) − (*Original sum*)
Number deleted = (*Original sum*) − (*New sum*)

Example:

The average of five numbers is 2. After one number is deleted, the new average is −3. What number was deleted?

Setup:

Find the original sum from the original average:

$$\textit{Original sum} = 5 \times 2 = 10$$

Find the new sum from the new average:

$$\textit{New sum} = 4 \times (-3) = -12$$

The difference between the original sum and the new sum is the answer.

$$\textit{Number deleted} = 10 - (-12) = 22$$

65. How to find an AVERAGE RATE

Convert to totals.

$$\textit{Average A per B} = \frac{\textit{Total A}}{\textit{Total B}}$$

Example:

If the first 500 pages have an average of 150 words per page, and the remaining 100 pages have an average of 450 words per page, what is the average number of words per page for the entire 600 pages?

Setup:

$$\begin{aligned} \textit{Total pages} &= 500 + 100 = 600 \\ \textit{Total words} &= (500 \times 150) + (100 \times 450) \\ &= 75{,}000 + 45{,}00 \\ &= 120{,}000 \end{aligned}$$

$$\textit{Average words per page} = \frac{120{,}000}{600} = 200$$

To find an average speed, you also convert to totals.

$$\textit{Average speed} = \frac{\textit{Total distance}}{\textit{Total time}}$$

Example:

Rosa drove 120 miles one way at an average speed of 40 miles per hour and returned by the same 120-mile route at an average speed of 60 miles per hour. What was Rosa's average speed for the entire 240-mile round trip?

Setup:

To drive 120 miles at 40 mph takes 3 hours. To return at 60 mph takes 2 hours. The total time, then, is 5 hours.

$$\textit{Average speed} = \frac{240 \text{ miles}}{5 \text{ hours}} = 48 \text{ mph}$$

66. How to solve a COMBINED WORK PROBLEM

In a combined work problem, you are given the rate at which people or machines perform work individually and you are asked to compute the rate at which they work together (or vice versa). The work formula states: *The inverse of the time it would take everyone working together equals the sum of the inverses of the times it would take each working individually.* In other words:

$$\frac{1}{r} + \frac{1}{s} = \frac{1}{t}$$

where r and s are, for example, the number of hours it would take Rebecca and Sam, respectively, to complete a job working by themselves, and t is the number of hours it would take the two of them working together. Remember that all these variables must stand for units of TIME and must all refer to the amount of time it takes to do the same task.

Example:

If it takes Joe 4 hours to paint a room and Pete twice as long to paint the same room, how long would it take the two of them, working together, to paint the same room, if each of them works at his respective individual rate?

Setup:

Joe takes 4 hours, so Pete takes 8 hours; thus:

$$\frac{1}{4}+\frac{1}{8} = \frac{1}{t}$$

$$\frac{2}{8}+\frac{1}{8} = \frac{1}{t}$$

$$\frac{3}{8} = \frac{1}{t}$$

$$t = \frac{1}{\left(\frac{3}{8}\right)} = \frac{8}{3}$$

So it would take them $\frac{8}{3}$ hours, or 2 hours and 40 minutes, to paint the room together.

67. How to determine a COMBINED RATIO

Multiply one or both ratios by whatever you need in order to get the terms they have in common to match.

Example:

The ratio of a to b is 7:3. The ratio of b to c is 2:5. What is the ratio of a to c?

Setup:

Multiply each member of $a:b$ by 2 and multiply each member of $b:c$ by 3, and you get $a:b = 14:6$ and $b:c = 6:15$. Now that the values of b match, you can write $a:b:c = 14:6:15$ and then say $a:c = 14:15$.

68. How to solve a DILUTION or MIXTURE problem

In dilution or mixture problems, you have to determine the characteristics of a resulting mixture when different substances are combined. Or, alternatively, you have to determine how to combine different substances to produce a desired mixture. There are two approaches to such problems—the straightforward setup and the balancing method.

Example:

If 5 pounds of raisins that cost \$1 per pound are mixed with 2 pounds of almonds that cost \$2.40 per pound, what is the cost per pound of the resulting mixture?

Setup:

The straightforward setup:

$(\$1)(5) + (\$2.40)(2) = \$9.80 =$ total cost for 7 pounds of the mixture

The cost per pound is $\frac{\$9.80}{7} = \$1.40$.

Example:

How many liters of a solution that is 10 percent alcohol by volume must be added to 2 liters of a solution that is 50 percent alcohol by volume to create a solution that is 15 percent alcohol by volume?

Setup:

The balancing method: Make the weaker and stronger (or cheaper and more expensive, etc.) substances balance. That is, (percent difference between the weaker solution and the desired solution) × (amount of weaker solution) = (percent difference between the stronger solution and the desired solution) × (amount of stronger solution). Make n the amount, in liters, of the weaker solution.

$$n(15 - 10) = 2(50 - 15)$$
$$5n = 2(35)$$
$$n = \frac{70}{5} = 14$$

So 14 liters of the 10 percent solution must be added to the original, stronger solution.

69. How to solve an OVERLAPPING SETS problem involving BOTH/NEITHER

Some GRE word problems involve two groups with overlapping members and possibly elements that belong to neither group. It's easy to identify this type of question because the words *both* and/or *neither* appear in the question. These problems are quite workable if you just memorize the following formula:

Group 1 + Group 2 + Neither − Both = Total

Example:

Of the 120 students at a certain language school, 65 are studying French, 51 are studying Spanish, and 53 are studying neither language. How many are studying both French and Spanish?

Setup:

$$65 + 51 + 53 - \textit{Both} = 120$$
$$169 - \textit{Both} = 120$$
$$\textit{Both} = 49$$

70 How to solve an OVERLAPPING SETS problem involving EITHER/OR CATEGORIES

Other GRE word problems involve groups with distinct "either/or" categories (male/female, blue-collar/white-collar, etc.). The key to solving this type of problem is to organize the information in a grid.

Example:

At a certain professional conference with 130 attendees, 94 of the attendees are doctors, and the rest are dentists. If 48 of the attendees are women and $\frac{1}{4}$ of the dentists in attendance are women, how many of the attendees are male doctors?

Setup:

To complete the grid, use the information in the problem, making each row and column add up to the corresponding total:

	Doctors	Dentists	Total
Male	55	27	82
Female	39	9	48
Total	94	36	130

After you've filled in the information from the question, use simple arithmetic to fill in the remaining boxes until you get the number you are looking for—in this case, that 55 of the attendees are male doctors.

71. How to work with FACTORIALS

You may see a problem involving factorial notation, which is indicated by the ! symbol. If n is an integer greater than 1, then n factorial, denoted by $n!$, is defined as the product of all the integers from 1 to n. For example:

$$2! = 2 \times 1 = 2$$
$$3! = 3 \times 2 \times 1 = 6$$
$$4! = 4 \times 3 \times 2 \times 1 = 24, \text{etc}$$

By definition, $0! = 1$.

Also note: $6! = 6 \times 5! = 6 \times 5 \times 4!$, etc. Most GRE factorial problems test your ability to factor and/or cancel.

Example:

$$\frac{8!}{6! \times 2!} = \frac{8 \times 7 \times 6!}{6! \times 2 \times 1} = 28$$

72: How to solve a PERMUTATION problem

Factorials are useful for solving questions about permutations (i.e., the number of ways to arrange elements sequentially). For instance, to figure out how many ways there are to arrange 7 items along a shelf, you would multiply the number of possibilities for the first position times the number of possibilities remaining for the second position, and so on—in other words: $7 \times 6 \times 5 \times 4 \times 3 \times 2 \times 1$, or 7!.

If you're asked to find the number of ways to arrange a smaller group that's being drawn from a larger group, you can either apply logic, or you can use the permutation formula:

$${}_nP_k = \frac{n!}{(n-k)!}$$

where $n =$ (the number in the larger group) and
$k =$ (the number you're arranging).

Example:

Five runners run in a race. The runners who come in first, second, and third place will win gold, silver, and bronze medals, respectively. How many possible outcomes for gold, silver, and bronze medal winners are there?

Setup:

Any of the 5 runners could come in first place, leaving 4 runners who could come in second place, leaving 3 runners who could come in third place, for a total of $5 \times 4 \times 3 = 60$ possible outcomes for gold, silver, and bronze medal winners. Or, using the formula:

$${}_5P_3 = \frac{5!}{(5-3)!} = \frac{5!}{2!} = \frac{5 \times 4 \times 3 \times \cancel{2} \times \cancel{1}}{\cancel{2} \times \cancel{1}}$$
$$= 5 \times 4 \times 3 = 60$$

73: How to solve a COMBINATION problem

If the order or arrangement of the smaller group that's being drawn from the larger group does NOT matter, you are looking for the numbers of combinations, and a different formula is called for:

$${}_nC_k = \frac{n!}{k!(n-k)!}$$

where $n =$ (the number in the larger group) and
$k =$ (the number you're choosing).

Example:

How many different ways are there to choose 3 delegates from 8 possible candidates?

Setup:

$${}_nC_k = \frac{8!}{3!(8-3)!} = \frac{8!}{3! \times 5!}$$
$$= \frac{8 \times 7 \times \cancel{6} \times \cancel{5} \times \cancel{4} \times \cancel{3} \times \cancel{2} \times \cancel{1}}{\cancel{3} \times \cancel{2} \times 1 \times \cancel{5} \times \cancel{4} \times \cancel{3} \times \cancel{2} \times \cancel{1}}$$
$$= 8 \times 7 = 56$$

So there are 56 different possible combinations.

74. How to solve PROBABILITY problems where probabilities must be multiplied

Suppose that a random process is performed. Then there is a set of possible outcomes that can occur. An event is a set of possible outcomes. We are concerned with the probability of events.

When all the outcomes are all equally likely, the basic probability formula is this:

$$\textit{Probability} = \frac{\textit{Number of desired outcomes}}{\textit{Number of total possible outcomes}}$$

Many more difficult probability questions involve finding the probability that several events occur. Let's consider first the case of the probability that two events occur. Call these two events *A* and *B*. The probability that both events occur is the probability that event *A* occurs multiplied by the probability that event *B* occurs given that event *A* occurred. The probability that *B* occurs given that

A occurs is called the conditional probability that *B* occurs given that *A* occurs. Except when events *A* and *B* do not depend on one another, the probability that *B* occurs given that *A* occurs is not the same as the probability that *B* occurs.

The probability that three events *A*, *B*, and *C* occur is the probability that *A* occurs multiplied by the conditional probability that *B* occurs given that *A* occurred multiplied by the conditional probability that *C* occurs given that both *A* and *B* have occurred.

This can be generalized to any number of events.

Example:

If 2 students are chosen at random to run an errand from a class with 5 girls and 5 boys, what is the probability that both students chosen will be girls?

Setup:

The probability that the first student chosen will be a girl is $\frac{5}{10} = \frac{1}{2}$, and since there would be 4 girls and 5 boys left out of 9 students, the probability that the second student chosen will be a girl (given that the first student chosen is a girl) is $\frac{4}{9}$. Thus, the probability that both students chosen will be girls is $\frac{1}{2} \times \frac{4}{9} = \frac{2}{9}$. There was conditional probability here because the probability of choosing the second girl was affected by another girl being chosen first. Now let's consider another example where a random process is repeated.

Example:

If a fair coin is tossed 4 times, what's the probability that at least 3 of the 4 tosses will be heads?

Setup:

There are 2 possible outcomes for each toss, so after 4 tosses, there are $2 \times 2 \times 2 \times 2 = 16$ possible outcomes.

We can list the different possible sequences where at least 3 of the 4 tosses are heads. These sequences are

HHHT
HHTH
HTHH
THHH
HHHH

Thus, the probability that at least 3 of the 4 tosses will come up heads is:

$$\frac{\textit{Number of desired outcomes}}{\textit{Number of total possible outcomes}} = \frac{5}{16}$$

We could have also solved this question using the combinations formula. The probability of a head is $\frac{1}{2}$, and the probability of a tail is $\frac{1}{2}$. The probability of any particular sequence of heads and tails resulting from 4 tosses is $\frac{1}{2} \times \frac{1}{2} \times \frac{1}{2} \times \frac{1}{2}$, which is $\frac{1}{16}$.

Suppose that the result of each of the four tosses is recorded in each of the four spaces.

____ ____ ____ ____

Thus, we would record an H for head or a T for tails in each of the 4 spaces.

The number of ways of having exactly 3 heads among the 4 tosses is the number of ways of choosing 3 of the 4 spaces above to record an H for heads.

The number of ways of choosing 3 of the 4 spaces is

$$_4C_3 = \frac{4!}{3!\,(4-3)!} = \frac{4!}{3!(1)!} = \frac{4 \times 3 \times 2 \times 1}{3 \times 2 \times 1 \times 1} = 4$$

The number of ways of having exactly 4 heads among the 4 tosses is 1.

If we use the combinations formula, using the definition that $0! = 1$, then

$$
\begin{aligned}
{}_4C_4 &= \frac{4!}{4!(4-4)!} = \frac{4!}{4!(0)!} \\
&= \frac{4 \times 3 \times 2 \times 1}{4 \times 3 \times 2 \times 1 \times 1} = 1
\end{aligned}
$$

Thus, ${}_4C_3 = 4$ and ${}_4C_4 = 1$. So the number of different sequences containing at least 3 heads is $4 + 1 = 5$.

The probability of having at least 3 heads is $\frac{5}{16}$.

75. How to deal with STANDARD DEVIATION

Like the terms *mean*, *mode*, *median*, and *range*, *standard deviation* is a term used to describe sets of numbers. Standard deviation is a measure of how spread out a set of numbers is (how much the numbers deviate from the mean). The greater the spread, the higher the standard deviation. You'll rarely have to calculate the standard deviation on Test Day (although this skill may be necessary for some high-difficulty questions). Here's how standard deviation is calculated:

- Find the average (arithmetic mean) of the set.
- Find the differences between the mean and each value in the set.
- Square each of the differences.
- Find the average of the squared differences.
- Take the positive square root of the average.

In addition to the occasional question that asks you to calculate standard deviation, you may also be asked to compare standard deviations between sets of data or otherwise demonstrate that you understand what standard deviation means. You can often handle these questions using estimation.

Example:

High temperatures, in degrees Fahrenheit, in two cities over five days:

September	1	2	3	4	5
City A	54	61	70	49	56
City B	62	56	60	67	65

For the five-day period listed, which city had the greater standard deviation in high temperatures?

Setup:

Even without trying to calculate them out, one can see that City A has the greater spread in temperatures and, therefore, the greater standard deviation in high temperatures. If you were to go ahead and calculate the standard deviations following the steps described above, you would find that the standard deviation in high temperatures for

City A $= \sqrt{\frac{254}{5}} \approx 7.1$ while the standard

deviation for City

B $= \sqrt{\frac{74}{5}} \approx 3.8$.

76. How to MULTIPLY/DIVIDE VALUES WITH EXPONENTS

Add/subtract the exponents.

Example:

$$x^a \times x^b = x^{a+b}$$
$$2^3 \times 2^4 = 2^7$$

Example:

$$\frac{x^a}{x^b} = x^{a-b}$$
$$\frac{2^8}{2^2} = 2^{8-2} = 2^6$$

77. How to handle a value with an EXPONENT RAISED TO AN EXPONENT

Multiply the exponents.

Example:

$$(x^a)^b = x^{ab}$$
$$(3^4)^5 = 3^{20}$$

78. How to handle EXPONENTS with a base of ZERO and BASES with an EXPONENT of ZERO

Zero raised to any nonzero exponent equals zero.

Example:

$$0^4 = 0^{12} = 0^1 = 0$$

Any nonzero number raised to the exponent 0 equals 1.

Example:

$$3^0 = 15^0 = (0.34)^0 = (-345)^0 = \pi^0 = 1$$

The lone exception is 0 raised to the 0 power, which is *undefined.*

79. How to handle NEGATIVE POWERS

A number raised to the exponent $-x$ is the reciprocal of that number raised to the exponent x.

Example:

$$n^{-1} = \frac{1}{n},\ n^{-2} = \frac{1}{n^2}, \text{ and so on.}$$

$$5^{-3} = \frac{1}{5^3} = \frac{1}{5 \times 5 \times 5} = \frac{1}{125}$$

80. How to handle FRACTIONAL POWERS

Fractional exponents relate to roots. For instance, $x^{\frac{1}{2}} = \sqrt{x}$.

Likewise, $x^{\frac{1}{3}} = \sqrt[3]{x}$, $x^{\frac{2}{3}} = \sqrt[3]{x^2}$, and so on.

Example:

$$\sqrt{x^{-2}} = (x^{-2})^{\frac{1}{2}} = x^{(-2)\left(\frac{1}{2}\right)} = x^{-1} = \frac{1}{x}$$

$$4^{\frac{1}{2}} = \sqrt{4} = 2$$

81. How to handle CUBE ROOTS

The cube root of x is just the number that, when used as a factor 3 times (i.e., cubed), gives you x. Both positive and negative numbers have one and only one cube root, denoted by the symbol $\sqrt[3]{\ }$, and the cube root of a number is always the same sign as the number itself.

Example:

$$(-5) \times (-5) \times (-5) = -125, \text{ so } \sqrt[3]{-125} = -5$$

$$\frac{1}{2} \times \frac{1}{2} \times \frac{1}{2} = \frac{1}{8}, \text{ so } \sqrt[3]{\frac{1}{8}} = \frac{1}{2}$$

82. How to ADD, SUBTRACT, MULTIPLY, and DIVIDE ROOTS

You can add/subtract roots only when the parts inside the $\sqrt{\ }$ are identical.

Example:

$$\sqrt{2} + 3\sqrt{2} = 4\sqrt{2}$$

$$\sqrt{2} - 3\sqrt{2} = -2\sqrt{2}$$

$\sqrt{2} + \sqrt{3}$ can not be combined.

To multiply/divide roots, deal with what's inside the $\sqrt{\ }$ and outside the $\sqrt{\ }$ separately.

Example:

$$\left(2\sqrt{3}\right)\left(7\sqrt{5}\right) = (2 \times 7)\left(\sqrt{3 \times 5}\right) = 14\sqrt{15}$$

$$\frac{10\sqrt{21}}{5\sqrt{3}} = \frac{10}{5}\sqrt{\frac{21}{3}} = 2\sqrt{7}$$

83. How to SIMPLIFY A RADICAL

Look for factors of the number under the radical sign that are perfect squares; then find the square root of those perfect squares. Keep simplifying until the term with the square root sign is as simplified as possible, that is, when there are no other perfect square factors (4, 9, 16, 25, 36, . . .) inside the $\sqrt{\ }$. Write the perfect squares as separate factors and "unsquare" them.

Example:

$$\sqrt{48} = \sqrt{16}\,\sqrt{3} = 4\sqrt{3}$$

$$\sqrt{180} = \sqrt{36}\,\sqrt{5} = 6\sqrt{5}$$

84. How to solve certain QUADRATIC EQUATIONS

Manipulate the equation (if necessary) so that it is equal to 0, factor the left side (reverse FOIL by finding two numbers whose product is the constant and whose sum is the coefficient of the term without the exponent), and break the quadratic into two simple expressions. Then find the value(s) for the variable that make either expression = 0.

Example:

$$\begin{aligned} x^2 + 6 &= 5x \\ x^2 - 5x + 6 &= 0 \\ (x-2)(x-3) &= 0 \\ x - 2 &= 0 \text{ or } x - 3 = 0 \\ x &= 2 \text{ or } 3 \end{aligned}$$

Example:

$$\begin{aligned} x^2 &= 9 \\ x &= 3 \text{ or } -3 \end{aligned}$$

85. How to solve MULTIPLE EQUATIONS

When you see two equations with two variables on the GRE, they're probably easy to combine in such a way that you get something closer to what you're looking for.

Example:

If $5x - 2y = -9$ and $3y - 4x = 6$, what is the value of $x + y$?

Setup:

The question doesn't ask for x and y separately, so don't solve for them separately if you don't have to. Look what happens if you just rearrange a little and "add" the equations:

$$\begin{aligned} 5x - 2y &= -9 \\ +[-4x + 3y &= 6] \\ \hline x + y &= -3 \end{aligned}$$

86. How to solve a SEQUENCE problem

The notation used in sequence problems scares many test takers, but these problems aren't as bad as they look. In a sequence problem, the nth term in the sequence is generated by performing an operation, which will be defined for you, on either n or on the previous term in the sequence. The term itself is expressed as a_n. For instance, if you are referring to the fourth term in a sequence, it is called a_4 in sequence notation. Familiarize yourself with sequence notation and you should have no problem.

Example:

What is the positive difference between the fifth and fourth terms in the sequence 0, 4, 18, . . . whose nth term is $n^2(n - 1)$?

Setup:

Use the definition given to come up with the values for your terms:

$$\begin{aligned} a_5 &= 5^2(5-1) = 25(4) = 100 \\ a_4 &= 4^2(4-1) = 16(3) = 48 \end{aligned}$$

So the positive difference between the fifth and fourth terms is $100 - 48 = 52$.

87. How to solve a FUNCTION problem

You may see function notation on the GRE. An algebraic expression of only one variable may be defined as a function, usually symbolized by f or g, of that variable.

Example:

What is the minimum value of x in the function $f(x) = x^2 - 1$?

Setup:

In the function $f(x) = x^2 - 1$, if x is 1, then $f(1) = 1^2 - 1 = 0$. In other words, by inputting 1 into the function, the output $f(x) = 0$. Every number inputted has one and only one output (although the reverse is not necessarily true). You're asked to find the minimum value, so how would you minimize the expression $f(x) = x^2 - 1$? Since x^2 cannot be negative, in this case $f(x)$ is minimized by making $x = 0$: $f(0) = 0^2 - 1 = -1$, so the minimum value of the function is -1.

88. How to handle GRAPHS of FUNCTIONS

You may see a problem that involves a function graphed onto the *xy*-coordinate plane, often called a "rectangular coordinate system" on the GRE. When graphing a function, the output, $f(x)$, becomes the *y*-coordinate. For example, in the previous example, $f(x) = x^2 - 1$, you've already determined 2 points, (1,0) and (0,−1). If you were to keep plugging in numbers to determine more points and then plotted those points on the *xy*-coordinate plane, you would come up with something like this:

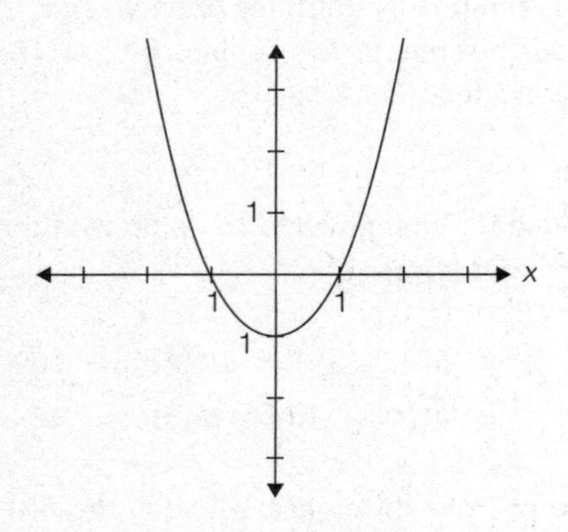

This curved line is called a *parabola*. In the event that you should see a parabola on the GRE (it could be upside down or narrower or wider than the one shown), you will most likely be asked to choose which equation the parabola is describing. These questions can be surprisingly easy to answer. Pick out obvious points on the graph, such as (1,0) and (0,−1) above, plug these values into the answer choices, and eliminate answer choices that don't work with those values until only one answer choice is left.

89. How to handle LINEAR EQUATIONS

You may also encounter linear equations on the GRE. A linear equation is often expressed in the form

$y = mx + b$, where

$m =$ the slope of the line $= \frac{rise}{run}$

$b =$ the *y*-intercept (the point where the line crosses the *y*-axis)

Example:

The graph of the linear equation

$$y = -\frac{3}{4}x + 3 \text{ is this:}$$

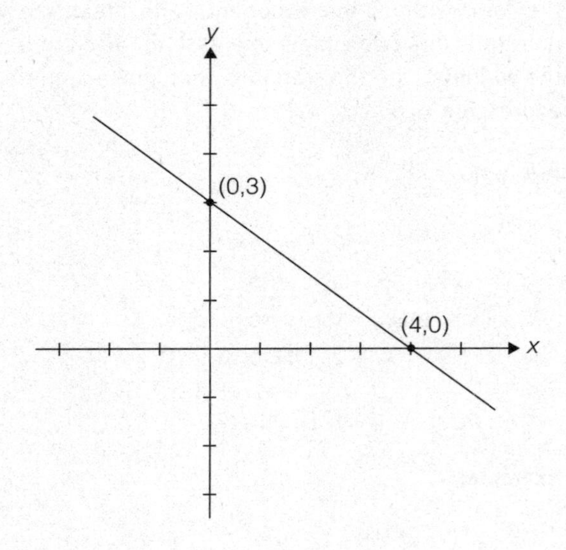

Note:

The equation could also be written in the form $3x + 4y = 12$, but this form does not readily describe the slope and *y*-intercept of the line.

To get a better handle on an equation written in this form, you can solve for *y* to write it in its more familiar form. Or, if you're asked to choose which equation the line is describing, you can pick obvious points, such as (0,3) and (4,0) in this example, and use these values to eliminate answer choices until only one answer is left.

90. How to find the *x*- and *y*-INTERCEPTS of a line

The *x*-intercept of a line is the value of *x* where the line crosses the *x*-axis. In other words, it's the value of *x* when $y = 0$. Likewise, the *y*-intercept is the value of *y* where the line crosses the *y*-axis (i.e., the value of *y* when $x = 0$). The *y*-intercept is also the value *b* when the equation is in the form $y = mx + b$. For instance, in the line shown in the previous example, the *x*-intercept is 4 and the *y*-intercept is 3.

91. How to find the MAXIMUM and MINIMUM lengths for a SIDE of a TRIANGLE

If you know the lengths of two sides of a triangle, you know that the third side is somewhere between the positive difference and the sum of the other two sides.

Example:

The length of one side of a triangle is 7. The length of another side is 3. What is the range of possible lengths for the third side?

Setup:

The third side is greater than the positive difference ($7 - 3 = 4$) and less than the sum ($7 + 3 = 10$) of the other two sides.

92. How to find the sum of all the ANGLES of a POLYGON and one angle measure of a REGULAR POLYGON

Sum of the interior angles in a polygon with n sides:

$$(n - 2) \times 180$$

The term *regular* means all angles in the polygon are of equal measure.

Degree measure of one angle in a regular polygon with n sides:

$$\frac{(n - 2) \times 180}{n}$$

Example:

What is the measure of one angle of a regular pentagon?

Setup:

Since a pentagon is a five-sided figure, plug $n = 5$ into the formula:

Degree measure of one angle:

$$\frac{(5 - 2) \times 180}{5} = \frac{540}{5} = 108$$

93. How to find the LENGTH of an ARC

Think of an arc as a fraction of the circle's circumference. Use the measure of an interior angle of a circle, which has 360 degrees around the central point, to determine the length of an arc.

$$\textit{Length of arc} = \frac{n}{360} \times 2\pi r$$

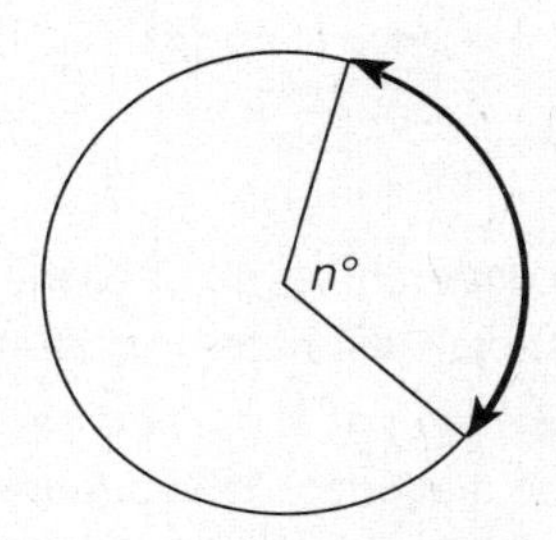

94. How to find the AREA of a SECTOR

Think of a sector as a fraction of the circle's area. Again, set up the interior angle measure as a fraction of 360, which is the degree measure of a circle around the central point.

$$\textit{Area of sector} = \frac{n}{360} \times \pi r^2$$

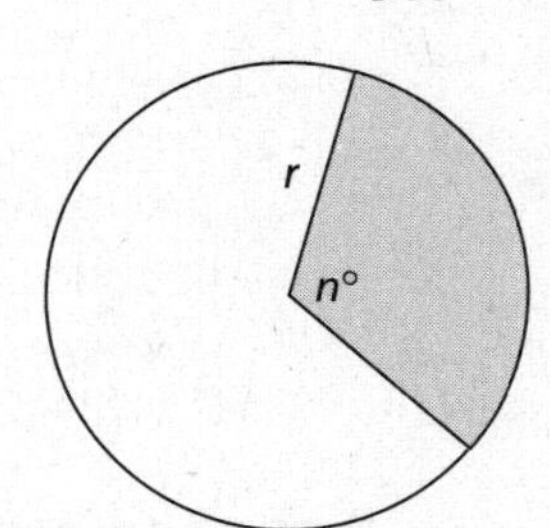

95. How to find the dimensions or area of an INSCRIBED or CIRCUMSCRIBED FIGURE

Look for the connection. Is the diameter the same as a side or a diagonal?

Example:

If the area of the square is 36, what is the circumference of the circle?

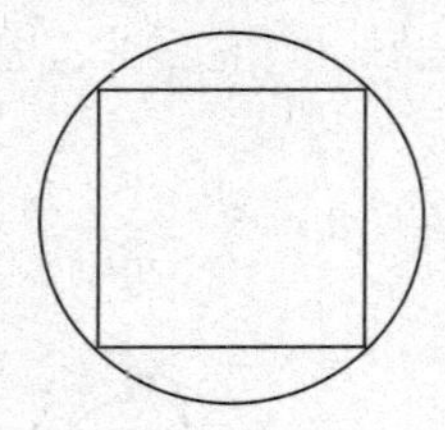

Setup:

To get the circumference, you need the diameter or radius. The circle's diameter is also the square's diagonal. The diagonal of the square is $6\sqrt{2}$. This is because the diagonal of the square transforms it into two separate $45° - 45° - 90°$ triangles (see #46). So, the diameter of the circle is $6\sqrt{2}$.

$$Circumference = \pi(Diameter) = 6\pi\sqrt{2}.$$

96. How to find the VOLUME of a RECTANGULAR SOLID

$$Volume = Length \times Width \times Height$$

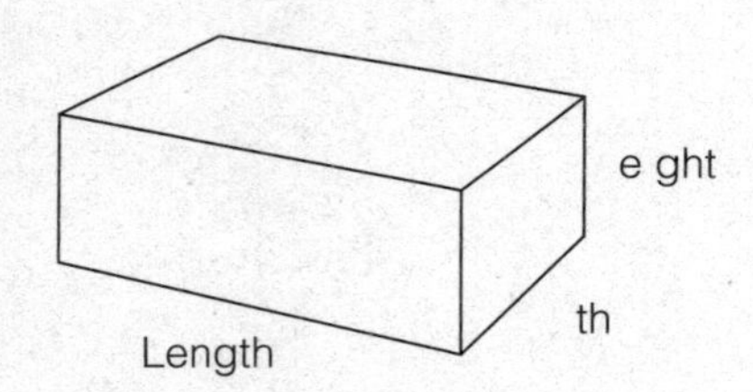

97. How to find the SURFACE AREA of a RECTANGULAR SOLID

To find the surface area of a rectangular solid, you have to find the area of each face and add the areas together. Here's the formula:

Let l = length, w = width, h = height:

$$Surface\ area = 2(lw) + 2(wh) + 2(lh)$$

98. How to find the DIAGONAL of a RECTANGULAR SOLID

Use the Pythagorean theorem twice, unless you spot "special" triangles.

Example:

What is the length of AG?

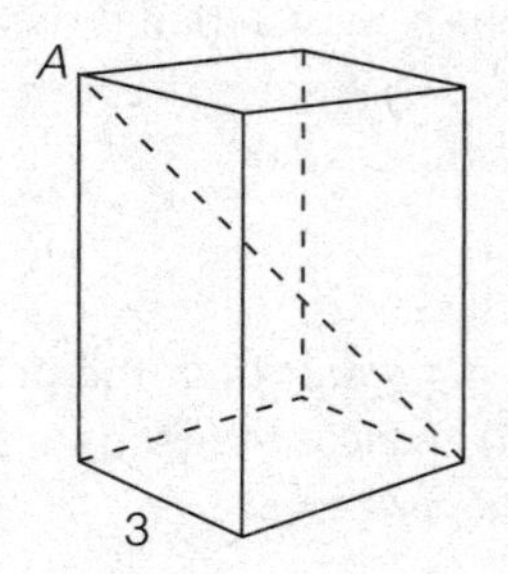

Setup:

Draw diagonal AC.

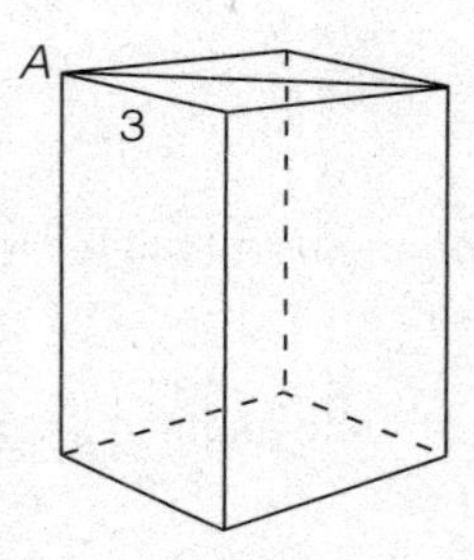

ABC is a 3:4:5 triangle, so $AC = 5$. Now look at triangle ACG:

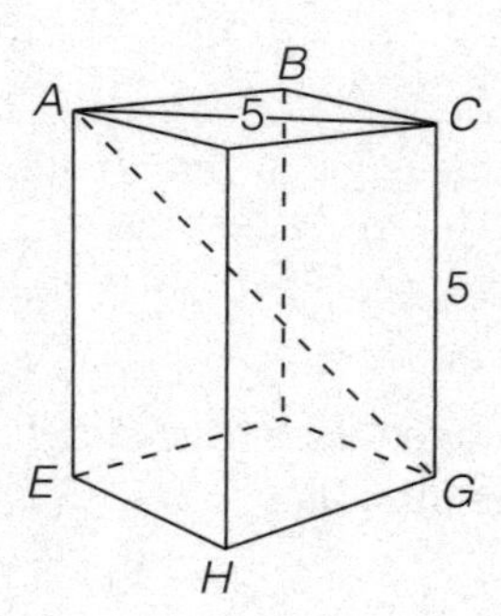

ACG is another special triangle, so you don't need to use the Pythagorean theorem. ACG is a $45° - 45° - 90°$ triangle, so $AG = 5\sqrt{2}$.

99. How to find the VOLUME of a CYLINDER

Volume = *Area of the base* × *Height* = $\pi r^2 h$

Example:

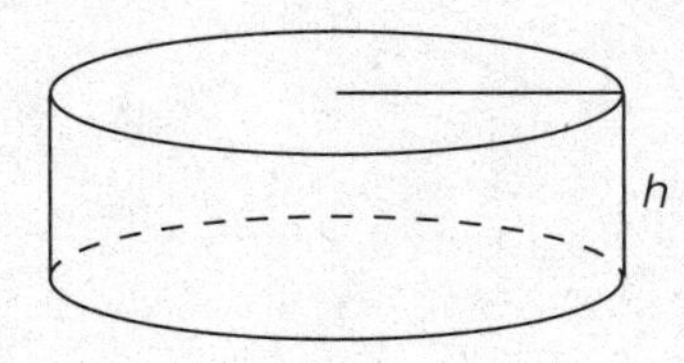

Let $r = 6$ and $h = 3$.

Setup:

Volume = $\pi r^2 h = \pi(6^2)(3) = 108\pi$

100. How to find the SURFACE AREA of a CYLINDER

Surface area = $2\pi r^2 + 2\pi rh$

Example:

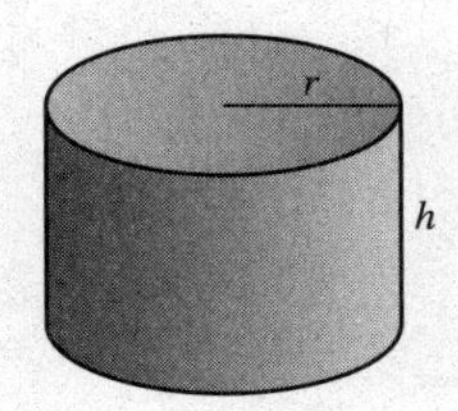

Let $r = 3$ and $h = 4$.

Setup:

$$\begin{aligned} \textit{Surface area} &= 2\pi r^2 + 2\pi rh \\ &= 2\pi(3)^2 + 2\pi(3)(4) \\ &= 18\pi + 24\pi = 42\pi \end{aligned}$$